Barry Cooper
Leeds 1994

Automorphisms of First-Order Structures

Automorphisms of First-Order Structures

Edited by

Richard Kaye

School of Mathematics and Statistics
The University of Birmingham

and

Dugald Macpherson

School of Mathematical Sciences
Queen Mary and Westfield College
London

CLARENDON PRESS • OXFORD
1994

Oxford University Press, Walton Street, Oxford OX2 6DP

Oxford New York Toronto
Delhi Bombay Calcutta Madras Karachi
Kuala Lumpur Singapore Hong Kong Tokyo
Nairobi Dar es Salaam Cape Town
Melbourne Auckland Madrid

and associated companies in
Berlin Ibadan

Oxford is a trade mark of Oxford University Press

Published in the United States
by Oxford University Press Inc., New York

A catalogue record for this book is available from the British Library

Library of Congress Cataloging in Publication Data
(Data available upon request)

ISBN 0 19 853468 X

Typeset by the authors—see Acknowledgements, p. ix.

Printed in Great Britain by
Biddles Ltd.,
Guildford & King's Lynn.

Preface

In 1986, Peter Neumann and Wilfrid Hodges organized a weekly joint seminar for model theorists and permutation group theorists, meeting at Queen Mary College, London (now Queen Mary and Westfield College) in the spring and Oxford in the summer. Peter Cameron, Jacinta Covington, David Evans, Wilfrid Hodges, Ian Hodkinson, Angus Macintyre, Dugald Macpherson, and Peter Neumann addressed the seminar, which was reported on in Wilfrid Hodges' article (Hodges, 1989). Following the success of these meetings, a symposium in Durham on Model Theory and Groups was held in August 1988 and supported by the London Mathematical Society. No formal proceedings of this conference were published, although Peter Cameron's series of lectures there were subsequently written up as the book *Oligomorphic permutation groups* (Cameron, 1990). Cameron's book is a discussion of many group-theoretical and combinatorial questions concerning ω-categoricity, one of the main themes of the Durham meeting.

After the conference, it was agreed to reconvene the Oxford–Queen Mary seminar as a series of one-day meetings (about three per year). Most have taken place in Oxford, at Jesus College, Queen's College, and the Mathematical Institute, but the new one-day meetings have attracted interest from other mathematicians across England. Meetings have also been held at Leeds, Manchester, and at Imperial College, London. The theme for these meetings began as 'Model theory and permutation groups', but the word 'permutation' was later dropped, and we have had many talks on the common ground between model theory and algebra.

Our reason for putting this volume together is that we feel that a book that could introduce a reader (whether he or she be a postgraduate student or more experienced researcher in model theory or group theory) to the many beautiful connections between the two subjects was very badly needed.

The present book is not a 'proceedings' of the 'Models and Groups' meetings as such, but most contributors to the book have spoken at the meetings and some of the contributions below are expanded versions of talks that were presented at the meetings. Indeed, this book is rather more integrated than a 'proceedings' volume would be, with a common index and bibliography, and many cross references between chapters. Any editorial royalties will be used to fund further meetings on the subject.

As an introduction to a mathematical discipline, this book is rather unusual. It will quickly be apparent that it is impossible to present all of the basic definitions and results from model theory, group theory and combinatorics relevant to permutation groups and models in a single book of this size. For this reason we have combined introductory material with guides to the more advanced literature in the journals, and also some samples of some recent (and previously unpublished) research work. We have encouraged authors to be more expansive and give more background than they may normally do when writing for journals. Thus, although there is clearly a suggested order in which most people will read the book, we invite any interested reader to dip in to the book at almost any point. We hope that the index will be particularly useful to the reader using the book in this way. There is a glossary of notation too, and pointers to definitions of key terms are highlighted in the index by page number an italic.

In commissioning and selecting papers for this volume, we have tried to bring together a representative collection of articles on topics involving both infinite permutation groups and model theory. Our intention was to put together a book that treats many of the topics that have been discussed at the Durham Symposium and the one-day 'Models and Groups' meetings. As such, it is broader in scope than Cameron's, but (we hope) complements it well. To keep the book as a reasonably coherent whole, we have stuck closely to the main theme of permutation groups. So, for example, stable groups arise in the book only in the permutation group context. Some of the articles (those by Evans, Macpherson, Hodges, Truss, Cameron) are essentially surveys of broad areas. The article by Kaye on Schmerl's Theorem concerning indiscernibles and Wagner on Hrushovski's constructions are expositions of important and difficult results and techniques which deserve as wide an audience as possible. The remaining articles (by Thomas, Ivanov, Adeleke, Lascar, Kaye, Borovik and Thomas, Nesin, and Epstein and Nesin) are essentially research papers, though in general they are self-contained and include a lot of background material. Many of the themes developed in the survey articles feature strongly in later research articles; in particular Ivanov's paper follows on naturally from Hodges' paper, and similarly Adeleke's paper is a follow-up to Macpherson's.

Although we have made some effort to see that notation and terminology are reasonably consistent from one chapter to another, this has not always been possible. In fact, terminology from permutation groups and that from model theory is occasionally mutually inconsistent. The index and glossary should provide a useful key to this potential trap. We have allowed authors a certain degree of stylistic freedom, and notation does change from one chapter to another, but not (we believe) in a confusing manner. All the 'named' chapters (articles, or papers, as they will sometimes be referred to) have been independently refereed. The other chapters, that is the first chapter of each part of the book, were written jointly by the editors.

Most of all, we hope that the selection of papers here will be enjoyed by a great number of mathematicians working in one or other of the two main subject areas, just as we have enjoyed the talks in the seminars and one-day meetings.

Birmingham R.K.
London D.M.

January, 1994

Acknowledgements

This book was inspired by the series of one-day conferences on Model Theory and Groups referred to in the Preface. Thanks are due to Jesus College, Oxford, The Queen's College, Oxford, The Mathematical Institute, Oxford, Imperial College, London, The School of Mathematics, Leeds, and The Department of Mathematics, Manchester for providing the venues for these meetings, and to the British Logic Colloquium and to the London Mathematical Society for providing financial support.

This book was typeset at Oxford and Birmingham Universities and the National Academic Typesetting Service, with the help of the Oxford University Computing Service.

Ali Nesin's research was partially supported by NSF grant DMS 92-04532. Research by Simon Thomas was also supported by NSF grants.

Contributors

Samson A. Adeleke
Mathematics Department, Western Illinois University, Macomb, IL61455, U.S.A.

Alexandre V. Borovik
Department of Mathematics, UMIST, PO Box 88, Manchester, M60 1QD, U.K.

Peter J. Cameron
School of Mathematical Sciences, Queen Mary and Westfield College, Mile End Road, London, E1 4NS, U.K.

David Epstein
Department of Mathematics, University of California at Irvine, Irvine, California, CA 92717, U.S.A.

David M. Evans
School of Mathematics, The University of East Anglia, Norwich, NR4 7TJ, U.K.

Wilfrid A. Hodges
School of Mathematical Sciences, Queen Mary and Westfield College, Mile End Road, London, E1 4NS, U.K.

Alexandre A. Ivanov
Institute of Mathematics, Wrocław University, pl. Grunwaldzki 2/4, 50–384 Wrocław, Poland.

Richard W. Kaye
The School of Mathematics and Statistics, University of Birmingham, Edgbaston, Birmingham, B15 2TT, U.K.

Daniel Lascar
Université Paris 7, URA 753, Tour 45-55, 2 Place Jussieu, 75251 Paris cédex 05, France.

H. Dugald Macpherson
School of Mathematical Sciences, Queen Mary and Westfield College, Mile End Road, London, E1 4NS, U.K. From 1st October, 1994: School of Mathematics, The University of Leeds, Leeds, LS2 9JT, U.K.

Ali Nesin
Department of Mathematics, University of California at Irvine, Irvine, California, CA 92717, U.S.A.

Simon R. Thomas
Department of Mathematics, Rutgers University, New Brunswick, NJ 08903, U.S.A.

John K. Truss
School of Mathematics, The University of Leeds, Leeds, LS2 9JT, U.K.

Frank O. Wagner
Mathematical Institute, 24-29 St. Giles, Oxford, OX1 3LB, U.K.

Contents

II Recursively saturated models

III Permutation groups of finite Morley rank

PART I

Automorphisms and permutation groups

Models and groups

This book is about automorphism groups, and the ways in which they bring model theory and permutation group theory together. Although the connections often help one to understand one or the other of the two subjects, published work is usually motivated and expressed either entirely in terms of model theory or entirely in terms of permutation group theory. We hope that the articles collected here will appeal to readers from both backgrounds. Part I of the book is intended to represent the core material connecting the two subjects. The present chapter should serve as a gentle introduction to the notation and terminology and some more elementary results.

Model theory can be thought of as the study of abstract mathematical structures, and especially their properties expressible in associated formal languages. Automorphism groups contribute to model theory in several ways. The automorphism group of a structure, considered for example as a permutation group, is often a useful invariant of the stucture. Many structures belonging to natural model-theoretic classes have large and interesting automorphism groups. Indeed the mathematical structures can often be classified via a classification of the automorphism groups. Also, groups of symmetries (at least on part of a structure) are often interpretable in the structure, and much of the structure is reflected in these groups; this seems to happen especially in that part of model theory called *stability theory*. Model theory also contributes to infinite permutation group theory: a permutation group can often be better understood if it is seen as a collection of permutations that preserve some sort of mathematical structure.

Applications of *sophisticated* techniques of one of these subjects to the other are quite rare. The main contribution of model theory to infinite permutation group theory has been through examples. Fraïssé's amalgamation theorem (see Theorem 2.4 on p. 42, and the discussion there) provides examples of infinite permutation groups with strong transitivity properties totally unlike any finite permutation groups. There have been a few applications of the results of Cherlin, Harrington and Lachlan (Cherlin *et al.*, 1985) to infinite permutation groups: for example, in Evans' proof (1987) that a primitive permutation group of countably infinite degree with fewer than $2^{\aleph_0}$ orbits on the power set is highly transitive, and in the theorem of Macpherson and Praeger (1990) that any subgroup of Sym($\mathbb{N}$) which is not highly transitive is contained in a maximal proper subgroup of Sym($\mathbb{N}$). And, of course, some questions on infinite permutation groups rapidly become very set-theoretic. In the other direction, the classification of the finite simple groups, via the O'Nan–Scott reduction theorem for finite primitive permutation groups, is a crucial part of the work of Cherlin and Lachlan (1986) on stable finitely homogeneous structures, and in more recent work on *smoothly approximated* structures (Kantor *et al.*, 1989). Also, many models are determined by their automorphism groups, as abstract groups or as permutation groups. Often this has been proved for a particular model via substantial permutation group theory.

Other connections between the subjects have been more subtle. In stability theory, they enter via, for example, Hrushovski's classification of groups acting transitively (in a stable structure) on a set of Morley rank one, and in certain techniques for defining a group in a structure. For a discussion of the latter in totally categorical structures, see the article by Hodges in this book (p. 111). More generally, the analogies between the language of types and orbits, seen most readily with recursively saturated and ω-categorical structures, make the language of permutation groups natural in model theory. But perhaps the richest connection of all has simply been in the wealth of intriguing questions which have appeared.

One final point. In this book, we are mostly concerned with permutation groups rather than abstract groups, that is, the groups have specified actions. But the distinction is artificial. For example, the chapters by Nesin and by Epstein and Nesin are on Frobenius groups of finite Morley rank, a natural class of permutation groups, but the arguments there are essentially about abstract groups. And there are many cases in stability theory where a group is defined in a structure as a permutation group, but then handled as an abstract group.

1 Models

We shall start our discussion here with the structures (or *models*) on which our groups will act. We will state a number of major theorems without proofs. In most cases, much fuller treatments can be found in, for example, Hodges (1993) or Chang and Keisler (1990).

A *structure* or *model* is a set M (called the model's *domain*) together with certain relations and functions on this domain, and named elements of this domain. For example, a group $(G, \cdot)$ is a pair consisting of a set G and a binary function $\cdot : G^2 \to G$, and the familiar linear order $(\mathbb{Q}, <)$ is the set $\mathbb{Q}$ together with a binary relation $<$. The words 'first-order' refer to the fact that these relations and functions are always subsets of M^n or functions $M^n \to M$ for some $n \in \mathbb{N}$ and thus many interesting properties can be expressed with the quantifiers $\forall x$ and $\exists y$ etc., where the variables x, y range over elements of M, rather than over subsets of M or other 'higher-order' notions. Thus, although topological notions do seem to be entering model theory slowly, topological structure (of a topological group, or of the usual topology on $\mathbb{Q}$, for example) cannot really be considered as 'first-order' in any natural way.

DEFINITION 1.1. A *model* with domain M is a tuple

$$\mathfrak{M} = (M; \ldots R_i \ldots; \ldots F_j \ldots; \ldots c_k \ldots)_{i \in I, j \in J, k \in K}$$

where $\mathscr{R} = \{R_i : i \in I\}$ is a family of relations, $R_i \subseteq M^{n_i}$, $\mathscr{F} = \{F_j : j \in J\}$ is a family of functions, $F_j \colon M^{m_j} \to M$, and $\mathscr{C} = \{c_k : k \in K\}$ is a family of constants, $c_k \in M$.

Some authors allow M to be empty, but most explicitly disallow this. Some identify $\mathfrak{M}$ with its domain M, using M for both, and some use $|\mathfrak{M}|$ or $\mathrm{Dom}\,\mathfrak{M}$ to denote M. The *cardinality* of the model $\mathfrak{M}$ (denoted $\mathrm{card}\,\mathfrak{M}$, but sometimes $|\mathfrak{M}|$ or even $||\mathfrak{M}||$) is the cardinality of its domain M.

Each model is associated with a formal language that describes it, and it is the interplay between the mathematical structure of the model and this logical language that forms the traditional part of model theory. First, associated with each model is its *similarity type* consisting of the following data:

$$\begin{array}{ll} I, J, K & \text{(assume that these are ordinals);} \\ \alpha : i \mapsto \mathrm{arity}\,(R_i); & \\ \beta : j \mapsto \mathrm{arity}\,(F_j). & \end{array}$$

(The *arity* of a relation R or function F is the $n \in \mathbb{N}$ for which $R \subseteq M^n$ or $F \colon M^n \to M$.) Then, for each similarity type we have the language $\mathscr{L}$ with symbols for

1. variables $v_0, v_1, \ldots$ (thought of as ranging over the domain of some given model),
2. logical operations $=$, $\wedge$ (or &, for 'and'), $\vee$ ('or'), $\to$ (or $\supset$ for 'implies'), $\neg$ ('not'), $\leftrightarrow$ ('iff'), $\forall$, $\exists$, and brackets ( );
3. relation symbols R_i for each $i \in I$, representing an $\alpha(i)$-ary relation;
4. function symbols F_j for each $j \in J$, representing a $\beta(j)$-ary function;
5. constant symbols c_k for each $k \in K$.

The distinction between the symbols R, F, etc., and the relations and functions they represent is important for some work; usually, however, the same letters can be used for each without any confusion, and this is the convention that most authors adopt. A language is a *relational language* if it has no function symbols or constant symbols.

Formulas of $\mathscr{L}$ are built up according to precise rules. But again the convention is that informal description of formulas is sufficient, with brackets being omitted when possible, $\wedge$ and $\vee$ being 'more binding than' $\to$ and $\leftrightarrow$, and quantifiers and $\neg$ being the most binding logical operators of all. The connections between models and languages is given by Tarski's definition of truth; if $\mathfrak{M}$ and $\mathscr{L}$ are of the same similarity type (we shall usually say '$\mathfrak{M}$ is a $\mathscr{L}$-structure', or '$\mathfrak{M}$ is a model for $\mathscr{L}$'), ϕ is a formula of $\mathscr{L}$, and $a_0, a_1, \ldots$ are elements of the domain of $\mathfrak{M}$ then we write

$$\mathfrak{M} \models \phi[a_0, a_1, \ldots] \qquad (*)$$

to mean ϕ is true in $\mathfrak{M}$ when v_0 is interpreted by a_0, v_1 by a_1, etc. More often, the free variables in ϕ will be indicated by writing ϕ as $\phi(v_0, v_1, \ldots, v_k)$ and we write $(*)$ as

$$\mathfrak{M} \models \phi(a_0, a_1, \ldots, a_k).$$

Formulas with no free variables are called *sentences*—these are the formulas which are true or false in a model independently of any choice of elements $a_0, a_1, \ldots$ for the variables $v_0, v_1, \ldots$ For a set of sentences S, we write $\mathfrak{M} \models S$ to mean $\mathfrak{M}$ satisfies each sentence in S. The structure $\mathfrak{M}$ is a *model of* a sentence σ or a set of sentences S if $\mathfrak{M} \models \sigma$ or S. (The 'correct' distinction between the words 'model' and 'structure' can now be explained: the word 'model' is used when it is a model *of something*, whether this 'something' is mentioned explicity or only implicit. Strictly, the word 'structure' should be used in all other cases. However, nearly all authors blur this distinction and use the two words interchangably.) A set of sentences S is *consistent* if it has a model, in which case it is often called a *theory*. If $\mathfrak{M}$ is an $\mathscr{L}$-structure, the *theory of* $\mathfrak{M}$, $\mathrm{Th}\,\mathfrak{M}$ is the set of all $\mathscr{L}$-sentences true in $\mathfrak{M}$. A consistent theory is *complete* if for every sentence σ, either σ or $\neg\sigma$ is in the theory; that is, if it is $\mathrm{Th}\,\mathfrak{M}$ for some $\mathfrak{M}$. (It is perhaps more usual to introduce these notions syntactically. Thus, a set of sentences is *consistent* if it does not prove a contradiction when the set of sentences is augmented by certain standard logical axioms and rules of derivation. The *Completeness Theorem* of Gödel—nothing to do with complete theories—shows that the above definition is the same.)

The most basic result in model theory is the compactness theorem.

THEOREM 1.2. (Compactness) A set of $\mathscr{L}$-sentences S has a model $\mathfrak{M}$ iff every finite subset of S has a model.

In simple applications, one often builds a structure $\mathfrak{M}$ for the language $\mathscr{L}$ by applying the compactness theorem to a set of sentences S in a larger language $\mathscr{L}'$, obtaining an $\mathscr{L}'$-structure $\mathfrak{M}'$, and then letting $\mathfrak{M}$ be the structure with the same domain as $\mathfrak{M}'$ but with fewer relations, functions, or constants. The words *expansion* and *reduct* refer exclusively to the processes of changing a language or structure by adding or removing some new relations, functions or constants, but (in the case of structures) keeping the same domain. For example, if A is a new set of constants (a subset of the domain of some $\mathscr{L}$-structure perhaps), $\mathscr{L}(A)$ denotes the expanded language when (symbols for) the new constants are added and $(\mathfrak{M}, a)_{a \in A}$ is the corresponding expanded structure. In practice though, we often abuse notation and denote $(\mathfrak{M}, a)_{a \in A}$ by $\mathfrak{M}$. The reduct of an $\mathscr{L}'$-structure $\mathfrak{N}$ to the smaller language $\mathscr{L}$ is denoted $\mathfrak{N}\restriction\mathscr{L}$.

Model theory also has notions of *extensions* and *substructures* or *submodels* of models. Here, the language (or, more properly, the similarity type) stays the same but the domain in made larger or smaller. A *submodel* or *substructure* $\mathfrak{N}$ of $\mathfrak{M}$ is determined by a set $N \subseteq M$ closed under all functions in $\mathscr{F}$ and containing all constants in $\mathscr{C}$. The relations and functions of $\mathfrak{N}$ are the restrictions of those from $\mathfrak{M}$. Thus, a submodel of a group $(G, \cdot)$ is a subsemigroup, whereas a submodel of $(G, \cdot, ^{-1}, 1)$ is a subgroup. An *extension* of $\mathfrak{M}$ is a model $\mathfrak{N}$ in which $\mathfrak{M}$ is a substructure. The notation $\mathfrak{N} \subseteq \mathfrak{M}$ denotes that $\mathfrak{N}$ is a substructure of $\mathfrak{M}$. A special case is that of *elementary* extensions, which preserve $\mathscr{L}$-formulas. If $\mathfrak{N}$ and $\mathfrak{M}$ are $\mathscr{L}$-structures with $\mathfrak{N}$ a substructure of $\mathfrak{M}$ then $\mathfrak{M}$ is an *elementary* extension of $\mathfrak{N}$, $\mathfrak{N} \prec \mathfrak{M}$, if for all $\mathscr{L}$-formulas $\phi(v_0, v_1, \ldots, v_k)$ in any number k of free variables, and all $a_0, a_1, \ldots, a_k$ from the domain of $\mathfrak{N}$,

$$\mathfrak{N} \models \phi(a_0, a_1, \ldots a_k) \quad \text{iff} \quad \mathfrak{M} \models \phi(a_0, a_1, \ldots a_k).$$

LEMMA 1.3. (The Tarski–Vaught Test) If $\mathfrak{N}$ is a substructure of $\mathfrak{M}$ then $\mathfrak{N} \prec \mathfrak{M}$ iff for all $a_0, \ldots, a_k$ in N and all $\phi(v_0, \ldots, v_k, v_{k+1})$,

$$\mathfrak{M} \models \exists v_{k+1}\, \phi(a_0, \ldots, a_k, v_{k+1}) \quad \text{implies} \quad \exists b{\in}N\ \mathfrak{M} \models \phi(a_0, \ldots, a_k, b).$$

There is no corresponding notion of an extension of a language and authors often refer to expansions of languages as extensions.

The two Löwenheim–Skolem Theorems (the first of which is an easy consequence of compactness and the second of the Tarski–Vaught Test) are

THEOREM 1.4. (Upward Löwenheim–Skolem Theorem) If $\mathfrak{M}$ is an $\mathscr{L}$-structure and κ is a cardinal with $\kappa \geqslant \max(\aleph_0, \operatorname{card} M, \operatorname{card} \mathscr{L})$ then there is an $\mathscr{L}$-structure $\mathfrak{N} \succ \mathfrak{M}$ with $\operatorname{card} N = \kappa$.

THEOREM 1.5. (Downward Löwenheim–Skolem Theorem) If $\mathfrak{M}$ is an $\mathscr{L}$-structure, $A \subseteq M$, and κ is a cardinal with

$$\operatorname{card}(M) \geqslant \kappa \geqslant \max(\aleph_0, \operatorname{card} A, \operatorname{card} \mathscr{L})$$

then there is an $\mathscr{L}$-structure $\mathfrak{N} \prec \mathfrak{M}$ with $A \subseteq N$ and $\operatorname{card}(N) = \kappa$.

These are proved in most books on model theory—see for example Hodges (1993, Corollaries 3.1.5 and 6.1.4).

There is also a good notion of *isomorphism*. Let $\mathfrak{M}$, $\mathfrak{N}$ be $\mathscr{L}$-structures. Then an *isomorphism* $\mathfrak{M} \to \mathfrak{N}$ is just a bijection $M \to N$ which preserves relations, functions, and constants in the usual algebraic or combinatorial sense. An *automorphism* of $\mathfrak{M}$ is an isomorphism of $\mathfrak{M}$ to itself. The set of all automorphisms of $\mathfrak{M}$ forms a group under composition, denoted $\operatorname{Aut} \mathfrak{M}$.

A *tuple* from a set M is an element of M^n for some $n \in \mathbb{N}$, that is, an n-tuple for some n. Model theorists commonly denote tuples by a bar or arrow over the letter (and without any other comment) thus: $\vec{a} \in M^n$ or $\bar{a} \in M^n$. Even the power 'n' is usually omitted. Given such a tuple $\bar{a} \in M$ from a model $\mathfrak{M}$, the *type* of $\bar{a}$, $\mathrm{tp}_{\mathfrak{M}}(\bar{a})$ or $\mathrm{t}_{\mathfrak{M}}(\bar{a})$ is the set of all formulas $\phi(v_0, v_1, \ldots, v_{k-1})$ for which $\phi(a_0, a_1, \ldots, a_{k-1})$ is true in $\mathfrak{M}$. (The $\mathfrak{M}$ in $\mathrm{tp}_{\mathfrak{M}}()$ or $\mathrm{t}_{\mathfrak{M}}()$ is omitted if clear.) A *k-type* is the type of a tuple $\bar{a}$ of length k. If $B \subseteq M$ then $\mathrm{tp}_{\mathfrak{M}}(\bar{a}/B)$ is the type of $\bar{a}$ in the expanded structure $(\mathfrak{M}, b)_{b\in B}$. A *type over a theory* T is a set of formulas that could be realized as $\mathrm{tp}_{\mathfrak{M}}(\bar{a})$ for some $\bar{a} \in \mathfrak{M} \models T$, i.e., it is a maximal set of formulas consistent with T when $v_0, v_1, \ldots, v_{k-1}$ are treated as new constant symbols. Since there are no additional parameters for such a type, it is also common usage to call it a *type over* $\varnothing$. It is an *n-type* if the tuple $\bar{a}$ is an n-tuple. A *type over a model* $\mathfrak{M}$ is a type over $\mathrm{Th}(\mathfrak{M}, a)_{a\in M}$. (Note, however, that some authors use 'type over T' in the more general sense of not necessarily being maximal, or in other words simply a subset of $\mathrm{tp}_{\mathfrak{M}}(\bar{a})$ for some $\bar{a} \in \mathfrak{M} \models T$. The definition just given of a type over a model is sometimes varied too.) If T is a complete theory, $\mathfrak{M} \models T$, and $A \subset M$, then $S_n(A)$ denotes the set of all complete n-types of $\mathrm{Th}(\mathfrak{M}, a)_{a\in A}$, and $S(A) := \bigcup\{S_n(A) : n \in \mathbb{N}\}$ (the theory T and the model $\mathfrak{M}$ should be clear from the context). A complete type $p(\bar{x})$ over a complete theory T is said to be *isolated* if there is a formula $\phi(\bar{x})$ in $p(\bar{x})$ such that, for all $\psi(\bar{x})$ in $p(\bar{x})$, $\forall\bar{x}\,(\phi(\bar{x}) \to \psi(\bar{x}))$ is true in all models of T. Observe that isolated types are realized in all models of T. A type will be called *algebraic* if it contains a formula which has just finitely many realizations. Any complete algebraic type over a complete theory is isolated.

The notion of type allows us to define some particularly well-behaved classes of models. A model $\mathfrak{M}$ is *atomic* if the type of each tuple $\bar{a} \in M^{<\omega}$ is isolated. Atomic models are usually thought of as 'small' since they realize as few types as possible. In 'large' models, many types are realized. If λ is an infinite cardinal, we say that a structure $\mathfrak{M}$ is *λ-saturated* if for every $A \subset M$ with $\mathrm{card}(A) < \lambda$, every type over $\mathrm{Th}(\mathfrak{M}, a)_{a\in A}$ is realized in M. An ω-saturated model is sometimes also called a *countably saturated* model. Also, $\mathfrak{M}$ is *saturated* if it is $\mathrm{card}(M)$-saturated. Thus, a countably infinite structure is saturated iff the following holds: for any $\bar{a} \in M^{<\omega}$ and any set of formulas $\Phi(\bar{a}, \bar{x})$ in $\mathscr{L}(\bar{a})$, if $\mathfrak{M}$ has an elementary extension $\mathfrak{N} \succ \mathfrak{M}$ with $\bar{c} \in N$ such that $\mathfrak{N} \models \phi(\bar{a}, \bar{c})$ for each $\phi \in \Phi$ then there is $\bar{b} \in M$ such that $\mathfrak{M} \models \phi(\bar{a}, \bar{b})$ for each $\phi \in \Phi$. A countably infinite model $\mathfrak{M}$ can be both saturated and atomic, in which case we say that it is *countably categorical* (*$\aleph_0$-categorical*, or *ω-categorical*). These models turn out to have several alternative characterizations, one of which is that for any countable $\mathfrak{N} \models \mathrm{Th}(\mathfrak{M})$ we have $\mathfrak{M} \cong \mathfrak{N}$. Another characterization involves the automorphism group; see Theorem 5.1 below for further discussion.

It is quite easy to build λ-saturated models, but they may turn out to have cardinality greater than λ, and hence not to be saturated. The key tool is the following general, and widely applicable, theorem.

THEOREM 1.6. (Elementary Chains Theorem) Let κ be an ordinal, and suppose that $(\mathfrak{M}_\lambda : \lambda < \kappa)$ is a chain of structures such that whenever $\mu < \lambda < \kappa$ we have $\mathfrak{M}_\mu \preceq \mathfrak{M}_\lambda$. Then the structure $\bigcup_{\lambda<\kappa} \mathfrak{M}_\lambda$ is an elementary extension of each $\mathfrak{M}_\lambda$ $(\lambda < \kappa)$.

The idea now is that, to build a λ-saturated model of some theory T, we build an elementary chain of models of T, taking unions at limit ordinals, and at successor ordinals using compactness to put in realizations of certain types over sets of size less than λ. By doing this with care, and then taking the union of the chain, we obtain the required λ-saturated model. But to obtain a *saturated* model, that is, to ensure that the λ-saturated model we build does not accidently have cardinality greater than λ, we need stronger hypotheses: either set-theoretic conditions on λ, or restrictions on the number of types over sets, that is, *stability conditions.*

For an infinite cardinal κ, a complete theory T is said to be *κ-categorical* if it has, up to isomorphism, only one model of cardinality κ. Thus, a complete theory is $\aleph_0$-categorical iff it has an $\aleph_0$-categorical model, according to the definition above. The notion of $\aleph_0$-categorical theory is perhaps slightly more 'primitive', and many authors use '$\aleph_0$-categorical model' simply to mean an infinite model of an $\aleph_0$-categorical theory—this is not quite the same because of the Löwenheim–Skolem Theorems. For countable languages, Morley's Theorem (proving a conjecture made by Łoś) says that a complete theory T is κ-categorical for some uncountable κ if and only if it is κ-categorical for all uncountable κ. Such theories are often referred to as *uncountably categorical*, or $\aleph_1$-categorical. A theory that is κ-categorical for all infinite κ is *totally categorical.*

It turns out that not all countably infinite models $\mathfrak{M}$ have an elementary extension which is countable and saturated. (In fact such an extension exists if and only if the set of types over $\mathrm{Th}\,\mathfrak{M}$ is countable.) Other notions of saturation have been devised for theories with no countable saturated models. See Part II for a discussion of one of the most useful of these.

If $\mathfrak{M}$ is a model for $\mathscr{L}$ and $A \subseteq M$, then $\mathrm{dcl}_{\mathfrak{M}}(A)$ (for 'definable closure') is the set of all $b \in M$ for which there is a formula $\phi(x)$ of $\mathscr{L}(A)$ such that

$$\mathfrak{M} \vDash \phi(b) \wedge \forall x\,(\phi(x) \to x = b),$$

i.e., b can be defined in $\mathfrak{M}$ using parameters from A. There is a related notion of *algebraic closure*, $\mathrm{acl}_{\mathfrak{M}}(A)$ which is the set of all $b \in M$ for which there are $\phi(x)$ in $\mathscr{L}(A)$, $n \in \mathbb{N}$, and $b_1, \dots, b_n \in M$ such that

$$\mathfrak{M} \vDash \phi(b) \wedge \forall x\,(\phi(x) \to x = b_1 \vee \dots \vee x = b_n).$$

The *closure of* A, $\mathrm{cl}_{\mathfrak{M}}(A)$ is usually defined to be one of $\mathrm{acl}_{\mathfrak{M}}(A)$ or $\mathrm{dcl}_{\mathfrak{M}}(A)$, depending on the context.

A set $A \subseteq M^n$ is *definable* over B if there is a formula $\phi(x_0, \ldots, x_{n-1})$ of $\mathscr{L}(B)$ such that

$$A = \{(a_0, \ldots, a_{n-1}) \in M^n : \mathfrak{M} \models \phi(a_0, \ldots, a_{n-1})\}.$$

It is *definable* if it is definable over some finite $B \subseteq M$, and it is 0-*definable* or $\varnothing$-definable if it is definable over $\varnothing$. In a similar way, a function $f: A \to B$ is definable if the sets A, B and $\{(\bar{a}, \bar{b}) : \bar{a} \in A \wedge \bar{b} = f(\bar{a})\}$ are all definable. Also, we say that a set A is *type-definable* over a set of parameters B if it is the set of solutions (in a model of an ambient theory) of a set of formulas with parameters in B, that is, a (possibly incomplete) type over B.

Algebra enters model theory in several ways. First, of course, the model considered may be an algebraic object—a group, a ring, a field, etc. Model theoretic methods can turn out to be quite powerful in handling such structures. Second, a model may *interpret* an algebraic structure, that is there may be a definable subset A of M^n, a definable equivalence relation $\equiv$ on A and definable functions $F_1, \ldots$, relations $R_1, \ldots$, etc., such that

$$(A/\equiv, R_1, \ldots, F_1, \ldots)$$

is a group, ring, or field, etc. Finally (and this is the main motivation for this book) we may investigate the properties of the group

$$G = \mathrm{Aut}(\mathfrak{M})$$

of automorphisms of $\mathfrak{M}$.

2 Stability

This is not a book about stability theory, but the subject is floating near the edge of several of the articles, notably those by Hodges, Wagner, Ivanov, and those in Part III. The article by Hodges, and the introduction to Part III, contain some introductory remarks about stability theory. Here, we just give a few definitions and state some major results. We will assume for the rest of the section that all languages are countable. And indeed, in this book, nearly all languages are countable.

Many of the key ideas of stability theory were first used in Morley's proof of Łoś's conjecture (see the discussion of categoricity above). Stability theory was developed in the 1970s, primarily through the work of Shelah. The classic reference is Shelah (1978), and in the next few paragraphs we will quote many results of Shelah without explicit reference. Most of the results can also be found in the books by Pillay (1983), Lascar (1987) and Baldwin (1988). One goal of stability theory was apparently to divide theories into two classes, the *good* ones, in which all models of the theory are classifiable by cardinal invariants, and the *bad* ones. Recall that by the Löwenheim–Skolem Theorems, a consistent theory over a countable language with infinite models has models of every infinite cardinality. A crude criterion for a theory to be bad is that it has 2^κ non-isomorphic models of size κ for any uncountable cardinal κ. It turns out that, with this criterion, the class of good theories is a small subclass of the stable theories.

If λ is an infinite cardinal, then a theory T is λ-*stable* if, for every $\mathfrak{M} \models T$ with $\operatorname{card} M = \lambda$, the number of 1-types over $\mathrm{Th}(\mathfrak{M}, a)_{a \in M}$ is at most λ. We say that T is *stable* if it is λ-stable for some infinite λ, and that it is *superstable* if it is λ-stable for all cardinals $\lambda \geqslant 2^{\aleph_0}$. Since T is over a countable language, it follows from results of Shelah that there are just four possibilities. The theory T could be: ω-stable; superstable but not ω-stable; stable but not superstable; or unstable. Furthermore, any ω-stable theory is λ-stable for all infinite λ (and hence is superstable), any superstable but not ω-stable theory is λ-stable if and only if $\lambda \geqslant 2^{\aleph_0}$ (and hence is stable), and any stable but unsuperstable theory is λ-stable if and only if $\lambda = \lambda^{\aleph_0}$. By a difficult result of Shelah, the unsuperstable theories are all bad under the criterion of the last paragraph.

There is an alternative combinatorial characterization of stability. A theory T is unstable if and only if there is $n \in \mathbb{N}$, a formula $\phi(\bar{x}, \bar{y})$ where $\bar{x}$ and $\bar{y}$ are n-tuples, a model $\mathfrak{M} \models T$, and a sequence $(\bar{a}_i : i < \omega)$ in M^n such that for all $i, j \in \omega$, $\mathfrak{M} \models \phi(\bar{a}_i, \bar{a}_j)$ if and only if $i < j$. We shall not here prove the equivalence of these definitions. However, by considering Dedekind cuts and using compactness, it is quite easy to see that any theory satisfying this combinatorial condition is unstable. Given this characterization of stability, it is not hard to see why unstable theories might have many models. If T is the complete theory of some total order, one can show that T has 2^κ models for any uncountable cardinal κ. Using the Ehrenfeucht–Mostowski construction (described on p. 259) to build models, one can then show that the same holds for *any* unstable theory T (over a countable language). The combinatorial details are intricate.

It is worth mentioning now that if T is a stable theory, then T has saturated models of arbitrarily large cardinality. This is essentially because, by the restriction on the number of types, one can use the Elementary Chains Theorem to build the model. It is a common convention, when considering a stable theory, to work with a fixed saturated model of the theory, known as the *monster model*, which is much larger than the sets being considered. Then all models of the theory which we handle can be regarded as elementary submodels of the monster model.

For stable theories, there are crucial notions of independence and rank. For convenience, we summarize them briefly. Suppose that T is a stable theory, and that $\mathfrak{C}$ is a large saturated model of T (or at least that $\mathfrak{C}$ is λ-saturated where λ is much larger than the cardinality of any of the sets of parameters we consider). All the models we consider will be elementary submodels of $\mathfrak{C}$. Suppose that A and B are (small) subsets of $\mathfrak{C}$, with $A \subseteq B$, and that $p \in S(A)$, that is, it is a type over the theory obtained from T by adding names for elements of A (we do not specify T when it is clear from the context). We consider types $q \in S(B)$ which contain p. We call such types *extensions of* p *over* B. In general there will be many such extensions q. We will define a privileged class of extensions, the *non-forking extensions*, whose realizations will be independent from B over A. The idea is, very roughly, that a realization c of q should be independent from B over A if there are no relationships between c and B other than those forced by the fact that c realizes p.

This is formalized as follows. First, we say that a formula $\phi(\bar{x}, \bar{y})$ is *represented* in a type $r \in S(D)$ (where D is some set of parameters), if there is some $\bar{d} \in D$ such that $\phi(\bar{x}, \bar{d}) \in r$. Then, with p, q as above, q is a *non-forking extension of* p if the following holds: for every formula $\phi(\bar{x}, \bar{y})$, if $\phi(\bar{x}, \bar{y})$ is represented in every extension of q to a model[1] of T containing B, then $\phi(\bar{x}, \bar{y})$ is also represented in every extension of p to a model of T containing A. In this situation, we also say that q *does not fork over* A. It turns out that if T is stable, then p will always have a non-forking extension. In fact, it will have at most $2^{\aleph_0}$ distinct non-forking extensions over B, and if T is ω-stable then p will have just finitely many non-forking extensions over B. A type over A is called *stationary* if it has a unique non-forking extension over any larger set of parameters.

If T is a stable theory in a countable language, then forking has the following properties (again, with all the sets living in some large model of T).

1. If $p \in S(A)$, then p does not fork over A.
2. If $A \subseteq B \subseteq C$ and $p \in S(C)$, then p does not fork over A if and only if p does not fork over B and the restriction of p to B does not fork over A.
3. If $A \subseteq B$ and $p \in S(A)$, then p has some non-forking extension $q \in S(B)$.
4. (forking symmetry) If A is a set and $\bar{b}$, $\bar{c}$ are tuples, then $\mathrm{tp}(\bar{b}/\bar{c} \cup A)$ forks over A if and only if $\mathrm{tp}(\bar{c}/\bar{b} \cup A)$ forks over A.
5. If $p \in S(B)$, then there is a countable set $A \subseteq B$ such that p does not fork over A.

It turns out that if T is superstable (and hence also if T is ω-stable), then in the last property (5), A can be chosen to be finite. This fact, Shelah's Finite Equivalence Relation Theorem, and an argument counting types yield an easy proof of a theorem of Lachlan (1974), that any ω-categorical superstable theory is ω-stable—see Proposition 5.16 of Pillay (1983). The article by Wagner in this volume makes implicit reference to the Finite Equivalence Relation Theorem, and to the notion of strong type, and again we refer the reader to Pillay (1983).

---

[1] An extension of q to a model of T is some $r \in S(M)$ containing q, such that $M \models T$

From forking, or rather non-forking, we derive a notion of independence. If A, B, C are sets, we say that B and C are *independent over* A if, for all $\bar{b} \in B$, $\text{tp}(\bar{b}/A \cup C)$ does not fork over A. By forking symmetry, this independence is symmetric between B and C. The notion is extended by saying that sets (or tuples, or elements) B_i $(i \in I)$ are *independent over* A if for all $i \in I$ and $\bar{b} \in B_i$, $\text{tp}(\bar{b}/A \cup \bigcup(B_j : j \in I \setminus \{i\}))$ does not fork over A. We also mention that a *Morley sequence* for a stationary n-type p over a set A of parameters is a sequence $(\bar{a}_\lambda : \lambda < \kappa)$ of realizations of p (where κ is some infinite ordinal) such that the following holds: for each $\lambda < \kappa$, the type of $\bar{a}_\lambda$ over $A \cup \{\bar{a}_\mu : \mu < \lambda\}$ is a non-forking extension of p. A Morley sequence of a stationary type over A will always be independent over A. More surprisingly, it will also be an *indiscernible sequence* over A (and in fact, by the order-theoretic characterization of stability, it will be an indiscernible set over A; that is, for any $r > 0$, any two r-tuples of distinct elements of A, regarded as rn-tuples of elements of $\mathfrak{C}$, have the same type).

In stability theory there are several notions of *rank* of a type or a formula. These mostly express the extent to which the set defined by the formula, or the set of realizations of the type, can be split up by other formulas. Typically, the ranks are ordinals in theories with sufficiently strong stability properties, and are ∞ otherwise. In this book only Morley rank arises. We give now a definition of Morley rank. In the article by Hodges, a game-theoretic definition valid for saturated structures is given for the property of having finite Morley rank.

Let M be a saturated structure, n a positive integer, and X a definable subset of M^n. Then the *Morley rank* $\text{RM}(X)$ (for 'rang de Morley') is defined as follows:

1. $\text{RM}(X) \geqslant 0$ if $X \neq \varnothing$.
2. If λ is a limit ordinal, then $\text{RM}(X) \geqslant \lambda$ if $\text{RM}(X) \geqslant \mu$ for all $\mu < \lambda$.
3. For any ordinal λ, $\text{RM}(X) \geqslant \lambda + 1$ if there are pairwise disjoint definable sets $X_i \subset M^n$ $(i < \omega)$ such that $\text{RM}(X \cap X_i) \geqslant \lambda$ for all $i < \omega$.
4. $\text{RM}(X) = \lambda$ if $\text{RM}(X) \geqslant \lambda$ but $\text{RM}(X) \not\geqslant \lambda + 1$.
5. If $\text{RM}(X) \geqslant \lambda$ for all ordinals λ, then $\text{RM}(X) = \infty$.

We may now define the Morley rank of a *formula* as the rank of the set it defines in a saturated model. If p is a type over a set A, then $\text{RM}(p)$ is defined to be the minimum of the ranks of the formulas in p. Finally, if X is a definable set of Morley rank λ, then it is easy to show that there is a greatest integer k such that there are disjoint definable sets $X_1, \ldots, X_k$ with $\text{RM}(X \cap X_i) = \lambda$ for each i. This number k is called the *Morley degree* of X.

Under our assumption that T is a complete theory over a countable language, it turns out that T is ω-stable if and only if the domain of any saturated model of T has ordinal Morley rank. Without the assumption of a countable language, theories with ordinal Morley rank are called *totally transcendental.* Furthermore, if T is a complete countable ω-stable theory, then forking can be defined in terms of Morley rank. In an ambient model of such a T, let A, B be sets with $A \subseteq B$, let $p \in S(A)$, and let $q \in S(B)$ with $p \subseteq q$. Then q is a non-forking extension of p if and only if $\mathrm{RM}(q) = \mathrm{RM}(p)$ (see for example Pillay (1983, Corollary 6.35)).

Finally, we mention the elegant interpretation of Morley rank and degree in algebraically closed fields. By a theorem of Tarski (1949), if F is an algebraically closed field and $n \in \mathbb{N}$, then any definable subset of F^n is definable by a quantifier-free formula. By inspection of the possible formulas, it follows easily that the definable subsets of F^n are exactly the *constructible sets*, that is, the Boolean combinations of Zariski closed sets. In particular, an algebraically closed field has Morley rank one and Morley degree one (so is *strongly minimal*). The Morley rank of a constructible set is exactly its Zariski dimension.

3 Permutation groups

If G is a group and Ω is a set then an *action* of G on Ω is given by a function $(\omega, g) \mapsto \omega^g$ from $\Omega \times G$ to Ω such that

1. $\alpha^1 = \alpha$ for all $\alpha \in \Omega$ (where 1 is the identity of G), and
2. for all $\alpha \in \Omega$ and $g, h \in G$, $\alpha^{(gh)} = (\alpha^g)^h$.

The *degree* of the action is the cardinality of the set Ω. There is no fixed convention on whether groups act on the right or the left. In this chapter they act on the right, but in some papers in this book the action is on the left. Some authors write α^g as αg (or $g\alpha$ if maps are on the left). If $\Gamma \subseteq \Omega$ and $g \in G$ then we extend the above action by writing Γ^g for $\{\gamma^g : \gamma \in \Gamma\}$. An action of G on Ω also induces actions of G on Ω^n and on $[\Omega]^n$, the set of *unordered* n-element subsets of Ω, for each $n \in \mathbb{N}$, in the obvious way.

The *kernel* of the group action is $\{g \in G : \forall \alpha{\in}\Omega\, (\alpha^g = \alpha)\}$. This is a normal subgroup of G; in fact it is the kernel of the homomorphism $G \to \mathrm{Sym}(\Omega)$ naturally induced by the action, where $\mathrm{Sym}(\Omega)$ is the symmetric group on Ω, i.e., the group of all permutations of Ω. The group action is *faithful* if it has trivial kernel. A *permutation group* is a pair (G, Ω) together with a faithful group action of G on Ω. For example, the group of automorphisms $\mathrm{Aut}\, M$ of a structure M may be considered as a permutation group, where a^g is defined to be the image of $a \in M$ under the automorphism g.

If G acts on two sets Γ and Δ, then the actions are said to be *isomorphic* or *G-isomorphic* if there is a bijection $\phi : \Gamma \to \Delta$ such that, for all $g \in G$

and all $\gamma \in \Gamma$,

$$(\gamma^g)\phi = (\gamma\phi)^g.$$

We shall say that two permutation groups (G, Ω) and (H, Δ) are *similar* if there is a bijection $\phi: \Omega \to \Delta$ such that $g \mapsto \phi^{-1}g\phi$ is an isomorphism of groups $G \to H$. There is an irritating inconsistency in terminology here which appears never to have been resolved: it could be that the same group G has two permutation representations which are similar but not isomorphic, if, say, an automorphism of G is required to exhibit the permutation group isomorphism. For example, if $n \geqslant 3$, F is a field, and $G = \mathrm{PGL}(n, F)$, then G has permutation representations on the set of points of $\mathrm{PG}(n-1, F)$ and this induces an action on the set of hyperplanes of $\mathrm{PG}(n-1, F)$; these actions are similar but not G-isomorphic. The use of 'similar' and 'isomorphic' which we are adopting is not common to all chapters of the book, but usually the meaning is clear from the context.

Let (G, Ω) be a permutation group. For $\bar{\alpha} \in \Omega^n$, $G_{\bar{\alpha}}$ denotes the stabilizer

$$G_{\bar{\alpha}} := \{g \in G : \bar{\alpha}^g = \bar{\alpha}\}.$$

For $\Gamma \subseteq \Omega$, the *setwise stabilizer* of Γ in G is the group

$$G_{\{\Gamma\}} := \{g \in G : \Gamma^g = \Gamma\}.$$

Also, the *pointwise stabilizer* of Γ in G is

$$G_{(\Gamma)} := \{g \in G : (\forall\gamma\in\Gamma)(\gamma^g = \gamma)\}.$$

We introduce some more standard terminology. If G acts on Ω, define the relation $\sim$ on Ω by $\alpha \sim \beta$ iff there is $g \in G$ such that $\alpha^g = \beta$. Then $\sim$ is an equivalence relation; equivalence classes are called *orbits*. The action is said to be *transitive* if it has just one orbit. If Γ is an orbit or a union of orbits of G acting on Ω, then G has (by restriction) a group action on Γ (which may not be faithful). We shall denote by G^Γ the permutation group induced on Γ by G in this action. So, as an abstract group,

$$G^\Gamma := G/\{g \in G : \gamma^g = \gamma \text{ for all } \gamma \in \Gamma\}.$$

These notions are extended to the action of G on Ω^k and $[\Omega]^k$ for non-zero $k \in \mathbb{N}$; the group (G, Ω) is said to be *k-transitive* if it is transitive on the set of ordered k-subsets of Ω, and it is *k-homogeneous* if it is transitive on the set of unordered k-subsets of Ω. We say that G is *highly transitive* (respectively *highly homogeneous*) if it is k-transitive (respectively k-homogeneous) for each positive integer k. Also, the permutation group (G, Ω) is said to be *regular* if it is transitive and $G_\alpha = 1$ for some (and hence, any) $\alpha \in \Omega$. It is an easy exercise to see that any faithful, transitive, abelian permutation group is regular, and that a regular permutation group (G, Ω) is G-isomorphic to the action of G on itself by right multiplication.

If (G, Ω) is a permutation group, then one can sometimes reduce it to a family of more manageable groups. First, if Ω is considered as the disjoint union of G-orbits $\{\Omega_i : i \in I\}$, then G is isomorphic to an (unrestricted) subdirect product of the groups $\{G^{\Omega_i} : i \in I\}$. Since the groups G^{Ω_i} are all transitive, we can expect to reduce questions about permutation groups to questions about transitive groups.

The next reduction works smoothly for finite groups, but is more problematic in the infinite case. We say that G is *primitive* on Ω if there is no G-invariant partition on Ω other than the trivial ones $\{\Omega\}$ and $\{\{\alpha\} : \alpha \in \Omega\}$. The group G is said to be *imprimitive* otherwise. We also call G-invariant equivalence relations *G-congruences* or simply *congruences* if G is clear. If (G, Ω) is transitive but imprimitive, and E is a non-trivial G-congruence on Ω, then the set of E-classes is called a *system of imprimitivity* and its elements are called *blocks of imprimitivity.* Suppose that (G, Ω) is transitive and has a system of imprimitivity $\{\Sigma_i : i \in I\}$. Then we may consider the groups $H_i := G^{\Sigma_i}$ (for any $i \in I$) and K, where K is the permutation group induced by G on the set $\{\Sigma_i : i \in I\}$. By the transitivity of K, the permutation groups (H_i, Σ_i) are all similar in the above sense. Many questions about G are reducible to questions about the H_i and K. In particular, there is a unique largest permutation group on Ω which preserves the above equivalence relation and induces H_i on each Σ_i and K on the set of classes. This is the wreath product $H \operatorname{Wr} K$ (where H is isomorphic to the H_i). It has a normal subgroup (the '*base* group') isomorphic to the cartesian product of $\operatorname{card} I$ copies of H, and is a split extension of this by the '*top* group', the group K. As an abstract group, the wreath product is determined by H, $\operatorname{card} I$, K and the action of K on I. For more on wreath products, see pp. 68–69 in the article by Evans in this volume. If (G, Ω) is an infinite permutation group then there may be infinite chains of G-congruences on Ω, and in particular, it may be that there is no maximal or minimal G-congruence. However, it is easily checked that if G has c orbits on the set of unordered pairs from Ω then there are at most $c - 1$ non-trivial G-congruences, in addition to the two trivial G-congruences.

If (G, Ω) is a transitive permutation group and $\alpha \in \Omega$ then (G, Ω) is G-isomorphic to the natural G-action on right cosets of G_α given by $(G_\alpha h)^g = G_\alpha(hg)$. (The isomorphism $\Omega \to G/G_\alpha$ is defined by $\beta \mapsto G_\alpha h$ where h satisfies $\alpha^h = \beta$.) Furthermore, there is a bijection between the lattice of G-congruences and the lattice $\{H : G_\alpha \leqslant H \leqslant G\}$. Here, if $G_\alpha < H < G$ then the set of right cosets of H gives a system of imprimitivity, and any system of imprimitivity arises in this way. In particular, it follows that if G is transitive on Ω, and $\alpha \in \Omega$, then G is primitive on Ω if and only if the point stabilizer G_α is a maximal (proper) subgroup of G.

Let us conclude with a result due to P. M. Neumann (1976, Lemma 2.3), really a version of an older result of B. H. Neumann (1954), which finds application in both permutation group theory and model theory. The proof here is copied from lectures given by P. M. Neumann and is based on the proof in Birch *et al.* (1976).

THEOREM 3.1. (The Separation Lemma) Let G be a group and suppose G acts on Ω. Suppose that all G-orbits of Ω are infinite, and that Γ and Δ are finite subsets of Ω. Then there is $g \in G$ such that $\Gamma^g \cap \Delta = \varnothing$.

Proof. By induction on $c = \operatorname{card}\Gamma$. The theorem is trivial for $c = 0$. Suppose it is true for all Γ' of size $c-1$ and all Δ'. Consider Γ and Δ and put $d = \operatorname{card}\Delta$. Without loss we may assume $\Gamma \not\subseteq \Delta$ (for otherwise find $a \in \Gamma$ and $g \in G$ with $a^g \notin \Delta$ and replace Γ by Γ^g). Let $\gamma_0 \in \Gamma \setminus \Delta$ and let $\Gamma_0 = \Gamma \setminus \{\gamma_0\}$. By the induction hypothesis there are $g_0, g_1, \ldots, g_d$ such that

$$\begin{aligned} \Gamma_0^{g_0} \cap \Delta &= \varnothing; \\ \Gamma_0^{g_1} \cap (\Delta \cup \Delta^{g_0}) &= \varnothing; \\ &\cdots \\ \Gamma_0^{g_d} \cap (\Delta \cup \Delta^{g_0} \ldots \cup \Delta^{g_{d-1}}) &= \varnothing. \end{aligned}$$

If $\gamma_0^{g_i} \notin \Delta$ for some i we are finished. But otherwise

$$\gamma_0^{g_0}, \gamma_0^{g_1}, \ldots, \gamma_0^{g_d} \in \Delta$$

hence $\gamma_0^{g_i} = \gamma_0^{g_j}$ for some $0 \leqslant i < j \leqslant d$. Let $g = g_j g_i^{-1}$. Then $\gamma_0^g = \gamma_0 \notin \Delta$, $\Gamma_0^g = \Gamma_0^{g_j g_i^{-1}}$, and $\Gamma_0^{g_j} \cap \Delta^{g_i} = \varnothing$ since $i < j$. Hence $\Gamma_0^g \cap \Delta = \varnothing$, so $\Gamma^g \cap \Delta = \varnothing$, as required. □

4 Groups of linear transformations

Here, we set up the notation for the classical matrix groups, and their infinite dimensional analogues. Let V be a vector space over a field F. Then $\mathrm{GL}(V)$, the *general linear group*, is the group of all non-singular linear transformations $V \to V$. If $\dim V$ is a natural number n, then the group may be denoted $\mathrm{GL}(n, F)$, or $\mathrm{GL}(n, q)$ if F is the finite field $\mathrm{GF}(q)$. The centre of $\mathrm{GL}(V)$ is the group of all scalar mappings $v \mapsto av$ (where $a \in F$), so is isomorphic to F^*, the multiplicative group of F. The quotient by the centre is denoted $\mathrm{PGL}(V)$. If $\dim(V)$ is finite, then we may talk of the *determinant* of a linear transformation. The *special linear group* $\mathrm{SL}(V)$ is the subgroup of $\mathrm{GL}(V)$ consisting of linear transformations of determinant one. The centre of $\mathrm{SL}(V)$ is the group of scalar linear transformations of determinant one, and the quotient by the centre is denoted $\mathrm{PSL}(V)$ (or $\mathrm{PSL}(n, F)$, and so on). It is well known that if n is finite, then $\mathrm{PSL}(n, F)$ is simple, except in the cases $(n, F) = (2, \mathrm{GF}(2))$ or $(n, F) = (2, \mathrm{GF}(3))$.

If V is infinite dimensional, then $\mathrm{GL}(V)$ has a normal subgroup with no finite analogue, the group of linear transformations whose fixed point space has finite codimension. However, apart from this, the normal subgroup structure is as one would expect—see Rosenberg (1958) for details.

A *semilinear* transformation of V is a mapping $\phi: V \rightarrow V$, determined by an automorphism σ of the field F, and satisfying $(au + bv)^\phi = a^\sigma u^\phi + b^\sigma v^\phi$ for all $a, b \in F$ and $u, v \in V$. The group of all semilinear transformations is denoted $\Gamma\mathrm{L}(V)$, and the corresponding quotient by the centre is $\mathrm{P}\Gamma\mathrm{L}(V)$. It can be checked that $\Gamma\mathrm{L}(V)$ is a split extension of $\mathrm{GL}(V)$ by the automorphism group of F. The *projective space* $\mathrm{PG}(V)$ corresponding to V is the incidence structure of *points, lines, planes*, etc., whose points are the one-dimensional subspaces of V, lines are two-dimensional subspaces of V, planes are three-dimensional subspaces, and hyperplanes are $\dim(V) - 1$ dimensional subspaces. Incidence is given by inclusion. The Veblen–Young Theorem asserts that the full automorphism group of $\mathrm{PG}(V)$ is $\mathrm{P}\Gamma\mathrm{L}(V)$. If $\dim(V) = n$ then $\mathrm{PG}(V)$ is said to have (projective) dimension $n - 1$, and is denoted $\mathrm{PG}(n - 1, F)$ (so $\mathrm{PGL}(n, F)$ acts naturally on $\mathrm{PG}(n - 1, F)$).

The *affine space* corresponding to V, denoted $\mathrm{AG}(V)$, is the incidence structure whose points are the vectors in V, lines are cosets of one-dimensional subspaces of V, planes are cosets of two-dimensional subspaces of V, etc., again with incidence given by inclusion. The vector space V, regarded as an abelian group, acts transitively on itself by addition as a group of automorphisms of the affine space. Also, $\mathrm{GL}(V)$ is a group of automorphisms of the affine space which stabilizes the zero vector and normalizes V. Together they generate their semidirect product, the *affine group* $\mathrm{AGL}(V)$, consisting of all maps $v \mapsto v^g + w$, where $w \in V$ and $g \in \mathrm{GL}(V)$. The full automorphism group of affine space is $\mathrm{A}\Gamma\mathrm{L}(V)$, with $\Gamma\mathrm{L}(V)$ acting naturally on V.

These groups all have important subgroups which preserve bilinear or quadratic forms, possibly twisted by a field automorphism. These are the symplectic, orthogonal, and unitary groups. We shall not define them here. The main theorem in the article by Thomas (p. 199) in this book concerns the countable dimensional symplectic group over a finite field, and we refer the reader to that chapter for the appropriate definitions.

5 Models and permutation groups

There is a standard way of regarding a permutation group (G, Ω) as a group of automorphisms of a relational structure with domain Ω. Introduce a relational language $\mathscr{L}$ which has, for each positive integer k and each G-orbit $\mathscr{O}$ of Ω^k, a unique k-ary relation symbol $R_\mathscr{O}$, and interpret $R_\mathscr{O}(\alpha_1, \ldots, \alpha_k)$ to be true just in case $(\alpha_1, \ldots, \alpha_k) \in \mathscr{O}$. The resulting structure $\mathfrak{M}$ is known as the *canonical structure* for (G, Ω), and the language $\mathscr{L}$ as the *canonical language*. For infinite Ω, it could be that G is a proper subgroup of Aut $\mathfrak{M}$, but G and Aut $\mathfrak{M}$ always have the same orbits on the set of finite ordered subsets of M. In fact Aut $\mathfrak{M}$ is the *closure*, $\overline{G}$, of G, where

$$\overline{G} := \{g \in \mathrm{Sym}(\Omega) : \forall n{\in}\mathbb{N}\, \forall \bar{a}{\in}\Omega^n\, \exists h{\in}G\, (\bar{a}^h = \bar{a}^g)\}.$$

(See below for more details on the topology from which this arises.)

A similar argument shows that if G is not 2-homogeneous on Ω then it is a group of automorphisms of some graph on Ω which is neither complete nor null—take as the edge set any orbit on unordered pairs. Here, G may not be the *full* automorphism group of the graph. Similarly, one can often investigate permutation groups (G, Ω) by finding G-invariant digraphs, tournaments, or k-uniform hypergraphs.

The canonical model $\mathfrak{M}$ for a permutation group (G, Ω) is atomic (in fact, $\mathrm{tp}_{\mathfrak{M}}(\bar{a})$ is determined by $R_{\mathscr{O}}(\bar{a})$ where $\mathscr{O}$ is the G-orbit of $\bar{a}$), and if it is countably infinite has a countable saturated elementary extension. (But perhaps these observations just serve to illustrate the fact that most model theoretic notions depend on rather more than just Aut $\mathfrak{M}$.) The canonical model seems to be of most use in the case when $\mathfrak{M}$ is a countable model of an $\aleph_0$-categorical theory. In this case we have the following result.

THEOREM 5.1. (Engeler, Ryll-Nardzewski, Svenonius) Let $\mathfrak{M}$ be a countably infinite $\mathscr{L}$-structure for a countable language $\mathscr{L}$, and let $G = \mathrm{Aut}\,\mathfrak{M}$, considered as a permutation group acting on $\Omega = \mathrm{Dom}(\mathfrak{M})$. Then the following are equivalent:

1. Th $\mathfrak{M}$ is $\aleph_0$-categorical;
2. for each $k \in \mathbb{N}$ there are only finitely many k-types over Th $\mathfrak{M}$;
3. for each k in $\mathbb{N}$, Ω^k has only finitely many G-orbits;
4. every model of Th $\mathfrak{M}$ is atomic;
5. every type over Th $\mathfrak{M}$ is isolated.

For a proof of this see Hodges (1993, Theorem 7.3.1). The most substantial ingredient in the proof is Vaught's Omitting Types Theorem (see for example Hodges (1993, Theorem 7.2.1)).

If card $\Omega = \aleph_0$ and (G, Ω) has the third condition here, it is said to be *oligomorphic*. Note that if (G, Ω) is an oligomorphic permutation group, then the corresponding canonical structure is $\aleph_0$-categorical.

Here is a simple application of model theoretic methods to oligomorphic permutation groups. Its proof is implicit in Theorem 2.1 on p. 259. (One applies that theorem to the canonical model to obtain that the proposition holds for some elementary extension, and then notes that the canonical model is $\aleph_0$-categorical, so the elementary extension is isomorphic to the original structure.)

PROPOSITION 5.2. Let (G, Ω) be oligomorphic and closed. Then there is an infinite $I \subset \Omega$ and a dense linear order $<$ on I such that any $g \in$

$\mathrm{Aut}(I,<)$ extends to some $g^* \in G$. The map $g \mapsto g^*$ can be taken to be an isomorphism of groups.

For a first-order structure M, the permutation group $\mathrm{Aut}\, M$ expresses certain 'language independent' features of the model. Sometimes the permutation group notions can be expressed in terms of model theory. For example, in many contexts in this book, *types* correspond to orbits.

DEFINITION 5.3. An $\mathscr{L}$-structure $\mathfrak{M}$ is *strongly κ-homogeneous* if whenever $f: M \to M$ is a partial elementary map with $\mathrm{card}(\mathrm{Dom}\, f) < \kappa$, the map f extends to an automorphism of $\mathfrak{M}$.

We warn that the word 'homogeneous' has many other related uses. For example, when talking about Fraïssé amalgamation, one often says that a countable structure is *homogeneous* if every isomorphism between finite substructures extends to an automorphism. In practice, the meaning of the word 'homogeneous' is always explained *in situ.*

Countable atomic structures, countable saturated structures, countable recursively saturated structures and (possibly uncountable) resplendent structures (see p. 246 and p. 250 respectively for the last two) are all homogeneous. But without saturation assumptions, one has to resort to infinitary languages. Given a language $\mathscr{L}$ (or, more properly, its similarity type), and given κ and λ which are infinite cardinals or ∞, by the language $\mathscr{L}_{\kappa\lambda}$ we understand the formal language in which we allow, as well as the usual 'finitary' formation rules of $\mathscr{L}$, the infinitary formation rules of conjunction and disjunction

> given $\mathscr{L}_{\lambda\kappa}$-formulas ϕ_i $(i < \alpha)$ where $\alpha < \lambda$ is an ordinal, both $\bigwedge_{i<\alpha} \phi_i$ and $\bigvee_{i<\alpha} \phi_i$ are $\mathscr{L}_{\lambda\kappa}$-formulas,

and infinitary quantification

> if ϕ is a $\mathscr{L}_{\lambda\kappa}$-formula and x_i $(i < \beta)$ are variables, where $\beta < \kappa$, both $\exists x_0,\ldots,x_i,\ldots\, \phi$ and $\forall x_0,\ldots,x_i,\ldots\, \phi$ are $\mathscr{L}_{\lambda\kappa}$-formulas.

The language $\mathscr{L}_{\omega\omega}$ is the usual first-order language. Tarski's definition of truth is extended to these languages in the obvious way.

For countable models, $\mathscr{L}_{\omega_1\omega}$ is important because of a result due to Scott (1964, 1965) which says that two countable $\mathscr{L}$-structures are isomorphic if and only if they satisfy the same $\mathscr{L}_{\omega_1\omega}$-sentences. It follows in particular that for a countable model $\mathfrak{M}$, $\mathrm{Aut}\,\mathfrak{M}$-orbits of M are $\mathscr{L}_{\omega_1\omega}$-definable. An interesting variation of Scott's Theorem (published in the same volume) is Karp's Theorem (Karp, 1965) that two $\mathscr{L}$-structures (whether countable or not) are back-and-forth equivalent if and only if they satisfy the same $\mathscr{L}_{\infty\omega}$- sentences. There are variations of these results for other cardinalities and other infinitary languages too; see Barwise (1975, pp. 292–303) for a discussion and proofs.

Quantifier elimination. A first-order theory T has *elimination of quantifiers* if for every formula $\phi(\bar{x})$ there is a formula $\psi(\bar{x})$ without quantifiers such that $\forall \bar{x}\,(\phi(\bar{x}) \leftrightarrow \psi(\bar{x}))$ is true in all models of T. Quantifier elimination is a useful property which, if the language is reasonable, enables one to get hold of the definable subsets of the structure. It is easily verified that the theory of the canonical structure for a permutation group has elimination of quantifiers. Also, if $\mathfrak{M}$ is a ω-categorical structure over a finite relational language then the theory of $\mathfrak{M}$ has elimination of quantifiers if and only if every isomorphism between finite substructures of $\mathfrak{M}$ extends to an automorphism of $\mathfrak{M}$.

Primitivity. By a criterion of Higman (1967) G is primitive iff each G-orbit on unordered pairs (regarded as the edge set of an undirected graph) is path-connected. If, for some k there are at most k 2-types then 'connected' is equivalent to 'of diameter $\leqslant k$' which is first-order. That is, if we regard (G, Ω) as a first-order structure (with unary predicates for G and Ω, and enough in the language to define the group structure and action), then, if G has finitely many orbits on Ω^2, primitivity is a first-order property of (G, Ω).

Double cosets. For a group G and a subgroup $H < G$, the double cosets $HgH := \{hgk : h, k \in H\}$ of H partition G in a similar way to ordinary cosets. If G is a permutation group on Ω and $H = G_{\bar{a}}$ these double cosets correspond precisely to orbits as follows: the map

$$G_{\bar{a}} g G_{\bar{a}} \mapsto \operatorname{orb}(\bar{a}, \bar{a}^g)$$

is well-defined and is a one-to-one correspondence from double cosets of $G_{\bar{a}}$ to orbits of tuples $(\bar{x}, \bar{y})$, where $\bar{x}$, $\bar{y}$, and $\bar{a}$ lie in the same G-orbit. Notice that any subgroup K with $H \leqslant K \leqslant G$ is a union of double cosets of H. It follows that if $G = \operatorname{Aut}\mathfrak{M}$ and $\mathfrak{M}$ is $\aleph_0$-categorical, then each subgroup $H = G_{\bar{a}}$ has at most finitely many such K since there are only finitely many types in $(\bar{x}, \bar{y})$ over $\mathfrak{M}$. Even if $\mathfrak{M}$ is not $\aleph_0$-categorical, this observation can sometimes give us useful information about such intermediate groups K.

We conclude this section with a straightforward result that says precisely when two countable structures have the same automorphism group (considered as a permutation group).

DEFINITION 5.4. Two $\mathscr{L}$-structures $\mathfrak{M}$ and $\mathfrak{N}$ are *bi-definable* if there is a bijection $f\colon \operatorname{Dom}\mathfrak{M} \to \operatorname{Dom}\mathfrak{N}$ mapping the 0-definable relations of $\mathfrak{M}$ onto those of $\mathfrak{N}$. Similarly, $\mathfrak{M}$ and $\mathfrak{N}$ are (ω_1, ω)-*bi-definable* if there is a bijection $f\colon \operatorname{Dom}\mathfrak{M} \to \operatorname{Dom}\mathfrak{N}$ mapping the parameter-free $\mathscr{L}_{\omega_1\omega}$-definable relations of $\mathfrak{M}$ (the analogue of 0-definable, but for the infinitary language) onto those of $\mathfrak{N}$.

PROPOSITION 5.5. Let $\mathfrak{M}$ and $\mathfrak{N}$ be countably infinite structures. Then Aut $\mathfrak{M}$ and Aut $\mathfrak{N}$, considered as permutation groups acting on Dom $\mathfrak{M}$ and Dom $\mathfrak{N}$ respectively, are similar iff $\mathfrak{M}$ and $\mathfrak{N}$ are (ω_1, ω)-bi-definable. If in addition $\mathfrak{M}$ and $\mathfrak{N}$ are $\aleph_0$-categorical, then Aut $\mathfrak{M}$ and Aut $\mathfrak{N}$ are similar (as permutation groups) iff $\mathfrak{M}$ and $\mathfrak{N}$ are bi-definable.

Proof. This is just chasing definitions round. Aut $\mathfrak{M}$ and Aut $\mathfrak{N}$ are similar iff there is a bijection $\phi: M \to N$ inducing an isomorphism of groups $g \mapsto \phi^{-1}g\phi$, which holds iff $\mathfrak{M}$ and $\mathfrak{N}$ are (ω_1, ω)-bi-definable (by taking the same ϕ) since unions of orbits are $\mathscr{L}_{\omega_1\omega}$-definable. The case of $\aleph_0$-categorical structures is the same except that, here, unions of orbits are actually 0-definable. □

6 Topological groups

A topological group is a group G with a topology on G for which the maps $(g, h) \mapsto gh$ and $g \mapsto g^{-1}$ are continuous. (For the first of these operations, the product topology on G^2 is used.) If G is a permutation group on an infinite set Ω, then there is a natural topology on G, making G into a topological group: basic open sets are cosets of the subgroups of the form $G_{(\Gamma)}$ where Γ is a finite subset of Ω. (These subgroups are sometimes called *basic open subgroups.*) Open sets are unions of basic open sets. (One can obtain other topologies by varying the choice of basic open subgroups. Provided the family of basic open subgroups is closed under finite intersections, the open sets obtained from them will give a topology making G into a topological group. For example, if Ω is uncountable, of cardinality κ say, and $\aleph_0 \leqslant \lambda \leqslant \kappa$ then the set of $G_{(\Gamma)}$ for $\Gamma \subseteq \Omega$ with card $\Gamma < \lambda$ is a suitable choice of basic open subgroups. However, most of the discussion here will be concerned with the countable case.)

This topology is always Hausdorff, but is not in general compact (and if (G, Ω) is oligomorphic, then the topology is not locally compact). It is easily verified that G is closed as a subgroup of the permutation group $(\mathrm{Sym}(\Omega), \Omega)$ in this topology iff $G = \overline{G}$ (as defined on p. 19 above) iff G is the full automorphism group of the canonical structure for G on Ω. Note too that a subgroup $H < G$ is open in this topology iff H contains $G_{(\Gamma)}$ for some finite Γ, and that each open subgroup is closed (since it is the complement of a union of open cosets). The open subgroups form a neighbourhood base of 1, since if U is an open set containing 1, then there is a coset $G_{(\Gamma)}g$ of a basic open subgroup with $1 \in G_{(\Gamma)}g \subseteq U$, and of course g must be 1 since $G_{(\Gamma)}$ is a subgroup.

If card $\Omega = \aleph_0$, then this topology is metrizable: enumerate $\Omega = \{x_n : n < \omega\}$, and write $d(f, g) = 1/2^{n+1}$ where n is least such that $x_n^f \neq x_n^g$ or $(x_n)^{f^{-1}} \neq (x_n)^{g^{-1}}$. Note that, if in addition G is closed, then G is a complete separable metric space, so is countable or has cardinality $2^{\aleph_0}$. In fact, in this situation, G is countable if and only if there is a finite $\Gamma \subset \Omega$ such that $G_{(\Gamma)} = 1$.

For the rest of this section we shall assume that all models are countable. It is possible to characterize precisely those topological groups that arise as $\operatorname{Aut}\mathfrak{M}$ for some $\mathfrak{M}$. Say that a sequence $a_0, a_1, \ldots$ from a topological group G is a *Cauchy sequence* if for any open subgroup $H \leqslant G$ there is $n_0 \in \mathbb{N}$ with

$$\forall n,m{>}n_0\,(a_n a_m^{-1}, a n_n^{-1} a_m \in H).$$

The Cauchy sequence *converges to a* if for any open $H \leqslant G$ there is $n_0 \in \mathbb{N}$ with

$$\forall n{>}n_0\,(a_n a^{-1}, a_n^{-1} a \in H).$$

Since G is Hausdorff, any Cauchy sequence converges to at most one point. We say G is *complete* if every Cauchy sequence converges to some point of G.

PROPOSITION 6.1. Let G be a topological group. Then $G = \operatorname{Aut}(M)$ for some countable structure M iff

1. G is Hausdorff;
2. G is complete;
3. if $H < G$ is open then $[G{:}H] \leqslant \aleph_0$;
4. there is a countable family $\{H_i : i \in \mathbb{N}\}$ of subgroups of G such that

$$\mathscr{B} = \{H_{i_0}^{g_0} \cap \ldots \cap H_{i_k}^{g_k} : k \in \mathbb{N}, g_0, \ldots, g_k \in G, i_0, \ldots, i_k \in \mathbb{N}\}$$

   is a base of open subgroups of G, i.e., the set of cosets of elements of $\mathscr{B}$ forms a base for the topology on G.

Proof. Suppose $G = \operatorname{Aut}\mathfrak{M}$ where M is countable and $H < G$ is open. Then H contains $G_{\bar{a}}$ for some $\bar{a} \in M^{<\omega}$ and the cosets of $G_{\bar{a}}$ are the non-empty sets of the form

$$U_{\bar{a},\bar{b}} := \{g \in G : \bar{a}^g = \bar{b}\}.$$

There are only countably many of these. This gives (3). For (4), take

$$\{H_i : i \in \mathbb{N}\} = \{G_{\bar{a}} : \bar{a} \in M^{<\omega}\}.$$

Conversely, let $\Omega = \{H_i g : i \in \mathbb{N}, g \in G\}$ and let G act on Ω by right-multiplication. The set Ω is countable by hypothesis. A group element h fixes each $H_{i_j} g_j$ for $j = 1, \ldots, k$ iff $g_j h g_j^{-1} \in H_{i_j}$ for each j, iff $h \in \bigcap_j H_{i_j}^{g_j}$. Thus the topology on G is the same as that given by the stabilizers $G_{(\Gamma)}$ for finite $\Gamma \subseteq \Omega$. This also shows that the action is faithful, since G is Hausdorff. Thus it suffices to check that G is closed in $\operatorname{Sym}\Omega$, and this follows from completeness. □

We shall say that a topological group is a *Polish* group if the underlying topology of the group is that of a Polish space, i.e., a complete separable metric space. It follows that any topological group satisfying the four conditions of the last proposition is a Polish group. Recall that, in a complete metric space, a set A is *meagre* or of *first category* if it is contained in a countable union of nowhere-dense sets, and that a set X has the *property of Baire* if it is $G \triangle P$ for some open G and some meagre set P (equivalently, if it is $F \triangle Q$ for some closed F and some meagre Q). (Here, $\triangle$ denotes the symmetric difference of two sets.)

PROPOSITION 6.2. (Lascar, 1991) If G is a Polish group and $H < G$ has the property of Baire then H is meagre or open.

Proof. Suppose $G = \mathrm{Aut}(M)$, and let O be open such that $H \triangle O$ is meagre. If O is empty we are done, so suppose otherwise. Then there is $\bar{a} \in M^{<\omega}$ and $f \in G$ such that $G_{\bar{a}} f \subseteq O$. By the theorem of Baire, there is $h \in G_{\bar{a}} f \setminus (H \triangle O)$, hence $h \in H \cap G_{\bar{a}} f$. But $Hh^{-1} = H$ and $G_{\bar{a}} f h^{-1} = G_{\bar{a}}$, so we have:

$$\begin{array}{rcl} G_{\bar{a}} f \setminus H \subseteq (H \triangle O) \cap G_{\bar{a}} f & \text{is meagre in} & G_{\bar{a}} f = G_{\bar{a}} h \\ G_{\bar{a}} h \setminus Hh & \text{is meagre in} & G_{\bar{a}} h \\ G_{\bar{a}} \setminus H & \text{is meagre in} & G_{\bar{a}} \\ G_{\bar{a}} \cap H & \text{is comeagre in} & G_{\bar{a}} \end{array}$$

so each coset of $G_{\bar{a}} \cap H$ is comeagre in $G_{\bar{a}}$. Intersections of countable families of comeagre sets are comeagre so $G_{\bar{a}} \cap H = G_{\bar{a}}$, i.e., $H \supseteq G_{\bar{a}}$ and hence is open. □

The property of Baire is preserved by continuous maps from one Polish space to another, and some very useful facts about Polish groups follow from this.

PROPOSITION 6.3. In Polish groups G_1 and G_2, if $\phi: G_1 \to G_2$ is a continuous homomorphism and $H < G_1$ is open then $H\phi$ has the property of Baire. If in addition $[G_2 : H\phi] \leqslant \aleph_0$ (for example, if ϕ is a surjection) then $H\phi$ is open.

Proof. See Kuratowski (1966). Note that if G_2 is the union of countably many cosets of $H\phi$, then $H\phi$ cannot be meagre and hence is open by the last result. □

It follows very simply from this that a continuous homomorphism from one Polish group onto another is actually an open map (i.e., the image of any open set is open) and a continuous isomorphism is actually a homeomorphism.

Hodges, Hodkinson, Lascar and Shelah (Hodges *et al.*, 1993) gave a tree-argument that proves the following result.

THEOREM 6.4. (Hodges *et al.*, 1993) Any meagre subgroup of a Polish group G has index $2^{\aleph_0}$ in G.

This theorem is a generalization of an observation due to Evans (which is in turn a version of a result known to model theorists as the Kueker–Reyes Theorem) which says that a closed subgroup H of a group $G = \operatorname{Aut}\mathfrak{M}$ is open iff it has index strictly less than $2^{\aleph_0}$ iff it has index at most $\aleph_0$. In practice, closed subgroups of $G = \operatorname{Aut}\mathfrak{M}$ often arise as $H = \operatorname{Aut}(\mathfrak{M}, R)$ for some new relation $R \subset \mathfrak{M}^n$. Images R^g of R under automorphisms g correspond to cosets Hg of H: $R^g = R^h$ iff $Hg = Hh$. This gives the Kueker–Reyes Theorem: R has countably many automorphic images R^g iff it has fewer than $2^{\aleph_0}$ such images iff H is open iff for some $\bar{a} \in \mathfrak{M}^{<\omega}$ there are no $g \in G$ with $\bar{a}^g = \bar{a}$ and $R^g \neq R$.

COROLLARY 6.5. A subgroup H of a Polish group G with the property of Baire is either open (and hence has index $[G{:}H] \leqslant \aleph_0$) or has index $2^{\aleph_0}$ in G.

Proof. By Proposition 6.2 and Theorem 6.4. □

The small index property. The last few results suggest a way that the topology on G may be obtained completely from G as an abstract group. Clearly the collection of subgroups of G is determined by the *abstract* group structure of G. Say that the topological group G (or the structure $\mathfrak{M}$ where $G = \operatorname{Aut}\mathfrak{M}$) has *the small index property* if for all subgroups $H < G$, H is open iff $[G{:}H] \leqslant \aleph_0$. One direction (that open subgroups have countable index) is true for all Polish groups G by separability. The other direction is true for all Polish groups G and all subgroups $H < G$ which have the property of Baire, by the last corollary. If the small index property is true for all subgroups H we could recognize open subgroups H by just looking at their index. We could then recover the topology completely, for in any topological group the open sets are unions of cosets of open subgroups.

Many familiar structures have the small index property. Examples include the set Ω with no structure at all, where $G = \operatorname{Sym}\Omega$ (Semmes, 1981; see Dixon *et al.* (1986) for a proof); the rationals with its usual linear order, with $G = \operatorname{Aut}(\mathbb{Q}, <)$ (Truss, 1989); the countable atomless boolean algebra (Truss, 1989); the random graph (Hodges *et al.*, 1993); vector spaces of countably infinite dimension over a field or skew field (Evans, 1986*c*); any countable saturated strongly minimal set (Lascar, 1992). However, using the axiom of choice examples of $\aleph_0$-categorical structures without the small index property can be given (see Evans' article here, p. 52 and p. 68 for an example). In his article in this volume, Lascar shows that sufficiently saturated countable models of arithmetic have the small index property.

If a structure $\mathfrak{M}$ or Polish group G has the small index property then this is reflected in the abstract group structure G alone. This follows from results already cited and the following simple observation.

PROPOSITION 6.6. If G and H are Polish groups, G has the small index property, and $\phi: G \to H$ is a homomorphism of abstract groups, then ϕ is continuous.

Proof. If $K < H$ is open, it has index at most $\aleph_0$ in H, so its inverse image $\phi^{-1}H$ has index at most $\aleph_0$ in G and hence is open. □

Now suppose G and H are Polish groups, G has the small index property, and $\phi: G \to H$ is an isomorphism of abstract groups, then ϕ is a homeomorphism, i.e., an isomorphism of topological groups by Proposition 6.3 and Proposition 6.6, so in particular H also has the small index property.

7 Interpretations

Much of the structural information about $\mathfrak{M}$ is carried by Aut $\mathfrak{M}$ as a topological group. Ahlbrandt and Ziegler (1986) gave some information here for $\aleph_0$-categorical structures. The following account is a slightly expanded version of this material due to Kaye and gives a model-theoretic description of the information in Aut $\mathfrak{M}$ for all countable structures $\mathfrak{M}$.

In the following, $\mathfrak{A}$ and $\mathfrak{B}$ denote countable structures for languages $\mathscr{L}_A$ and $\mathscr{L}_B$ respectively. If $U \subseteq A^n$ for some n, we say that U is *quasidefinable* (in $\mathfrak{A}$) if it is the union of Aut $\mathfrak{M}$-orbits of A^n. Note that if A is $\aleph_0$-categorical this notion corresponds precisely to being 0-definable, but in general U is quasidefinable in an $\mathscr{L}$-structure iff it is definable by an $\mathscr{L}_{\omega_1\omega}$-formula with no parameters from A.

DEFINITION 7.1. We shall say that $\mathbf{f}: \mathfrak{A} \rightsquigarrow \mathfrak{B}$ is an (infinitary) *interpretation* of $\mathfrak{B}$ in $\mathfrak{A}$ if $\mathbf{f}$ is a countable set of functions $f_i: U_i \to B$ ($U_i \subseteq A^{n_i}$, $n_i \in \mathbb{N}$) such that

1. $\bigcup_i \operatorname{Im} f_i = B$;
2. for all i, j,
$$\{(\bar{u}, \bar{v}) \in U_i \times U_j : f_i(\bar{u}) = f_j(\bar{v})\}$$
is quasidefinable in $\mathfrak{A}$;
3. for all k, all quasidefinable $R \subseteq B^k$, and all i,
$$f_i^{-1}R =_{\text{def}} \{(\bar{u}_1, \ldots, \bar{u}_k) \in U_i^k : (f_i(\bar{u}_1), \ldots, f_i(\bar{u}_1)) \in R\}$$
is quasidefinable in $\mathfrak{A}$.

This terminology is slightly non-standard: the word 'interpretation' is usually reserved for 'finitary interpretation' (as defined below). It will be convenient, however, to use 'interpretation' in its infinitary sense unless explicitly stated otherwise for the rest of this chapter. See for example Evans' article (p. 39) for the standard definition of 'interpretation'.

Countable structures with interpretations form a category $\mathfrak{I}$. The identity interpretation is given by $U_i = A$ and $f_i = \mathrm{id}_A$ for all i, and the composition of $\mathbf{f}: \mathfrak{A} \rightsquigarrow \mathfrak{B}$ as above and $\mathbf{g}: \mathfrak{B} \rightsquigarrow \mathfrak{C}$ given by the data $g_i: V_i \to C$ ($V_i \subseteq B^{m_i}$) is defined by taking $W_{i,j} = (f_j^{m_i})^{-1} V_i \subseteq U_i^{m_i}$ where $f_j^{m_i}(\bar{u}_1, \ldots, \bar{u}_{m_i}) = (f_j \bar{u}_1, \ldots, f_j \bar{u}_{m_i})$ and the map $h_{i,j}: W_{i,j} \to C$ is the composition of $f_j^{m_i}$ and g_i. In this setting, we can regard Aut as a functor from $\mathfrak{I}$ to $\mathfrak{G}$, the category of Polish groups with continuous homomorphisms as arrows. Given $\mathbf{f}: \mathfrak{A} \rightsquigarrow \mathfrak{B}$ and $\sigma \in \operatorname{Aut} \mathfrak{A}$ we define $\rho = \operatorname{Aut} \mathbf{f}(\sigma)$ in such a way that

$$\begin{array}{ccc} & f_i & \\ U_i & \longrightarrow & B \\ \sigma \downarrow & & \downarrow \rho \\ U_i & \longrightarrow & B \\ & f_i & \end{array}$$

commutes. (Note that σ is well-defined on U_i since U_i, being the inverse image of a quasidefinable set, is itself quasidefinable.) This means that, for $b \in B$, we should define b^ρ to be $f_i(\bar{u}^\sigma)$ for some $\bar{u} \in U_i$ with $b = f_i(\bar{u})$. We first show that this definition does not depend on the choice of $\bar{u}$ within U_i, and then we show that it does not depend on i either.

If $b = f_i(\bar{u}) = f_i(\bar{v})$ with $\bar{u}, \bar{v} \in U_i$, consider $R \subseteq B^2$, the orbit of (b, b) in B^2; R is quasidefinable, so its inverse image under f_i is, and $(\bar{u}, \bar{v}) \in f_i^{-1} R$. Thus $(\bar{u}^\sigma, \bar{v}^\sigma) \in f_i^{-1} R$ so $(f_i(\bar{u}^\sigma), f_i(\bar{v}^\sigma)) \in R$, and hence $f_i(\bar{u}^\sigma) = f_i(\bar{v}^\sigma)$.

Given $b = f_i(\bar{u}) = f_j(\bar{v})$ where $\bar{u} \in U_i$, $\bar{v} \in U_j$ and j, i, we must show $f_i(\bar{u}^\sigma) = f_j(\bar{v}^\sigma)$. This is identical to the argument in the last paragraph except that we use condition 2 in the definition of an interpretation to see that

$$\{(\bar{u}, \bar{v}) \in U_i \times U_j : f_i(\bar{u}) = f_j(\bar{v})\}$$

is quasidefinable.

To see that ρ is an automorphism, it suffices to check that ρ is onto and preserves quasidefinable sets. It is onto because $\bigcup_i \operatorname{Im} f_i = B$ and σ is a permutation of each U_i. It preserves quasidefinable sets since if $(b_1, \ldots, b_n) \in B^n$ and R is the $\operatorname{Aut} B$-orbit of $(b_1, \ldots, b_n)$ then the inverse image of R is preserved by σ so $(b_1, \ldots, b_n)^\rho \in R$.

It is immediate that $\operatorname{Aut}(\mathbf{f})$ is a homomorphism from its definition. To see that $\operatorname{Aut}(\mathbf{f})$ is continuous, note that if we are looking for $\rho = \operatorname{Aut} \mathbf{f}(\sigma)$ such that $b_1^\rho = b_2$ it suffices to consider any automorphism σ such that $\bar{u}_1^\sigma = \bar{u}_2$ where $b_1 = f_i(\bar{u}_1)$ and $b_2 = f_i(\bar{u}_2)$.

It is now straightforward to check that Aut is a functor as previously indicated. Our main theorem, to be presented next, is a converse to this: all continuous homomorphisms arise as $\operatorname{Aut}(\mathbf{f})$ for some $\mathbf{f}$.

THEOREM 7.2. Let $\phi: \operatorname{Aut} \mathfrak{A} \to \operatorname{Aut} \mathfrak{B}$ be a homomorphism. Then ϕ is continuous iff $\phi = \operatorname{Aut}(\mathbf{f})$ for some $\mathbf{f}: \mathfrak{A} \rightsquigarrow \mathfrak{B}$.

Proof. Set $U_0 = \varnothing$, $n_0 = 0$, and let $b_1, \ldots, b_i, \ldots$ be an enumeration of B. Inductively, suppose we have defined U_i to be the union of the $\operatorname{Aut}\mathfrak{A}$-orbits of

$$\begin{gathered} a_1, a_1, a_2, \ldots, a_i \\ a_2, a_1, a_2, \ldots, a_i \\ \ldots \\ a_i, a_1, a_2, \ldots, a_i \end{gathered}$$

for some $\bar{a}_j \in A^{<\omega}$ (but for simplicity we will not use a bar above the letter), and that

$$f_i \colon (a_j, a_1, \ldots, a_i)^g \mapsto b_j^{g\phi}$$

is well-defined and satisfies all the required conditions in the definition of 'interpretation'. Notice that $\operatorname{Im} f_i$ is the union of the $\operatorname{Im}\phi$-orbits of $b_1, \ldots, b_i$. Consider $b_{i+1} \in B$; since ϕ is continuous we can find $a_{i+1} \in A^{<\omega}$ such that the image of $(\operatorname{Aut}\mathfrak{A})_{a_{i+1}}$ under ϕ is contained in $(\operatorname{Aut}\mathfrak{B})_{b_{i+1}}$. Clearly we can assume $\operatorname{len}(a_{i+1}) \geqslant \operatorname{len}(a_j)$ for each $j \leqslant i$. Next, we increase the length of the tuples a_j so that

$$\operatorname{len}(a_j) = \operatorname{len}(a_{i+1}) \qquad \text{for all } j$$

and

$$a_{i+1} \notin \bigcup_{j \leqslant i} \operatorname{orb}(a_j).$$

We do this by replacing a tuple $a_j = x_1, \ldots, x_n$ with $x_1, \ldots, x_n, x_n, \ldots, x_n$, and extending a_{i+1} in some other way, if necessary. For simplicity, we continue to denote the new tuples by $a_1, \ldots, a_i, a_{i+1}$.

Let U_{i+1} be the union of the $\operatorname{Aut}\mathfrak{A}$-orbits of

$$\begin{gathered} a_1, a_1, a_2, \ldots, a_{i+1} \\ a_2, a_1, a_2, \ldots, a_{i+1} \\ \ldots \\ a_{i+1}, a_1, a_2, \ldots, a_{i+1} \end{gathered}$$

(in the new sense) and let

$$f_{i+1} \colon (a_j, a_1, \ldots, a_{i+1})^g \mapsto b_j^{g\phi}.$$

If $(a_j, a_1, \ldots, a_{i+1})^g = (a_k, a_1, \ldots, a_{i+1})^h$ then either $k, j \leqslant i$ and $b_j^{g\phi} = b_k^{h\phi}$ since f_i is well-defined, or $j = k = i + 1$ (since a_j and a_{i+1} lie in different orbits for $j \leqslant i$) and

$$a_{i+1}^g = a_{i+1}^h.$$

In this second case, $gh^{-1} \in (\operatorname{Aut}\mathfrak{A})_{a_{i+1}}$, so $gh^{-1}\phi \in (\operatorname{Aut}\mathfrak{B})_{b_{i+1}}$, i.e., $b_{i+1}^{g\phi} = b_{i+1}^{h\phi}$, since ϕ is a homomorphism. Thus f_{i+1} is well-defined.

There are some properties that we must check hold for f_{i+1}. First, if

$$S := \{(a_j^g, a_1^g, \ldots, a_i^g, a_k^h, a_1^h, \ldots, a_{i+1}^h) \in U_i \times U_{i+1} : b_j^{g\phi} = b_k^{h\phi}\}$$

and $\sigma \in \operatorname{Aut}\mathfrak{A}$ then

$$b_j^{g\phi} = b_k^{h\phi} \Leftrightarrow b_j^{g\phi\sigma\phi} = b_k^{h\phi\sigma\phi} \Leftrightarrow b_j^{(g\sigma)\phi} = b_k^{(h\sigma)\phi}$$

so

$$\begin{aligned} S^\sigma &= \{(a_j^{g\sigma}, a_1^{g\sigma}, \ldots, a_i^{g\sigma}, a_k^{g\sigma}, a_1^{g\sigma}, \ldots, a_{i+1}^{g\sigma}) : b_j^{g\phi} = b_k^{h\phi}\} \\ &= \{(a_j^{g\sigma}, a_1^{g\sigma}, \ldots, a_i^{g\sigma}, a_k^{g\sigma}, a_1^{g\sigma}, \ldots, a_{i+1}^{g\sigma}) : b_j^{(g\sigma)\phi} = b_k^{(h\sigma)\phi}\} \\ &= S. \end{aligned}$$

Thus S is quasidefinable. Since $\operatorname{Im} f_j \subseteq \operatorname{Im} f_{j+1}$ for $j \leqslant i$, and projections of quasidefinable sets are quasidefinable, it follows that

$$\{(\bar{u}, \bar{v}) \in U_j \times U_k : f_j(\bar{u}) = f_k(\bar{v})\}$$

is quasidefinable for all $j \neq k \leqslant i+1$.

Second, if R is a quasidefinable subset of $\mathfrak{B}$ we must show

$$f_{i+1}^{-1} R := \{(x_1, x_2, \ldots, x_n) \in U_{i+1}^n : (f_{i+1}(x_1), \ldots, f_{i+1}(x_n)) \in R\}$$

is quasidefinable. It suffices to show that each

$$T := \left\{ (x_1^{g_1}, x_2^{g_2}, \ldots, x_n^{g_n}) : \begin{array}{l} (f_{i+1}(x_1^{g_1}), \ldots, f_{i+1}(x_n^{g_n})) \in R, \\ g_1, \ldots, g_n \in \operatorname{Aut}\mathfrak{A} \end{array} \right\}$$

is quasidefinable, for fixed $x_1, \ldots, x_{i+1} \in U_{i+1}$. But since ϕ is a homomorphism, if $c_k := f_{i+1}(x_k)$ we have $f_{i+1}(x_k^h) = c_k^{h\phi}$ and so

$$T = \{(x_1^{g_1}, x_2^{g_2}, \ldots, x_n^{g_n}) : (c_1^{g_1\phi}, \ldots, c_n^{g_n\phi})) \in R, g_1, \ldots, g_n \in \operatorname{Aut}\mathfrak{A}\}$$

which is quasidefinable since R is.

Since $\mathfrak{B}$ is countable, this induction gives an interpretation $\mathbf{f}$ of $\mathfrak{B}$ in $\mathfrak{A}$. Thus it is only necessary to show that $\phi = \operatorname{Aut}(\mathbf{f})$. This is almost immediate: if $h \in \operatorname{Aut}\mathfrak{A}$ and $f_i(a_i^g, a_1^g, \ldots, a_i^g) = b_i^{g\phi}$ then

$$\begin{aligned} (b_i^{g\phi})^{h\,\operatorname{Aut}(\mathbf{f})} &= f_i(a_i^{gh}, a_1^{gh}, \ldots, a_i^{gh}) \\ &= b_i^{gh\phi} \\ &= (b_i^{g\phi})^{h\phi} \end{aligned}$$

as required. □

Notice that if $\mathfrak{B}$ has only finitely many $\operatorname{Im}\phi$-orbits, the construction above can be carried out in such a way that $B = \operatorname{Im} f_i$ for some i. Say that $\mathbf{f}\colon \mathfrak{A} \rightsquigarrow \mathfrak{B}$ is a *finitary interpretation* if $B = \operatorname{Im} f_i$ for some i and all the conditions in the definition of 'interpretation' hold when 'quasidefinable' is replaced thoughout by '0-definable'. (This is essentially the usual meaning of 'interpretation'.) Note also that if $\mathfrak{A}$ is $\aleph_0$-categorical and $\mathbf{f}: \mathfrak{A} \rightsquigarrow \mathfrak{B}$ is finitary then $\mathfrak{B}$ is also $\aleph_0$-categorical. The following result of Ahlbrandt and Ziegler (1986) follows from this and the proof of the last theorem.

THEOREM 7.3. (Ahlbrandt and Ziegler) If $\mathfrak{A}$ is $\aleph_0$-categorical,

$$\phi: \operatorname{Aut} \mathfrak{A} \to \operatorname{Aut} \mathfrak{B}$$

is a continuous homomorphism and $\mathfrak{B}$ has finitely many $\operatorname{Im} \phi$-orbits, then $\phi = \operatorname{Aut}(\mathbf{f})$ for some finitary interpretation $\mathbf{f}$. In particular, it follows from this that $\mathfrak{B}$ is $\aleph_0$-categorical.

To complete the story, we give a model-theoretic notion equivalent to $\operatorname{Aut} \mathbf{f} = \operatorname{Aut} \mathbf{g}$ for interpretations $\mathbf{f}: \mathfrak{A} \rightsquigarrow \mathfrak{B}$ and $\mathbf{g}: \mathfrak{A} \rightsquigarrow \mathfrak{B}$.

DEFINITION 7.4. If $\mathbf{f}: \mathfrak{A} \rightsquigarrow \mathfrak{B}$ and $\mathbf{g}: \mathfrak{A} \rightsquigarrow \mathfrak{B}$ and interpretations, where $\mathbf{f}$ is a set of $f_i: U_i \to B$ and $\mathbf{g}$ is a set of $g_j: V_j \to B$, we say that they are *homotopic* ($\mathbf{f} \sim \mathbf{g}$) if for all i, j

$$\{(\bar{a}, \bar{b}) \in U_i \times V_j : f_i(\bar{a}) = g_j(\bar{b})\}$$

is quasidefinable in $\mathfrak{A}$.

THEOREM 7.5. For interpretations $\mathbf{f}: \mathfrak{A} \rightsquigarrow \mathfrak{B}$ and $\mathbf{g}: \mathfrak{A} \rightsquigarrow \mathfrak{B}$, $\mathbf{f} \sim \mathbf{g}$ iff $\operatorname{Aut} \mathbf{f} = \operatorname{Aut} \mathbf{g}$.

Proof. Let $\phi = \operatorname{Aut} \mathbf{f}$ and $\psi = \operatorname{Aut} \mathbf{g}$. Recall that, from the definitions,

$$f_i(\bar{a})^{h\phi} = f_i(\bar{a}^h)$$

and

$$g_j(\bar{b})^{h\psi} = g_j(\bar{b}^h)$$

for any $h \in \operatorname{Aut} \mathfrak{A}$. So, for one direction, if $\mathbf{f} \sim \mathbf{g}$ then, for all $c \in B$ and $\bar{a} \in U_i$, $\bar{b} \in V_j$, if $c = f_i(\bar{a}) = g_j(\bar{b})$ then $f_i(\bar{a}^h) = g_j(\bar{b}^h)$ so $c^{h\phi} = c^{h\psi}$. For the other direction, if $h\phi = h\psi$ for all h,

$$X =_{\text{def}} \{(\bar{a}, \bar{b}) \in U_i \times V_j : f_i(\bar{a}) = g_j(\bar{b})\}$$

and $(\bar{a}, \bar{b}) \in X$ then

$$f_i(\bar{a}^h) = f_i(\bar{a})^{h\phi} = g_j(\bar{b})^{h\phi} = g_j(\bar{b}^h)$$

so $(\bar{a}^h, \bar{b}^h) \in X$, that is, X is a union of orbits, as required. □

DEFINITION 7.6. Structures $\mathfrak{A}$ and $\mathfrak{B}$ are *(infinitarily) bi-interpretable* if there are (infinitary) interpretations $\mathbf{f}: \mathfrak{A} \rightsquigarrow \mathfrak{B}$ and $\mathbf{g}: \mathfrak{A} \rightsquigarrow \mathfrak{B}$ such that $\mathbf{f} \circ \mathbf{g} \sim \mathrm{id}_{\mathfrak{A}}$ and $\mathbf{g} \circ \mathbf{f} \sim \mathrm{id}_{\mathfrak{B}}$. They are *finitarily bi-interpretable* if there are finitary interpretations as above.

COROLLARY 7.7. Let $\mathfrak{A}$ and $\mathfrak{B}$ be countable structures.

1. $\operatorname{Aut}\mathfrak{A} \cong \operatorname{Aut}\mathfrak{B}$ as topological groups iff $\mathfrak{A}$ and $\mathfrak{B}$ are (infinitarily) bi-interpretable.
2. If $\mathfrak{A}$ and $\mathfrak{B}$ are both $\aleph_0$-categorical then $\operatorname{Aut}\mathfrak{A} \cong \operatorname{Aut}\mathfrak{B}$ as topological groups iff they are finitarily bi-interpretable.

Examples of $\aleph_0$-categorical structures

David M. Evans

School of Mathematics,
The University of East Anglia,
Norwich,
NR4 7TJ,
U.K.
Electronic mail address: D.Evans@uea.ac.uk

1 Introduction

1.1 About the paper

This paper is an attempt to survey the known examples of countable $\aleph_0$-categorical structures. Inevitably, this is a selective and subjective account, and few details and proofs are given. The emphasis throughout the paper is on *constructions.*

It is convenient to identify an example by the method used for its construction. Thus, the paper has three main sections, each of which first outlines a particular construction technique and then uses it to produce specific examples. The techniques discussed are the following.

- Amalgamation methods: Fraïssé's Theorem and variations.
- Boolean power constructions: the theorem of Waszkiewicz and Węglorz, and generalizations.
- Canonical structures: describing a structure by specifying its automorphism group.

These constructions suffice to produce the vast majority of the known examples of $\aleph_0$-categorical structures. There are two further techniques which are also useful, and which we mention later in this introduction: direct constructions of some of the more familiar $\aleph_0$-categorical structures, and showing that a structure is $\aleph_0$-categorical by interpreting it in a known $\aleph_0$-categorical structure.

By concentrating on examples, I have underplayed the theorems in the area. However, many of these are classification theorems, and of course the accumulation of examples and the pursuit of classification results are complementary activities: it is impossible to do one without keeping an eye on the other. So the reader will notice that although our main theme is the construction of examples, we mention in passing a number of classification results (this is particularly noticeable in Section 2.2).

I have tried to avoid too much overlap with existing surveys in the area. In particular, I have tried to emphasize topics different from those covered in Cameron's recent book (1990), and the reader is directed to this book for much of the background material and motivation. To obtain the broader picture on recent work, as well as reading the books by Cameron (1990) and Hodges (1993), the reader can consult the survey articles: Lachlan (1987) for work on classification of homogeneous structures; Lascar (1994) for work on automorphism groups; Cherlin (1984) and Hrushovski (1989*c*) for surveys of results on totally categorical structures; and Hodges (1989) for work on model theory and permutation groups.

Thanks are due to Gregory Cherlin for some very helpful comments on an earlier version of this paper.

1.2 Notation and terminology

We shall work throughout in ZFC: Zermelo–Fraenkel set theory with the Axiom of Choice.

We shall assume that the reader is familiar with the basic notions of model theory as summarized in the previous chapter. All first-order languages will have equality as a binary relation symbol, and it will always be assumed that this is interpreted as true equality. Our notation is hopefully either standard or self-explanatory. We use a capital Roman letter such as A to denote a structure, and we usually abuse this notation by using the same symbol to denote the domain (that is, the underlying set) of the structure. We stray a little from standard practice in model theory and allow empty structures. The automorphism group of a structure A will be denoted by $\mathrm{Aut}(A)$.

If A is a set, $\mathrm{Sym}(A)$ is the *symmetric group* on A; *permutation groups* G on A are thought of as subgroups of $\mathrm{Sym}(A)$, and we sometimes refer to the permutation group $(G; A)$. We shall assume that the reader is familiar with the basic notions of permutation group theory, such as that contained in the previous chapter. Our notation for permutation groups is essentially the same as elsewhere in this book except that our permutations (and functions) act on the left: thus if $a \in A$ and $g \in \mathrm{Sym}(A)$, then the image of a under g is denoted by ga, not ag or a^g. In particular, if $(G; A)$ is a permutation group and $a \in A$ then $G_a = \{g \in G : ga = a\}$ is the stabilizer in G of a, and if $X \subseteq A$ then

$$G_{(X)} = \{g \in G : \forall x{\in}X \; gx = x\}$$

is the pointwise stabilizer in G of X.

We also use the following notation. Suppose n is a natural number. Denote the set of n-tuples from A by A^n; the set of subsets of size n from A by $A^{\{n\}}$; and the set of n-tuples of distinct elements from A by $A^{(n)}$. It is clear that any permutation of A induces a permutation of each of these three sets. So an action of a group G on A induces an action of G on each of these sets.

We shall assume that the reader is familiar with the idea of regarding the symmetric group on a countable set A as a complete topological group in which closed subgroups are automorphism groups of structures with domain A. See the discussion starting on p. 22 for details. An alternative exposition of this can be found in Cameron (1990), Section 2.4. For the argument as to why the category of topological groups and continuous homomorphisms is the 'correct' category in which to study automorphism groups of $\aleph_0$-categorical structures, see the section on interpretations (from p. 26 above) especially Theorem 7.3 on p. 30, or the introduction to Ahlbrandt and Ziegler (1986).

Finally, it should be mentioned that what we have chosen to call '$\aleph_0$-categorical structures' are often referred to as 'ω-categorical structures', 'ω_0-categorical structures' or 'countably categorical structures'.

1.3 $\aleph_0$-categorical structures

If L is a first-order language and A is an L-structure, the *theory* of A, denoted by $\mathrm{Th}(A)$, is the set of all L-sentences which are true in A. In what follows, the language L will always be countable, and we will be interested in $\aleph_0$-categorical structures. Recall that a countably infinite L-structure A is $\aleph_0$-categorical if and only if it is determined up to isomorphism amongst countable L-structures by its first-order theory.

Theorem 5.1 on p. 19 gives a remarkable characterization of countable $\aleph_0$-categorical structures. This result was proved independently, but almost simultaneously, by Engeler (1959), Ryll-Nardzewski (1959), and Svenonius (1959). We shall refer to it as the *Ryll-Nardzewski Theorem.* This is standard practice, although rather unfair on Engeler and Svenonius. The reader can consult Chapter 3 of Chang and Keisler (1990) for a proof.

Condition 3 of the Ryll-Nardzewski Theorem motivates the following definition.

DEFINITION 1.1. Suppose G is a permutation group on a set Ω. We say that $(G;\Omega)$ is *oligomorphic*, or that G acts *oligomorphically* on Ω, if G has finitely many orbits on Ω^n for every natural number n.

Thus a countable L-structure A (where L is countable) is $\aleph_0$-categorical if and only if $\mathrm{Aut}(A)$ is oligomorphic as a permutation group on the domain of A.

It is straightforward to show that for a permutation group $(G;\Omega)$, each of the following conditions is equivalent to $(G;\Omega)$ being oligomorphic:

1. G has finitely many orbits on $\Omega^{\{n\}}$ for all n;
2. G has finitely many orbits on $\Omega^{(n)}$ for all n;

3. for every finite subset X of Ω, the pointwise stabilizer in G of X has only finitely many orbits on Ω.

The proof of the Ryll-Nardzewski Theorem gives us a 'dictionary' which, for a countable $\aleph_0$-categorical L-structure A, provides a translation between model-theoretic terminology and group-theoretic terminology. Here are some examples.

1. There is a 1–1 correspondence between the n-types of $\mathrm{Th}(A)$ and the $\mathrm{Aut}(A)$-orbits on A^n. Two n-tuples $\bar{a}$ and $\bar{b}$ from A realize the same n-type if and only if they lie in the same $\mathrm{Aut}(A)$-orbit.
2. If X is a finite subset of A, a subset D of A^m is X-definable if and only if any element of $\mathrm{Aut}(A)$ which fixes each element of X stabilizes the set D.
3. If X is a finite subset of A, the algebraic closure $\mathrm{acl}(X)$ of X in A (that is, the union of the finite X-definable subsets of A) is the union of the finite orbits on A of the pointwise stabilizer in $\mathrm{Aut}(A)$ of X.

See Section 5 of the previous chapter for further examples and discussion of these points.

1.4 Some straightforward examples

The examples of $\aleph_0$-categorical structures which we give here are 'straightforward' in the sense that they are familiar objects from mathematics in general, and require no additional work to construct them.

1.4.1 *A pure set*

Let L be the first-order language with just equality as a relation symbol, and with no function and constant symbols. So an L-structure A in which the equality symbol is interpreted as true equality is a set with no additional structure (a *pure* set). If A is infinite, then it is $\aleph_0$-categorical.

1.4.2 *The rationals as an ordered set*

Let L be a first-order language with equality which has a binary relation symbol $\leqslant$ (and no other constant, relation and function symbols). An L-structure A is a *partial order* (or a *partially ordered set* or *poset*) if it satisfies:

1. $\forall x, y\ (x \leqslant y) \wedge (y \leqslant x) \rightarrow (x = y)$;
2. $\forall x, y, z\ (x \leqslant y) \wedge (y \leqslant z) \rightarrow (x \leqslant z)$.

It is a *linear* order (or *total* order) if additionally

3. $\forall x, y\ (x \leqslant y) \vee (y \leqslant x)$.

A partial order is *dense* if

4. $\forall x, y\ (x < y) \rightarrow (\exists z\ x < z < y)$,

where $x < y$ is shorthand for $(x \leqslant y) \wedge (x \neq y)$. A linear order is *without endpoints* if

5. $\forall x \exists y, z\ (x < y) \wedge (z < x)$.

THEOREM 1.2. (Cantor) There exists a countably infinite dense, linear order without endpoints. It is unique up to isomorphism (and is therefore $\aleph_0$-categorical).

Proof. Existence is straightforward: the rational numbers are a countably infinite set, and together with their usual ordering, they satisfy axioms 1–5 above.

The uniqueness statement is via a *back-and-forth* argument (this style of proof is apparently due to Hausdorff: cf. Cameron (1990, Section 5.2)). Suppose A, B are countably infinite, dense linear orders without endpoints. We construct an increasing chain $f_1 \subseteq f_2 \subseteq \cdots$ of functions with the following properties:

1. each f_i is an order-preserving bijection from a finite subset of A to a finite subset of B;
2. for every $a \in A$ there exists an i with a in the domain of f_i;
3. for every $b \in B$ there exists an i with b in the image of f_i.

The union $f = \bigcup_{i<\omega} f_i$ is then an isomorphism from A to B.

The partial isomorphisms f_i are constructed inductively. Enumerate A and B as $a_1, a_2, \ldots$ and $b_1, b_2, \ldots$. Suppose we have constructed $f_1, \ldots, f_{2n}$. Using axioms 1–5 in A there exists $a \in A$ such that $f_{2n+1} = f_{2n} \cup \{(a, b_n)\}$ is a partial isomorphism from A to B. Similarly, using 1–5 in B there exists $b \in B$ such that $f_{2n+2} = f_{2n+1} \cup \{(a_n, b)\}$ is a partial isomorphism. This eventually achieves all tasks in 1, 2, and 3. □

1.4.3 *The 'random' graph*

Let L be a first-order language (as always, with equality) which has a single binary relation symbol R. An L-structure A is a *graph* if R is a symmetric, irreflexive relation on A. We refer to the elements of A as *vertices*, and if $A \models R(a, b)$ for some $a, b \in A$, then we say that a, b are *adjacent* in A and that $\{a, b\}$ is an *edge* of A. Consider the following property that a graph A may or may not have:

$(*)$ if U, V are disjoint, finite sets of vertices in A, then there exists a vertex x of A which is adjacent to all vertices in U and to no vertices in V.

Note that property $(*)$ is equivalent to the satisfaction of an infinite collection of L-sentences (so it is a property of the first-order theory of A). The graph in the following theorem is usually referred to as the random graph, or Rado's graph (after its discoverer, R. Rado).

THEOREM 1.3. There exists a countably infinite graph A satisfying property $(*)$. It is unique up to isomorphism (and therefore is $\aleph_0$-categorical).

Proof. Uniqueness follows from a back-and-forth argument, as in the previous subsection.

For existence, one can build a graph A satisfying $(*)$ in a countable number of stages as the union of a chain of finite graphs (this method will be formalized in Fraïssé's Theorem (see Theorems 2.4 and 2.10 of Section 2.1). A different approach—due to Erdős and Rényi (1963)—is the following. We are going to build a graph with vertex set $\mathbb{N}$, the natural numbers. Enumerate all pairs $\{1,2\}$, $\{1,3\}$, $\{2,3\}, \ldots$; these are the potential edges in the graph. Now go along this sequence, deciding independently, with probability $1/2$, whether each pair is an edge or not. With probability 1, the resulting graph has property $(*)$. This is the existence proof.

The $\aleph_0$-categoricity of A follows from the uniqueness statement, and the first-order nature of property $(*)$. □

1.4.4 *Vector spaces over finite fields*

Let F be a finite field. Let L be a first-order language with one binary function symbol $+$, one constant symbol 0, and for each $\alpha \in F$ a unary function symbol μ_α. If V is a vector space over F, consider V as an L-structure (where μ_α is interpreted as scalar multiplication by α).

THEOREM 1.4. If V is an infinite dimensional F-vector space, then V is $\aleph_0$-categorical.

Proof. A model of $Th(V)$ of cardinality $\aleph_0$ is a countably infinite dimensional F-vector space. Any two of these are isomorphic (by taking bases). □

REMARK. This result remains true if we extend the language L and equip V with a non-degenerate quadratic form.

1.4.5 *Boolean algebras*

Introductory material on boolean algebras is given in Section 3.1.1 (which the reader should consult for terminology). In this subsection, we merely observe the following.

THEOREM 1.5. There exists a countably infinite, atomless boolean algebra B^1. This is unique up to isomorphism, and is $\aleph_0$-categorical.

Proof. Uniqueness (and $\aleph_0$-categoricity) follow from a back-and-forth argument.

For existence, consider the rationals $\mathbb{Q}$ embedded as a subset of the real numbers $\mathbb{R}$. Let $\alpha \in \mathbb{R} \setminus \mathbb{Q}$, and let B be the subalgebra of the field of subsets of $\mathbb{Q}$ generated by sets of the form

$$\{x \in \mathbb{Q} : a + \alpha < x < b + \alpha\}$$

where $a < b$ are rationals. Then B is countably infinite and atomless. □

1.5 Interpretations

DEFINITION 1.6. We say that a structure B is *interpretable* in a structure A if there exist

- a 0-definable subset D of A^n,
- a 0-definable equivalence relation E on D,
- a bijection $\alpha: B \to D/E$

satisfying: for every 0-definable m-ary relation R on B there is a 0-definable mn-ary relation $\hat{R}$ on A such that

$$B \models R(b_1, \ldots, b_m)$$

holds if and only if whenever $\bar{a}_i \in \alpha(b_i)$ for $i = 1, \ldots, m$ we have

$$A \models \hat{R}(\bar{a}_1, \ldots, \bar{a}_m).$$

So the domain of B can be identified with a 0-definable subset of A^n factored by a 0-definable equivalence relation, and with this identification, relations in B can all be derived from relations in A.

If the classes of E are all singletons in the above we say that B is *definable* in A. If, additionally, we have $D = A$ then B is called a *reduct* of A. So in the latter case, B consists of A possibly with some of its structure forgotten.

Suppose that A and B are structures and that B is interpretable in A as in the above definition. Any automorphism of A preserves D and E and so induces a permutation of D/E. Translating this via α into a permutation of B gives an *automorphism* of B. Now suppose further that A is countable and $\aleph_0$-categorical. Then by the Ryll-Nardzewski Theorem, $\mathrm{Aut}(A)$ has finitely many orbits on A^{mn} for all $m \in \mathbb{N}$, and so the group of automorphisms of B induced by $\mathrm{Aut}(A)$ has finitely many orbits on B^m for all $m \in \mathbb{N}$. So by the Ryll-Nardzewski Theorem again we have:

THEOREM 1.7. Suppose A is a countable, $\aleph_0$-categorical structure and B is a structure interpretable in A. Then B is $\aleph_0$-categorical.

Of course, the most familiar example of an interpretation of one structure in another is the standard construction of the field of rational numbers from equivalence classes of ordered pairs of integers. We now give two examples of ways in which interpretations can be used to construct explicit examples of $\aleph_0$-categorical structures.

1.5.1 *A non-abelian $\aleph_0$-categorical group*

Suppose R is a countable (commutative) $\aleph_0$-categorical ring with identity and $n \in \mathbb{N}$. Then $\mathrm{GL}(n, R)$, the group of invertible $n \times n$-matrices with entries from R, is interpretable in R, and so is an $\aleph_0$-categorical group (by Theorem 1.7). Furthermore, if n is at least 2 then $\mathrm{GL}(n, R)$ is non-abelian. So for example, taking R to be the countable atomless boolean algebra constructed in Section 1.4.5 and $n = 2$ gives what is probably the easiest construction of an infinite non-abelian $\aleph_0$-categorical group.

1.5.2 *Reducts*

In Theorem 1.7, it is not necessarily the case that every automorphism of B is induced from an automorphism of A (via the interpretation). Indeed, the reducts of A up to interdefinability (that is, changing to an equivalent language) are in one-to-one correspondence with the closed subgroups of $\mathrm{Sym}(A)$ which contain $\mathrm{Aut}(A)$. So we can either describe a reduct B of a countable $\aleph_0$-categorical structure A by giving an interpretation as above, or we can describe its automorphism group as a permutation group on A. Perhaps the most straightforward illustration of this is where A is the rationals as an ordered set, and B is the reduct to the language with the ternary 'betweenness' relation

$$R(a, b, c) \leftrightarrow (a < b < c) \vee (c < b < a).$$

So $\mathrm{Aut}(B)$ consists of the order-preserving or order-reversing permutations of A.

Rather more interesting is the case where A is the random graph. Thomas (1991) has determined all reducts of this structure.

Reducts and interpretations occasionally provide straightforward counterexamples to superficially appealing conjectures. The following instance of this is taken from Cherlin and Lachlan (1986, p. 819).

Let A be a countably infinite set, with no structure other than equality. Let $D \subseteq A^4$ consist of 4-tuples of distinct elements, and say that two elements $(a_1, \ldots, a_4)$ and $(b_1, \ldots, b_4)$ of D are related by E if and only if

$$\{\{a_1, a_2\}, \{a_3, a_4\}\} = \{\{b_1, b_2\}, \{b_3, b_4\}\}.$$

So D/E, the domain of B, can be regarded as the set of unordered pairs of disjoint subsets of size 2 from A. The relations on B are all those induced by 0-definable relations on A^n (for all $n \in \mathbb{N}$). The automorphism group of B consists of the permutations of D/E induced by elements of $\mathrm{Sym}(A)$.

Cherlin and Lachlan observe that B is not homogeneous for a finite language (see Definition 2.3 for terminology), but is nevertheless totally categorical and coordinatized by disintegrated strictly minimal sets. So the class of stable, $\aleph_0$-categorical structures which are homogeneous for a finite relational language is *properly* contained in the class of $\aleph_0$-stable, $\aleph_0$-categorical structures coordinatized by disintegrated strictly minimal sets.

This example makes a further appearance in Thomas (1991). Macpherson pointed out that if in the above construction we take the rationals as an ordered set in place of the pure set A, then the resulting interpreted structure *is* homogeneous for a finite relational language. Thus B is an example of a reduct of a structure homogeneous for a finite relational language which is not itself homogeneous for a finite language.

2 Amalgamation

2.1 The Constructions

2.1.1 *Fraïssé's Theorem*

Constructing a structure by glueing together (amalgamating) smaller pieces is a basic theme throughout mathematics. Of course, if the structure is to have reasonable properties then the amalgamation of the pieces has to be done in a special way. The construction we describe in this section is the most widely used method of producing $\aleph_0$-categorical structures (or indeed, structures with a large supply of automorphisms). It admits many refinements and variations, but in its basic form originates in work of Fraïssé (1954*a*, 1953), and Jónsson (1956, 1960).

DEFINITION 2.1. Suppose L is a first-order language and $\mathscr{C}$ is a collection of L-structures. We are really only interested in the isomorphism types of elements of $\mathscr{C}$, thus when we say that an L-structure A is in $\mathscr{C}$ (or write $A \in \mathscr{C}$) we mean that A is isomorphic to an element of $\mathscr{C}$. Regard two collections as equal if they represent the same isomorphism types. With this in mind, consider the following properties that a collection $\mathscr{C}$ of L-structures may (or may not) have:

Hereditary Property. (HP) If $A \in \mathscr{C}$ and B is a substructure of A then $B \in \mathscr{C}$.
The Joint Embedding Property. (JEP) If $A, B \in \mathscr{C}$ there exists $C \in \mathscr{C}$ such that A, B are isomorphic to substructures of C.
The Amalgamation Property. (AP) Suppose $A, B_1, B_2 \in \mathscr{C}$ and that $\alpha_i : A \to B_i$ are embeddings. Then there exists $C \in \mathscr{C}$ and embeddings $\beta_i : B_i \to C$ such that $\beta_1\alpha_1 = \beta_2\alpha_2$.

The reader unfamiliar with these concepts might like to look at some of the examples in Section 2.2 before continuing with this section. The following is a useful exercise.

EXERCISE 2.2. 1. The collection of all fields (regarded as structures in the language $(+, -, \cdot, ^{-1}, 0, 1)$) has AP and HP but not JEP.

2. The collection of all trees (i.e., connected graphs with no cycles) has AP, JEP, but not HP. The collection of all forests (i.e., graphs with no cycles) has HP and JEP, but not AP.

3. If $\mathscr{C}$ has AP and there exists a structure in $\mathscr{C}$ which embeds in all others, then $\mathscr{C}$ has JEP.

DEFINITION 2.3. (i) If M is an L-structure the *age* of M (denoted by Age(M)) is the collection of finitely generated substructures of M. Note that Age(M) has HP and JEP.

(ii) A collection of finitely generated L-structures is an *amalgamation class* if it has HP, JEP and AP.

(iii) An L-structure M is *homogeneous* if any partial isomorphism between finitely generated L-substructures of M extends to an automorphism of M.

The following theorem will be a consequence of a more general result to be proved in Section 2.1.3.

THEOREM 2.4. (Fraïssé's Theorem) Let L be a first-order language. Suppose $\mathscr{C}$ is an amalgamation class of finite L-structures. Suppose that the number of isomorphism types of structures in $\mathscr{C}$ of the same size is always finite. Then there exists a countable L-structure M such that:

(a) the age of M is $\mathscr{C}$;

(b) if $A, B \in \mathscr{C}$ and we have embeddings $\alpha: A \to M$ and $\beta: A \to B$ then there exists an embedding $\gamma: B \to M$ such that $\alpha = \gamma\beta$.

The structure M is determined up to isomorphism amongst countable L-structures by properties (a) and (b). It is homogeneous for the language L. Furthermore, if the number of isomorphism types of n-generator structures in $\mathscr{C}$ is finite for each $n < \omega$, then M is $\aleph_0$-categorical.

DEFINITION 2.5. If $\mathscr{C}$ is an amalgamation class satisfying the hypotheses of the above theorem, we refer to the homogeneous structure M guaranteed by the theorem as the *Fraïssé limit* of $\mathscr{C}$, and denote it by H($\mathscr{C}$).

As a straightforward consequence of Fraïssé's Theorem, we have

THEOREM 2.6. Let $\mathscr{C}$ be a class of finite L-structures in which, for every $n < \omega$, the number of isomorphism types of n-generator structures is finite. The following are equivalent:

(a) $\mathscr{C}$ is the age of a countable homogeneous L-structure;

(b) $\mathscr{C}$ is an amalgamation class.

2.1.2 *Strong amalgamation bases and algebraic closure*

DEFINITION 2.7. Let $\mathscr{C}$ be a class of L-structures and let $A \in \mathscr{C}$. We say that A is *an amalgamation base* if for all $B_1, B_2 \in \mathscr{C}$ and embeddings $\alpha_i: A \to B_i$ there exists $C \in \mathscr{C}$ and embeddings $\beta_i: B_i \to C$ such that $\beta_1\alpha_1 = \beta_2\alpha_2$. We say that A is a *strong amalgamation base* if we can arrange additionally that $\beta_1(B_1) \cap \beta_2(B_2) = \beta_1\alpha_1(A)$.

Let $\mathscr{C}_S$ consist of the strong amalgamation bases in the amalgamation class $\mathscr{C}$. In many of the examples in Section 2.2 we have $\mathscr{C}_S = \mathscr{C}$ (in which case we say that $\mathscr{C}$ is a *strong amalgamation class*). This is not too surprising: proving the amalgamation property for something which is not a strong amalgamation base involves identifying more than just the images of A in B_1 and B_2, and this is usually a difficult problem. The link with algebraic closure is this:

LEMMA 2.8. Suppose $\mathscr{C}$ is an amalgamation class of finite L-structures in which the number of isomorphism types of n-generator structures is finite, for all $n \in \mathbb{N}$. Then $A \in \mathscr{C}$ is a strong amalgamation base in $\mathscr{C}$ if and only if A is isomorphic to an algebraically closed substructure of the Fraïssé limit $\mathrm{H}(\mathscr{C})$ of $\mathscr{C}$, equivalently if all substructures $A' \cong A$ of $\mathrm{H}(\mathscr{C})$ are algebraically closed.

Proof. First, assume $A' \subseteq \mathrm{H}(\mathscr{C})$ is algebraically closed, $A' \cong A$, and $\alpha_i\colon A \to B_i$ $(i = 1, 2)$ are embeddings. Without loss of generality (using homogeneity) $A = A' \subseteq B_1 \subseteq \mathrm{H}(\mathscr{C})$ and $A \subseteq B_2 \subseteq \mathrm{H}(\mathscr{C})$, and α_1, α_2 are inclusion maps. Since $G_{(A)}$ has no finite orbits outside A the separation lemma (Theorem 3.1 on p. 17) gives $g \in G_{(A)}$ with $g(B_1 \setminus A) \cap (B_2 \setminus A) = \varnothing$. Then $C = gB_1 \cup B_2$ is the required amalgam of B_1, B_2 with embeddings g and the inclusion map.

For the converse, let $A \cong A' \subseteq \mathrm{H}(\mathscr{C})$ be arbitrary, and $x \in \mathrm{H}(\mathscr{C}) \setminus A'$. We prove that x is not in the algebraic closure of A'. Let $\{x\} \cup A' \subseteq B_1 \subseteq \mathrm{H}(\mathscr{C})$ where B_1 is isomorphic to a structure in $\mathscr{C}$, let $B_2 = B_1$, and let $\alpha_i\colon A' \to B_i$ be the inclusion maps. Since A is a strong amalgamation base there are $C \in \mathscr{C}$ and $\beta_i\colon B_i \to C$ such that $\beta_1 B_1 \cap \beta_2 B_2 = \beta_1\alpha_1 A'$. By homogeneity we may assume $C \subseteq \mathrm{H}(\mathscr{C})$ and β_1 is the inclusion map. Then $\beta_2 x \mapsto x$ extends to an isomorphism $\beta_2 B_2 \to B_1$ which is constant on A', so x and $\beta_2 x$ lie in the same $G_{(A)}$-orbit, by homogeneity. But $x \neq \beta_2 x$. By repeating this argument with C in place of B_1, and so on, we can show that the $G_{(A)}$-orbit of x is infinite, as required. □

2.1.3 *A variation on the Fraïssé construction*

We introduced the Fraïssé construction by referring to it as a way of glueing structures together. A slightly more sophisticated viewpoint is to see it as a way of piecing *embeddings* together. One advantage to doing this is that one can sometimes prove versions of Fraïssé's Theorem for collections of L-structures which are not amalgamation classes for general embeddings, but which are amalgamation classes when 'embedding' is read throughout in some restricted sense. This technique has been used extremely effectively in recent work of Hrushovski (1988*b*) (see also the article by Wagner in this volume), providing counterexamples to Lachlan's Conjecture on stable $\aleph_0$-categorical structures (Lachlan, 1974). The art (and the difficulty) is in finding a suitably restricted notion of 'embedding' which will allow an amalgamation property to be proved. We shall illustrate this in Example 5 of Section 2.2.2 on p. 50 and in Example 2 of Section 2.2.5 on p. 56. The following formalism allows us to go ahead with an 'amalgamation of embeddings' version of Fraïssé's Theorem. It is straightforward, if somewhat *ad hoc.* A slightly different viewpoint on the same construction can be found in Kueker and Laskowski (1992).

DEFINITION 2.9. Let $\mathscr{C}$ be a class of L-structures. A *class of $\mathscr{C}$-embeddings* is a collection $\mathscr{E}$ of embeddings $\alpha\colon A \to B$ (where $A, B \in \mathscr{C}$) such that

(i) any isomorphism is in $\mathscr{E}$;
(ii) $\mathscr{E}$ is closed under composition;
(iii) if $\alpha\colon A \to B$ is in $\mathscr{E}$ and $C \subseteq B$ is a substructure in $\mathscr{C}$ such that $\alpha(A) \subseteq C$, then the map obtained by restricting the range of α to C is also in $\mathscr{E}$.

Suppose $\mathscr{E}$ is a class of $\mathscr{C}$-embeddings. If M is an L-structure and A a finite substructure which is in $\mathscr{C}$ then we say that A is *$\mathscr{E}$-embedded* in M if whenever B is a finite substructure of M which is in $\mathscr{C}$ and contains A, the inclusion map $A \hookrightarrow B$ is in $\mathscr{E}$. In this case we write $A \leqslant M$. Note that if $\alpha\colon A \to B$ is in $\mathscr{E}$ then $\alpha(A) \leqslant B$, and if $A, B, C \in \mathscr{C}$ and $A \leqslant B \leqslant C$, then $A \leqslant C$.

As before, we are interested in embeddings up to isomorphism: say that embeddings $\alpha\colon A \to B$ and $\alpha'\colon A' \to B'$ are *isomorphic* if there exist isomorphisms $\gamma\colon A \to A'$ and $\delta\colon B \to B'$ such that $\delta\alpha = \alpha'\gamma$.

Let $\mathscr{E}$ be a class of $\mathscr{C}$-embeddings. Consider the following modifications of the joint embedding and amalgamation properties:

(JEP$'$) If $A, B \in \mathscr{C}$ there exists $C \in \mathscr{C}$ and embeddings $\alpha\colon A \to C$ and $\beta\colon B \to C$ such that $\alpha, \beta \in \mathscr{E}$.

(AP$'$) Suppose $A, B_1, B_2 \in \mathscr{C}$ and $\alpha_i\colon A \to B_i$ are embeddings in $\mathscr{E}$. Then there exists $C \in \mathscr{C}$ and embeddings $\beta_i\colon B_i \to C$ in $\mathscr{E}$ such that $\beta_1\alpha_1 = \beta_2\alpha_2$.

The new version of Fraïssé's Theorem follows. (Note that we can recover Theorem 2.4 by taking $\mathscr{E}$ to be the class of all embeddings between structures in $\mathscr{C}$.)

THEOREM 2.10. Suppose $\mathscr{C}$ is a collection of finite L-structures in which the number of isomorphism types of any finite size is finite. Suppose $\mathscr{E}$ is a class of $\mathscr{C}$-embeddings which satisfies JEP$'$ and AP$'$. Then there exists a countable L-structure M with the following properties:

(a) the class of $\mathscr{E}$-embedded substructures of M is equal to $\mathscr{C}$;
(b) M is a union of finite $\mathscr{E}$-embedded substructures;
(c) if $A \leqslant M$ and $\alpha\colon A \to B$ is in $\mathscr{E}$ then there exists $C \leqslant M$ containing A and an isomorphism $\beta\colon B \to C$ such that $\beta\alpha(a) = a$ for all $a \in A$.

Any two countable structures with properties (a), (b) and (c) are isomorphic, and any isomorphism between $\mathscr{E}$-embedded finite substructures of M extends to an automorphism of M. Moreover, suppose the following property holds:

Uniform boundedness. There is a function f on the natural numbers such that if C is an n-generator substructure of an element of $\mathscr{C}$ then there exists a structure $A \in \mathscr{C}$ which contains C, has at most $f(n)$ elements, and has the property that whenever $B \in \mathscr{C}$ and $\beta: A \to B$ is an embedding, then β is an $\mathscr{E}$-embedding.

then M is $\aleph_0$-categorical.

Proof. First, we give the construction. We find $B_i \in \mathscr{C}$ and $\mathscr{E}$-embeddings $\beta_i: B_i \to B_{i+1}$ such that:

1. any member of $\mathscr{C}$ with at most i elements is $\mathscr{E}$-embedded in B_i;
2. if $A \leqslant B_i$ and $\alpha: A \to B$ is an $\mathscr{E}$-embedding, and B has at most $i+1$ elements, then there exists $C \leqslant B_{i+1}$ containing $\beta_i(A)$ and an isomorphism $\beta: B \to C$ such that $\beta\alpha(a) = \beta_i(a)$ for all $a \in A$.

To do this, first choose $B_0 \in \mathscr{C}$ arbitrarily. Suppose B_i has been constructed. List the finitely many substructures $A_1, \dots, A_n$ which are $\mathscr{E}$-embedded in B_i and have at most i elements. For each $k \leqslant n$ list the isomorphism types of embeddings $A_k \to C_j$ in $\mathscr{E}$, where the C_j have at most $i+1$ elements (there are finitely many of these). Amalgamate these, and the embedding $A_k \hookrightarrow B_i$, over A_k to obtain the diagram of embeddings (in $\mathscr{E}$) into a new structure $B_i^k \subset \mathscr{C}$:

$$\begin{array}{ccc} & A_k & \\ \swarrow & \downarrow & \searrow \\ C_1 \cdots & C_r & B_i. \\ \searrow & \downarrow & \swarrow \\ & B_i^k & \end{array}$$

Do this for each k, and then amalgamate the embeddings $B_i \to B_i^k$ into a new structure $B_{i+1} \in \mathscr{C}$ by using AP$'$:

$$\begin{array}{ccc} & B_i & \\ \swarrow & & \searrow \\ B_i^1 & \cdots & B_i^n. \\ \searrow & & \swarrow \\ & B_{i+1} & \end{array}$$

It should be clear that the embedding $B_i \to B_{i+1}$ has the second of the desired properties. By using JEP$'$, we can ensure that B_{i+1} also has the first.

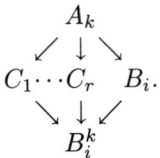

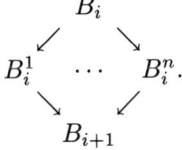

Let M be the direct limit of the embeddings $B_i \to B_{i+1}$, and to ease notation, think of each B_i as a substructure of M. Properties (a) and (b) are clear once we observe that $B_i \leqslant M$ for each i. For (c), let A and α be as stated there. By (a) we can assume $A \leqslant B_i$ and we can take i larger than the cardinality of B. By construction, there exists C with $A \leqslant C \leqslant B_{i+1}$ such that the embeddings $A \to B$ and $A \hookrightarrow C$ are isomorphic. The statement in (c) now follows.

The uniqueness statement and its generalization follow from a straightforward back-and-forth argument. By using uniform boundedness, the $\aleph_0$-categoricity then follows from the Ryll-Nardzewski Theorem (alternatively, the condition can be used to produce first-order axioms which guarantee properties (a), (b), (c) above). □

2.2 Examples

2.2.1 *Ordered structures*

Many examples of ordered $\aleph_0$-categorical structures are most easily constructed directly (such as the rationals), or constructed from existing ordered structures (cf. the construction of the 2-homogeneous trees from the rationals described by Droste (1985)). Consequently, this subsection will be rather brief, and the reader wanting to see a plethora of amalgamation classes should look instead at Section 2.2.2 on graphs and digraphs.

1. *Linear orders.* Let $\mathscr{C}$ be the class of finite linear orders. As a substructure of a finite linear order is a linear order, it follows that $\mathscr{C}$ has the Hereditary Property. The empty structure is in $\mathscr{C}$, and so to show that $\mathscr{C}$ is an amalgamation class, it suffices to prove the Amalgamation Property. So let $A, B_1, B_2 \in \mathscr{C}$ and let $\alpha_i : A \to B_i$ be embeddings (that is, order-preserving injections). For ease of notation, regard A as a subset of both B_1 and B_2, and let C be the disjoint union over A of B_1 and B_2. So we regard B_1 and B_2 as subsets of C intersecting in A. We now define an ordering on C which agrees with the original orderings on B_1 and B_2. Let $c_1, c_2 \in C$. Then:

1. if $c_1, c_2 \in B_i$ (for some i) and $c_1 \leqslant c_2$ in B_i, then we set $c_1 \leqslant c_2$ in C;
2. if (say) $c_i \in B_i \setminus A$, and there does not exist $a \in A$ greater than one of them and less than the other, then we set $c_1 \leqslant c_2$;
3. if (say) $c_i \in B_i \setminus A$ (for $i = 1, 2$), and there exists $a \in A$ with $c_i \leqslant a$ in B_i and $c_j \geqslant a$ in B_j, then set $c_i \leqslant c_j$.

The reader can verify that this makes C a linear order which agrees with the existing linear orders on B_1 and B_2, and so we have that $\mathscr{C}$ is a (strong) amalgamation class.

Let M be the Fraïssé limit of $\mathscr{C}$. Then M is a dense linear order without endpoints. One way to see this is to use property (b) in Theorem 2.4. Let

$a, b \in M$ with $a < b$. Let B be the three element linear order $b_1 < b_2 < b_3$, and consider the embedding β with $\beta(a) = b_1$ and $\beta(b) = b_3$. Let $\gamma: B \to M$ be as given by Theorem 2.4. Then $a < \gamma(b_2) < b$. So the ordering on M is dense. Similarly M is without endpoints. As M is countable, it follows from Cantor's Theorem that M is isomorphic to the rationals with their usual ordering.

Another way to do this is to observe that homogeneity of M implies that its automorphism group is transitive (so there are no endpoints), and transitive on unordered pairs of elements of M (so the ordering is dense). A variation on this argument can be used to determine all countable $\aleph_0$-categorical linear orders (see Rosenstein (1969)).

2. *Partial orders.* The class $\mathscr{C}$ of finite partial orders is a strong amalgamation class (the proof of this is similar to the proof for linear orders above), and so there exists a (unique) homogeneous partially ordered set $(P, <)$ which embeds any finite (indeed, any countable) partially ordered set. There does not appear to be a 'concrete' description of this, and the Fraïssé method seems to be the only known way to construct $(P, <)$. (It is worth mentioning in passing that Albert and Burris (1986) give a finite axiomatization of $(P, <)$.) Schmerl (1979) gives a description of all countable, homogeneous partially ordered sets. This does not give any 'new' $\aleph_0$-categorical structures.

3. *Other homogeneous ordered structures.* The countable 2-homogeneous trees considered by Droste (1985) are $\aleph_0$-categorical partially ordered sets which are not homogeneous. However, by enlarging the language and introducing a ternary and a 4-ary relation symbol, they can be made homogeneous. The reader should consult Droste's memoir for details.

Thomas (1986) shows that for every prime power q there is a linear order on the elements of a countably infinite dimensional vector space V over the field with q elements, with the property that for every pair A, B of subspaces of V of the same finite dimension, there exists an order preserving automorphism of V taking A to B. The construction is a rather non-trivial application of Fraïssé's Theorem.

2.2.2 *Graphs and digraphs*

1. *Homogeneous graphs.* The random graph mentioned in the introduction is the first member of a family of countable, homogeneous graphs, the other members of which were were first described by Henson (1971).

Let n be a natural number. A graph is called *K_n-free* if it has no complete n-vertex subgraph. Let $\mathscr{C}^{(n)}$ be the class of finite K_n-free graphs. (We also allow n to be ∞ here, and $\mathscr{C}^{(\infty)}$ is the class of all finite graphs.)

Clearly $\mathscr{C}^{(n)}$ has the Hereditary Property. To see that it has the Amalgamation Property, suppose $A, B_1, B_2 \in \mathscr{C}^{(n)}$ and $\alpha: A \to B_i$ are embeddings. To ease notation, regard A as a subset of B_i, and let C be the disjoint union of B_1 and B_2 over A. Consider B_1 and B_2 as subsets of C. Make C into a graph by setting $\{a, b\}$ to be an edge if and only if $a, b \in B_i$ for some i, and $\{a, b\}$ is an edge in B_i. Then C is K_n-free and the inclusion maps of B_1 and B_2 into C are embeddings. (We refer to C as the *free amalgam* of B_1 and B_2 over A.) It follows that $\mathscr{C}^{(n)}$ is a strong amalgamation class.

Let G_n denote the Fraïssé limit of $\mathscr{C}^{(n)}$. So G_n is a countable, homogeneous K_n-free graph with age $\mathscr{C}^{(n)}$ and G_n is $\aleph_0$-categorical. Note that the complement $\overline{G_n}$ of G_n is also homogeneous and $\aleph_0$-categorical. Lachlan and Woodrow (1980) show that, apart from some degenerate examples, these account for all countably infinite homogeneous graphs (there are also some finite examples).

The reader can show that the class of all finite bipartite graphs is not an amalgamation class. To remedy this, instead of thinking of a bipartite graph as a set with a binary relation on it, think of it as also having an equivalence relation which partitions the vertices into two sets, with no edge between vertices in the same equivalence class. Let $\mathscr{B}$ denote the class of finite structures of this type. Then $\mathscr{B}$ is an amalgamation class and its Fraïssé limit is $\aleph_0$-categorical.

2. *The Henson digraphs.* A (simple) *digraph* is a structure with a single asymmetric, irreflexive binary relation (which we will denote by R). Henson (1972) constructs $2^{\aleph_0}$ non-isomorphic countable, homogeneous digraphs. We recall his construction here. First, a digraph A is called a *tournament* if for any distinct $a, b \in A$ we have $R(a, b)$ or $R(b, a)$. Now suppose $\mathscr{T}$ is a class of finite tournaments which has the Hereditary Property. Let $\mathscr{C}(\mathscr{T})$ be the class of finite digraphs A satisfying the following condition:

any sub-digraph of A which is a tournament is in $\mathscr{T}$.

It is not hard to show that $\mathscr{C}(\mathscr{T})$ is an amalgamation class (use the same amalgamation procedure as was used for graphs in 1 above), and the Fraïssé limit $\Gamma(\mathscr{T})$ has age $\mathscr{C}(\mathscr{T})$.

To see that $2^{\aleph_0}$ examples arise here, it suffices to find $2^{\aleph_0}$ classes of finite tournaments with the Hereditary Property. One way to do this is as follows.

For a natural number n (at least 3), let A_n be the tournament with vertices $\{0, \ldots, n+2\}$ and directed edges

$$\{(i, i+r), (i+1, i), (n+2, n+1) : 0 \leqslant i \leqslant n, 2 \leqslant r \leqslant n+2-i\}.$$

Let T_n be the tournament consisting of A_n with one new vertex α and new directed edges

$$(0, \alpha), (\alpha, 1), \ldots, (\alpha, n+1), (n+2, \alpha).$$

CLAIM. If $m \neq n$ then T_m does not embed in T_n.

Proof. Suppose to the contrary that $f: T_m \to T_n$ is an embedding. Label each directed edge in T_k according to the number of directed 3-cycles in which it is contained. Each of these labels is 0, 1 or 2, except that the edge $(k+2, \alpha)$ has label k. Now, f cannot decrease the labelling of any edge, and so as $m, n \geqslant 3$ we have $f(m+2) = n+2$ and $f(\alpha) = \alpha$. Apart from α, $n+2$ has a unique predecessor, so $f(m+1) = n+1$. Similarly, $f(m) = n$, ..., $f(0) = n - m$. By consideration of $(0, \alpha)$ we then get $n = m$. □

So for a subset S of $\mathbb{N} \setminus \{1, 2\}$, let $\mathscr{T}(S)$ be the class of tournaments which embed in T_n for some $n \in S$. Clearly $\mathscr{T}(S)$ has the Hereditary Property, and different choices for S give rise to different classes $\mathscr{T}(S)$.

There is a slightly different viewpoint on this construction which is worthy of note. Suppose we have an amalgamation class $\mathscr{C}$ of finite structures (of the same signature), and suppose that the amalgamation procedure is fixed (for example, when verifying the amalgamation property for K_n-free graphs on p. 48, we gave a rule for performing the amalgamation, which was only one of many possible choices). Call this fixed type of amalgamation *free* amalgamation. Say that $A \in \mathscr{C}$ is *a-indecomposable* if it cannot be written as a non-trivial free amalgamation in $\mathscr{C}$. Suppose S is a set of pairwise non-embeddable a-indecomposable elements of $\mathscr{C}$. Let $\mathscr{C}(S)$ be the set of elements of $\mathscr{C}$ which do not embed any $A \in S$. Then $\mathscr{C}(S)$ is an amalgamation class. Moreover, different choices for S give rise to different Fraïssé limits.

3. *Homogeneous tournaments and digraphs.* There are other countable homogeneous digraphs besides the $\Gamma(\mathscr{T})$ just described. Included amongst these are the homogeneous tournaments. These have been classified by Lachlan (1984) (and this has been reworked by Cherlin (1988)). A linear order is a special form of tournament, so the homogeneous linear order gives an example here. Likewise the universal, homogeneous local order can be thought of as a homogeneous tournament. The class of all finite tournaments is an amalgamation class, and so there is a countable, homogeneous tournament which embeds every finite tournament. This is the analogue for tournaments of the random graph (indeed, it can be described as the tournament obtained with probability 1 by randomly (and independently) orienting directed edges between a countable vertex set).

The class of finite digraphs is a strong amalgamation class, so there is a universal, countable homogeneous digraph. Countable homogeneous digraphs have recently been classified by Cherlin (as yet unpublished).

4. *Constructing* $2^{\aleph_0}$ $\aleph_0$*-categorical graphs.* As mentioned in 1, there are only countably many countable, homogeneous graphs. However, the Fraïssé

technique can be used to construct continuum-many non-isomorphic $\aleph_0$-categorical graphs, and we shall now describe one way of doing this. The basic idea here is to consider amalgamation classes of finite structures which consist of a graph together with some extra structure. The Fraïssé limit of each class is then a graph with some extra structure. So then one ignores the extra structure and obtains an $\aleph_0$-categorical graph. The trick is to choose the 'extra structure' so that different amalgamation classes give non-isomorphic graphs (not just non-isomorphic graphs with extra structure). The examples we give are taken from Droste and Macpherson (1991).

Let S be a subset of the natural numbers and consider a language L_S having a binary relation symbol E (for the edge relation in the graph) and for each $i \in S$ an i-ary relation symbol R_i. Let $\mathscr{C}_S$ be the class of finite L_S-structures A such that:

(i) E is a symmetric and irreflexive relation on A (so A is a graph with E the adjacency relation);

(ii) $A \models R_i(a_1, \ldots, a_i)$ if and only if

  (a) the a_i are distinct

  (b) the induced subgraph on $\{a_1, \ldots, a_i\}$ is an i-cycle

  (c) no vertex of A is adjacent to all elements of $\{a_1, \ldots, a_i\}$.

It is easy to see that $\mathscr{C}_S$ is an amalgamation class (for the amalgamation property, we can use a 'free' amalgamation as on p. 48). Let Γ_S^+ be the Fraïssé limit of $\mathscr{C}_S$. So Γ_S^+ is $\aleph_0$-categorical. Let Γ_S be the reduct of Γ_S^+ to the graph language (so forget about the R_i, and keep the graph relation E). By the Ryll-Nardzewski Theorem, Γ_S is $\aleph_0$-categorical (of course, it is not homogeneous unless $S = \varnothing$). We claim that if $S \neq S'$ then Γ_S and $\Gamma_{S'}$ are non-isomorphic. To see this, we prove the following.

Claim. For any natural number n, we have $n \in S$ if and only if there is an n-cycle in Γ_S with no vertex of Γ_S adjacent to all of its vertices.

Proof. If n is in S, then we can find $a_1, \ldots, a_n \in \Gamma_S^+$ such that the relation $R_n(a_1, \ldots, a_n)$ holds. For the converse, suppose there is such an n-cycle $a_1, \ldots, a_n$ in Γ_S^+ yet $n \notin S$. Let A be the induced L_S-structure on $\{a_1, \ldots, a_n\}$ in Γ_S^+. So $A \in \mathscr{C}_S$. Let B be the L_S-structure with point set $\{a_1, \ldots, a_n, b\}$ and relations on $\{a_1, \ldots, a_n\}$ as in A, together with new edges $\{a_i, b\}$. Thus A is a substructure of B. The point to notice is that $B \in \mathscr{C}_S$: for any $i \leqslant n$ with $i \in S$ we have $R_i[B] = \varnothing$, so the defining conditions for $\mathscr{C}_S$ are satisfied vacuously. So now, using property (b) of the Fraïssé limit in Theorem 2.4, there exists $b \in \Gamma_S^+$ adjacent to each a_i. This contradiction establishes the claim. □

5. *An example due to Hrushovski.* This example is taken from Hrushovski (1988*b*). It is used there to give a straightforward example of an $\aleph_0$-categorical pseudoplane. We include it here to illustrate the modification of the Fraïssé construction given in Theorem 2.10.

If A is a finite graph, let $x(A)$ and $e(A)$ denote respectively the number of vertices and the number of edges in A. Let $y(A) = 2x(A) - e(A)$. Let $\mathscr{C}$ be the class of finite graphs A such that if B is a subgraph of A then $x(B) \leqslant \exp(y(B) - 3) + 1$, that is, $e(B) \leqslant 2x(B) - 3 - \log(x(B) - 1)$. (In particular, the smallest cycle in $\mathscr{C}$ is a 5-cycle.)

If $B \in \mathscr{C}$ we say that a subgraph A of B is a *self-sufficient* subgraph (and write $A \leqslant B$) if for every B' with $A \subset B' \subseteq B$ we have $y(A) < y(B')$. An embedding $\alpha: A \to B$ is *self-sufficient* if $\alpha(A) \leqslant B$. The class $\mathscr{E}$ of self-sufficient embeddings into elements of $\mathscr{C}$ is a class of $\mathscr{C}$-embeddings. (Parts (i) and (iii) in the definition are trivial; (ii) takes some work, but is elementary and uses only the definition of y.) It is not too hard to show that $\mathscr{E}$ satisfies AP$'$. (In the notation of Example 1 of Section 2.2.2 one takes the amalgamated structure C to be the 'free' amalgam of B_1 and B_2 over A; see Lemma 6 of Hrushovski (1988*b*)). See Macpherson's article in this volume for a related construction also due to Hrushovski. Both constructions are disscussed further in the article by Wagner below.

Thus, we can form the Fraïssé limit M of $\mathscr{E}$, as in Theorem 2.10. To see that M is $\aleph_0$-categorical, we now show uniform boundedness of $\mathscr{E}$. Let $A \in \mathscr{C}$ have n elements. Let $C \in \mathscr{C}$ contain A and be maximal subject to the constraint $y(C) \leqslant y(A)$. Note that this is possible as such a C has at most $\exp(2n - 3) + 1$ elements. Then if $B \in \mathscr{C}$ and $C \subset B$ we have $y(C) \leqslant y(A) < y(B)$, so $C \leqslant B$, as required for uniform boundedness.

2.2.3 *Other relational structures*

The first two types of examples in this subsection are homogeneous for an infinite language rather than for a finite language.

1. *Arbitrarily fast growth rates of number of n-types.* If M is a countable $\aleph_0$-categorical structure, the number $S_n(M)$ (repectively $s_n(M)$) of orbits of Aut(M) on n-tuples of distinct elements (respectively, subsets of size n) of M is finite. The reader should consult Cameron (1990) and references given there for results on the fine detail of the behaviour of the sequences $(S_n(M))$ and $(s_n(M))$. As is well-known, we can use Fraïssé's Theorem to show that these sequences can grow arbitrarily quickly.

The number of sets of k-tuples from a set of size n is 2^{n^k}. Thus, if the finite language L has r relation symbols of arities $k_1 \leqslant \cdots \leqslant k_r$, then the number of isomorphism types of n-element L-structures is at most $\prod_{i=1}^{r} 2^{n^{k_i}} \leqslant 2^{rn^{k_r}}$. So if a structure M is homogeneous for a *finite* relational language the function $n \mapsto S_n(M)$ is bounded by a function which is the exponential of a polynomial in n.

Given a sequence (a_n) and a natural number n_0 we construct a homogeneous structure M with $S_n(M) \geqslant a_n$ for all $n \geqslant n_0$ as follows. The language L has a_k k-ary relation symbols $R_1^k, \ldots, R_{a_k}^k$ for $k \geqslant n_0$. Let $\mathscr{C}$

be the class of finite L-structures satisfying the axioms

$$\forall x_1,\ldots,x_k\,(R_i^k(x_1,\ldots,x_k) \to \bigwedge_{j\neq l} x_j \neq x_l).$$

So there are finitely many, and at least a_k, isomorphism types of k-element structures in $\mathscr{C}$. Now, $\mathscr{C}$ is an amalgamation class (a 'free' amalgamation works very nicely). If M is the Fraïssé limit, then M is $\aleph_0$-categorical and has the required growth rate of $S_n(M)$.

2. *An example due to Cherlin and Hrushovski.* The following example was proposed by Cherlin in response to a question of Lascar (1982). It was subsequently rediscovered by Hrushovski, and is typical of $\aleph_0$-categorical structures known not to have the small index property (see the previous chapter).

Let L be a first-order language having for each n a $2n$-ary relation symbol E_n. Let $\mathscr{C}$ be the class of finite L-structures in which, for each n, $E_n[A]$ is an equivalence relation on the set $A^{(n)}$ of n-tuples of distinct elements of A, with at most two classes. Then $\mathscr{C}$ is an amalgamation class. Let M be the Fraïssé limit of $\mathscr{C}$. So M is $\aleph_0$-categorical.

Let Γ denote the automorphism group of M and let Γ_0 be the subgroup of Γ consisting of automorphisms which fix each of the equivalence classes of every equivalence relation $E_n[M]$. Then $\Gamma_0 \trianglelefteq \Gamma$ and it can be shown that the quotient group Γ/Γ_0 is isomorphic to Z_2^ω, the full direct product of countably many copies of the cyclic group of order 2 (the equivalence classes in each E_n can be interchanged independently). In particular, Γ has $2^{2^{\aleph_0}}$ subgroups of index 2.

The construction is quite flexible. In Evans and Hewitt (1990) it is used to show that any separable profinite group is a homomorphic image of some automorphism group of a countable $\aleph_0$-categorical structure. With a further variation, the profinite quotient can be 'hidden' in point stabilizers. Consider $(2n+1)$-ary relations F_n on structures A so that for any point $c \in A$, $F_n(c,*,*)$ is an equivalence relation on $(A\setminus\{c\})^{(n)}$ with two equivalence classes. This is an amalgamation class and if M is the Fraïssé limit then $\mathrm{Aut}(M)$ has no proper, closed subgroup of finite index, but the stabilizer in $\mathrm{Aut}(M)$ of any point of M has Z_2^ω as a homomorphic image.

3. *Tree-like structures.* In contrast to the previous two types of examples, the structures we describe now are homogeneous for a finite relational language (and they also involve some non-trivial verification of the amalgamation property). These structures were discovered by Cameron, and are described in detail (and with variations) in Cameron (1983*a*) and Cameron (1987). They are related to the 2-homogeneous trees described in Droste (1985).

The class of finite trees has the amalgamation property, but not the hereditary property, so the Fraïssé construction cannot be used directly. Of course, we could apply the variation on Fraïssé's Theorem, Theorem 2.10 of Section 2.1, but we do not have uniform boundedness. Nevertheless, it is desirable to find some way of utilizing the amalgamation property for trees to produce an $\aleph_0$-categorical structure. One way of doing this (due to Droste) has already been mentioned: consider the (rooted) trees as a partially ordered set with some extra structure, and prove the amalgamation property for these (Droste, 1985). Another way, due to Cameron, is as follows.

If A is a finite tree, a vertex of A of valency 1 is called a *leaf*. Suppose a, b, c, d are four distinct leaves of A. We write $ab|cd$ if the (unique) shortest path from a to b in A is disjoint from the shortest path from c to d. We denote by $\tilde{A}$ the set of leaves of A, together with the (possibly empty) 4-ary relation $|$. Can we recover A from $\tilde{A}$? Not quite: if we change A by introducing a new vertex (of valency 2) into an existing edge then we do not alter $\tilde{A}$. To avoid this problem, we consider the class of *series-reduced* trees, that is, trees with no vertices of valency 2. Note that from each tree A we obtain a unique series-reduced tree: if x is a vertex of valency 2 in A adjacent to the two vertices y and z, then in the series-reduced tree x has been deleted and $\{y, z\}$ is an edge. We then have:

LEMMA 2.11. Suppose A_1 and A_2 are series-reduced trees. Let $\theta: \tilde{A}_1 \to \tilde{A}_2$ be an isomorphism. Then there exists a unique isomorphism $\tau: A_1 \to A_2$ such that if x is a leaf of A_1 then $\tau(x) = \theta(x)$.

Proof. See Cameron (1987, Proposition 3.1). Alternatively, show that if A is series-reduced then A can be interpreted in $\tilde{A}$ in a uniform fashion. □

Now let $\widetilde{\mathcal{T}_\infty}$ be the class of relational structures $\tilde{A}$ obtained from finite series-reduced trees A. We claim that $\widetilde{\mathcal{T}_\infty}$ is an amalgamation class. We leave HP and JEP to the reader, and verify the amalgamation property (cf. Cameron, 1987, Proposition 3.2). Suppose $\tilde{B}_1$ and $\tilde{B}_2$ are in $\widetilde{\mathcal{T}_\infty}$ and have a common substructure $\tilde{A}$. Let B_i be the series-reduced tree giving rise to $\tilde{B}_i$. Let A_i be the subtree of B_i consisting of points on shortest paths between pairs of vertices in $\tilde{A}$. By the above lemma, the series-reduced versions of A_1 and A_2 are isomorphic via an isomorphism which is the identity on the leaves: call the series-reduced version A (this is consistent!). So each of B_1 and B_2 can be obtained from A by adding branches, either at existing vertices, or at new vertices 'in the middle of' an existing edge. If we add both sets of branches at once, we get a tree C where $\tilde{C}$ provides the required amalgamation of $\tilde{B}_1$ and $\tilde{B}_2$ over $\tilde{A}$.

Let T_∞ be the Fraïssé limit of $\widetilde{\mathcal{T}_\infty}$. Then T_∞ is homogeneous for the 4-ary language and is $\aleph_0$-categorical. In particular, the automorphism group of T_∞ is 3-transitive, and has two orbits on 4-sets.

Cameron (1987, p. 159 *et passim*) gives some variations on this construction. One of these is to consider instead of all finite trees, just those of valency at most n. The corresponding class of 'leaf-structures' $\tilde{\mathscr{T}}_n$ is still an amalgamation class. Cameron denotes the corresponding Fraïssé limit by T_n.

2.2.4 *Algebraic structures*

So far, the classes of structures we have considered have been relational, that is, they have not involved functions. If we wish to consider amalgamation classes of algebraic objects (such as lattices, rings or groups) then the situation changes: we should have function symbols in the language, so as to control the nature of the substructures over which we are amalgamating. (So, for example, we do not want to consider a group operation as being given by a ternary relation, as then giving an amalgamation class of groups would involve amalgamating over arbitrary subsets, rather than subgroups, of a group.) Showing that a particular class of algebraic structures has the amalgamation property is ususally a major piece of work: so in this section we usually just give references, and omit virtually all the details. First though, we give some genuinely straightforward examples. Groups are given a section to themselves later on.

1. *Vector spaces over a fixed finite field.* Let F be a finite field and let $\mathscr{V}_F$ be the class of all finite-dimensional F-vector spaces. (It is not very important what we take for the language here, but the usual thing to do is to have a binary function symbol $+$ for addition, a constant symbol 0 for the zero, and for each $\alpha \in F$ a unary function symbol μ_α for scalar multiplication by α.)

To show that $\mathscr{V}_F$ is an amalgamation class, suppose $A, B_1, B_2 \in \mathscr{V}_F$, and $\alpha_i\colon A \to B_i$ are embeddings. Let D be the direct sum $B_1 \oplus B_2$ and let E be the subspace $\{(\alpha_1 a, -\alpha_2 a) : a \in A\}$. Let $C = D/E$. Define maps $\beta_i\colon B_i \to C$ by $\beta_1(b) = (b, 0) + E$ and $\beta_2(b) = (0, b) + E$. These are embeddings, and for $a \in A$ we have $\beta_1\alpha_1(a) = \beta_2\alpha_2(a)$. This gives the amalgamation property. (We call C the *free amalgam* of B_1 and B_2 over A.)

The Fraïssé limit of $\mathscr{V}_F$ is a countably infinite dimensional vector space over F (of course, we did not need Fraïssé's Theorem to manufacture this). As F is finite, this is $\aleph_0$-categorical.

2. *Bounded abelian groups.* It follows easily from the Ryll-Nardzewski Theorem that an $\aleph_0$-categorical group must have finite exponent. For any natural number n let $\mathscr{A}_n$ be the class of finite abelian groups (written additively) of exponent dividing n (so $A \in \mathscr{A}_n$ if and only if $nA = 0$).

As above, a 'free' amalgamation shows that $\mathscr{A}_n$ is an amalgamation class. The reader can verify (using the characteristic properties in Theorem 2.4) that the Fraïssé limit of $\mathscr{A}_n$ is isomorphic to $Z_n^{(\omega)}$, the direct sum of countably many copies of the cyclic group of order n.

There are various preservation theorems of the form 'do this construction to a collection of $\aleph_0$-categorical groups and you obtain another $\aleph_0$-categorical group'. We will see a very general one of these in the section on boolean powers (3.1). A very straightforward one (due to Grzegorczyk) is that the direct product of a finite number of (countable) $\aleph_0$-categorical groups is $\aleph_0$-categorical. (Exercise: verify this using the Ryll-Nardzewski Theorem.) So now the reader can show that an abelian group is $\aleph_0$-categorical if and only if it has finite exponent (in which case it is isomorphic to a direct sum of cyclic groups, by a theorem of Prüfer (see Fuchs, 1960)).

3. *Boolean algebras.* It has already been mentioned in Section 1.4.5 that, up to isomorphism, there is a unique countable atomless boolean algebra B^1. In fact, this is homogeneous (in the algebra language): the proof of the uniqueness goes by a back-and-forth argument which can also be used to give homogeneity. So by the converse to Fraïssé's Theorem (Theorem 2.6 of Section 2.1), the class of finite boolean algebras is an amalgamation class. We leave it as an amusing exercise for the reader to verify the amalgamation property directly, using the fact that a finite boolean algebra is isomorphic to a field of sets (see Section 3.1.1).

By taking products of B^1 with finite boolean algebras, we see that any countable boolean algebra with finitely many atoms is $\aleph_0$-categorical. The converse of this follows from the Ryll-Nardzewski Theorem.

4. *Lattices.* The reader can consult Grätzer (1978) for information about amalgamation classes in various classes of lattices. In general, finding such things is an extremely difficult problem, and this does not appear to be a good source of $\aleph_0$-categorical structures. However, there is one example to be noted: the class $\mathscr{D}$ of finitely generated distributive lattices has the amalgamation property, and an n-generator distributive lattice has at most 2^{2^n-2} elements. The Fraïssé limit of $\mathscr{D}$ is a distributive lattice so we have

THEOREM 2.12. There exists a countable $\aleph_0$-categorical homogeneous distributive lattice with age $\mathscr{D}$.

Droste and Macpherson (1991) attribute this observation to Schmerl.

2.2.5 *Non-abelian groups*

One of the most straightforward ways to construct a rich supply of non-abelian $\aleph_0$-categorical groups is to take boolean powers of finite groups, as in Section 3.2.3. There are, however, ways of using the Fraïssé construction (and variants of it) to produce $\aleph_0$-categorical groups. The difficulty with all of these ways is of course to prove a suitable amalgamation property. One class where one does have control over amalgamation is in the class of groups which are nilpotent of class 2, and such groups provide a plentiful supply of $\aleph_0$-categorical groups.

1. *Homogeneous groups.* It is well-known that the class of finite groups is an amalgamation class. The Fraïssé limit of the class (more properly known as Philip Hall's countable, universal homogeneous locally finite group (Hall, 1959)) is not $\aleph_0$-categorical (it is not of finite exponent, and so there are infinitely many 1-types).

Cherlin and Felgner (1992) give a description of all countable, homogeneous *solvable* groups. Apart from abelian groups and extensions of abelian groups by involutions, the infinite $\aleph_0$-categorical examples are nilpotent of class 2 and have exponent 4.

Saracino and Wood (1982) construct $2^{\aleph_0}$ examples of such groups. Their method is, at least in spirit, similar to Henson's construction of $2^{\aleph_0}$ homogeneous digraphs (see Example 2 of Section 2.2.2), although the technical difficulties are much more substantial. They construct a countable family $\{G_i : i < \omega\}$ of pairwise non-embeddable finite nil-2 (i.e., nilpotent of class 2), exponent 4 groups and an amalgamation class $\mathscr{C}$ which contains it. They then show that for every $S \subseteq \omega$, if $\mathscr{C}_S$ denotes the class of members of $\mathscr{C}$ not embedding any G_i with $i \in S$, then $\mathscr{C}_S$ is also an amalgamation class (this is the difficult part). Let G^S denote the Fraïssé limit of $\mathscr{C}_S$. Easily, different choices of S give non-isomorphic groups G^S, and as an r-generator nil-2, exponent 4 group has size at most $4^r 2^{r(r-1)/2}$, each G^S is $\aleph_0$-categorical. More recently, Cherlin, Saracino and Wood (Cherlin *et al.*, in press) give a new framework for these constructions, and produce further examples (see also the remarks below).

2. *Non-homogeneous examples.* If $\mathscr{G}$ is a class of groups, an element $G \in \mathscr{G}$ is called *existentially closed* (or *e.c.*) in $\mathscr{G}$ if whenever a finite system of equations (and inequations) $\mathscr{S}$ involving parameters from G has a solution in some overgroup $H \in \mathscr{G}$ of G, it has a solution in G. This can be seen as a property similar to that in Fraïssé's Theorem: if A is the subgroup of G generated by the parameters in $\mathscr{S}$ and $B \in \mathscr{G}$ a finitely generated subgroup of H containing A and solutions to the equations in $\mathscr{S}$, then an embedding $B \to G$ over A would provide solutions to $\mathscr{S}$ in G. It should come as no surprise therefore that constructing existentially closed groups involves solving amalgamation problems, and that examples of $\aleph_0$-categorical groups can be found amongst countable existentially closed groups.

Saracino and Wood (1979) show that if m is a natural number and $\mathscr{N}_m$ is the class of groups which are nil-2 and have exponent (dividing) m, then there is a unique countable existentially closed model in $\mathscr{N}_m$, and furthermore, this is $\aleph_0$-categorical. Apps (1983*a*, Theorem C) has a further variation on this theme, where the equations have a restricted form.

Hodges' book (1985) is to be recommended for an enjoyable presentation of material on existentially closed nil-2 groups (and for much else).

The perceptive reader will have noticed that this subsection has been long on assertions and references, and short on proofs. To remedy this a little, we give an amalgamation procedure for a class of groups which are nil-2, exponent p, where p is an odd prime. We will then use the modification of Fraïssé's Theorem to build an $\aleph_0$-categorical group which is nil-2 and exponent p. (In fact, the group will be existentially closed in this class, and so isomorphic to one of the groups of Saracino and Wood (1979).) The presentation of the amalgamation procedure is taken from Baudisch (1992) (where it is used to construct a new $\aleph_1$-categorical group). It is not dissimilar to the construction used by Cherlin *et al.* (in press) to produce homogeneous nilpotent rings of characteristic p.

Let $p > 2$ be a prime, and let $\mathscr{C}$ be the class of finite groups A satisfying:

1. for all $x, y, z \in A$ we have $[[x, y], z] = 1$ and $x^p = 1$;
2. $[A, A] = Z(A)$.

Condition 1 says that A is nilpotent of class 2 and exponent p, and 2 says that the derived subgroup of A is equal to the centre of A.

For $A \in \mathscr{C}$ let $\overline{A}$ denote $A/Z(A)$. This is an elementary abelian p-group, therefore it can be regarded as a vector space over the field with p elements. We write $\overline{A}$ additively, and denote the coset $aZ(A)$ by $\overline{a}$.

Note that if $A, B \in \mathscr{C}$ and $A \leqslant B$ then $Z(A) = [A, A] \leqslant [B, B] \leqslant Z(B)$. Also, $Z(A) \geqslant A \cap Z(B)$, and so we get

$$Z(A) = Z(B) \cap A.$$

In particular, $\overline{A}$ embeds naturally in $\overline{B}$, via the map $aZ(A) \mapsto aZ(B)$ for $a \in A$.

By standard commutator identities, commutation in A gives a skew-symmetric bilinear map from $\overline{A}$ to $Z(A)$: for $a, a' \in A$ let $\langle \overline{a}, \overline{a'} \rangle_A = [a, a']$. By 2, this is surjective. Moreover, this map determines A completely: A is isomorphic to $\overline{A} \times Z(A)$ with multiplication

$$(\overline{a}, z)(\overline{a'}, z') = (\overline{a} + \overline{a'}, z + z' + \langle \overline{a}, \overline{a'} \rangle).$$

Now, this map factors through the alternating square of $\overline{A}$ (see Lang (1965), p. 424); there exists a (essentially unique) linear map $\phi \colon \bigwedge^2 \overline{A} \to Z(A)$ such that $\langle u, v \rangle_A = \phi(u \wedge v)$ for all $u, v \in \overline{A}$. Let $N(A)$ denote the kernel of ϕ. So A is determined by $\overline{A}$ and the subspace $N(A)$ of $\bigwedge^2 \overline{A}$. We identify A with $\overline{A} \times (\bigwedge^2 \overline{A})/N(A)$, with multiplication as above.

We now give the amalgamation procedure. Let $A, B_1, B_2 \in \mathscr{C}$ and suppose $\alpha_i \colon A \to B_i$ are embeddings. As above, these induce embeddings $\overline{\alpha_i} \colon \overline{A} \to \overline{B_i}$. Let C be the free amalgam of $\overline{B_1}$ and $\overline{B_2}$ over $\overline{A}$, as in Example

1 of Section 2.2.4, and let $\overline{\beta_i}: \overline{B_i} \to C$ be as given there. The $\overline{\alpha_i}$ and $\overline{\beta_i}$ induce embeddings $\hat{\alpha}_i: \bigwedge^2 \overline{A} \to \bigwedge^2 \overline{B}_i$ and $\hat{\beta}_i: \bigwedge^2 \overline{B} \to \bigwedge^2 C$, and $\hat{\beta}_1\hat{\alpha}_1 = \hat{\beta}_2\hat{\alpha}_2$.

Let $N(C)$ be the subspace of $\bigwedge^2 C$ that is generated by $\hat{\beta}_1(N(\overline{B_1}))$ and $\hat{\beta}_2(N(\overline{B_2}))$. We leave the reader to show that $\hat{\beta}_i(\bigwedge^2 \overline{B_i}) \cap N(C) = \hat{\beta}_i(N(\overline{B_i}))$. Let $Z = (\bigwedge^2 C)/N(C)$ and define the map $\langle , \rangle : C \times C \to Z$ by $\langle c, c' \rangle = c \wedge c' + Z$. This is a surjective, skew-symmetric bilinear map and so determines a group $G \in \mathscr{G}$ given by $C \times Z$ with multiplication

$$(c, z)(c', z') = (c + c', z + z' + \langle c, c' \rangle).$$

Identifying B_i with $\overline{B_i} \times (\bigwedge^2 \overline{B_i}/N(\overline{B_i}))$ we get maps $\beta_i : B_i \to G$ given by $\beta_i(b, f + N(\overline{B_i})) = (\overline{\beta_i}b, \hat{\beta}_i f + N(C))$. By the reader's exercise, these are embeddings. As $\beta_1\alpha_1 = \beta_2\alpha_2$, we have the amalgamation procedure.

If we let $\mathscr{E}$ be the class of all embeddings between elements of $\mathscr{C}$ then, in the notation of Theorem 2.10 of Section 2.1, $\mathscr{E}$ is a class of $\mathscr{C}$-embeddings which satisfies AP′. Let G_p be the Fraïssé limit of $\mathscr{E}$ as given by Theorem 2.10. To show that G_p is $\aleph_0$-categorical, we have to show uniform boundedness. In this context, this amounts to proving the following:

> Any r-generator subgroup A of G_p is contained in a $\mathscr{C}$-subgroup of G_p of size at most $p^{4r+4r(4r-1)/2}$.

We sketch a proof of this.

Firstly, note that (by (b) of Theorem 2.10 in Section 2.1) A is contained in some $\mathscr{C}$-subgroup B of G_p. We claim that there is a $\mathscr{C}$-subgroup C of G_p which contains B, and a $\mathscr{C}$-subgroup C_1 of C which contains A and has at most $4r$ generators. This will suffice to prove the above.

STEP 1. There exists a subgroup A_1 of B with $A \subseteq A_1$, having at most $2r$ generators and such that $Z(A_1) = A_1 \cap Z(B)$.

Proof. First, note that the dimension of $Z(A)/(Z(B) \cap A)$ is at most r. If this is zero, there is no problem. Otherwise, let $t \in Z(A) \setminus Z(B)$. Let $y \in B$ be such that $[t, y] \neq 1$, and let $A_1 = \langle A, y \rangle$. We claim that the dimension of $Z(A_1)/(Z(B) \cap A_1)$ is less that the dimension of $Z(A)/(Z(B) \cap A)$. (To see this note that

$$Z(A_1) = (A \cap Z(A_1))[A_1, A_1]$$

and

$$Z(B) \cap A_1 = (A \cap Z(B))[A_1, A_1],$$

and compute dimensions using the fact that $[A_1, A_1] \leqslant Z(A_1), Z(B)$.) Now repeat this procedure (at most r times) to get A_1 with $Z(A_1) = Z(B) \cap A_1$.

STEP 2. Let $x_1, \ldots, x_s$ be a basis over $[A_1, A_1]$ for $Z(A_1)$. Then $s \leqslant r$, and each $x_i \in Z(B)$. Take central products of B with extraspecial exponent

p groups $\langle y_i, z_i \rangle$, where $[y_i, z_i]$ is identified with x_i. Let C be the resulting group. Then $C \in \mathscr{C}$ and $C_1 = \langle A_1, y_1, \ldots, y_s, z_1, \ldots, z_s \rangle \in \mathscr{C}$ contains A.

This finishes the proof of uniform boundedness of G_p, so now we know that these groups are $\aleph_0$-categorical.

3 Boolean powers

3.1 The construction

3.1.1 *Facts about boolean rings*

We collect some facts about boolean rings and algebras. For more details, the reader could consult Sikorski (1964) or Bell and Slomson (1969).

A *boolean ring* is a ring B (not necessarily having an identity element) which satisfies $x^2 = x$ for all $x \in B$. If B has an identity element, it is called a boolean algebra. If B is a boolean ring then $x + x = 0$ for all $x \in B$ and B is commutative.

The most straightforward examples of boolean rings are *fields of sets*: if Ω is a set define operations $+$ and $\cdot$ on the power set $\mathscr{P}(\Omega)$ by $A + B = A \triangle B$ (where $\triangle$ denotes symmetric difference) and $A \cdot B = A \cap B$. Any finite boolean ring is isomorphic to one of these.

We can move between boolean rings with and without an identity element in the following way. If I is an ideal in the boolean ring B then I can also be considered as a boolean ring, and if I is non-principal then I has no identity element. Conversely, if B_0 is a boolean ring, we can form a boolean algebra $1 \oplus B_0$ in which B_0 is a maximal ideal.

An *atom* in a boolean ring B is a non zero element a such that $a \cdot b = 0$ if $b \neq a$. By the Ryll-Nardzewski Theorem, an $\aleph_0$-categorical boolean ring can have only finitely many atoms (if $a_1, a_2, \ldots \in B$ are atoms, then $a_1, a_1 + a_2, a_1 + a_2 + a_3, \ldots$ have different 1-types over the empty set). The converse is true, and we have:

THEOREM 3.1. There are two countable, atomless boolean rings B^0 and B^1, one without an identity element (B^0) and the other with an identity element. These are $\aleph_0$-categorical. Any countable $\aleph_0$-categorical boolean ring is isomorphic to a product of one of these with a finite field of sets.

Proof. Existence of B^1 has been remarked on in Section 1.4.5. Existence of B^0 follows from this and remarks above. A back-and-forth argument gives uniqueness and $\aleph_0$-categoricity. The final statement follows from the Ryll-Nardzewski Theorem. □

A Zorn's Lemma argument shows that any proper ideal in B is contained in a maximal ideal. We let $S(B)$ denote the set of maximal ideals of B. For $b \in B$ let $[b] = \{I \in S(B) : b \notin I\}$. Say that $X \subseteq S(B)$ is *open* if for every $I \in X$ there exists $b \in B$ such that $I \in [b] \subseteq X$.

THEOREM 3.2. (The Stone Representation Theorem) Let B be a boolean ring.

1. The set $S(B)$, together with its open sets, is a Hausdorff topological space. The compact open subsets of $S(B)$ are precisely the sets $[b]$ for $b \in B$.
2. The map $B \to \mathscr{P}(S(B))$ given by $b \mapsto [b]$ for $b \in B$ is a ring (or algebra) embedding of B into the field of sets of $S(B)$.

The topological space $S(B)$ is called the *Stone space* of B. Any automorphism of B induces a homeomorphism from $S(B)$ to itself (an autohomeomorphism), and by 1 in the above theorem, any autohomeomorphism arises in this way. Thus B and the Stone space $S(B)$ have isomorphic automorphism groups.

We now focus on the case of the countable atomless boolean algebra B^1. As B^1 is an algebra, $S(B^1)$ is compact and totally disconnected. It is clearly separable, and (as B^1 is atomless) it has no isolated points. Thus $S(B^1)$ is homeomorphic with the Cantor ternary set ξ. We remark that $S(B^0)$ is homeomorphic to ξ with a point removed.

If T is a clopen subset of ξ then $T = [t]$ for some $t \in B^1$. We write $B^1|T$ for the ideal of B^1 generated by t. The Stone space of this is naturally homeomorphic to T with the relative topology. Of course, in this case $B^1|T$ is isomorphic to B^1.

3.1.2 *The theorem of Waszkiewicz and Węglorz*

Let L be a first-order language and A a non-empty L-structure. Let B be a boolean algebra with Stone space S. We define an L-structure $A[B]$, the *boolean power* of A by B as follows:

1. the domain of $A[B]$ is the set of continuous functions $S \to A$, where A is regarded as having the discrete topology;
2. if c is a constant symbol in L interpreted by the element c in A, then we interpret c in $A[B]$ as the constant function which takes the value c everywhere;
3. if R is an n-ary relation symbol in L and $f_1, \ldots, f_n \in A[B]$ then $A[B] \models R(f_1, \ldots, f_n)$ if and only if $A \models R(f_1(x), \ldots, f_n(x))$ for each $x \in S$;
4. if F is an n-ary function symbol in L and $f_1, \ldots, f_n, f \in A[B]$ then $A[B] \models F(f_1, \ldots, f_n) = f$ if and only if $A \models F(f_1(x), \ldots, f_n(x)) = f(x)$ for each $x \in S$.

(It requires a moment's thought to see that F as given above is indeed interpreted by a function on $A[B]$ which is everywhere-defined.)

THEOREM 3.3. (Waszkiewicz and Węglorz, 1969) If B is a countable $\aleph_0$-categorical boolean algebra and A a countable $\aleph_0$-categorical structure for the language L, then the boolean power $A[B]$ is countable and $\aleph_0$-categorical.

Proof. The boolean algebra B is the direct product of B^1 with a finite field of sets, so $S(B)$ is the Cantor set ξ with a finite number of isolated points adjoined. Thus $A[B]$ is a direct product of a finite number of copies of A together with $A[B^1]$. Thus it will suffice to assume that $B = B^1$.

Let $f \in A[B]$. Then f has compact (so finite) image, and therefore there is a partition $\{T_1, \ldots, T_n\}$ of ξ into clopen subsets and elements $a_1, \ldots, a_n \in A$ such that $f(x) = a_i$ if and only if $x \in T_i$. We write $f = (a_1)_{T_1} \cdots (a_n)_{T_n}$. It is clear from this that $A[B]$ is countable.

Consider the following types of automorphisms of $A[B]$:

1. Let $g \in \mathrm{Aut}(A)$ and T a clopen subset of ξ. For $f \in A[B]$ define $f^{(g,T)}$ by
$$f^{(g,T)}(x) = \begin{cases} f(x) & \text{if } x \notin T \\ g(f(x)) & \text{if } x \in T. \end{cases}$$
(These automorphisms generate the boolean power $(\mathrm{Aut}(A))[B]$.)
2. Let $g \in \mathrm{Aut}(B)$ and $f \in A[B]$. Define f^g by
$$f^g(x) = f(gx).$$
(Here we are identifying $\mathrm{Aut}(B)$ with the group of autohomeomorphisms of ξ.)

Now we can prove the theorem. We do this by showing that if $G = \mathrm{Aut}(A[B])$ and $f_1, \ldots, f_n \in A[B]$ then the pointwise stabilizer $G_{f_1,\ldots,f_n}$ has finitely many orbits on $A[B]$. The result then follows from the Ryll-Nardzewski Theorem. First, we do the case where $n = 0$. Let $\Delta_1, \ldots, \Delta_k$ be the orbits of $\mathrm{Aut}(A)$ on A. Consider $f = (a_1)_{T_1} \cdots (a_m)_{T_m} \in A[B]$, where $a_1, \ldots, a_m$ are distinct. Using automorphisms of type 1, we can assume that $a_1, \ldots, a_m$ lie in distinct Δ_i's. So $m \leqslant k$. We can use automorphisms of type 2 to move the partition $\{T_1, \ldots, T_m\}$ around, and so using automorphisms of type 1 again, we get that G has at most $(k+1)^k$ orbits on $A[B]$.

Now take $f_1, \ldots, f_n \in A[B]$. Each of these defines a partition of ξ. Let $\{T_1, \ldots, T_m\}$ be the common refinement of these. Let $f_i(T_j) = \{a_{ij}\}$. Then $G_{f_1,\ldots,f_n}$ contains (g, T_i), where $g \in (\mathrm{Aut}(A))_{a_{1i} \cdots a_{ni}}$. If $f \in A[B]$, then the orbit of $G_{f_1,\ldots,f_n}$ in which f lies is determined by each $f|T_i$. Now, $B|T_i$ is isomorphic to B and $(\mathrm{Aut}(A))_{a_{1i} \cdots a_{ni}}$ has finitely many orbits on A for all i, and so the first part of the proof shows that there are only a finite number of orbits on the $f|T_i$'s for each i, and so there are only a finite number of $G_{f_1,\ldots,f_n}$-orbits on $A[B]$. □

REMARK. 1. There is no assumption in the above that A be infinite. Indeed, many of the interesting examples of boolean powers are where A is taken to be finite.

2. In the context of the above theorem, iteration of the boolean power operation produces nothing new. It is known (Burris, 1975) that, for boolean algebras B_1 and B_2, the iterated boolean power $(A[B_1])[B_2]$ is isomorphic to $A[B_1 \star B_2]$, where $B_1 \star B_2$ is the boolean free product (the Stone space of $B_1 \star B_2$ is the topological product of the Stone spaces of B_1 and B_2). As $B^1 \star B^1$ is isomorphic to B^1, we get that $(A[B^1])[B^1]$ is isomorphic to $A[B^1]$.

3. The proof of the above theorem shows that the subgroup of $A[B^1]$ generated by automorphisms of types 1 and 2 acts oligomorphically on $A[B^1]$. We leave it as an exercise to show that this subgroup is actually a semidirect product

$$(\mathrm{Aut}(A))[B^1] \cdot \mathrm{Aut}(B^1).$$

3.1.3 *Filtered boolean powers*

Let L be a first-order language and $P_0, \ldots, P_{n-1}$ unary predicate symbols not in L. Let $\tilde{L}$ be the language $L \cup \{P_0, \ldots, P_{n-1}\}$. Let A be an L-structure and let $A_0, \ldots, A_{n-1}$ be L-substructures of A. The $\tilde{L}$-structure $\tilde{A} = \langle A; A_0, \ldots, A_{n-1}\rangle$ has P_i interpreted as A_i.

Let B be a boolean algebra and $I_0, \ldots, I_{n-1}$ ideals in B. Consider the structure $\tilde{B} = \langle B; I_0, \ldots, I_{n-1}\rangle$. (Macintyre and Rosenstein (1976) call this an *augmented* boolean algebra of rank n.) Let $S = S(B)$ and let $X_i = \{I \in S : I_i \subseteq I\}$. So X_i is a closed subset of S. The *filtered boolean extension* $\tilde{A}[\tilde{B}]$ is the substructure of the boolean power $A[B]$ consisting of (continuous) functions $f\colon S \to A$ such that $f(X_i) \subseteq A_i$.

THEOREM 3.4. (Schmerl, 1978) If $\tilde{A}$ and $\tilde{B}$ are countable and $\aleph_0$-categorical, then so is $\tilde{A}[\tilde{B}]$.

For a proof of this, see Schmerl (1978). The proof is similar to the proof of the theorem of Waszkiewicz and Węglorz given above, making use of the Ryll-Nardzewski Theorem. Macintyre and Rosenstein (1976, Theorem 5), give a different proof of this in the case where A is finite.

REMARK. 1. The determination of which augmented boolean algebras are $\aleph_0$-categorical is a non-trivial problem involving the structure of Heyting algebras. The reader should consult Macintyre and Rosenstein (1976, Section 3) for information about this.

2. There is a particular case of the construction which is worthy of special mention. Suppose $n = 1$, $B = B^1$, I_0 is a maximal ideal, and A_0 is a singleton $\{a\}$. Then $\tilde{A}[\tilde{B}]$ consists of the continuous functions $f\colon \xi \to A$ where $f(I_0) = \{a\}$. Equivalently, these are the continuous functions $\hat{f}\colon C \setminus \{I_0\} \to A$ which take the value a outside of a compact set. In

the case where A is a group and a the identity element, we see that $\tilde{A}[\tilde{B}]$ is naturally isomorphic to the (bounded) boolean power $A[B^0]$ defined by Apps (1982) and others, that is: $A[B^0]$ is the set of continuous functions $f: S(B^0) \to A$ which have compact support.

3.2 Examples

Once we have described the construction of a boolean power of an $\aleph_0$-categorical structure, there is no additional work to be done to show that the result is $\aleph_0$-categorical. This is in contrast to the Fraïssé construction, where we have to provide an amalgamation procedure for a class of finite structures. It is, however, worthwhile to check whether the boolean power gives a 'new' structure. Indeed the following problem arises:

> Let $\mathscr{K}$ be a collection of countable $\aleph_0$-categorical structures with the same signature. For $A_1, A_2 \in \mathscr{K}$, does isomorphism of $A_1[B^1]$ and $A_2[B^1]$ imply isomorphism of A_1 and A_2?

We say that $\mathscr{K}$ is *booleanly separated* if this is so. This problem is closely related to the problem of determining the full automorphism group of a boolean power. Suppose A is a finite structure. Then, as in the proof of the Waszkiewicz–Węglorz Theorem (Theorem 3.3 of Section 3.1), we have automorphisms of $A[B^1]$ given by $(\mathrm{Aut}(A))[B^1]$ and $\mathrm{Aut}(B^1)$. If these generate all automorphisms of $A[B^1]$ then we say that the boolean power *splits*. (There is a generalization of this notion if A is countable and $\aleph_0$-categorical: see Archer (1993).)

3.2.1 *Graphs, digraphs, and partially ordered sets*

There appears to have been little work done on boolean powers of structures of this type. One exception is some recent work of Archer (1993) on splitting of boolean powers of finite graphs. We content ourselves here with making some trivial obsevations.

LEMMA 3.5. (a) A boolean power of a graph (respectively, digraph) is also a graph (respectively, digraph).

(b) A boolean power of a partially ordered set is a partially ordered set, whereas a boolean power of a linear ordering is a linear ordering only under trivial circumstances.

(c) If G is a countable graph which is neither null nor complete, then $G[B^1]$ is not homogeneous.

Proof. Either prove (a) and the first part of (b) directly, or note that the boolean power construction preserves validity of strict universal Horn sentences (cf. Hodges, 1985, Section 3.1). For the second part of (b), suppose $(A, <)$ is a linear ordering and $a, b \in S$ with $a < b$. Let C be a proper clopen subset of the Cantor set ξ and consider the following two elements

f_1 and f_2 of $A[B^1]$:

$$f_1(x) = \begin{cases} a & \text{if } x \in C \\ b & \text{if } x \notin C \end{cases}$$

and

$$f_2(x) = \begin{cases} b & \text{if } x \in C \\ a & \text{if } x \notin C. \end{cases}$$

Then f_1 and f_2 are incomparable in $A[B^1]$.

For (c) let a, b be vertices in G and let C be a clopen subset of ξ. Let $f_1, f_2, f_3 \in G[B^1]$ be given by $f_1(x) = a$, $f_2(x) = b$ for all $x \in \xi$ and $f_3(x) = a$ if $x \in C$ and $f_3(x) = b$ otherwise. Then any vertex $f \in G[B^1]$ which is adjacent to both f_1 and f_2 is adjacent to f_3. This is enough to prevent homogeneity of $G[B^1]$, unless G is null or complete. □

3.2.2 *Algebraic structures: lattices and rings*

A boolean power of a ring is a ring (it is a subring of a direct power), and clearly if the ring R has no nilpotent elements, then neither has the boolean power $R[B^1]$ (note, however, that $R[B^1]$ always has zero-divisors). Macintyre and Rosenstein (1976) show that a countable $\aleph_0$-categorical ring without nilpotent elements is isomorphic to a finite direct product of filtered boolean extensions $\tilde{A}[\tilde{B}]$, where A is a finite field, and $\tilde{B}$ is an $\aleph_0$-categorical augmented boolean algebra. The reader should consult Macintyre and Rosenstein (1976) for details of this, and for examples of $\aleph_0$-categorical augmented boolean algebras. This result of Macintyre and Rosenstein is complemented by a result of Cherlin (1980): the Jacobson radical of an $\aleph_0$-categorical ring is nilpotent.

The reader should consult Burris (1975) for results on boolean powers of lattices.

3.2.3 *Algebraic structures: groups*

The $\aleph_0$-categorical groups produced by Fraïssé's construction are generally nilpotent. Other examples of $\aleph_0$-categorical groups can be constructed from $\aleph_0$-categorical rings (see Berline and Cherlin (1981)). Boolean powers of finite groups provide convenient examples of non-solvable $\aleph_0$-categorical groups (if the group A is non-solvable, then neither is $A[B^1]$). This is no accident: Apps (1983*b*, Theorem C) proves the following (previously unpublished) result of J. S. Wilson:

THEOREM 3.6. If Γ is a countable, $\aleph_0$-categorical, characteristically simple group, then Γ is of one of the following types:

(a) finite;

(b) a boolean power (by B^0 or B^1) of a finite simple group;

(c) an infinite p-group, for some prime p.

REMARK. 1. A subgroup H of a group G is *characteristic* if it is invariant under all automorphisms of G. We say that G is characteristically simple if its only characteristic subgroups are G and $\{1\}$. If G is countable and $\aleph_0$-categorical then by the Ryll-Nardzewski Theorem G has only finitely many characteristic subgroups. Thus G has a sequence of characteristic subgroups $G = G_1 > G_2 > \cdots > G_r = \{1\}$ such that each G_i/G_{i+1} is characteristically simple and $\aleph_0$-categorical.

2. Case (b) includes elementary abelian p-groups.

3. J. S. Wilson conjectures that there does not exist a characteristically simple, locally finite, non-abelian p-group (in which case in (c) above the group must be elementary abelian).

Apps (1982) investigates the automorphism groups of boolean powers of (finite) groups, and shows that the class of finite indecomposable groups is booleanly separated. (This extends earlier work of Jónsson (1957).) As a warm-up exercise to reading Apps' paper, the reader might like to show that the class of non-abelian finite simple groups is booleanly separated. Here is a hint, taken from Apps (1982, Proposition 5.1). Suppose G is a finite non-abelian simple group and let $\Gamma = G[B^1]$. Show that B^1 is isomorphic to the lattice of normal closures of elements of Γ. Now show that the union of a maximal ideal in the latter is a normal subgroup of Γ and that the quotient group is isomorphic to G.

Apps (1983*c*) uses boolean powers to show that there exists a countable $\aleph_0$-categorical group G with a subgroup H of index 2, such that H is not $\aleph_0$-categorical (in fact G can be taken as $S_5[B^1]$, where S_5 denotes the symmetric group on 5 symbols). Also in this paper, Apps constructs an extension of groups $G \rhd H$ with G/H finite and H but not G being $\aleph_0$-categorical.

4 Canonical structures

Suppose Ω is a countable set and G a subgroup of $\mathrm{Sym}(\Omega)$, the symmetric group on Ω. Suppose further that the number N_k of orbits of G on Ω^k is finite, for all $k < \omega$. (So G acts oligomorphically on Ω.)

Let L be a first-order language which has N_k k-ary relation symbols, for each $k < \omega$. Consider Ω as the domain of an L-structure where the k-ary relation symbols are interpreted as the G-orbits on Ω^k (in some order). As in the previous chapter, this L-structure is called the *canonical structure* afforded by the permutation group $(G; \Omega)$ and will be denoted here $\mathrm{CS}(G; \Omega)$. (We will continue to denote its domain by Ω).

Clearly $G \leqslant \mathrm{Aut}(\mathrm{CS}(G; \Omega))$, and there is equality here if and only if G is a *closed* permutation group on Ω. It follows from the Ryll-Nardzewski Theorem that $\mathrm{CS}(G; \Omega)$ is $\aleph_0$-categorical. In this section, we give some ways of constructing oligomorphic permutation groups directly (that is, without first constructing an $\aleph_0$-categorical structure). In most cases, this will give new examples of $\aleph_0$-categorical structures when we consider the associated canonical structures.

4.1 Boolean powers again

Let A be a countable $\aleph_0$-categorical structure and let $\Omega = A[B^1]$ be the domain of the boolean power $A[B^1]$ of A by the countable atomless boolean algebra B^1, as in Section 3.1. Then, as in the proof of the theorem of Waszkiewicz and Węglorz (Theorem 3.3 of Section 3.1), we have that the groups $(\mathrm{Aut}(A))[B^1]$ and $\mathrm{Aut}(B^1)$ act on Ω in the following way:

1. if $f \in \Omega$ and $g \in (\mathrm{Aut}(A))[B^1]$ define $f^g(x) = g(x)(f(x))$ for $x \in \xi$ (remember, ξ denotes the Cantor ternary set);
2. if $f \in \Omega$ and $g \in \mathrm{Aut}(B^1)$ define $f^g(x) = f(g(x))$, for $x \in \xi$ (remember, $\mathrm{Aut}(B^1)$ can be identified with the group of autohomeomorphisms of ξ).

Let $G(A, B^1)$ be the subgroup of $\mathrm{Sym}(\Omega)$ generated by $(\mathrm{Aut}(A))[B^1]$ and $\mathrm{Aut}(B^1)$ with the above actions (in fact, as already remarked, this is the semidirect product of $(\mathrm{Aut}(A))[B^1]$ and $\mathrm{Aut}(B^1)$). The proof of the Waszkiewicz and Węglorz Theorem gives

THEOREM 4.1. With the above notation, $G(A, B^1)$ acts oligomorphically on Ω.

So the canonical structure $\mathrm{CS}(G(A, B^1); \Omega)$ is $\aleph_0$-categorical. It is interdefinable with the boolean power $A[B^1]$ if and only if $G(A, B^1) = \mathrm{Aut}(A[B^1])$.

Here is an application of this construction due to Evans (1986*a*), answering the following question.

QUESTION 4.2. ((Macpherson, 1986*a*, question 2)) Is it true that for every function f on the natural numbers there is an $\aleph_0$-categorical structure M such that $a_k(M) \geqslant f(k)$ for sufficiently large k?

Here, $a_k(M)$ is the maximal size of the algebraic closure of a subset of size k from M. (This is finite, by the Ryll-Nardzewski Theorem.)

A straightforward computation (along the lines of that used in the proof of Theorem 3.3 of Section 3.1) shows:

THEOREM 4.3. Let A be a countably infinite $\aleph_0$-categorical structure and let M denote the canonical structure $\mathrm{CS}(G(A, B^1); \Omega)$. Let $G = G(A, B^1)$. Denote by r_k the minimal number of $G_{(X)}$-orbits on $\Omega \setminus X$, where X is any k-element subset of Ω. Then $a_k(M) \geqslant k^{r_{k-1}}$.

Now, the structures in Example 1 of Section 2.2.3 have arbitrarily fast growth rate of the sequence (r_k), so using these as the structures A in the above theorem gives arbitrarily fast growth rate of the sequence $(a_k(M))$, and so Macpherson's question has an affirmative answer. As an additional refinement, we can take the structure A to be doubly transitive (take n_0 of

Example 1 in Section 2.2.3 to be at least 2). This makes the structure M in the above theorem primitive (and there are 3 2-types over $\varnothing$).

REMARK. Independently, Macpherson gave an affirmative answer to his question by using Fraïssé's Theorem. The examples he produces are not primitive. Hrushovski (personal communication) suggested that his construction in Example 4 of Section 2.2.2 could be modified to: (i) produce $\aleph_0$-categorical graphs giving an affirmative answer to Macpherson's question; (ii) this could also be modified to produce multiply transitive $\aleph_0$-categorical structures with arbitrarily fast growth rate of algebraic closure. The details for (i) have been provided by Emanuela Pantano (personal communication).

4.2 Coset spaces

Given a group G and a (not necessarily normal) subgroup H, G/H is the set $\{gH : g \in G\}$ of left cosets of H in G. G acts on G/H in a natural way: $g'(gH) = (g'g)H$ for $g' \in G$ and $gH \in G/H$.

As is well-known (see p. 16 for example) any transitive permutation action is equivalent to an action of this form. However, I know of no construction of an oligomorphic permutation group which proceeds by giving a group G and a subgroup H and verifying that the action of G on the coset space G/H is oligomorphic, *unless* G is already given as an oligomorphic permutation group. Here is a simple application of the coset space construction where the latter verification is straightforward. It is adapted from Evans and Hewitt (1990), where it is used to produce two countable $\aleph_0$-categorical structures whose automorphism groups are isomorphic as groups, but not as topological groups (Evans and Hewitt, 1990, Theorem 3.4).

Let G be a permutation group on a set A and let H be a subgroup of G. Let $C = G/H$ and define a map $\rho: G \to \mathrm{Sym}(A \sqcup C)$ by $\rho(g)(a) = ga$ if $a \in A$ and $\rho(g)(g'H) = gg'H$ if $g'H \in C$. (Here, $\sqcup$ denotes disjoint union.)

LEMMA 4.4. Suppose H is a normal subgroup of finite index in G.

(i) If $(G; A)$ is oligomorphic then $(\rho(G); A \sqcup C)$ is also oligomorphic.

(ii) Suppose G is closed on A. Then $\rho(G)$ is closed on $A \sqcup C$ if and only if H is a closed subgroup of G.

(iii) The closure in $\mathrm{Sym}(A \sqcup C)$ of $\rho(G)$ is isomorphic to the semidirect product $(\overline{H}/H) \cdot \rho(G)$, where $\overline{H}$ denotes the closure in G of H.

Proof. (i) Clearly $\rho(H)$ is the pointwise stabilizer in $\rho(G)$ of C. So it will suffice to show that H has finitely many orbits on A^k (for all k). But each G-orbit on A^k is the union of a finite number of H-orbits which are permuted transitively by G/H.

(ii) This will follow from (iii).

(iii) Let F be the closure in $\mathrm{Sym}(A \sqcup C)$ of $\rho(G)$. The restriction map $\theta: F \to \mathrm{Sym}(A)$ has image G (as G is closed on A) and kernel $F_{(A)}$. Furthermore, ρ is a splitting for this. So F is the semidirect product of $F_{(A)}$ and $\rho(G)$. Thus it remains to show that $F_{(A)}$ is isomorphic to $\overline{H}/H$.

As C is finite, any permutation of C induced by an elment of F can also be induced by an element of $\rho(G)$. This gives a map $\gamma: F \to G/H$. We show that $\gamma(F_{(A)}) = \overline{H}/H$. As γ is injective on $F_{(A)}$, this will be enough. Let $f \in F_{(A)}$. So there exists a sequence (g_i) of elements of G such that $\rho(g_i) \to f$ in $\mathrm{Sym}(A \sqcup C)$. Now, (as C is finite), the sequence of restrictions $\rho(g_i)|C$ is eventually constant. So we can assume that there exists $g \in G$ such that $(\rho(g_i))(c) = (\rho(g))(c)$ for all $c \in C$, and all i. Thus $g^{-1}g_i \in H$ for all i. As $g_i \to 1$ in G, it follows that $g^{-1} \in \overline{H}$. This shows that $\gamma(F_{(A)}) \leqslant \overline{H}/H$. Equality here is left to the reader. □

As an application of this, here is a weaker (and simpler) version of Theorem 3.4 of Evans and Hewitt (1990) (this was pointed out to us by Lascar).

THEOREM 4.5. There exist countable $\aleph_0$-categorical structures M_1 and M_2 and an isomorphism $\phi: \mathrm{Aut}(M_1) \to \mathrm{Aut}(M_2)$ which is not continuous.

Proof. Let M be the structure due to Cherlin and Hrushovski in Example 2 of Section 2.2.3. So $\Gamma = \mathrm{Aut}(M)$ has a closed normal subgroup Γ_0 such that the quotient group Γ/Γ_0 is isomorphic to the profinite group Z_2^{ω}. Let H be a dense subgroup of index 2 in Γ which contains Γ_0 (there are $2^{2^{\aleph_0}}$ of these, assuming the axiom of choice). Form the structure $M_1 = M \sqcup \Gamma/H$ as above. Then by the lemma, $\mathrm{Aut}(M_1)$ is isomorphic to $\Gamma \times Z_2$, and the first factor here is dense.

Now let M_2 be the structure $M \sqcup \{0,1\}$ with automorphism group $\Gamma \times Z_2$, where the two factors act independently on M and $\{0,1\}$ respectively. In particular, the factor Γ is closed in $\mathrm{Aut}(M_2)$. It is now clear what the map ϕ should be, and that it has the required properties. □

It is clear from this that both $\mathrm{Aut}\, M_1$ and $\mathrm{Aut}\, M_2$ have open subgroups not having the small index property; in fact, by Proposition 6.6 on p. 26, neither M_1 nor M_2 can have the small index property.

4.3 Wreath products and imprimitive actions

Recall that a (transitive) permutation group $(G;\Omega)$ is *imprimitive* if there exists $\Delta \subset \Omega$, with at least two elements, such that for all $g \in G$ we have $\Delta \cap g\Delta = \Delta$ or $\varnothing$. Such a Δ is called a *block* of imprimitivity. If G is transitive on Ω then the translates $\Sigma = \{g\Delta : g \in G\}$ of Δ form a partition of Ω and there is an induced action of G on Σ. Denote the kernel of this action by K and the permutation group induced on Σ by F; so G is an extension of K by F.

We give below a few examples where oligomorphic permutation groups $(G;\Omega)$ are constructed by first specifying $(F;\Sigma)$ and then giving suitable kernels.

4.3.1 *The full wreath product*

Suppose $(F;\Sigma)$ and $(H;\Delta)$ are permutation groups. Let $\Lambda = \Delta \times \Sigma$ and $B = H^{\Sigma}$, the set of functions from Σ to H. Define actions of F and B on Λ by

$$f(\delta,\sigma) = (\delta, f\sigma)$$

and

$$g(\delta,\sigma) = (g(\sigma)\delta,\sigma)$$

where $\delta \in \Delta$, $\sigma \in \Sigma$, $f \in F$ and $g \in B$.

These give embeddings of F and B into $\mathrm{Sym}(\Lambda)$, and the images generate their semidirect product inside $\mathrm{Sym}(\Lambda)$ (so $B \trianglelefteq \langle B, F\rangle \leqslant \mathrm{Sym}(\Lambda)$, and $B \cap F = \{1\}$). The permutation group on Λ so generated is called the *wreath product* of $(F;\Sigma)$ and $(H;\Delta)$, and is denoted by $H \operatorname{Wr} F$. This group is imprimitive (if Δ has size at least 2) with blocks of imprimitivity (also called the *fibres*) $\Delta_\sigma = \Delta \times \{\sigma\}$ for $\sigma \in \Sigma$. Note that if $(F;\Sigma)$ and $(H;\Delta)$ are oligomorphic, then so is $(H \operatorname{Wr} F;\Lambda)$ (and conversely).

There is a straightforward interpretation of this in terms of relational structures. Suppose S is a relational structure on Σ with automorphism group F and D a relational structure on Δ with automorphism group H. Replace each point of S with a copy of D, and have as relations on the resulting set those induced by relations in S, and those coming from relations in each copy of D. The resulting structure has automorphism group $H \operatorname{Wr} F$.

The wreath product is a 'universal' imprimitive group in the following sense. Suppose $(G;\Omega)$ is a transitive permutation group and Δ is a block of imprimitivity. Let Σ be the set of G-translates of Δ and let F be the permutation group induced on Σ by G. Let H be the permutation group induced on Δ by the setwise stabilizer in G of Δ. Then there is a bijection from Ω to $\Lambda = \Delta \times \Sigma$ such that the resulting map from $\mathrm{Sym}(\Omega)$ to $\mathrm{Sym}(\Lambda)$ embeds G as a subgroup $H \operatorname{Wr} F$.

This is sometimes called the *full* wreath product. This is to distinguish it from the *restricted wreath product* of H and F, notated $H \operatorname{wr} F$, which is defined as for $H \operatorname{Wr} F$ except that we take B to be the restricted cartesian power of H by Σ, that is the set of functions $\Sigma \to H$ which has value 1 except for finitely many arguments.

4.3.2 *Imprimitive actions with abelian kernel*

Suppose $(F;\Sigma)$ and $(H;\Delta)$ are permutation groups, and let G be a subgroup of $H \operatorname{Wr} F$. We say that G is a *full* subgroup (of the wreath product $H \operatorname{Wr} F$) if G induces F on the blocks of imprimitivity $\{\Delta_\sigma : \sigma \in \Sigma\}$.

Suppose this throughout this section. Let K be the kernel of the action on these blocks of imprimitivity. If K is *abelian*, conjugation in G gives an action (by automorphisms) of F on K. In this case we can partially specify G by describing K, F, and the action of F on K.

To render this useful for actually constructing oligomorphic permutation groups, we explore the action of F on K a little further. Let K_σ be the permutation group induced on the fibre Δ_σ by K. So there is a natural embedding of K into the product $P = \prod_{\sigma\in\Sigma} K_\sigma$ (and the latter is naturally a permutation group on $\Delta \times \Sigma$). The action of F on K induces an action of F on P which is consistent with this embedding:

> if $p, q \in P$ and $f \in F$ then $fp = q$ if and only if for all $\sigma \in \Sigma$ and $\delta \in \Delta$ we have
>
> $$q(\sigma)(\delta, \sigma) = (p(f^{-1}\sigma)\delta, \sigma).$$

Clearly K embeds as an F-invariant subgroup of P. Conversely, if we have a (closed) abelian permutation group P on a set $\Delta \times \Sigma$ as above, and an action of a group F on P which permutes the factors in P, then F-invariant subgroups of P will provide us with a supply of imprimitive permutation groups. We illustrate this in the following.

EXAMPLE 4.6. Let V be a countably infinite dimensional vector space over the field with 2 elements and let F be the group of invertible linear transformations on V. So F acts transitively (indeed, 2-transitively) on V^0, the non-zero vectors from V. Let $(H; \Delta)$ be the cyclic group of order 2 acting regularly on a set with 2 elements. Ahlbrandt and Ziegler (1991*a*, 1991*b*) determine all closed, full subgroups of the wreath product $H \operatorname{Wr} F$. (Note that such a subgroup is necessarily oligomorphic.) The proof proceeds in two steps: first, describe the possible kernels for such subgroups (Ahlbrandt and Ziegler, 1991*a*); second, decide how the kernel fits together with F to give a closed subgroup (this is the *extension problem*, and is done in Ahlbrandt and Ziegler (1991*b*)). The first step amounts to a determination of all closed subgroups of the group $P = H^{V^0}$ which are invariant under the action of F: $(fp)(v) = p(f^{-1}v)$, for $f \in F$, $p \in H^{V^0}$ and $v \in V^0$. These are of the following forms.

1. Con, the subgroup consisting of the two constant functions.
2. Pol_n, consisting of all maps $p \in P$ of the following form: Let $x_1, x_2, \ldots$ be a basis for V, and let S be a set of monomials of degree n in variables $t_1, t_2, \ldots$. For a monomial $\underline{s} = t_{m_1} \cdots t_{m_n}$ and a vector $v = \sum_i a_i x_i$ let $\underline{s}(v) = a_{m_1} \cdots a_{m_n}$. Now define $p(v) = \sum_{\underline{s}\in S} \underline{s}(v)$ (there are only finitely many non-zero terms in this sum). Then Pol_n is an F-invariant subgroup of P.
3. $\mathrm{Con} + \mathrm{Pol}_n$.

4. The trivial subgroup, and the whole of P.

We remark that there is an alternative definition of Pol_n: the set of $p \in P$ with the property that $\sum_{v \in X \setminus \{0\}} p(v) = 0$ for every n-dimensional subspace X of V, is Pol_{n-1}.

Ahlbrandt and Ziegler phrase their results mentioned above in terms of *finite covers.* The reader can consult Ahlbrandt and Ziegler (1991*b*) for this terminology and the more general notion of an *affine cover*, and also Evans and Hrushovski (1993) for general results on finite covers of $\aleph_0$-categorical structures. Hodges and Pillay (in press) answer questions left outstanding from Ahlbrandt and Ziegler (1991*b*), and give some general results on the extension problem mentioned above. For more information about covers and the extension problem, see the papers by Hodges (p. 111) and Ivanov (p. 215) in this volume.

4.3.3 *Duality*

Suppose $(F; \Sigma)$ and $(H; \Delta)$ are closed permutation groups of countable degree, and let G be a full subgroup of $H \operatorname{Wr} F$. Suppose that the kernel K of G is abelian. As above, we are interested in the possibilities for K, and we consider K as a subgroup of the group $P = \prod_{\sigma \in \Sigma} K_\sigma$ on which F acts as in the previous section. So P is a closed subgroup of $\mathrm{Sym}(\Delta \times \Sigma)$, and K is a closed subgroup of P. Suppose that G is oligomorphic on $\Delta \times \Sigma$. Then it can be shown that $K_\sigma \leqslant \mathrm{Sym}(\Delta_\sigma)$ is of bounded exponent, and is either a discrete group or a profinite group, depending on whether it is countable or not. In either case P is a product of locally compact groups. We are going to apply (a simple instance of) Pontrjagin duality to the case where the K_σ are profinite. (Of course, this includes the case where these groups are finite).

Recall (Pontrjagin, 1966) that if A is topological abelian group then the *dual* of A, denoted by $\hat{A}$, is the group of continuous homomorphisms from A into the multiplicative group of the complex numbers. This has a natural topology, and every continous homomorphism $\theta: A \to B$ gives rise to a continuous homomorphism $\hat{\theta}: \hat{B} \to \hat{A}$. The operation of dualizing interchanges compact and discrete abelian groups, and provides a faithful contravariant functor between the categories of compact and discrete abelian groups. To put this more simply, if A is a compact (respectively, discrete) abelian group then its dual $\hat{A}$ is a discrete (respectively, compact) abelian group and the dualizing operation gives an order reversing bijection between the closed subgroups of A and $\hat{A}$: if $B \leqslant A$ then the bijection sends B to $\{f \in \hat{A} : f(b) = 1 \text{ for all } b \in B\}$.

So now suppose that each K_σ is compact. Then P is compact, and the dual of P is $\hat{P} = \coprod_{\sigma \in \Sigma} \hat{K}_\sigma$ (here, $\coprod$ denotes direct sum). The action of F on P induces an action of F on $\hat{P}$ and the duality gives a 1–1 correspondence between F-invariant closed subgroups of P and $\hat{P}$. The simplest case of this is where the groups K_σ are finite. In this case $\hat{K_\sigma}$ is isomorphic to K_σ, and the original problem of describing the possible kernels K is equivalent to finding F-invariant subgroups of the countable abelian group $\coprod_{\sigma\in\Sigma} \hat{K_\sigma}$.

An example of this duality occurred in Section 4.3.2. Recall that there we were describing F-invariant subgroups of $H^{V\setminus\{0\}}$, where H was a cyclic group of order 2, V a countably infinite vector space over the field Φ_2 with 2 elements, and F the automorphism group of V. By the duality, this is equivalent to finding the F-invariant subgroups of the Φ_2-vector space $\Phi_2(V \setminus \{0\})$ with basis $V \setminus \{0\}$ (cf. the alternative description of Pol_n given above). In a similar way, if K_σ is cyclic of prime order p, and F_σ (the stabilizer in F of σ) induces no non-trivial automorphisms of K_σ, then the problem of determining the closed F-invariant subgroups of P is equivalent to finding all $\Phi_p F$-submodules of the permutation module $\Phi_p\Sigma$, where Φ_p denotes the field with p elements.

A survey of Jordan groups

Dugald Macpherson

School of Mathematical Sciences,[1]
Queen Mary and Westfield College,
Mile End Road,
London,
E1 4NS,
U.K.
Electronic mail address: H.D.Macpherson@qmw.ac.uk

1 Introduction

This article surveys recent results on a rather ancient topic in permutation group theory. In the 1870s, Jordan developed the early theory of the permutation groups now known as Jordan groups, that is, permutation groups in which, in some non-degenerate way, the pointwise stabilizer of a subset is transitive on the complement. One such reference is Jordan (1871). Over the last 20 years a much clearer picture of them has developed. Finite primitive Jordan groups were classified in the early 1980s. In a slightly vague sense, there is now a classification of infinite primitive Jordan groups (Adeleke and Neumann, in press*b*; Adeleke and Macpherson, 1994). Some of these results, in particular the classification of finite primitive Jordan groups, have had important applications in model theory, and model-theoretic methods have been used to construct new Jordan groups. The purpose of this paper is to describe the known examples of Jordan groups and the classification theorem, and to sketch some of its applications to permutation group theory and to model theory. I also consider how some of the relational structures associated with Jordan groups arise in other contexts. The article is essentially a survey of known results.

Throughout this paper, G will denote a permutation group on a set Ω. A subset Γ of Ω is called a *Jordan set* for G if $|\Gamma| > 1$ and, for all $\alpha, \beta \in \Omega$, there is $g \in G$ such that $\alpha g = \beta$ and g fixes each element of $\Omega \setminus \Gamma$. We say that Γ is a *proper* Jordan set if, in addition, if $k \in \mathbb{N}$ and G is $(k+1)$-transitive on Ω then $|\Omega \setminus \Gamma| > k$. The permutation group (G, Ω) is called a *Jordan group* if it has a proper Jordan set.

[1] Address from 1st October, 1994: School of Mathematics, The University of Leeds, LS2 9JT, U.K.

For history of the work on finite Jordan groups in the last century, I refer to Neumann (1985). In particular, a proof is given there of the following theorem.

THEOREM 1.1. (Jordan) Every finite primitive Jordan group is 2-transitive.

Finite doubly transitive permutation groups have been classified. This was an immediate consequence of the classification of the finite simple groups, together with information already available on finite simple groups and on subgroups of affine groups. It is discussed in the survey paper by Cameron (1981) on applications of the classification of finite simple groups to permutation group theory. A list of the finite 2-transitive groups can be found in Liebeck (1987) (see also Section 4 from p. 186 in Cameron's paper in this volume). Since, by Jordan's Theorem, all finite primitive Jordan groups are doubly transitive, it is not surprising that finite primitive Jordan groups were classified in the early 1980s. A classification was given by Neumann (1985) and Kantor (1985), in independent work. Up to finitely many exceptions (which was all that was needed for the model-theoretic application) a classification is also given by Cherlin, Harrington and Lachlan (Cherlin *et al.*, 1985) (and a similar route to the classification of ω-categorical strictly minimal sets was taken by Mills). The examples of finite primitive Jordan groups are projective and affine linear groups over finite fields, the Mathieu groups $M_{22}, M_{23}, M_{24}, \mathrm{Aut}\, M_{22}$ (where Jordan sets are complements of blocks) and A_7 embedded in PGL(4, 2) and its transitive extension embedded in AGL(4, 2). There is further discussion of the projective and affine examples in Section 4.

There are examples of infinite Jordan groups with no finite analogues. For example, let G be the automorphism group of the countable dense linear order $(\mathbb{Q}, \leqslant)$. Then G is primitive and any proper non-empty open interval of $(\mathbb{Q}, \leqslant)$ is a Jordan set, but unlike in the finite examples G is not 2-transitive. Indeed, this example is not remotely like any finite permutation group, since finite total orders have no non-trivial automorphisms. There are many other examples of infinite Jordan groups. These include automorphism groups of certain Steiner systems, circular orders, trees and related structures, and various homeomorphism groups. It is not clear what form a full classification theorem should take. In particular, it does not at present seem feasible to classify the Steiner systems whose automorphism groups are Jordan groups; and there is little information at present on those Jordan groups which are k-transitive for all positive integers k. However there is now a classification, though some of the families arising are still not well understood. Many of the essential ideas in its proof can be found in Adeleke and Neumann (in press *b*). The classification is given in Adeleke and Macpherson (1994), using results from Adeleke and Neumann (in press *b*).

I now describe the contents of the rest of this paper. The results of Adeleke, Neumann and myself are discussed in Section 2, and the structure of the proof of the classification is sketched, very roughly, in Section 3. Arguments similar to those in these papers can be found in Adeleke's article in this volume. In Section 4, I examine the connections between Jordan groups, homogeneous geometries, and strongly minimal sets. In particular I describe the application to the Cherlin–Mills–Zil'ber classification of ω-categorical strictly minimal sets.

It is natural to ask whether, for every positive integer k, there is an infinite Jordan group which is k-transitive but not $(k+1)$-transitive. The familiar examples involving treelike structures or projective and affine spaces, are not even 4-transitive. However, in 1987, work of Hrushovski yielded a positive answer to this question. Hrushovski (1993*a*, 1993*a*) came up with two counterexamples to old model-theoretic conjectures: Zil'ber's Conjecture that any non locally modular strongly minimal set interprets (and is interpretable in) an algebraically closed field; and Lachlan's Conjecture that every ω-categorical stable theory is ω-stable. The two constructions of Hrushovski are quite similar, and the method has since had other applications. This technique is the subject of Wagner's article in this volume; there, a uniform approach is given to the constructions of Hrushovski, Baldwin and others. Evans pointed out that the construction gives, for every positive integer k, a Jordan group which is k-transitive but not $(k+1)$-transitive. This is described in Section 5, shorn of all model theory except for Fraïssé amalgamation.

In the last three sections, some topics related to Jordan groups are discussed. We consider applications to permutation group theory in Section 6; in particular, to work by Adeleke and Neumann on permutation groups on uncountable sets with elements of small support, and to work by myself and Praeger on the question: which cycle types are realizable in primitive permutation groups which are not highly transitive? Then, in Section 7, I consider some applications to model theory. There are connections between Jordan groups, the existence of prime models over sets of parameters, and the definability of types. This has led to work by Haskell, Steinhorn and myself on certain variations on o-minimality.

In Section 8 I discuss briefly some topics only faintly related to Jordan groups. The point here is to emphasize how structures related to Jordan groups occur in many different contexts. Topics mentioned include: the work of Cameron (1990) and McLeish (1994) on when the 'forth' part of the back-and-forth argument suffices to build an automorphism; some remarks on highly transitive Jordan groups; other contexts in which the relational stuctures associated with Jordan groups arise.

This paper does not assume much background in model theory, but most model-theoretic notions we consider are examined in Hodges (1993) or Pillay (1983). For further information on Jordan groups, together with a description of the classification of the finite Jordan groups, I recommend Neumann (1985).

I conclude with some further pieces of terminology. If Γ is an infinite set, then a *moiety* of Γ is a subset Δ of Γ such that $|\Gamma| = |\Delta| = |\Gamma \setminus \Delta|$. A permutation group (G, Ω) is said to be *k-primitive* if it is k-transitive and the stabilizer of $k-1$ points acts primitively on the remaining points. Also, (G, Ω) is *simply primitive* if it is primitive but not 2-transitive. If (G, Ω) is a permutation group with a Jordan set Γ, and $\mathscr{P}$ is a property of permutation groups (such as *primitivity, k-homogeneity*), then we say that Γ is a *$\mathscr{P}$–Jordan set* if $(G_{(\Omega\setminus\Gamma)}, \Gamma)$ has property $\mathscr{P}$. Also, following Droste (1985), if $k \in \mathbb{N}$ then we say that a structure M is *k-homogeneous* if every isomorphism between substructures of size k extends to an automorphism of M. Unfortunately, there is a common usage whereby 'k-transitivity' for a structure is weaker than 'k-homogeneity', in conflict with the convention for permutation groups. A *chain* is just a totally ordered set.

I thank Peter Neumann and Samson Adeleke for many valuable conversations on Jordan groups, and for several suggestions on the present paper.

2 The classification of infinite primitive Jordan groups

In this section I will describe the classification results of Adeleke, Neumann and myself on infinite Jordan groups. I first give an informal account of the structures preserved by Jordan groups. The proof will be described in Section 3. For further information on some of the structures discussed here, see the chapter by Evans.

The most familiar examples of Jordan groups are projective and affine linear groups. If V is a vector space over a field F, and U is a proper subspace of V, then every basis of U can be extended to a basis of V. Since $\mathrm{GL}(V)$ is transitive on the set of ordered bases of V, it follows that $V \setminus U$ is a Jordan set for the action of $\mathrm{GL}(V)$ on V. This action is not transitive (as 0 is fixed), and the action on $V \setminus 0$ is not primitive if $|F| > 2$, since one-dimensional subspaces of V are blocks of imprimitivity. However, $\mathrm{PGL}(V)$, in its action on the corresponding projective space $\mathrm{PG}(V)$, is a primitive Jordan group, and indeed is 2-transitive. Similarly the affine group $\mathrm{AGL}(V)$, which consists of all maps $v \mapsto vg + w$ determined by pairs (w, g) where $w \in V$ and $g \in \mathrm{GL}(V)$, is a doubly transitive Jordan group. It is a semidirect product of the translation group $(V, +)$, acting regularly, by the stabilizer $\mathrm{GL}(V)$ of the zero vector. Note that in both these examples, if V and F are finite then we obtain finite Jordan groups. If F is finite but V is infinite dimensional, then we obtain examples in which every finite set is contained in the complement of a cofinite Jordan set.

Recall that a *Steiner system* is an incidence structure consisting of a set $\mathscr{P}$ of points, a set $\mathscr{L}$ of lines, and an incidence relation between them (a subset of $\mathscr{P} \times \mathscr{L}$). It has parameters t and k, where t is an integer greater than 1, and k is a cardinal (possibly infinite) with $t \leqslant k$, such that

- if $\alpha_1, \ldots, \alpha_t$ are distinct elements of $\mathscr{P}$ then there is a unique element of $\mathscr{L}$ incident with each of $\alpha_1, \ldots, \alpha_t$, and
- any element of $\mathscr{L}$ is incident with exactly k elements of $\mathscr{P}$.

We also call such a structure a *Steiner t-system.* The Steiner system is *non-trivial* if $t < k$. There is a natural Steiner system on the projective space $\mathrm{PG}(V)$ over a field F with $t = 2$, $k = |F| + 1$; the lines are the two-dimensional subspaces. This is preserved by $\mathrm{PGL}(V)$. Similarly there is a Steiner system on $\mathrm{AG}(V)$ (affine space over a field F) invariant under $\mathrm{AGL}(V)$: if $|F| \neq 2$ then the lines are the cosets of the one-dimensional subspaces (so $t = 2$, $k = |F|$) and if $F = \mathrm{GF}(2)$ then the lines are the cosets of the two-dimensional spaces (so $t = 3, k = |F|^2 = 4$). There are other examples of Jordan groups which preserve Steiner systems. These arise, for example, as automorphism groups of saturated strongly minimal sets and regular types, and are the subject of Sections 4 and 5. We shall say that a Jordan group is of *geometric type* if it preserves a Steiner system. There are many examples of such groups, and at present a classification does not seem feasible.

As mentioned in the introduction, the automorphism group $\mathrm{Aut}(\mathbb{Q}, \leqslant)$ is a Jordan group. For let $\alpha, \beta \in \mathbb{Q}$ with $\alpha < \beta$, and put $I := (\alpha, \beta)$. Choose $\mu, \nu \in I$. By the ω-categoricity of the theory of dense linear orders without endpoints, there are isomorphisms $\phi: (\alpha, \mu) \cap \mathbb{Q} \to (\alpha, \nu) \cap \mathbb{Q}$ and $\psi: (\mu, \beta) \cap \mathbb{Q} \to (\nu, \beta) \cap \mathbb{Q}$. Let x be the permutation of $\mathbb{Q}$ extending ϕ and ψ, taking μ to ν, and fixing the rest of $\mathbb{Q}$ pointwise. Then $x \in \mathrm{Aut}(\mathbb{Q}, \leqslant)_{(\mathbb{Q} \setminus I)}$, and $\mu x = \nu$, as required.

There are other relational structures closely related to total orders. First, there is the *linear betweenness relation.* This is a ternary relation $B(\alpha; \beta, \gamma)$, interpreted naturally on a total order $(I, <)$ by

$$B(\alpha; \beta, \gamma) \longleftrightarrow ((\beta \leqslant \alpha \leqslant \gamma) \vee (\gamma \leqslant \alpha \leqslant \beta)).$$

It is easily checked that $\mathrm{Aut}(\mathbb{Q}, <)$ is a subgroup of index two of $\mathrm{Aut}(\mathbb{Q}, B)$, and that $\mathrm{Aut}(\mathbb{Q}, B)$ is 2-transitive but not 2-primitive, since the stabilizer of 0 has two blocks, the positive rationals and the negative rationals. It is also possible to twist a total order around to obtain a circular order. The automorphism group of the circular order induced from the rationals is 2-transitive, indeed 2-primitive, but not 3-transitive. Finally, there is the group of permutations of a circular order which preserve or reverse it, that is, which preserve the induced quaternary separation relation. The automorphism group of the separation relation induced from the rationals is 3-transitive, but not 3-primitive, and hence not 4-transitive. These structures are all described in more detail by Cameron (1976). Since all these groups contain the automorphism group of the underlying linear order, there are examples of Jordan groups of each of these types. Furthermore, under reasonable hypotheses (such as that the underlying linear ordering is countable, dense and without end points), the groups are all highly homogeneous. The following beautiful theorem, due to Cameron (1976), is the fundamental result on these permutation groups.

THEOREM 2.1. Let G be a permutation group on an infinite set Ω, and suppose that G is highly homogeneous but not highly transitive. Then G preserves one of the following structures (and has the same orbits on finite ordered sets as the full automorphism group): a dense linear order; a dense circular order; a dense linear betweenness relation; or a dense separation relation.

There are some other easily described examples of Jordan groups, arising as automorphism groups of treelike objects. They are examined in detail by Adeleke and Neumann (to appear). Since they are discussed in the literature in several places (Adeleke and Neumann, to appear; Adeleke and Macpherson, 1994; Cameron, 1987) the treatment here will be brief.

By a *semilinear order* we mean a partially ordered set (poset) $(P, \leqslant)$ such that

- for all $\alpha, \beta \in P$ there is $\gamma \in P$ such that $\alpha \leqslant \gamma$ and $\beta \leqslant \gamma$,
- for all $\alpha \in P$, $\{x \in P : \alpha \leqslant x\}$ is totally ordered,
- $(P, \leqslant)$ is not a total order.

Adeleke and Neumann call these posets are called *upper* semilinear orders, to distinguish them from the structures with the ordering reversed.

Droste (1985) introduced the following terminology. First, a semilinear order $(P, \leqslant)$ has a natural *Dedekind completion* $(P^+, \leqslant)$. An element of P^+ is called a *ramification point* if it is of the form $\sup(a, b)$ for some incomparable $a, b \in P$. If $(P, \leqslant)$ is a semilinear order and $\alpha \in P^+$ then there is an equivalence relation E_α on $\{x \in P : x < \alpha\}$ defined as follows:

$$u\, E_\alpha\, v \longleftrightarrow \exists w{\in}P\, (u \leqslant w < \alpha) \wedge (v \leqslant w < \alpha).$$

The equivalence classes are called *cones at* α.

Semilinear orders can have large automorphism groups; for example, they can be 1- and 2-homogeneous as posets, in the sense that any isomorphism between subposets of size at most 2 extends to an automorphism. There is a classification of countable 2-homogeneous semilinear orders in Droste (1985).

In a countable 2-homogeneous semilinear order, all the maximal chains are densely ordered without endpoints. The isomorphism type of a 2-homogeneous upper semilinear order is determined by two 'parameters':

whether or not they have *positive type*, that is, whether or not, for incomparable pairs (α, β), $\sup\{\alpha, \beta\} \in P$; and the number of cones at a ramification point. Note that both these parameters are independent of the choice of nodes, by the assumption of 2-homogeneity.

The automorphism group of any countable 2-homogeneous semilinear order is a primitive Jordan group: the Jordan sets are precisely unions of sets of cones at a ramification point, and unions of chains of cones (the chain being ordered under inclusion). There are further countable semilinear orders, indeed $2^{\aleph_0}$ of them, which are not 2-homogeneous but whose automorphism groups are still primitive Jordan groups.

There is a natural ternary relation, called a B-relation, which is induced on a semilinearly ordered set $(P, \leqslant)$. First, the notation $\alpha || \beta$ (where $\alpha, \beta \in P$) means that α and β are incomparable. We write $B(\alpha; \beta, \gamma)$ if $\alpha, \beta, \gamma \in P$ and one of the following holds.

$$\beta \leqslant \alpha \leqslant \gamma$$
$$\gamma \leqslant \alpha \leqslant \beta$$
$$\beta \leqslant \alpha \wedge \alpha || \gamma$$
$$\gamma \leqslant \alpha \wedge \alpha || \beta$$
$$(\beta \leqslant \alpha) \wedge (\gamma \leqslant \alpha) \wedge \nexists \sigma \in \Sigma \setminus \{\alpha\} \, (\beta \leqslant \sigma \wedge \gamma \leqslant \sigma \wedge \sigma < \alpha).$$

A relation obtained in this way is called a *B-relation*. Adeleke and Neumann (to appear) axiomatize B-relations without reference to a semilinear order, but it is shown that given any B-relation on a set, there is a semilinear order on the set from which it is induced as above.

We call the relation B above a *general betweenness relation* if in addition it satisfies the following axiom:

$$\neg B(\alpha; \beta, \gamma) \longrightarrow \exists \delta {\neq} \alpha \, (B(\delta; \alpha, \beta) \wedge B(\delta; \alpha, \gamma)).$$

In fact, in this paper, we shall always work with general betweenness relations rather than B-relations. Given any 0-definable B-relation which is dense in the natural sense, there is a 0-definable general betweenness relation on the same set. These relations are discussed in detail in Adeleke and Neumann (to appear).

Clearly $\mathrm{Aut}(P, B) \geqslant \mathrm{Aut}(P, \leqslant)$. If (P, B) is a countable 2-homogeneous semilinear order then the containment is strict and $\mathrm{Aut}(P, B)$ is 2-transitive. If in addition $(P, \leqslant)$ is of positive type, with r cones at a node (for some $r \leqslant \aleph_0$) then for $\alpha \in P$ we have $\mathrm{Aut}(P, B)_\alpha \cong \mathrm{Aut}(P, \leqslant) \,\mathrm{Wr}\, S_{r+1}$, for there is a natural semilinear order induced on each cone, and $\mathrm{Aut}(P, B)_\alpha$ induces the symmetric group on the set of cones. If $(P, \leqslant)$ is of negative type (i.e. not of positive type) then $\mathrm{Aut}(P, B)_\alpha \cong \mathrm{Aut}(P, \leqslant) \,\mathrm{Wr}\, S_2$, for any $\alpha \in P$. From this description we can see what Jordan sets can arise.

If $(P, \leqslant)$ is a semilinear order then $\mathrm{Aut}(P, \leqslant)$ acts on the set of maximal chains of $(P, \leqslant)$. It is easily seen that this action is not 3-transitive (or even

3-homogeneous), since it preserves the ternary relation $C(\alpha;\beta,\gamma)$ which says

$$(\alpha \neq \beta = \gamma) \vee (\alpha \cap \beta \subset \beta \cap \gamma) \qquad (*)$$

(where we regard α, β, γ as subsets of P). Adeleke and Neumann say that a ternary relation C is a *C-relation* if it satisfies the following five universal axioms:

C1 $C(\alpha;\beta,\gamma) \to C(\alpha;\gamma,\beta)$;
C2 $C(\alpha;\beta,\gamma) \to \neg C(\beta;\alpha,\gamma)$;
C3 $C(\alpha;\beta,\gamma) \to (C(\delta;\beta,\gamma) \vee C(\alpha;\delta,\gamma))$;
C4 $\exists \alpha\, C(\alpha;\beta,\gamma)$;
C5 $(\alpha \neq \beta) \to \exists \gamma{\neq}\beta\, C(\alpha;\beta,\gamma)$.

They also show that if C is a C-relation on a set M then there is a semilinear order $(P,\leqslant)$ such that M can be identified with a dense set of maximal chains of $(P,\leqslant)$, dense in the sense that $\bigcup M = P$, and C is interpreted as in $(*)$. Indeed, $(P,\leqslant)$ is first-order interpretable without parameters in (M,C) (it lives on a quotient of M^2). If $\alpha \in P$ then it is natural also to define the equivalence relation E_α on the set of chains through α which lie in M, putting two such chains equivalent if their intersection contains a point strictly below α. Hence we may regard a *cone at* α as being a subset of M. If $(P,\leqslant)$ is 2-homogeneous and $\alpha \in P$, then for any set of cones at α, the union of that set, regarded as a subset of M, is a Jordan set for $\operatorname{Aut} M$. It can be checked that the automorphism group of a C-relation can be 2-transitive, but never 2-primitive. We remark that sometimes there are Jordan sets other than the union of a set of cones at a node. For example, the set of all chains *not* passing through a node is often a Jordan set.

Finally, there is a quaternary relation, known as the D-relation, which corresponds to the B-relation just as the C-relation corresponds to a semilinear order. The elements of the D-relation can be thought of as *directions* of the B-relation. They are closely related to the ends of a graph, and indeed Möller (1992) has shown that if Γ is an infinite graph with infinitely many ends whose automorphism group is transitive on the vertices and the ends, and whose vertices all have just finitely many neighbours, then there is a natural D-relation on which $\operatorname{Aut}\Gamma$ acts. Again, Adeleke and Neumann axiomatize the relation and show that it must come from a betweenness relation in the natural way. The automorphism group can be 3-transitive, but not 3-primitive. The (universal) axioms for a D-relation are as follows:

D1 if $D(\alpha,\beta;\gamma,\delta)$ then $D(\beta,\alpha;\gamma,\delta)$ and $D(\gamma,\delta;\alpha,\beta)$;
D2 if $D(\alpha,\beta;\gamma,\delta)$ then $\neg D(\alpha,\gamma;\beta,\delta)$;
D3 if $D(\alpha,\beta;\gamma,\delta)$ then $D(\alpha,\beta;\gamma,\epsilon)$ or $D(\alpha,\epsilon;\gamma,\delta)$;
D4 if α,β,γ are distinct then there is $\delta \in \Omega \setminus \{\alpha,\beta,\gamma\}$ with $D(\alpha,\beta;\gamma,\delta)$.

Note that if (Ω, D) is a D-relation and $\alpha \in \Omega$, then there is a α-definable C-relation induced on $\Omega \setminus \{\alpha\}$.

Relational structures which are essentially B-relations, C-relations and D-relations were introduced by Cameron (1983*b*), where the homogeneity and transitivity properties of their automorphism groups were investigated. Various constructions were there given, using sets of sequences and using Fraïssé's Theorem. Cameron (1987) examined, for several of these structures, the growth rate of the sequence counting the number of orbits of the automorphism group on the set of k-element subsets. The general theory of these structures was developed in a much more elaborate form by Adeleke and Neumann in the mid-1980s. They axiomatized these structures, described precisely the relationships between them, and gave a number of constructions and classification theorems (in terms of symmetry properties). For a while it seemed likely that any primitive infinite Jordan group which is not highly transitive or of geometric type must preserve one of these relations (a linear order, linear betweenness relation, circular order, separation relation, semilinear order, general betweenness relation, or C- or D-relation). Indeed, it is shown by Adeleke and Neumann (in press*b*) that this is true if there is a proper *primitive* Jordan set. However, when attempts were made to prove this in general, problems arose with Jordan groups which are 2-primitive but not 3-transitive. Finally, in 1988, Adeleke came up with a beautiful construction of a 2-primitive but not 3-transitive Jordan group which is not of any of the above types. Essentially, his example is built as a limit of general betweenness relations. Later, he constructed another example, built as a limit of D-relations. These examples are the subject of Adeleke (to appear).

I now state the main theorem of Adeleke and Macpherson (1994).

THEOREM 2.2. (Classification of infinite primitive Jordan groups) Let (G, Ω) be an infinite Jordan group which is primitive but not highly transitive. Then G preserves on Ω one of the following structures:

(a) a dense linear order;
(b) a dense circular order;
(c) a dense linear betweenness relation;
(d) a dense separation relation;
(e) a semilinear order;
(f) a general betweenness relation;
(g) a C-relation;
(h) a D-relation;
(i) a non-trivial Steiner t-system (with G acting t-transitively on Ω); or
(j) a structure which is not one of (a)–(i) but is a limit of structures of type (f), (h), or (i).

Remarks. 1. If G is not 2-transitive on Ω then it is of type (a), (e) or (g), and if not of types (e) or (g) then it is highly homogeneous (see Theorem 3.2 below). In his chapter in this volume, Adeleke introduces the notion of a *c-Jordan* group, a slight generalization of a Jordan group, and shows that every primitive but not 2-homogeneous c-Jordan group is of type (a), (e) or (g) above.

2. If (G, Ω) is of type (i) or (j) then it has no proper primitive Jordan set, and if it is of type (b), (c), (d), (e), or (f) but not of any of the types (a), (e), (g), (h), or (i) then it has a proper primitive Jordan set.

3. A simple-minded argument shows that a t-transitive group of automorphisms G of a non-trivial Steiner t-system on Ω could not have a proper *primitive* Jordan set. For let Γ be a proper Jordan set of such a Jordan group and pick $\alpha_0, \dots, \alpha_{t-1} \in \Omega \setminus \Gamma$. Let $\alpha_t \in \Gamma$ and let L denote the Steiner line through $\alpha_1, \dots, \alpha_t$.

We claim that $\Gamma \supseteq L \setminus \{\alpha_1, \dots, \alpha_{t-1}\}$. For suppose that there is $\beta \in L \setminus \Gamma$ with $\beta \neq \alpha_1, \dots, \alpha_{t-1}$. Then $G_{(\Omega \setminus \Gamma)}$ fixes L, so as Γ is a Jordan set, $\Gamma \subset L$. Now pick $\gamma \in \Omega \setminus L$. Then $\gamma \notin \Gamma$. If L' is the line through $\gamma, \alpha_2, \dots, \alpha_t$, then as $|L \cap L'| = t - 1$, $L' \cap \Gamma = \{\alpha_t\}$. It follows that L' contains at least t points outside Γ. Hence $G_{(\Omega \setminus \Gamma)}$ fixes L', so it also fixes $L' \cap \Gamma = \{\alpha_t\}$. Hence $\Gamma = \{\alpha_t\}$, contradicting the assumption that it is a Jordan set.

Now let M be the line through $\alpha_0, \dots, \alpha_{t-2}, \alpha_t$. By the last paragraph $\alpha_0 \notin L$, so $M \neq L$. Again by the last paragraph, $\Gamma \supseteq M \setminus \{\alpha_0, \dots, \alpha_{t-2}\}$. Hence there is $\delta \in (M \cap \Gamma) \setminus L$. It follows that $L \cap \Gamma$ is a proper subset of Γ, so is a block of imprimitivity of $(G_{(\Omega \setminus \Gamma)}, \Gamma)$.

4. If (G, Ω) is of type (g) but not of any of the types (a)–(f), (h), (i), then G preserves a C-relation on Ω with one of the following properties:

(a) every cone is a Jordan set for G;
(b) for every node a of the underlying semilinear order, the set of all elements of Ω which do not pass through a is a Jordan set for G.

5. There are some interesting proper expansions of C-relations whose automorphism groups are primitive Jordan groups. Several of these are discussed by Cameron (1987). For example, let $(P, \leqslant)$ be a countable 1- and 2-homogeneous semilinear order, and suppose that its nodes are coloured densely red and green. Pick a dense set M of maximal chains of $(P, \leqslant)$, and interpret a C-relation C in the usual way on M. Also define a graph structure on M, putting two members of M adjacent if, regarding them as chains, the infimum of their intersection is a red point of P. Then the C-relation is interpretable without parameters in the graph on M, and the automorphism group of the graph is a primitive Jordan group with three orbits on ordered pairs and each cone is a Jordan set. The graph here is

of independent interest: it is a universal *N-free* graph, that is, a graph not embedding a path with 4 vertices, and universal among countable graphs with this property. An alternative construction, using Fraïssé's Theorem in an expanded language, was given by Covington (1989).

In Section 6 of Cameron (1987) there is another proper expansion of a C-relation, involving iterated wreath products, for which the automorphism group is a primitive Jordan group. It is also possible to combine cases (g) and (i), as mentioned by Macpherson and Steinhorn (to appear, Example 4.1). Let M be the set of all sequences of zeros and ones indexed by $\mathbb{Q}$, each with finitely many ones. This carries naturally the structure of an affine space of dimension $\aleph_0$ over GF(2). Interpret a C-relation on M, putting $C(\alpha;\beta,\gamma)$ if the value of $\mathbb{Q}$ at which the sequences α and β differ is smaller than that at which β and γ differ. The subgroup of $\mathrm{AGL}(\aleph_0, 2)$ preserving the C-relation is a doubly transitive Jordan group in which, for each node of the underlying semilinear order, the set of all chains not passing through the node is a Jordan set.

6. I give more detail on groups of type (j). First, when I say that G preserves on Ω a limit of general betweenness relations or D-relations, I mean that the following holds: there is a linearly ordered set $(J, \leqslant)$ with no greatest element, a strictly increasing chain $(\Gamma_i : i \in J)$ of subsets of Ω, and an increasing chain $(H_i : i \in J)$ of subgroups of G satisfying the following conditions:

(i) for each $i \in J$, $H_i = G_{(\Omega \setminus \Gamma_i)}$, and H_i is transitive on Γ_i and has a unique maximal proper congruence σ_i on Γ_i;

(ii) for each i, $(H_i, \Gamma_i/\sigma_i)$ is a 2-transitive but not 2-primitive Jordan group preserving a general betweenness relation or a D-relation (so in this sense G preserves a limit of general betweenness relations or D-relations);

(iii) $\bigcup(\Gamma_i)_{i\in J} = \Omega$;

(iv) $(\bigcup(H_i)_{i\in J}, \Omega)$ is a 2-primitive but not 3-transitive Jordan group;

(v) $\sigma_i \supseteq \sigma_j \restriction_{\Gamma_i}$ if $i < j$;

(vi) $\bigcap(\sigma_i : i \in J)$ is equality in Ω (where we regard $\Omega \setminus \Gamma_i$ as a single σ_i-class);

(vii) $\forall g{\in}G\, \exists i_0{\in}J\, \forall i{>}i_0\, \exists j{\in}J\, (\Gamma_i g = \Gamma_j \wedge g^{-1}H_i g = H_j)$;

(viii) for any $\alpha \in \Omega$, there is a C-relation on $\Omega \setminus \{\alpha\}$ which is G_α-invariant.

The only situation in which the above holds but none of (a)–(i) occurs is when (G, Ω) is 2-primitive but not 3-transitive.

The following is meant when I say that G preserves a limit of Steiner systems on Ω: for some $k \geqslant 3$, (G, Ω) is k-transitive but not k-primitive, and there is a linearly ordered set $(J, \leqslant)$ with no greatest element and an increasing chain $(\Pi_i : i \in J)$ of subsets of Ω such that conditions (a)–(e) below hold.

(a) $\bigcup(\Pi_i : i \in J) = \Omega$.

(b) For each $i \in J$, $G_{\{\Pi_i\}}$ is $(k-1)$-transitive on Π_i and preserves a non-trivial Steiner $(k-1)$-system on Π_i.

(c) If $i < j$ then Π_i is a subset of a line of the Steiner system invariant under $G_{\{\Pi_j\}}$.

(d) For all $g \in G$ there is $i_0 \in J$, dependent on g, such that for every $i > i_0$ there is $j \in J$ so that $\Pi_i g = \Pi_j$ and the image under g of every line of Π_i is a line of Π_j.

(e) The stabilizer in G of any distinct $k-2$ points of Ω preserves a C-relation on the remaining points in which, for any node of the underlying semilinear order, the set of chains not passing through the node is a Jordan set. Furthermore, we may obtain the set $(\Pi_j)_{j \in J}$ from the C-relation and the semilinear ordering as follows. Let $(a_j : j \in J^*)$ be a maximal chain of nodes in the semilinear ordering, chosen as an element of the C-structure. For each $j \in J$, we take Π_j to be the set of chains passing through a_j, together with the $(k-2)$ points determining the C-relation (so Π_j is the complement of one of the Jordan sets mentioned above).

Adeleke (1992) has constructed a 3-transitive not 3-primitive Jordan group which preserves a limit of Steiner 2-systems and is not of any of the other types in Theorem 2.2. It is not known if there are similar constructions of a higher degree of transitivity.

3 Description of the proof

This section consists of a very rough sketch of the proof of the classification of infinite primitive Jordan groups. For details we refer to Adeleke and Neumann (in press*b*) (which handles the case when some Jordan set is primitive) and Adeleke and Macpherson (1994) (which handles the general case). In both papers, the proof is by induction on the degree of transitivity, and in both cases the hard part is to get the induction started, that is, to handle Jordan groups whose degree of transitivity is less than 4.

Adeleke and Neumann's paper (in press*b*) is eminently readable, and there is little point in adding further explanation to it. The authors prove that a primitive permutation group with a proper primitive Jordan set is highly transitive or preserves a relational structure of one of the following kinds: a linear or circular order or a linear betweenness or separation relation, or a semilinear order, a general betweenness relation or C- or D-relation.

One of Adeleke and Neumann's main ideas is to examine the possible intersections between sets of Jordan sets. First, note the following lemma.

LEMMA 3.1. (i) If Σ_1, Σ_2 are Jordan sets for (G, Ω) and $\Sigma_1 \cap \Sigma_2 \neq \varnothing$, then $\Sigma_1 \cup \Sigma_2$ is a Jordan set.

(ii) If $(\Sigma_i)_{i\in I}$ is a chain of Jordan sets for (G,Ω) ordered under inclusion, then $\bigcup(\Sigma_i)_{i\in I}$ is a Jordan set.

In fact, Adeleke and Neumann prove more. It is shown (Adeleke and Neumann, in press*b*) that for any of the properties $\mathscr{P}$ such as primitivity, k-primitivity, k-homogeneity, and k-transitivity, if the Σ_i in (i) or (ii) are $\mathscr{P}$-Jordan sets (in the sense of Section 1), then so is their union. From this and (ii), it follows quite easily that if (G,Ω) is a primitive Jordan group with a Jordan set having $\mathscr{P}$, then (G,Ω) has $\mathscr{P}$. We remark that by (ii) above, if $\mathscr{P}$ is one of the above properties, $\Gamma\subset\Omega$, and there is a $\mathscr{P}$-Jordan set in Ω disjoint from Γ, then there is a maximal such one.

Adeleke and Neumann (in press*b*) call a pair of subsets Σ_1,Σ_2 of Ω a *typical pair* if each of $\Sigma_1\setminus\Sigma_2$, $\Sigma_2\setminus\Sigma_1$, $\Sigma_1\cap\Sigma_2$ is non-empty. They show that if (Σ_1,Σ_2) is a typical pair of proper Jordan sets, such that Σ_1 is primitive and Σ_2 is k-homogeneous where $k\geqslant 1$ and $2k\leqslant|\Sigma_2|$, then $\Sigma:=\Sigma_1\cup\Sigma_2$ is $(k+1)$-homogeneous. It follows in particular, by the last paragraph, that G is $(k+1)$-homogeneous. This line of argument clearly gives good possibilities for induction (since, for example, if G is transitive enough then Σ will be typical with respect to one of its G-translates).

These arguments suggest that one should investigate G-invariant families of Jordan sets containing few typical pairs. Several lemmas on such families are proved by Adeleke and Neumann (in press*b*) and are used also by Adeleke and Macpherson (1994); in fact, for some of these lemmas the sets need not even be Jordan sets. For example, if (G,Ω) is primitive and Σ is a non-empty proper subset of Ω such that for all $g\in G$, $\Sigma\subseteq\Sigma^g$ or $\Sigma^g\subseteq\Sigma$, then there is a G-invariant linear order on Ω, of which Σ is an initial segment. Similarly, if (G,Ω) is primitive on Ω with a proper Jordan set Σ, and there are distinct $\alpha,\beta\in\Omega$ such that every G-translate of Σ which contains β also contains α, then there is a G-invariant linear or semilinear order on Ω: put $\gamma\leqslant\delta$ if and only if

$$\forall g{\in}G\,(\delta\in\Sigma g\longrightarrow\gamma\in\Sigma g).$$

If on the other hand (G,Ω) is transitive, $\Sigma\subset\Omega$ with $|\Sigma|>1$ and $|\Omega\setminus\Sigma|\geqslant 2$, the family $\{\Sigma g : g\in G\}$ contains no typical pairs, and

$$\forall\gamma,\delta{\in}\Omega\,(\gamma\neq\delta\longrightarrow\exists g{\in}G\,(\gamma\in\Sigma g\wedge\delta\notin\Sigma g)),$$

then there is a G-invariant C-relation on Ω: put $C(\alpha;\beta,\gamma)$ if for some $g\in G$ we have $\beta,\gamma\in\Sigma g$ and $\alpha\notin\Sigma g$. Finally, under certain conditions on a G-invariant family $\mathscr{F}$ of subsets of Ω, there is a G-invariant D-relation on Ω. The main condition here is that typical pairs in $\mathscr{F}$ exist, but the union of any such typical pair is Ω. Arguments of this nature yield the classification of infinite primitive Jordan groups with proper primitive Jordan sets. Also, they yield the following result of Adeleke and Neumann (in press*b*), though it is not explicitly stated there.

THEOREM 3.2. If (G, Ω) is a simply primitive Jordan group, then G preserves on Ω a linear order, a semilinear order, or a C-relation. Furthermore, if it does not preserve a semilinear order or a C-relation then it is highly homogeneous.

A more detailed version of this theorem is given as Theorem 3.1.1 in Adeleke and Macpherson (1994). As remarked in Section 2, a version of this theorem with a slight weakening of the notion of Jordan set is proved by Adeleke in this volume.

The second paper (Adeleke and Macpherson, 1994) essentially starts with the assumption that (G, Ω) is a 2-transitive infinite Jordan group with no proper primitive Jordan set, since the other cases are handled in Adeleke and Neumann (in press*b*). Most of the problems arise when G is not 3-transitive, so for the moment we shall consider only this case. A crucial first result is the following. (This is Theorem 5.1.2 of Adeleke and Macpherson (1994).)

THEOREM 3.3. If (G, Ω) is a 2-transitive Jordan group, all of whose proper Jordan sets are imprimitive, and there is a G-invariant family $\mathscr{F}$ of proper Jordan sets and distinct $\alpha, \beta, \gamma \in \Omega$ such that no $\Gamma \in \mathscr{F}$ contains β and excludes α and γ, then there is a G-invariant C-relation or Steiner 2-system on Ω.

I remark that in the above theorem, if a C-relation is preserved, then the Jordan sets in $\mathscr{F}$ are not cones but are upper sections of the underlying upper semilinear order; that is, if a is a node of the semilinear order, then the set of all chains in Ω which do not pass through a is a Jordan set. An example of this situation is the combined affine structure and C-structure in Remark 5 of Section 2 on p. 82. The strategy for proving Theorem 3.3 is to fix α and then define a preorder $\preceq$ on $\Omega \setminus \{\alpha\}$, putting $\beta \preceq \gamma$ if every proper Jordan set which contains γ also contains β. The various combinatorial possibilities are then examined. For example, if $\preceq$ is symmetric then there is a G-invariant Steiner 2-system: a typical line is $\{\alpha\} \cup \{\mu : \beta \preceq \mu\}$ for some $\beta \neq \alpha$. Also, $\preceq$ cannot be antisymmetric, for otherwise a short argument shows that $(\Omega \setminus \{\alpha\}, \preceq)$ is essentially a disjoint union of semilinearly ordered sets, and it follows easily that there is a primitive Jordan set.

Given this result, if there is no G-invariant C-relation or Steiner 2-system, we have that for any distinct $\alpha, \beta, \gamma \in \Omega$ there is a unique Jordan set, denoted $\mathrm{MJ}(\alpha, \beta/\gamma)$, which is maximal subject to containing γ and excluding α and β. Note that as we are assuming that G is not 3-transitive, $\mathrm{MJ}(\alpha, \beta/\gamma)$ is a *proper* Jordan set. Furthermore, as in the remark before Theorem 3.2, we may suppose that there is a typical pair of proper Jordan sets, for otherwise there is a G-invariant C-relation. Furthermore, if the union of every typical pair of proper Jordan sets is Ω, then there is a G-invariant C- or D-relation on Ω. We may suppose that this does not hold. It follows that there are distinct $\alpha, \beta, \gamma \in \Omega$ such that $\{\mathrm{MJ}(\alpha, \beta/\gamma), \mathrm{MJ}(\alpha, \gamma/\beta)\}$ is a typical pair. A key lemma now is the following.

LEMMA 3.4. *Let (G, Ω) be an infinite Jordan group and let $\{\Gamma_1, \Gamma_2\}$ be a typical pair of proper Jordan sets with union Γ. Then there is a unique maximal $G_{(\Omega \setminus \Gamma)}$-congruence ρ on Γ. Furthermore, either each of $\Gamma_1 \setminus \Gamma_2, \Gamma_2 \setminus \Gamma_1$ is a ρ-block, in which case $G_{(\Omega \setminus \Gamma)}$ is 2-transitive on Γ/ρ, or $G_{(\Omega \setminus \Gamma)}$ preserves a linear order or linear or general betweenness relation on Γ/ρ.*

We can describe ρ partly as follows. Let $\alpha_1 \in \Gamma_1 \setminus \Gamma_2$, $\alpha_2 \in \Gamma_2 \setminus \Gamma_1$. A typical ρ-class is the set of $\delta \in \Gamma \setminus \Gamma_2$ such that the maximal Jordan set in Γ which contains α_2 and omits α_1 is the same as that which contains α_2 and omits δ.

DEFINITION 3.5. If $\alpha, \beta, \gamma \in \Omega$ are distinct, we say that the pair of Jordan sets $\{\mathrm{MJ}(\alpha, \beta/\gamma), \mathrm{MJ}(\alpha, \gamma/\beta)\}$ has (P) if

(a) the pair $\mathrm{MJ}(\alpha, \beta/\gamma)$, $\mathrm{MJ}(\alpha, \gamma/\beta)$ is typical,
(b) G does not induce the symmetric group on $\{\alpha, \beta, \gamma\}$,
(c) $\mathrm{MJ}(\alpha, \beta/\gamma) \setminus \mathrm{MJ}(\alpha, \gamma/\beta)$ is a block of the unique maximal congruence (promised by Lemma 3.4) on $\mathrm{MJ}(\alpha, \beta/\gamma) \cup \mathrm{MJ}(\alpha, \gamma/\beta)$.

In a series of technical lemmas, it is shown that either G preserves a structure on Ω of known type, or (P) does indeed hold for some triples, or a situation rather similar to (P) often arises, but with a general betweenness relation living on the quotient of $\mathrm{MJ}(\alpha, \beta/\gamma) \cup \mathrm{MJ}(\alpha, \gamma/\beta)$. Eventually the following lemma is proved.

LEMMA 3.6. *Let (G, Ω) be an infinite 2-transitive but not 3-transitive Jordan group, and suppose that there is no G-invariant C- or D-relation, linear or general betweenness relation, circular order, separation relation or non-trivial Steiner 2-system. Then there is a linearly ordered set $(I, <)$ and a chain $\{\Sigma_i : i \in I\}$ of Jordan sets of (G, Ω) with the following properties, where $i, j \in I$ with $i < j$.*

(i) *There is $\alpha_i \in \Omega$ such that the pair*

$$\{\mathrm{MJ}(\alpha_i, \beta/\gamma), \mathrm{MJ}(\alpha_i, \gamma/\beta)\}$$

satisfies (P) (or a related condition with general betweenness relations) and has union Σ_i;

(ii) $\Sigma_i \subset \Sigma_j$;

(iii) *there is a unique non-trivial maximal G-congruence ρ_i of $G_{(\Omega \setminus \Sigma_i)}$ on Σ_i;*

(iv) $\rho_i \supseteq \rho_j \restriction \Sigma_i$;
(v) $\bigcup(\Sigma_k : k \in I) = \Omega$;
(vi) $\bigcap(\rho_k : k \in I)$ is equality on Ω (where for convenience we regard $\Omega \setminus \Sigma_k$ as a single ρ_k-class);

Furthermore, under the above conditions, either G preserves a limit of general betweenness relations, where the betweenness relations live on the sets Σ_j/ρ_j as j ranges through a cofinal subset of I, or in addition the following conditions are satisfied by the Σ_i.

- $G_{(\Omega \setminus \Sigma_i)}$ is not 3-transitive on Σ_i/ρ_i;
- the complement in Ω of any ρ_i-class is a Jordan set;
- Δ_i, which is

$$\{\omega \in \Omega : \mathrm{MJ}(\omega, \beta/\gamma) \cup \mathrm{MJ}(\omega, \gamma/\beta) = \mathrm{MJ}(\alpha_i, \beta/\gamma) \cup \mathrm{MJ}(\alpha_i, \gamma/\beta)\}$$

equals $\bigcap(\Sigma_j : j > i) \setminus \Sigma_i$.

We are now (in the second case above) quite close to the situation described in case (j) of Theorem 2.2 and in more detail in Remark 6 of Section 2—the case where there is a G-invariant limit of general betweenness relations or D-relations. In particular, it is shown that the situation of Lemma 3.6 cannot arise if (G, Ω) is 2-transitive but not 2-primitive. Thus, the possibilities when (G, Ω) is 2-transitive but not 3-transitive are described by the following theorem.

THEOREM 3.7. Let (G, Ω) be an infinite 2-transitive but not 3-transitive Jordan group.

(a) If (G, Ω) is not 2-primitive, then G preserves on Ω a linear or general betweenness relation, a C- or D-relation, or a Steiner 2-system.
(b) If (G, Ω) is 2-primitive then it preserves on Ω a circular order, a D-relation, or a limit of general betweenness relations or D-relations.

The classification of Jordan groups of a larger degree of transitivity is inductive, for if (G, Ω) is k-transitive then the stabilizer in G of a point is $(k-1)$-transitive on the remaining points. Hence we only have to examine transitive extensions of already known permutation groups. The known results can be summarized in the following theorem. The case $k = 3$ still involves a substantial amount of work, with arguments like those in Lemma 3.6 being applied to the point stabilizer.

THEOREM 3.8. Suppose that for some integer $k \geqslant 3$, G is k-transitive but not $(k+1)$-transitive. Then G is not k-primitive, and G preserves a Steiner k-system, a limit of Steiner $(k-1)$-systems, or a D-relation or separation relation (and the last two cases can only occur if $k = 3$).

In the last theorem, 'limit of Steiner systems' is meant in the sense of Remark 6 in Section 2. For all $k \geqslant 3$, Steiner systems admitting k-transitive Jordan groups are constructed in Section 5, and Adeleke (1992) has also built Jordan groups involving limits of Steiner 2-systems.

It is worth remarking that some parts of the above classification go through without the full assumption that (G, Ω) is a Jordan group. The paper by Adeleke in this volume, which classifies simply primitive permutation groups satisfying a condition slightly weaker than the Jordan condition, gives an example of this (and also exemplifies the kinds of arguments used in the classification). Also, in Adeleke and Neumann (in press *b*, to appear) much effort is made to recover G-invariant relational structures on Ω just from the existence of G-invariant families of subsets of Ω satisfying certain intersection properties, without the assumption that there is a proper Jordan set.

In the early attempts at the classification of infinite Jordan groups, several alternative partial arguments were found. For example, it is possible to classify the 2-transitive, not 2-primitive Jordan groups without going via Lemma 3.6, and also to give a more direct proof that every 3-primitive Jordan group is 4-transitive. These arguments, however, gave no insights into the structures of type (j), but did give as a by-product the following result, which might have independent interest. Recall that a transitive permutation group (G, Ω) is said to be a *transitive extension* of (H, Σ), if there is $\alpha \in \Omega$ such that $\Sigma = \Omega \setminus \{\alpha\}$ and the permutation groups $(G_\alpha, \Omega \setminus \{\alpha\})$ and (H, Σ) are isomorphic. Parts (iv) and (v) follow immediately from Cameron's result, Theorem 2.1 above.

PROPOSITION 3.9. Permutation groups of the following kinds have no transitive extensions:

(i) a simply primitive Jordan group preserving a semilinear order;

(ii) a doubly transitive Jordan group preserving a general betweenness relation;

(iii) a triply transitive Jordan group preserving a D-relation;

(iv) a doubly primitive highly homogeneous permutation group preserving a circular order;

(v) a triply transitive Jordan group preserving a separation relation.

4 Jordan groups and stability theory

First, I emphasize that this section is about Jordan groups of geometric type. For, by Theorem 7.8, if M is a stable saturated structure whose automorphism group is a primitive Jordan group, then $(\operatorname{Aut} M, M)$ is of geometric type.

Recall that if M is a first-order structure, $\bar{a}$ is a tuple from M, and $\phi(x, \bar{a})$ is a formula, then the set $X := \{x \in M : M \models \phi(x, \bar{a})\}$ is called *strongly minimal* if for every N such that $M \prec N$ and every formula $\psi(x, \bar{b})$ with $\bar{b}$ from N, the set $\{x : N \models \psi(x, \bar{b}) \wedge \phi(x, \bar{a})\}$ is finite or cofinite in the set $\{x : N \models \phi(x, \bar{a})\}$. Similarly, the formula $\phi(x, \bar{a})$ is called a *strongly minimal formula*. By the results of Baldwin and Lachlan (1971), strongly minimal sets are the building blocks of $\aleph_1$-categorical structures. In a little more detail, if M is an uncountable model of an uncountably categorical theory T, then there is a sequence $\bar{a}$ in M whose type is isolated over $\varnothing$ and a strongly minimal formula $\phi(x, \bar{a})$ such that if N is an elementary submodel of M containing $\bar{a}$ then the isomorphism type of N is determined by the cardinality of a maximal algebraically independent (over $\bar{a}$) subset I of $\phi(x, \bar{a}) \cap N$ (and N is prime over $I \cup \mathrm{Ran}(\bar{a})$)). Note that $\aleph_1$-categorical theories (over, as always in this paper, countable languages) have saturated models of every infinite cardinal. Also, every strongly minimal *structure* has $\aleph_1$-categorical theory.

Recall that saturated models of any theory are *homogeneous*, in the model-theoretic sense that any partial elementary map from the structure to itself with domain of cardinality less than that of the structure extends to an automorphism. These observations yield the following well-known lemma.

LEMMA 4.1. Let X be an $\bar{a}$-definable strongly minimal set in a saturated structure M, and let $G := \mathrm{Aut}\, M$. Then for every $A \subset M$ such that $|A| < |M|$ and $\mathrm{Ran}(\bar{a}) \subseteq A$, $X \setminus \mathrm{acl}(A)$ is a Jordan set for the action of $G_{\bar{a}}$ on X.

Proof. It suffices to note that, by strong minimality, any two elements of $X \setminus \mathrm{acl}(A)$ have the same type over $A \cup \bar{a}$. □

We shall now consider the situation of the last lemma in the special case where $X = M$ and $\bar{a} = \varnothing$. Here, (G, M) is a Jordan group. Possibly, of course, G is intransitive on M. However, $A := \mathrm{acl}(\varnothing)$ is a subset of M of size at most $\aleph_0$ which is setwise invariant under $\mathrm{Aut}\, M$, and by the lemma $\mathrm{Aut}\, M_{(A)}$ is transitive on $M \setminus A$, so $(\mathrm{Aut}\, M_{(A)}, M \setminus A)$ is a transitive Jordan group. This group may not be primitive. However, if we define $\sim$ on $M \setminus A$ by putting $a \sim b$ if $b \in \mathrm{acl}(a)$, then $\sim$ is an $\mathrm{Aut}\, M$-invariant equivalence relation, and it is easily checked that $\mathrm{Aut}\, M_{(A)}$ is 2-transitive on $(M \setminus A)/\sim$ and is a Jordan group. This is how projective space is obtained from a vector space.

For this section and the next we recall some combinatorial notions. A *pregeometry* is a set X equipped with a closure operator

$$\mathrm{cl}\colon \mathscr{P}(X) \to \mathscr{P}(X)$$

satisfying the following axioms (where $A, B \subseteq X$ and $a, b \in X$).

1. $A \subseteq \mathrm{cl}(A)$;

2. if $A \subseteq B$ then $\mathrm{cl}(A) \subseteq \mathrm{cl}(B)$;
3. if $A \subseteq \mathrm{cl}(B)$ and $B \subseteq \mathrm{cl}(C)$ then $A \subseteq \mathrm{cl}(C)$;
4. if $a \in \mathrm{cl}(A \cup \{b\}) \setminus \mathrm{cl}(A)$ then $b \in \mathrm{cl}(A \cup \{a\})$;
5. $\mathrm{cl}(A) = \bigcup(\mathrm{cl}(F) : F \subseteq A, |F| < \aleph_0)$.

A *geometry* is a pregeometry in which all singletons are closed. If (X, cl) is a pregeometry and $A \subseteq X$, we say that A is *independent* if for all $a \in A$, $a \notin \mathrm{cl}(A \setminus \{a\})$. Furthermore, if $A \subseteq X$ and B is a closed subset of X, we say that A is a *basis* for B if $A \subseteq B$, A is independent, and $\mathrm{cl}(A) = B$. Closed subsets are called *subspaces* or *flats*. If B is a subspace of X then by condition (4) any two bases of B have the same size. This size is called the *dimension* of B. Thus, a k-*flat* is just a k-dimensional subspace..

It is well-known and elementary to verify that on a strongly minimal set, algebraic closure gives a pregeometry. Similarly, there is a pregeometry on a stationary regular type (see Pillay (1983) for definitions) p, once we have introduced constants for the parameters over which the type is defined. Here, if A is a set of realizations of p and a realizes p, then $a \in \mathrm{cl}(A)$ if $\mathrm{tp}(a/A)$ forks over the empty set. Lemma 4.1 tells us that in a saturated strongly minimal set, the complements of the k-flats ($k \in \mathbb{N}$) are Jordan sets, and a similar lemma holds for regular types in saturated structures.

In the strongly minimal case it is easily seen that, once we have factored out the equivalence relation $\sim$ defined above, the doubly transitive Jordan group obtained is either highly transitive or of geometric type. I now list some familiar examples. In the next section I hint at Hrushovski's construction of more complicated examples.

(i) Let V be an infinite vector space over a field F, and regard V as a first-order structure over the language $(+, 0, f_i)_{(i \in I)}$ where the f_i are unary functions defining multiplication by corresponding field elements. This structure has elimination of quantifiers in the usual language for modules (where there is a unary function symbol for multiplication by each field element) so is strongly minimal. Its automorphism group is $\mathrm{GL}(V)$. We have $\mathrm{acl}(\varnothing) = \{0\}$, and the equivalence relation $\sim$ is just linear dependence. The quotient $(V \setminus \{0\})/\sim$ is the corresponding projective space, and the group induced on it by $\mathrm{Aut}\, V$ is the 2-transitive group $\mathrm{PGL}(V)$.

(ii) Any infinite dimensional affine space over a field F is strongly minimal, essentially because it is just a reduct of a vector space. An appropriate language for affine space was given by Givant (1979). It has a ternary function symbol $f(x, y, z)$ interpreted by putting $f(x, y, z) = x + y - z$, and, for each $\lambda \in F$, a binary function symbol f_λ, where $f_\lambda(x, y) = \lambda x + (1 - \lambda)y$.

(iii) Let K be an algebraically closed field of transcendence degree at least two, and let k be the algebraic closure of the prime subfield of

K. Then by Tarski's elimination of quantifiers for the theory of algebraically closed fields, K is strongly minimal, and, as model-theoretic and field-theoretic algebraic closure coincide, k is just $\mathrm{acl}(\varnothing)$. For $x, y \in K \setminus k$, we now have that $x \sim y$ if and only if y is in the field-theoretic algebraic closure of $k(x)$. As above, $\mathrm{Gal}(K/k)$ is 2-transitive on $(K \setminus k)/\sim$.

Suppose now that M is an ω-categorical strongly minimal set. Such sets arise in attempts to classify totally categorical theories. By the Ryll-Nardzewski Theorem we no longer need to worry about the choice of language, but may simply choose the *canonical* language, with a relation symbol corresponding to each orbit on M^n. Now $\mathrm{acl}(\varnothing)$ is finite, as are the classes of the equivalence relation $\sim$. It is natural to replace M by the set $(M \setminus \mathrm{acl}(\varnothing))/\sim$, viewed as a structure over the canonical language. This structure is a *strictly minimal set*, that is, an ω-categorical strongly minimal set with primitive automorphism group. There are three obvious examples of strictly minimal sets, namely a pure set, and infinite dimensional projective and affine spaces over finite fields. (Projective and affine spaces over infinite fields could not arise here, since they have infinitely many 3-types, so, by the Ryll-Nardzewski Theorem, cannot be ω-categorical.) By the following important theorem proved independently by Cherlin, Mills and Zil'ber, these are essentially the only possibilities.

THEOREM 4.2. Let M be a strictly minimal set, and $G = \mathrm{Aut}\, M$. Then one of the following holds.

(i) $G = \mathrm{Sym}(M)$.
(ii) $\mathrm{PGL}(\aleph_0, q) \leqslant G \leqslant \mathrm{P\Gamma L}(\aleph_0, q)$, and G acts naturally on the projective space $\mathrm{PG}(\aleph_0, q)$ (q a prime power).
(iii) $\mathrm{AGL}(\aleph_0, q) \leqslant G \leqslant \mathrm{A\Gamma L}(\aleph_0, q)$, and G acts naturally on the affine space $\mathrm{AG}(\aleph_0, q)$ (q a prime power).

This result is at the heart of the structure theory due to Cherlin *et al.* (1985) for ω-categorical, ω-stable theories, and its subsequent refinements by Ahlbrandt and Ziegler (1986, Ahlbrandt and Ziegler (1991*b*)), Hrushovski (1989*b*), and others. The key result from Cherlin *et al.* (1985) is the Coordinatization Theorem, which asserts that if M is ω-categorical and ω-stable then there is a rank one set A, 0-definable in M, such that for any $x \in M$, $\mathrm{acl}(x) \cap A \neq \varnothing$. Theorem 4.2 is also an ingredient of Zil'ber's proof (Zil'ber, 1979, 1984*b*) that totally categorical theories are not finitely axiomatizable (which is also proved by Cherlin *et al.* (1985)).

The proof of Theorem 4.2 due to Cherlin (and independently, Mills) is given in Cherlin *et al.* (1985). The idea is as follows: if M is a strictly minimal set, then as noted above, $\mathrm{Aut}\, M$ is 2-transitive on M. If we give M the canonical structure over the canonical language L and pick a finite algebraically closed subset X of M with the induced relations, then if $\{x_1, \ldots, x_n\}$, $\{y_1, \ldots, y_n\}$ are maximal algebraically independent (in M) subsets of X, there is $g \in \operatorname{Aut} M$ such that $x_i g = y_i$ for $i = 1, \ldots, n$, and such a g must fix X setwise and induce an automorphism of X. It follows easily that $(\operatorname{Aut} X, X)$ is a doubly transitive Jordan group. In fact, if $A \subset X$ is algebraically closed, then $\operatorname{Aut} X_{(A)}$ is transitive on $X \setminus A$. Using the classification of finite doubly transitive groups, Cherlin and Mills classified sufficiently large finite doubly transitive Jordan groups, and showed that they are symmetric or alternating groups, or projective or affine groups. By examining algebraic closure in the infinite limit, they then proved Theorem 4.2.

The proof by Zil'ber (1979, 1984*b*) is more geometrical. Later, a combinatorial proof was given by Evans (1986*b*), who combines some of Zil'ber's geometrical ideas with some linear algebraic arguments involving coherent configurations. Finally, a beautiful proof was given by Hrushovski (1992*b*). Here, something more general is proved (essentially, that every *unimodular* strongly minimal set is locally modular).

5 Multiply transitive Jordan groups

In this section I will describe how a construction of Hrushovski (1993*a*) yields, for any positive integer $k \geqslant 2$, a k-transitive Jordan group of geometric type which is not $(k+1)$-transitive. The construction here is much easier than that in Hrushovski's work, since we do not need to worry about either stability or ω-categoricity. It was first pointed out to me by David Evans that the construction yields such Jordan groups. Since the method of construction is described in detail in Wagner's article in this volume, treatment is brief but self-contained. The approach here is taken from Goode (1989). My intention is that it be readily accessible to a permutation group theorist not interested in the connections with stability theory. The theorem we shall prove is the following.

THEOREM 5.1. For any integer $k \geqslant 2$ there is a Jordan group which is k-transitive but not $(k+1)$-transitive and preserves a Steiner k-system.

To prove this, we shall build a geometry (M, cl) of dimension $\aleph_0$ (in the sense of Section 4) such that if $G = \operatorname{Aut}(M, \mathrm{cl})$, then the following hold:

(i) G is k-transitive but not $(k+1)$-transitive on M;
(ii) the complement in M of any finite dimensional subspace is a Jordan set for G;
(iii) $\dim M = \aleph_0$.

The lines of the Steiner system will then be the closed sets of dimension k.

Fix $k \geqslant 2$. A $(k+1)$-*hypergraph* is just a pair (Y, S) where Y is a set and S is a set of $(k+1)$-element subsets of Y (called the *edges* of Y). If

$H = (Y, S)$ is a $(k+1)$-hypergraph then $v(H)$ and $e(H)$ denote respectively $|Y|$ and $|S|$. For any $(k+1)$-hypergraph H, define $\delta(H) := v(H) - e(H)$. We use the word *subhypergraph* just for substructure of a hypergraph in the model-theoretic sense, and write $K \leqslant H$ if K is a subhypergraph of H (note that this differs from Wagner's notation in this volume). Let $\mathscr{H}_{k+1}$ be the class of all $(k+1)$-hypergraphs. Also let

$$\mathscr{A}_k := \{H \in \mathscr{H}_{k+1} : \forall K \leqslant H\, (\delta(K) \geqslant \min(|K|, k))\}$$

If $A, M \in \mathscr{A}_k$ with $A \leqslant M$, we say that A is *self-sufficient* in M if for any finite B with $A \leqslant B \leqslant M$ we have $\delta(B) \geqslant \delta(A)$.

CLAIM 1. If $A, M \in \mathscr{A}_k$ with $A \leqslant M$ then there is a finite $B \in \mathscr{A}_k$ such that $A \leqslant B \leqslant M$ and B is self-sufficient in M.

Proof. Choose B such that $A \leqslant B \leqslant M$ and $\delta(B)$ is as small as possible. □

CLAIM 2. If A is self-sufficient in B and B is self-sufficient in M then A is self-sufficient in M.

Proof. Suppose for a contradiction that there is $C \in \mathscr{A}_k$ such that $A \leqslant C \leqslant M$ and $\delta(C) < \delta(A)$. Clearly we cannot have $C \leqslant B$ or $B \leqslant C$. Also, $\delta(C \cap B) \geqslant \delta(A)$, so adjoining $C \setminus B$ to $C \cap B$ adjoins more edges than vertices. Hence adjoining $C \setminus B$ to B adds more edges than vertices, so $\delta(C \cup B) < \delta(B)$, a contradiction. □

CLAIM 3. If $B_1, B_2, M \in \mathscr{A}_n$ and B_1, B_2 are self-sufficient in M then $B_1 \cap B_2$ is self-sufficient in M.

Proof. Suppose that $B_1 \cap B_2 \leqslant C \leqslant M$, $\delta(C) < \delta(B_1 \cap B_2)$, and that $|C|$ is minimal subject to this. Suppose $C \nsubseteq B_1$. Then, by the minimality of $|C|$, adjoining $C \setminus B_1$ to $C \cap B_1$ adds more edges than vertices. Hence adjoining $C \setminus B_1$ to B_1 adds more edges than vertices, so $\delta(B_1 \cup C) < \delta(B_1)$, contradicting that B_1 is self-sufficient in M. So $C \leqslant B_1$. Similarly $C \leqslant B_2$, so $C \leqslant B_1 \cap B_2$, a contradiction. □

Note that by Claim 3, if $A, M \in \mathscr{A}_k$ with $A \leqslant M$, and if A is finite, then there is a unique smallest self-sufficient set B with $A \leqslant B \leqslant M$. Such a B is called the self-sufficient closure of A (in M), and is denoted $\mathrm{SSC}_M(A)$. Note that this is *not* the closure operator giving the geometry. Also, we define $d_M(A) := \delta(\mathrm{SSC}_M(A))$. When M is clear from the context, we drop the subscript. Clearly we always have $d_M(A) \leqslant \delta(A)$.

Next, we define our closure operator on any $M \in \mathscr{A}_k$. For $M \in \mathscr{A}_k$ and finite $A \leqslant M$, put

$$\mathrm{cl}_M(A) := \{x \in M : d_M(A) = d_M(A \cup \{x\})\}.$$

For infinite $A \leqslant M$, put

$$\mathrm{cl}_M(A) := \bigcup(cl_M(F) : F \leqslant A, |F| < \aleph_0).$$

CLAIM 4. If $M \in \mathscr{A}_k$, then (M, cl_M) is a geometry.

Proof. We verify that the exchange property holds. Let $A \subseteq M$, $b, c \in M$, and $c \in \mathrm{cl}_M(A \cup \{b\}) \setminus \mathrm{cl}_M(A)$. Then $d_M(A \cup \{b, c\}) = d_M(A \cup \{b\})$ and $d_M(A \cup \{c\}) = d_M(A) + 1$. It follows that $d_M(A \cup \{b\}) = d_M(A) + 1$; for otherwise $d_M(A \cup \{b\}) = d_M(A)$, so $d_M(A \cup \{c\}) \leqslant d_M(A \cup \{b, c\}) = d_M(A \cup \{b\}) = d_M(A)$, which is a contradiction. Putting this together, $d_M(A \cup \{b, c\}) = d_M(A \cup \{c\})$, so $b \in \mathrm{cl}_M(A \cup \{c\})$, as required. □

The dimension function associated with this geometry is d_M. Note that $\mathrm{SSC}_M(A)$ and $\mathrm{cl}_M(A)$ are usually very different: for finite $A \leqslant M$, the set $\mathrm{SSC}_M(A)$ is always finite but $\mathrm{cl}_M(A)$ is usually infinite. Note too that if $A, M \in \mathscr{A}_k$ with $A \leqslant M$ then $\mathrm{SSC}_M(A) \subseteq \mathrm{cl}_M(A)$ and $\mathrm{cl}_M(A)$ is self-sufficient in M. We call subsets of M which are closed in M *subspaces* of M.

We shall build the countable geometry of the theorem by a slight variant of Fraïssé's Theorem. For this, we need a notion of amalgamation. If $A, B, C, D \in \mathscr{A}_k$ with $A \leqslant B$ and $A \leqslant C$ and $D = B \cup C$, we say that D is a *free composite* of B and C over A if $B \cap C = A$ and there is no edge of D which meets both $B \setminus A$ and $C \setminus A$. The following lemma makes a version of amalgamation possible.

LEMMA 5.2. Let $A, B, C \in \mathscr{A}_k$ be finite with $A \leqslant B$ and $A \leqslant C$, and suppose that A is self-sufficient in B and in C. Suppose that $B \cap C = A$, and let D be an n-hypergraph with vertex set $B \cup C$, such that $B \leqslant D$, $C \leqslant D$ and no edge of D meets both $B \setminus A$ and $C \setminus A$. Then $D \in \mathscr{A}_k$, and B, C are both self-sufficient in D.

Proof. We first show that B is self-sufficient in D, that is, that $\delta(B) \leqslant \delta(E)$ for all E with $B \leqslant E \leqslant D$. Suppose that this is false for some E. Then adjoining $E \setminus B$ to B adds more edges than vertices. Hence, as the composition is free, adjoining $E \setminus B$ to A adds more edges than vertices. This contradicts that A is self-sufficient in C.

A similar argument shows that $D \in \mathscr{A}_k$. □

The above lemma has the following partial converse.

LEMMA 5.3. Let $M \in \mathscr{A}_k$, let X be a finite dimensional subspace of M, let $B \leqslant M$ be finite and self-sufficient in M, put $A = B \cap X$, and suppose that $d_M(A) = d_M(X)$. Then X and B form a free composite over A.

Proof. Since B and X are both self-sufficient in M, by Claim 3 A is self-sufficient in M. Since $\delta(A) \geqslant d_M(A) = d_M(X)$ we have $\delta(A) = \delta(X)$. Let

C be finite with $A \leqslant C \leqslant X$. Put $C' := \mathrm{SSC}_X(C)$. Then $C \leqslant C' \leqslant X$ and so $\delta(C') = \delta(X) = \delta(A)$. If there was a hyperedge of $B \cup C'$ meeting both $B \setminus A$ and $C' \setminus A$ then, since adding $C' \setminus A$ to A adds the same number of edges as vertices, adding $C' \setminus A$ to B would add more edges than vertices; hence we would have $\delta(B \cup C') < \delta(B)$, contradicting that B is self-sufficient in M. □

Using Lemma 5.2 we can build a countably infinite structure $M \in \mathscr{A}_k$, such that

(∗) whenever $A, B \in \mathscr{A}_k$ are finite and $f: A \to B$, $g: A \to M$ are hypergraph embeddings such that $f(A)$ is self-sufficient in B and $g(A)$ is self-sufficient in M, there is an embedding $h: B \to M$ such that $h \circ f \restriction A = g \restriction A$ and $h \circ f(B)$ is self-sufficient in M.

We build M as a union of an ω-chain $(M_n : n < \omega)$ of finite structures. At the nth stage we have, say, embeddings $f: A \to B$, $g: A \to M_n$ with $f(A)$ self-sufficient in B and $g(A)$ self-sufficient in M_n. Let M_{n+1} be the free composite of M_n and B over $g(A)$ (where A and $g(A)$ are identified in the natural way). There are countably many steps in the chain, and countably many amalgams to handle, so we may ensure that M satisfies (∗).

Put $G := \mathrm{Aut}\, M$. We shall show first that the complement of any finite dimensional subspace of (M, cl) is a Jordan set for G. It follows that G is k-transitive, and it is immediate that it is not $(k+1)$-transitive.

First, by a back-and-forth argument using (∗), we have:

Claim 5. Any isomorphism between finite self-sufficient substructures of M extends to an element of G.

Let X be a finite dimensional subspace of M and let $a, b \in M \setminus X$. To show that $M \setminus X$ is a Jordan set, we must find $g \in \mathrm{Aut}\, M_{(X)}$ with $ag = b$. For any finite $A \leqslant X$, $d_M(A \cup \{a\}) = d_M(A) + 1 = d_M(A \cup \{b\})$. Hence no edge of M lies in $X \cup \{a\}$ and contains a, or in $X \cup \{b\}$ and contains b. Thus, the function f fixing X pointwise and taking a to b is a k-hypergraph isomorphism. Note that for any finite A self-sufficient in X, since $a \notin \mathrm{cl}(A)$, $A \cup \{a\}$ is self-sufficient in M, and similarly $A \cup \{b\}$ is self-sufficient in M. We shall build an element g of G fixing X pointwise and taking a to b, by a back-and-forth argument. We build g as the union of a chain $(g_n : n < \omega)$ of partial automorphisms, where $g_0 \subseteq g_1 \subseteq \ldots$, each g_i induces the identity on X, and $|\mathrm{Dom}(g_i) \setminus X| < \aleph_0$. At the nth stage we will have built finite sets $A_0 \subseteq A_1 \subseteq \ldots \subseteq A_n$, where A_n is self-sufficient in M, $\mathrm{cl}(A_n \cap X) = X$, and for any finite B with $A_n \cap X \leqslant B \leqslant X$, $B \cup A_n$ forms a free composite over $B \cap A_n$. For each n we will have $\mathrm{Dom}(g_n) = X \cup A_n$. We may choose g_0 as id_X, A_0 finite and containing some basis of X, g_1 as the element f above, and A_1 as $A_0 \cup \{a\}$. Suppose that we have such g_n, A_n. Let $x \in M \setminus (X \cup A_n)$. We must show that we can extend g_n to g_{n+1} so that $x \in \mathrm{Dom}(g_{n+1})$; (this is the 'forth' stage in the argument; the 'back' stage is similar). Put $A_{n+1} := \mathrm{SSC}_M(A_n \cup \{x\})$. By Lemma 5.3, for any finite B with $A_{n+1} \cap X \leqslant B \leqslant X$, B and A_{n+1} form a free composite over $A_{n+1} \cap X$. Let $C := X \cap A_{n+1}$.

CLAIM 6. $A_n \cup C$ is self-sufficient in M.

Proof. First note that as A_{n+1} and X are both self-sufficient in M, it follows by Claim 3 that C is self-sufficient in M. Similarly, $A_n \cap X$ is self-sufficient in M. Hence $\delta(A_n \cap X) \leqslant \delta(X)$ and $\delta(C) \leqslant \delta(X)$. Since A_n contains a basis for X, $\delta(A_n \cap X) \geqslant \delta(X)$ and $\delta(C) \geqslant \delta(X)$. It follows that $\delta(A_n \cap X) = \delta(X) = \delta(C)$. Hence, as A_n and C form a free composite over $A_n \cap C$, $\delta(A_n \cup C) = \delta(A_n)$. Since A_n is self-sufficient in M, the claim follows. □

Let $\tilde{g}_n$ be the restriction of g_n to $A_n \cup C$. By Claims 5 and 6, $\tilde{g}_n$ extends to some $h_n \in G$. Put $k_n := h_n \restriction_{X \cup A_{n+1}}$. Since k_n fixes pointwise a basis for X, it fixes X setwise, so induces an automorphism l_n of X. Also, k_n fixes C pointwise. Since X and A_{n+1} form a free composite over $X \cap A_{n+1}$, the mapping $\tilde{l}_n$ fixing A_{n+1} pointwise and inducing l_n^{-1} on X is a hypergraph isomorphism. Put $g_{n+1} := k_n \tilde{l}_n^{-1}$. Then g_{n+1} satisfies the required conditions, so the construction goes through.

REMARK. As noted in Section 6 of Wagner's paper in this volume (p. 153), the structure we have built is ω-stable of Morley rank ω. It is not ω-categorical.

6 Group-theoretic applications of Jordan groups

In this section, I discuss some applications of the results on Jordan groups to general permutation group theory. I also discuss very briefly some group-theoretic weakenings of the definition of a Jordan set.

Recall that the *support*, supp(g), of a permutation g is the set of elements moved by g. The *minimal degree* of a finite permutation group is the least cardinality of support of a non-identity element of the group. One of the early results on permutation groups, with the flavour of Jordan groups, is the following.

THEOREM 6.1. (Jordan, 1875) The minimal degree of a primitive subgroup of S_n not containing A_n tends to infinity as n tends to infinity.

Wielandt (1959) pointed out that this theorem has the following analogue for infinite permutation groups.

THEOREM 6.2. Every primitive subgroup of $\mathbb{N}$ which contains a non-identity element of finite support must contain the finitary alternating group on $\mathbb{N}$ (i.e., the permutation group on $\mathbb{N}$ consisting of even permutations of finite support).

Cameron (1990) gives a short proof of Theorem 6.2 using the Separation Lemma, Theorem 3.1 on p. 17. (Cameron attributes this proof to Dixon.) In Neumann (1975, 1976) an elegant theory was developed for finitary permutation groups, that is, permutation groups of infinite degree all of whose elements have finite support. This has been applied in recent work of Hall, Phillips, Wehrfritz and others on finitary linear groups (where irreducible imprimitive linear groups give finitary permutation groups, induced on the set of direct summands of an invariant direct sum decomposition).

There are several natural ways in which one might try to generalize these results. The theory of Jordan groups is the fruit of one of them. Others include: uncountable analogues of Wielandt's Theorem; attempts to classify those cycle types which can be realized by primitive permutation groups which are not highly transitive; and finitary extensions of the above theorem of Jordan. I will discuss here the first two topics. The third has also spawned much recent work (see for example Pyber (in press), where it is shown that any primitive subgroup of S_n not containing A_n has minimal degree at least $(\sqrt{n}-1)/2$).

1. *Cycle types.* I first consider briefly the question: which cycle types (on a countable set) are realizable by a primitive permutation group of countably infinite degree which is not highly transitive? The results on this are very incomplete. On one side of the question, the best results I know are those due to Truss (1985), who classifies the cycle types which are realized by automorphisms of the k-coloured random graph, that is, the universal homogeneous object in a language with k binary relations ($k \leqslant \aleph_0$), all of them symmetric and irreflexive. I remark that any cycle type which includes infinitely many infinite cycles is realized by the automorphism group of the (1-coloured) random graph: simply build the graph on a permutation of that cycle type, ensuring in the process that for any two finite disjoint sets U, V of vertices there is a vertex adjacent to everything in U and to nothing in V.

In the other direction, some initial results have been made in joint work with Praeger (Macpherson and Praeger, in press *a*). There, we make the following conjectures.

CONJECTURE 6.3. (i) Any primitive but not highly transitive permutation group with a non-identity element having finitely many non-trivial finite cycles, infinitely many fixed points, and a finite number of infinite cycles is contained in a Jordan group which is not highly transitive.

(ii) Any primitive permutation group with an element having finitely many infinite cycles, a finite number greater than zero of non-trivial finite cycles, and infinitely many fixed points, is highly transitive.

Some evidence for these conjectures is given in Macpherson and Praeger (in press *a*). First, it is immediate that any permutation group containing an element with infinitely many fixed points whose support is just a single infinite cycle is a Jordan group. Such a cycle type is realized by certain 2-transitive groups, namely the automorphism groups of some C-relations (Ω, C) in which the maximal chains of the underlying semilinear order have order type $(\mathbb{Z}, <)$. The point here is that an infinite cycle can act regularly on a cone K in such a C-relation; for it is possible to define a translation-invariant relation C satisfying axioms C1, C2, C3, and C5 and isomorphic to the relation induced by C on K: put $C(i; j, k)$ if for some $n \in \mathbb{N}$, some coset of $2^n.\mathbb{Z}$ in $\mathbb{Z}$ contains j and k but not i. Indeed, by this trick, $\mathrm{Aut}(\Omega, C)$ (and the automorphism group of the corresponding D-relation) can realize any cycle type having finitely many infinite cycles, infinitely many fixed points, and no other finite cycles. Such a group cannot be oligomorphic, since as the maximal chains are discrete there are infinitely many orbits on triples. Using results on Jordan groups it is quite easy to check that any primitive permutation group which is oligomorphic and contains an element with infinitely many fixed points, an infinite cycle, and no other cycles, is highly transitive.

Furthermore, we prove (Macpherson and Praeger, in press *a*) the following theorem.

THEOREM 6.4. If (G, Ω) is a primitive permutation group containing an element consisting of a single infinite cycle, infinitely many fixed points, and a finite number greater than zero of non-trivial finite cycles, then G is highly transitive on Ω.

In the proof of Theorem 6.4, we first show, by induction on the degree of transitivity, that (G, Ω) is contained in a Jordan group of the same degree of transitivity. The result follows by inspection of the list of Jordan groups. Most of the difficulties are in finding a Jordan set when G is not 2-transitive.

2. *Bounded groups.* On the second topic, uncountable analogues of Wielandt's Theorem, substantial progress was made by Adeleke and Neumann (in press *a*). The key result here is the following.

THEOREM 6.5. If (G, Ω) is a permutation group of uncountable degree κ, and G contains an element whose support has cardinality $\lambda < \kappa$, then if $\Sigma \subseteq \Omega$ with $|\Sigma| < \kappa$, there is a proper Jordan set $\Gamma \subset \Omega$ with $\Sigma \subseteq \Gamma$ and $|\Gamma| \leqslant \max\{\lambda, |\Sigma|\}$.

Using refinements of this and results from (Adeleke and Neumann, in press *b*), many results on infinite bounded permutation groups are proved in Adeleke and Neumann (in press *a*). For example, suppose that (G, Ω) is a primitive permutation group such that for all $g \in G$, $|\mathrm{supp}(g)| \leqslant \lambda$, and

$\kappa \geqslant \lambda^{++}$. Then, by Theorem 3.2 of Adeleke and Neumann (in press*a*), G is highly transitive on Ω.

3. *Piecewise patching.* I conclude this section with some further thoughts on possible generalizations of Jordan groups. One of the most familiar examples of an infinite Jordan group is $\mathrm{Aut}(\mathbb{Q}, \leqslant)$. There is a highly developed theory of automorphism groups of total orders—see for example Glass (1981). The fact that open intervals of $(\mathbb{Q}, \leqslant)$ are Jordan sets is really a special case of the following observation:

REMARK. Suppose that $(I, \leqslant)$ is a total order with partitions into intervals $I = \bigcup(A_j : j \in J) = \bigcup(B_j : j \in J)$ such that for all $i, j \in J$ with $i < j$, if $a \in A_i$ (respectively B_i) and $b \in A_j$ (respectively B_j) then $a < b$. Then, if $f_j \colon A_j \to B_j$ $(j \in J)$ are order preserving bijections, the function $\bigcup(f_j : j \in J)$ is an automorphism of $(I, \leqslant)$.

This condition is known as the *piecewise patching* condition. It is at the heart of the analysis of conjugacy, normal subgroup structure, and subgroups of small index in such automorphism groups. Indeed, much of the theory is preserved by subgroups which are sufficiently transitive and are, in the natural sense, closed under piecewise patching—see for example Droste and Truss (1991). There is a similar notion of piecewise patching for automorphism groups of semilinear orders. This is used by Droste, Holland and Macpherson (1989*a*, 1989*b*) in the study of the normal subgroup structure and subgroups of small index of the automorphism group. Clearly, the notion can be defined also for betweenness relations and C- and D-relations. Projective and affine spaces also have a natural notion of piecewise patching. Many other structures, whose automorphism groups are never Jordan groups, have such a notion. The list includes projective and affine spaces carrying a symplectic (or orthogonal or unitary) geometry. It also includes a certain class of partial orders examined by Warren (1992), the *cycle-free* partial orders. I shall not define Warren's class here, but comment that the class includes semilinear orders. It should be possible to define a general notion of piecewise patching for a first-order structure. It would be interesting to try to classify structures admitting piecewise patching in some sense, and perhaps for such structures to produce general arguments on small index and normal subgroup structure.

7 Model-theoretic applications

The best known and most important application of Jordan groups to model theory is the Cherlin–Mills approach to classifying strictly minimal sets, decribed in Section 4. However, the definition of a Jordan group is quite a natural one for automorphism groups of saturated structures, and other applications have arisen.

1. *Model-theoretic conditions giving a Jordan set.* First, I give a slight generalization of some work from Macpherson and Steinhorn (to appear), to be described shortly.

THEOREM 7.1. Let M be a saturated structure over a countable language L, with $|M| > \aleph_0$. Suppose

(i) for all $A \subseteq M$, $\mathrm{acl}(A) = A$,
(ii) for every set $A \subseteq M$ which is A-definable, every M-definable subset of A^n is A-definable.

Then every infinite definable subset of M contains an infinite Jordan set for $\mathrm{Aut}\, M$.

REMARK. Note that condition (ii) would hold if $\mathrm{Th}(M)$ is stable, even without the assumption that A is A-definable. However, condition (i) is quite unnatural in the context of stability. For example, it is shown by Macpherson (1991) that any stable structure satisfying (i) is an indiscernible set or has a 0-definable non-trivial equivalence relation (so has imprimitive, possibly intransitive, automorphism group).

The following lemma is the main ingredient for Theorem 7.1.

LEMMA 7.2. Let M be as in Theorem 7.1, and let A be an A-definable moiety of M. Then there is a chain

$$A = A_0 \subset A_1 \subset \ldots \subset A_i \subset \ldots \qquad (i < \omega)$$

such that the following hold, where $A_\omega := \bigcup(A_i : i < \omega)$.

(i) For all $i < \omega$, A_i is an A_i-definable moiety of M.
(ii) A_ω is a moiety of M.
(iii) if $x, y \in M \setminus A_\omega$ then $\mathrm{tp}(x/A_\omega) = \mathrm{tp}(y/A_\omega)$.

Proof. Fix $b \in M \setminus A$. Let $\{\phi_i(x,\bar{y}) : i < \omega\}$ enumerate all the L-formulas with distinguished variable x. Let $f\colon \omega \to \omega$ be a function so that each $i \in \omega$ has infinite preimage. Suppose that we have constructed $A_0, \ldots, A_i$ satisfying (i) above. Let $\phi(x,\bar{y})$ be the formula $\phi_{f(i+1)}(x,\bar{y})$. By hypothesis (ii) in the theorem, there is $\bar{a}_\phi \subseteq A_i$ and a formula $d\phi(\bar{y}) \in L(\mathrm{Ran}(\bar{a}_\phi))$ such that for all $\bar{c} \subseteq A_i$, $M \models \phi(b,\bar{c})$ if and only if $M \models d\phi(\bar{c})$. Let

$$B_{i+1} := \{x \in M \setminus A_i : M \models \forall \bar{y} \subseteq A_i\, (\phi(x,\bar{y}) \longleftrightarrow d\phi(\bar{y}))\}.$$

Then put $A_{i+1} := M \setminus B_{i+1}$. Note that as A_i is A_i-definable and $\bar{a}_\phi \subseteq A_i$, the sets B_{i+1} and A_{i+1} are A_i-definable. Also, any two elements of B_{i+1} satisfy the same formulas $\phi(x,\bar{a})$ ($\bar{a} \subseteq A_i$). By saturation and the assumption that algebraic closure is trivial in M, we have part (i) of the lemma. Part (ii) follows by saturation, and part (iii) drops out of the construction. □

Proof of Theorem 7.1. Let $(A_i : i \leqslant \omega)$ be as in Lemma 7.2, and put $B := M \setminus A_\omega$. We claim that B is a Jordan set for $\operatorname{Aut} M$. This is proved by a back-and-forth argument. A typical step in the construction is the following.

Let f be a partial elementary map $M \to M$ such that

(a) $\operatorname{Dom}(f) \supseteq A_\omega$ and $f \restriction_{A_\omega} = \operatorname{id}_{A_\omega}$,

(b) $|\operatorname{Dom}(f) \setminus A_\omega| < |M|$.

Let $c \in B \setminus \operatorname{Dom}(f)$. For the back-and-forth step, we must show that there is $d \in B \setminus \operatorname{Ran}(f)$ such that $f \cup \{(c,d)\}$ is an elementary map.

Let $|\operatorname{Dom}(f) \setminus A_\omega| = \lambda$ and let $(b_\mu : \mu < \lambda)$ enumerate $\operatorname{Dom}(f) \setminus A_\omega$. Let

$$p_1 := \{\phi(x, \bar{b}) : \phi(x, \bar{y}) \in L, \bar{b} = (b_{\lambda_1}, \ldots, b_{\lambda_n}), M \vDash \phi(c, \bar{b})\},$$

where the λ_i are listed in increasing order. For each $\bar{b} \in \operatorname{Dom}(f)$ and each L-formula $\psi(\bar{x}, \bar{y})$ with $l(\bar{x}) = l(\bar{b}) + 1$ and $l(\bar{y}) > 0$, and each $i < \omega$ there is $\bar{e}_i \subseteq A_i$ and $d_{i,\bar{b}}\psi(\bar{y}) \in L(\operatorname{Ran}(\bar{e}_i))$ such that for all $\bar{m} \in A_i^{l(\bar{y})}$,

$$M \vDash \psi(c\bar{b}, \bar{m}) \longleftrightarrow d_{i,\bar{b}}\psi(\bar{m}).$$

Also by Lemma 7.2 (i), for each $i < \omega$ there is $\bar{a}_i \subseteq A_i$ and a formula $\chi_i(z, \bar{w})$ such that $A_i = \{z : M \vDash \chi_i(z, \bar{a}_i)\}$. Put

$$p_2 := \{\forall \bar{y}\, [\chi_i(y_1, \bar{a}_i) \wedge \ldots \wedge \chi_i(y_{l(\bar{y})}, \bar{a}_i) \to (\psi(x\bar{b}, \bar{y}) \leftrightarrow d_{i,\bar{b}}\psi(\bar{y}))] : i < \omega\},$$

where $\bar{b}$ and ψ range as above. Let $p = p_1 \cup p_2$. Then p determines the type $\operatorname{tp}(c/\operatorname{Dom}(f))$ and has $\operatorname{Max}(\lambda, \aleph_0)$ parameters. Since f is elementary, the set $f(p)$ obtained by replacing b by $f(b)$ for all $b \in \operatorname{Dom}(f)$ is consistent. Hence, as M is saturated, the partial type $f(p)$ has a realization $d \in M$. We now can extend f in the required way by putting $f(c) = d$. □

The hypothesis (ii) in Theorem 7.1 is reminiscent of other conditions familiar in stability theory. The connections between these conditions are examined in detail in work of Kueker and Steitz (unpublished *b*, unpublished *a*). One observation, which appears to be folklore, is the following. It follows from Corollary 1.9 of Kueker and Steitz (unpublished *b*).

THEOREM 7.3. Let M be an ω-categorical structure and let A be a moiety of M which is A-definable. Then the following are equivalent.

(i) M is prime over A;

(ii) every type of $\operatorname{Th}(M, a)_{a \in A}$ which is realized in M is isolated.

(iii) $(\operatorname{Aut} M_{\{A\}}, A)$ is a closed permutation group;

(iv) every M-definable subset of A^n $(n < \omega)$ is A-definable.

Conditions of this sort have arisen in several places recently, for example in work on covers of totally categorical structures, and in the work of Cherlin and Hrushovski (described in Hrushovski (1994)) on the coordinatization by Lie geometries of smoothly approximated structures. The proof of the above theorem is straightforward, the only non-trivial implication being (iv) implies (ii), which is handled by an argument of Mati Rubin (see the proof of Lemma 2.10 of Rubin (in press)). The conditions above are related to Theorem 7.1 by the following result of Macpherson and Steinhorn (to appear).

THEOREM 7.4. Let M be an ω-categorical structure such that $\mathrm{acl}(S) = S$ for all $S \subseteq M$, and suppose that whenever A is an infinite A-definable subset of M, M is prime over A. Then every infinite definable subset of M contains an infinite Jordan set for $\mathrm{Aut}\, M$.

PROBLEM 7.5. Obtain analogues of Theorem 7.1 under weaker assumptions (such as that the lattice of algebraically closed sets is distributive).

Questions of the above nature have led to work in several directions, not all involving Jordan groups. Kueker and Steitz (unpublished*b*, unpublished*a*) examined conditions similar to those of Theorem 7.3, usually in an uncountable and saturated setting. One question that was asked by several people was the following: if M is ω-categorical and ω-stable and $A \subseteq M$, must $\mathrm{Aut}\, M$ induce a closed permutation group on A? The answer is easily yes if A is definable, and Bouscaren and Laskowski (1993) gave an affirmative answer for any A, and indeed, with 'ω-categorical' replaced by 'countable and saturated'. Related questions were examined in Macpherson and Woodrow (1992). There it is shown, for example, that if (H, Σ) is a closed permutation group of countable degree, then there is a moiety Δ of the vertex set of the random graph Γ such that $\mathrm{Aut}\, \Gamma_{(\Delta)} = 1$ and the permutation groups (H, Σ) and $(\mathrm{Aut}\, \Gamma_{\{\Delta\}}, \Delta)$ are isomorphic. The common theme in all this is that there is a saturated structure M, and an elementary partial map $f: M \to M$ whose domain is a moiety of M, and one must try to extend f to an automorphism of M. This is much more problematical than when $|\mathrm{Dom}(f)| < |M|$.

2. *o-Minimality and variations.* Recall from Pillay and Steinhorn (1986) that a totally ordered structure $(M, <, \ldots)$ is *o-minimal* if every parameter-definable subset of M is a finite union of intervals with endpoints in $M \cup \{\pm\infty\}$. By Theorem 5.1 of Pillay and Steinhorn (1986), o-minimal structures satisfy a strengthening of Theorem 7.3 (a); indeed, if M is o-minimal and $A \subseteq M$ then $\mathrm{Th}(M, a)_{a \in A}$ has a unique prime model. This, and the rich supply of Jordan groups, suggested that it would be interesting to replace the total order in the definition of o-minimality by another relational structure arising from Jordan groups, and see whether interesting variants of o-minimality arise.

This was begun in Macpherson and Steinhorn (to appear) and continued by Haskell and Macpherson (1994). The idea is as follows. Fix a relational language L and an L-theory T (not necessarily complete—T could be the set of axioms of all total orders, or of all C-relations). Let L^+ be an extension of L, and M be an L^+-structure whose reduct to L is a model of T. We say that M is *T-minimal* if for every $N \equiv M$, every definable subset of N is quantifier-free definable in the language L. Thus, if L is empty and T is the theory of infinite sets we have strong minimality, and if L has a single binary relation and T is the theory of total orders then we have o-minimality (but this uses Theorem 0.2 of Knight *et al.* (1986), which says that the class of o-minimal structures is closed under elementary equivalence). In Macpherson and Steinhorn (to appear), possibilities considered for the theory T include circular orders, trees, the random graph, the universal poset and the C-relation. The case examined in most detail in Macpherson and Steinhorn (to appear) and Haskell and Macpherson (1994) is the C-relation. We define the notion of C-minimality in the natural way, taking T above to be the common theory of all C-relations (except that we no longer require that a C-relation must satisfy axiom C4).

With C-minimal structures, the connection with Jordan groups is rather tenuous. Unlike with o-minimality, we do not have good general prime model theorems for C-minimal structures. The only general result is a theorem of Macpherson and Steinhorn (to appear), that if $(M, C, \ldots)$ is a C-structure which is C-minimal and in which cones are Jordan sets for $\operatorname{Aut} M$, then over any set of parameters there is a prime model, and if in addition (M, C) is a *pure* C-structure then these prime models over sets are unique.

Our main reason for investigating C-minimality is that there are several nice examples. If a set has a definable chain of equivalence relations $(E_i : i < \omega)$ such that E_i refines E_j whenever $i > j$, then we can define C by requiring $C(\alpha; \beta, \gamma)$ to hold whenever there is $i \in I$ such that some E_i-class contains β and γ but not α. Under easily formulated extra hypotheses this will give a C-relation. Hence, by considering certain abelian groups with a definable descending chain of subgroups with trivial intersection, we can obtain examples of C-minimal groups, that is, groups equipped with a C-relation preserved by left and right group multiplication, whose theory is C-minimal. In Macpherson and Steinhorn (to appear), partial descriptions of C-minimal groups are obtained.

It is easily seen that if F is a field equipped with a valuation to an ordered group Γ, then there is a C-relation on F given by $C(x; y, z)$ if and only if $v(y-x) < v(y-z)$. Furthermore, addition by any field element and multiplication by non-zero elements preserves the C-relation, so we may call (F, C) a *C-field*. It is noted in Macpherson and Steinhorn (to appear) that any C-field arises from a valuation in this way. Furthermore, the affine group $F^+ \cdot F^*$ preserves this C-relation, so a C-relation which comes from a valuation always has a doubly transitive automorphism group. From this it follows easily that the union of any set of cones at a node is a Jordan set for the automorphism group of the pure C-structure. These observations, together with 2.1 in Ch.VIII of Shelah (1978) give one way of seeing that for any uncountable κ there are 2^κ non-isomorphic C-structures of size κ whose automorphism groups are doubly transitive Jordan groups.

It is shown in Macpherson and Steinhorn (to appear) that if (F, v, Γ) is an *algebraically closed* non-trivially valued field, then the corresponding C-field is C-minimal. The converse, that any C-minimal C-field is algebraically closed, is also true. This is proved in Haskell and Macpherson (1994), via a cell-decomposition theorem for definable subsets of M^n for any C-minimal structure M.

There is also a connection with the D-relation, as was noticed in a conversation with Haskell (and, independently, by Adeleke and Neumann). If an extra element is added to a field you obtain, in some sense, the projective line over the field. If you add a point to a C-relation you can obtain a D-relation. It is now easy to see how to define a D-relation on the projective line over a valued field. Put $D(\alpha, \beta; \gamma, \delta)$ if one of the following holds:

1. $\alpha, \beta, \gamma, \delta$ are distinct and the cross-ratio $\mathrm{cr}(\alpha, \gamma; \beta, \delta)$ lies in the maximal ideal of the valuation ring;
2. $\alpha, \beta, \gamma, \delta$ are not all distinct, but $\{\alpha, \beta\} \cap \{\gamma, \delta\} = \varnothing$.

Here, cross-ratio is as defined on p. 41 of Hughes and Piper (1973): if P_1, P_2, P_3, P_4 are four distinct points on the projective line with parametric coordinates t_1, t_2, t_3, t_4 then the cross-ratio of the ordered quadruple (P_1, P_2, P_3, P_4) is $(t_3 - t_1)(t_4 - t_2)/(t_4 - t_1)(t_3 - t_2)$. Over, for example, the p-adics, this construction enables one to recover the natural tree on which $\mathrm{SL}(2, \mathbb{Q}_p)$ acts (see Serre (1980) for a full account).

3. *Further classification.* Finally, I consider a couple of possible extensions of the classification of primitive Jordan groups, motivated by model theory.

PROBLEM 7.6. Classify primitive closed oligomorphic Jordan groups.

PROBLEM 7.7. Classify saturated stable structures whose automorphism groups are primitive Jordan groups.

Neither question is motivated by a very specific model-theoretic question, though the first is suggested by Theorem 7.4. We consider both questions by running down the list in Theorem 2.2.

On Problem 7.6 we have very little information. First, consider oligomorphic primitive Jordan groups of geometric type. The obvious examples are infinite dimensional projective and affine spaces over finite fields. There are proper expansions of some of these examples admitting oligomorphic

2-transitive Jordan groups, as noted in Section 2, Remark 5. However, we do not know of any essentially different geometric examples. Note, here, that if (G, Ω) is oligomorphic and preserves a Steiner t-system with *finite* lines, then the algebraic closure of any sufficiently non-degenerate finite set is a finite Steiner t-system with lines of the same size.

All the familiar relational structures admitting Jordan groups, such as those associated with linear orders, and semilinear orders, general betweenness relations and C- and D-relations, yield oligomorphic examples, and a complete classification seems quite feasible. Indeed, some partial results have already been obtained by Adeleke and Neumann. It is not known if there are oligomorphic examples which are limits of betweenness relations, D-relations, or Steiner systems.

We now turn to Problem 7.7. Here, Hrushovski's construction of a strongly minimal set (see Wagner's paper in this book, and also Section 5 above) suggests that there is no hope of a classification in the geometric case. It is clear that if any of the structures of types (a)–(h) in Theorem 2.2 is interpretable in a structure, then it is unstable. For example, if (M, C) is a C-relation, then it is possible to interpret in M a semilinear order. Essentially, this lives on M^2/E for some 0-definable equivalence relation E on M^2; see Adeleke and Neumann (to appear) for details. It turns out that for primitive Jordan groups of relational type, that is, those not of geometric type, we have the strongest possible answer to Problem 7.7.

THEOREM 7.8. Let M be a saturated stable structure over a countable language, and suppose that $\mathrm{Aut}\, M$ is a primitive but not highly transitive Jordan group. Then $(\mathrm{Aut}\, M, M)$ is of geometric type.

I shall not give full details of the proof, as the arguments are all very similar, but I shall do the hardest cases. Suppose first that case (g) of Theorem 2.2 holds, so G preserves a C-relation on M. In the simply primitive case, we have by Theorem 3.1.1 of Adeleke and Macpherson (1994) (see also Theorem 3.2 above) that all cones are Jordan sets. In the doubly transitive case, by Remark 4 in Section 2, we may suppose that either all cones are Jordan sets, or, for any node of the semilinear order, the set of all chains not passing through that node is a Jordan set. First suppose that all cones are Jordan sets. We claim that for any finite $A \subset M$, $\mathrm{acl}(A) = A$; for if $x \in M \setminus A$, then some cone contains x but is disjoint from A, so as the cone is a Jordan set, $x \notin \mathrm{acl}(A)$. Given this claim, we have a contradiction, for by Proposition 1.3 of Macpherson (1991), any stable structure with no 0-definable equivalence relation and with trivial algebraic closure in this sense is an indiscernible set.

Next, suppose that some (and hence all) cones are not Jordan sets, so 'upper sections' of chains are Jordan sets. Fix $a \in M$. By the above Jordan property, all pairs (x_1, x_2) which satisfy $C(x_1; x_2, a)$ lie in the same G_a-orbit, so have the same type over a, $p(x_1, x_2, a)$ say. However, since $C(x_1;x_2,a) \longrightarrow \neg C(x_2;x_1,a)$, if $C(x_1;x_2,a)$ then $\mathrm{tp}(x_1,x_2,a) \neq \mathrm{tp}(x_2,x_1,a)$. Hence there is a formula $\phi(x_1,x_2,a)$ such that $\phi(x_1,x_2,a) \wedge \neg\phi(x_2,x_1,a) \in p(x_1,x_2,a)$. The formula $\phi(x_1,x_2,a)$ is an unstable formula.

Two other points in the proof need to be made. First, if $(\mathrm{Aut}\, M, M)$ is 2-primitive and preserves a D-relation or a limit of general betweenness relations or D-relations, then a one point stabilizer is a primitive Jordan group preserving a C-relation, so by the above argument M is unstable; similarly, if $\mathrm{Aut}\, M$ is k-transitive ($k \geqslant 3$) and preserves a limit of Steiner $(k-1)$-systems, then the stabilizer of $k-2$ points is a doubly transitive group of automorphisms of a C-relation. Second, if $\mathrm{Aut}\, M$ is 2-transitive, not 2-primitive and preserves a D-relation but no Steiner system or relational structure of other type, then by the proof of Theorem 5.4 of Adeleke and Neumann (in press$b$), G has no proper primitive Jordan set. Hence, by Theorem 3.3, the following holds: for any proper Jordan set A and any distinct $x, y, z \in M$ there is $g \in G$ such that Ag contains x and excluding y and z. Using this together with the structure of the D-relation, it follows that given finite $A \subset M$ and $b \in M \setminus A$, there is a Jordan set containing b and disjoint from A. Hence, $\mathrm{acl}(A) = A$ for all $A \subset M$. This case is now eliminated as with the C-relation above.

8 Concluding remarks

I consider in this section several topics peripherally related to Jordan groups.

1. *The 'forth' part of the back-and-forth construction.* I first consider the question, raised by Mathias and discussed by Cameron (1990), of when 'forth suffices' for a countable structure M. When Cantor proved that any two countable dense total orders without endpoints are isomorphic, he did not give the now usual back-and-forth argument, but just gave the 'forth' part of the proof. More precisely, he argued as follows. Let $(M, \leqslant)$, $(N, \leqslant)$ be countable dense total orders without endpoints and let $(a_i : i < \omega)$, $(b_i : i < \omega)$ enumerate their respective domains. Define an injection $\phi\colon M \to N$ inductively. At the ith stage we have defined $\phi(a_0), \ldots, \phi(a_i)$, and we define $\phi(a_{i+1})$ to be b_n where n is least such that $\mathrm{tp}(b_n, \phi(a_0), \ldots, \phi(a_i)) = \mathrm{tp}(a_{i+1}, a_0, \ldots, a_i)$. The construction clearly embeds M into N, and it turns out that in this particular case ϕ is also surjective. Cameron's question (raised earlier by Mathias) was: for which countable ω-categorical structures M does the above construction, starting with *any* two enumerations of M, give an automorphism of M? If the construction gives an automorphism for all enumerations of M, we say that *forth suffices* for M. In fact, as mentioned by Cameron (1990), the property can be defined (in terms of orbits) for any closed permutation group acting on a countably infinite set. The countability assumption can also

be dispensed with, and we can ask whether forth suffices for an arbitrary saturated structure. Cameron (1990) gives a necessary condition for forth to suffice, and a sufficient condition, but unfortunately there is a gap between them. It emerges from his work that there is a strong connection with Jordan groups. For example, if (G, Ω) is a closed permutation goup of countably infinite degree such that for every finite $\Gamma \subseteq \Omega$, each orbit of $G_{(\Gamma)}$ is a Jordan set for (G, Ω) (possibly singleton or improper), then forth suffices for (G, Ω). The primitive groups satisfying this condition have been classified by Neumann (see Cameron (1990)) and all are highly transitive or preserve one of: a linear order, circular order, linear betweenness relation or separation relation, a C- or D-relation, an affine space over a finite field, a projective space over $GF(2)$. There is also a partial reduction of permutation groups with this condition to primitive groups.

McLeish (1994) has extended some of these results. He has sharpened Cameron's necessary and sufficient conditions, shown that forth suffices for countable saturated structures of Morley rank one but that it doesn't for certain ω-categorical structures of Morley rank two, and also examined saturated o-minimal structures for which forth suffices. To the best of my knowledge, for all the known examples of structures for which forth suffices (other than indiscernible sets) the automorphism group is contained in a Jordan group which is not highly transitive.

2. *Highly transitive Jordan groups.* In the classification of infinite Jordan groups, nothing is said about highly transitive Jordan groups. So far, little has been done on this, though several examples are given in Adeleke and Neumann (in press*b*). These include the following: finitary symmetric groups (or more generally bounded symmetric groups); homeomorphism groups of homogeneous 0-dimensional topological spaces (for which the open sets are Jordan sets); homeomorphism groups of d-dimensional path-connected manifolds without boundary, for $d \geqslant 2$ (for $d = 1$ we just recover the group preserving the linear betweenness relation or the separation relation on the reals).

3. *Other properties of relational structures associated with Jordan groups.* The familiar relational structures such as semilinear orders, general betweenness relations, and C- and D-relations had already arisen in permutation group theory before the recent work on infinite Jordan groups. Automorphism groups of semilinear orders were considered in Droste (1985). In his work on partial orders with strong transitivity properties, Droste classified countable 2-homogeneous semilinear orders (see Section 2 above). Independently, Cameron (1983*b*) examined permutation groups whose degree of transitivity is exceeded by their degree of homogeneity, and certain Jordan groups arise here. There are several variations on Cameron's theorem that any highly homogeneous but not highly transitive permuta-

tion group of infinite degree preserves or reverses a linear or circular order (Theorem 2.1 above). For example, there is the result of John MacDermott (see Cameron (1976)) that any 3-homogeneous but not 2-transitive group preserves a linear order. Similarly, any 3-homogeneous, 2-transitive but not 2-primitive permutation group of infinite degree preserves a linear betweenness relation or a C-relation derived from a semilinear order in which the number of cones at each node is two (Cameron, 1983*b*). And the transitive extension of this last group, which preserves a D-relation, can also be characterized, by the following result of Macpherson (1986*b*): any 5-homogeneous, 3-transitive, not 3-primitive permutation group of infinite degree preserves a separation relation or a D-relation induced from a C-relation of the above type. The proofs of these results have much in common, and they are discussed in more detail by Cameron (1990). The key point is that if (G, Ω) is k-homogeneous but not k-transitive then by Ramsey's Theorem there is a subset $\{a_i : i < \omega\}$ such that if $i_1 < \ldots < i_k$ and $j_1 < \ldots < j_k$ then $(a_{i_1}, \ldots, a_{i_k})$ and $(a_{j_1}, \ldots, a_{j_k})$ lie in the same G-orbit. If G is also $(k+1)$-transitive, this gives us extra information on the structure induced by G on a $(k+1)$-set, and this enables one often to axiomatize an invariant relational structure. Using these arguments together with the classification of finite simple groups and an extensive case-by-case analysis, it was shown (Macpherson, 1986*b*) that if $k \geqslant 5$ and G is $(k-1)$-transitive but not k-transitive then G is not $(k+3)$-homogeneous.

I now mention two other, possibly related, contexts in which the treelike structures arise. The first involves growth rates for orbits on k-sets, and is discussed by Cameron (1990). If (G, Ω) is an oligomorphic permutation group, let $f_k(G)$ denote the number of G-orbits on the set of k-sets for each positive integer k. It is well known that the sequence (f_k) is non-decreasing in k, and it was shown (Macpherson, 1985) that if G is primitive but not highly homogeneous then the sequence grows at least exponentially. The question is, for which such groups is the growth no faster than exponential? All the known examples are subgroups of closed Jordan groups which are not highly transitive. In particular, there are semilinear orders, general betweenness relations and C- and D-relations which have growth no faster than exponential. Some of these examples are discussed in detail by Cameron (1987), but we do not have a classification of such examples.

PROBLEM 8.1. Is it true that if (G, Ω) is an oligomorphic primitive but not highly transitive permutation group realizing growth no faster than exponential, then there is some Jordan group which is not highly transitive and contains G?

I close with a possibly related question. If M is a homogeneous structure (in Fraïssé's sense) over a finite relational language then one can regard its age as a partially ordered set, ordered under inclusion. It is of interest to pin down the cases in which the age has no infinite antichains. I will not give details here, but this condition is connected with a number of other conditions on M, such as the following: G-finiteness of M; growth rate of algebraic closure in M; nice enumerations of M. I pose the following problem. The examples suggest it has a positive answer.

PROBLEM 8.2. Is it true that if M is homogeneous over a finite relational language and its age has no infinite antichains, then $\operatorname{Aut} M$ is contained in a Jordan group which is not highly transitive?

In fact, the condition we are really interested in is a slight refinement of the above. Let L be the language of M, and let c be a constant symbol not in L. Let $\mathscr{A}$ be the class of all structures isomorphic to finite substructures of M, expanded by interpreting c in all possible ways. Then the condition that $\mathscr{A}$ has no infinite antichains is useful, and implies, for example, that every enumeration of M is nice (in the sense of Camina and Evans (1991) and Ahlbrandt and Ziegler (1986)). Problem 8.2 could also be raised for this class $\mathscr{A}$. These notions were also examined by Pouzet (1978).

I conclude the paper with an open-ended problem.

PROBLEM 8.3. Investigate further those closed Jordan groups of type (j), that is, those which preserve limits of general betweenness relations, D-relations, and Steiner systems. In particular, find 'natural' structures of which they are the automorphism groups.

The structure of totally categorical structures

Wilfrid Hodges

School of Mathematical Sciences,
Queen Mary and Westfield College,
Mile End Road,
London,
E1 4NS,
U.K.
Electronic mail address: W.Hodges@qmw.ac.uk

Among the many programmes that go to make up research in model theory today, one of the most elegant is the programme *to catalogue all the countable models of totally categorical theories.* Morley (1965, 1967) and Baldwin and Lachlan (1971) laid the foundations. Between about 1976 and 1984, Zil'ber made enormous progress, but a large part of his work was unavailable in English until the recent translation of his doctoral thesis (Zil'ber, 1993). In the meantime, hints were published (Zil'ber, 1979, 1984*b*, and elsewhere), enough for workers in the West to carry the work forward. Hrushovski, Ahlbrandt and Ziegler all published papers that cast fresh light. More is on its way into print, and indeed I know that there are people who know many things not yet known to me. So this account makes no claim to completeness. I aim only to tell a good story.

Most of the story is available in other places. Zil'ber's own fullest account is in Zil'ber (1993). Hrushovski (1989*c*) gave a much briefer report, but with many insights. Here I have concentrated on things that can be said in terms of automorphism groups, and I have added some recent news on covers. I warmly thank the editors for their advice and patience, and David Evans for various improvements.

1 The structures to be classified

Henceforth L is a fixed countable first-order language and T is a complete theory in L. We say that T is *κ-categorical* if T has up to isomorphism exactly one model of cardinality κ. A *κ-categorical* structure is by definition an infinite structure whose complete theory is κ-categorical. *Totally categorical*, or *t.c.* for short, means κ-categorical for all infinite cardinals κ.

T.c. theories are interesting because they come as close as possible to the ideal of pinning down an infinite structure by first-order statements about the structure. They represent the closest possible liaison between syntax and semantics. So one suspects that their models will form a particularly tidy collection.

In fact we can recognize t.c. theories purely from their countable models, thanks to the following theorem. (Here and henceforth, *countable* means of cardinality $\aleph_0$. A *definable k-ary relation* in a structure M is a set $\phi(M^k)$ where $\phi(x_0, \ldots, x_{k-1})$ is a formula of L, possibly with parameters from M, and $\phi(M^k)$ means $\{\bar{a} : M \models \phi(\bar{a})\}$; if there are no parameters in the formula, we say $\varnothing$*-definable.* A *definable set* is a definable unary relation. In this and other notation I follow Hodges (1993).)

THEOREM 1.1. Suppose M is a countable structure. Then M is t.c. if and only if M has the following three properties.

Svenonius property. The automorphism group $\mathrm{Aut}(M)$, considered as a permutation group on the domain $\mathrm{Dom}(M)$ of M, is oligomorphic (i.e., for every positive integer n, the number of orbits of $\mathrm{Aut}(M)$ in its induced action on n-tuples of elements of $\mathrm{Dom}(M)$ is finite).

Non-two-cardinal property. If X is an infinite definable set in M, then every elementary embedding $e: M \to M$ which is the identity on X is an isomorphism.

Finite rank property. Given a positive integer k and a definable k-ary relation $X = X_0$ in M, we define a cut-and-choose game for players $\forall$ and $\exists$: at step 0, player $\forall$ finds countably many pairwise disjoint definable relations which are subsets of X_0, say $X_0^0, X_0^1, \cdots$, and player $\exists$ replies by choosing one of them, say $X_1 = X_0^{i_0}$; at step 1, player $\forall$ finds countably many pairwise disjoint definable subsets of X_1, say $X_1^0, X_1^1, \cdots$, and player $\exists$ replies by choosing one of them, say $X_2 = X_1^{i_1}$; etc. Player $\exists$ wins when she chooses a relation which is empty. The condition on M is that for each definable relation X in M there is a finite m such that player $\exists$ has a strategy which will win the game for her by the end of step m. The least such m is called the *rank* of the relation X.

Sketch proof. The Svenonius property is equivalent to categoricity in cardinality $\aleph_0$, and it implies that M is saturated. The rank defined by the game is Cantor–Bendixson rank, but it is equal to the better known but more complicated Morley rank whenever the structure is saturated. A theorem of Baldwin and Lachlan (1971) states that a countable complete theory is categorical in uncountable cardinalities if and only if its countable models have the non-two-cardinal property and all their relations have Morley rank; Baldwin and Zil'ber independently showed that the rank must be finite. □

By a theorem of Hrushovski (1989*b*), every totally categorical theory T is definitionally equivalent to a single sentence together with countably many sentences which say that there are infinitely many elements. This confirmed a conjecture of Vaught. In fact part of the impetus to classify the models of t.c. theories was to prove this conjecture of Vaught. Hrushovski's argument sidestepped the classification; but it would still be interesting if the classification led to a proof of the conjecture that doesn't involve the heavy combinatorics of Hrushovski's argument. (A forthcoming book by Pillay has a chapter analysing Hrushovski's proof.)

Hrushovski's result immediately implies the following theorem.

THEOREM 1.2. Up to choice of language, there are only countably many totally categorical theories.

This increases our chances of finding a manageable classification.

2 Examples

At this point we can call a halt for some examples. In each case we describe the unique countable model M.

EXAMPLE 2.1. M is a countable set Ω with no further structure. $\text{Aut}(M)$ is the symmetric group $\text{Sym}(\Omega)$ on Ω.

EXAMPLE 2.2. We write F_q for the finite field with q elements. M is the vector space of dimension ω over F_q. The elements of M are the vectors; the symbols of M are the abelian group symbols and for each element σ of the field, a 1-ary function symbol S_σ to represent scalar multiplication by σ. $\text{Aut}(M) = \text{GL}(\omega, F_q)$.

EXAMPLE 2.3. Write V for the vector space of Example 2.2. The projective space M is formed by taking as elements the one-dimensional subspaces of V, and adding symbols for the orbits of the projective linear group $\text{PGL}(\omega, F_q)$ on n-tuples of elements, for each $n < \omega$. For example if σ is an element of F_q, then there will be a 4-ary relation symbol to pick out the 4-tuples whose harmonic ratio is σ. $\text{Aut}(M) = \text{PGL}(\omega, F_q)$.

EXAMPLE 2.4. Write P for the projective space of Example 2.3. The projective geometry M is formed as follows. The elements of M are the same as those of P. For each positive integer n there is an $(n+1)$-ary relation R_n:

$$R_n(a_0, \ldots, a_n) \Leftrightarrow a_n \text{ is in the projective flat spanned by } a_0, \ldots, a_{n-1}.$$

Unlike P, M is unaffected by automorphisms of the field F_q. $\text{Aut}(M)$ is the projective semilinear group $\text{P}\Gamma\text{L}(\omega, F_q)$.

EXAMPLE 2.5. Let V be the vector space of Example 2.2. We form the affine space M: the elements of M are those of V, but instead of the vector space operations we have relations expressing

$$a_0 + a_1 = a_2 + a_3,$$

$$\sum_{i<n} \sigma_i a_i = 0 \quad (\text{where } \sum_{i<n} \sigma_i = 0).$$

$\mathrm{Aut}(M)$ is the affine linear group $\mathrm{AGL}(\omega, F_q)$.

EXAMPLE 2.6. This example M comes from Example 2.5 by exactly the same process that led from Example 2.3 to Example 2.4, except that we use affine flats instead of projective flats. Then M is an affine geometry. $\mathrm{Aut}(M)$ is the affine semilinear group $\mathrm{A\Gamma L}(\omega, F_q)$.

EXAMPLE 2.7. A finite group G is given, and for each $i < \omega$, G_i is a new copy of G. The symbols of L are 1-ary function symbols F_g ($g \in G$). The domain of M is the union of the G_i ($i < \omega$), and if g' in G_i is the element corresponding to g in G, and h is in G, then $F_h^M(g')$ is the element in G_i corresponding to hg. $\mathrm{Aut}(M)$ is the wreath product $G \,\mathrm{Wr}\, \mathrm{Sym}(\omega)$.

EXAMPLE 2.8. A countable t.c. structure N is given, together with a positive integer k. The domain of M is the set of all k-element subsets of $\mathrm{Dom}(N)$. The relations on M are those induced from N; they include a binary relation R^M which relates a to b if and only if $a \cap b \neq \varnothing$. We can use R to recover the elements of N, coding them by pairs of elements of M whose intersection is a singleton. So $\mathrm{Aut}(M) = \mathrm{Aut}(N)$.

EXAMPLE 2.9. M is an abelian group of the form $A + B$ where A is a finite abelian group and B the direct sum of countably many cyclic groups of order p^n, for some prime p and some positive integer n.

Some more examples will appear in Section 8.

3 How to classify?

When we catalogue 'up to choice of language', we have some freedom about how much of the language to throw away. It's a matter of defining an equivalence relation between structures, so that two equivalent structures will be regarded as the same for purposes of the catalogue. There are two natural options that are worth mentioning. Let M and N be countable t.c. structures.

(1) We can count M and N as equivalent if there is a bijection β from $\mathrm{Dom}(M)$ to $\mathrm{Dom}(N)$ which takes $\varnothing$-definable relations to $\varnothing$-definable relations (in both directions). Thanks to ω-categoricity, these bijections β are exactly those which induce an isomorphism of automorphism groups, $\mathrm{Aut}(M) \cong \mathrm{Aut}(N)$, as permutation groups. For permutation group theorists this is quite a comfortable level of equivalence. When M and N are equivalent in this sense, we say they are *bi-definable*, and we call the map β a *bi-definition* from M to N.

(2) We can count M and N as equivalent if $\mathrm{Aut}(M)$ and $\mathrm{Aut}(N)$ are isomorphic as topological groups. The topology is defined as follows: a

subset J of $\mathrm{Aut}(N)$ is open if and only if J contains the pointwise stabilizer $\mathrm{Aut}(N)_{(X)}$ of some finite set X of elements of N. (This topology has the property that a subgroup G is closed if and only if $G = \mathrm{Aut}(M)$ for some expansion M of N—we shall use this fact. Also every open subgroup is closed, but in general not vice versa.)

Ahlbrandt and Ziegler (1986) showed that $\mathrm{Aut}(M)$ and $\mathrm{Aut}(N)$ are isomorphic as topological groups if and only if there are interpretations Γ of M in N and Δ of N in M, both without parameters, such that the natural isomorphisms $M \cong \Gamma\Delta(M)$ and $N \cong \Delta\Gamma(N)$ are definable in M and N respectively (again without parameters). (See the section starting from p. 26 in this volume for more details.) When M and N are related in this way, we say that M and N are *bi-interpretable*. If M and N are bi-interpretable countable structures and one is t.c., then so is the other.

Bi-definability implies bi-interpretability. One obvious weakening of bi-interpretability is for M and N to be each interpretable in the other, but without the further condition on the natural isomorphisms. In practice this turns out to be a less natural equivalence; for example it need not preserve the property of being t.c. I shall not pursue it.

Another weaker relation, or so it seems at first sight, is that $\mathrm{Aut}(M)$ and $\mathrm{Aut}(N)$ are isomorphic as abstract groups. However, the following theorem of Lascar (Hodges *et al.*, 1993) will show that in fact this is exactly the same relation as (2). We say that a countable structure N has the *small index property* if the open subgroups of $\mathrm{Aut}(N)$ are exactly those of at most countable index in $\mathrm{Aut}(N)$. If N has the small index property, then the topology on $\mathrm{Aut}(N)$ can be recovered from $\mathrm{Aut}(N)$ as abstract group.

THEOREM 3.1. Countable t.c. structures have the small index property.

Let us examine and compare these two equivalence relations (1) and (2) for a moment.

One can show that an oligomorphic permutation group G on a countable set Ω is the automorphism group of some ω-categorical structure with domain Ω if and only if it is closed in the symmetric group $\mathrm{Sym}(\Omega)$ on Ω. (The topology on $\mathrm{Sym}(\Omega)$ is defined just as for automorphism groups; $\mathrm{Sym}(\Omega)$ is the automorphism group of the set Ω regarded as a structure.) The required ω-categorical structure can be found by introducing relation symbols for the orbits on n-tuples, for each positive integer n; moreover this structure differs only in language from all the other structures whose automorphism group is G. So the structure can be reconstructed very easily from its automorphism group as a permutation group. The non-two-cardinal property and the finite rank property can be translated routinely into group-theoretic properties of G, but these properties are artificial from a group-theoretic point of view. It would be very interesting to find a purely 'group-theoretic' characterization of those permutation groups which are automorphism groups of t.c. structures. But pessimistically, it may be necessary to complete the classification in order to get such a characterization.

Turning to (2), a topological group is the automorphism group of a countable structure provided that the topology is Hausdorff and complete, and the family of open neighbourhoods of the identity has a countable base of open subgroups, each of which has index at most ω (see Hodges, 1989 or the section from p. 22 in this volume).

When we reconstruct the structure from its topological automorphism group, we have to invent the elements. The natural way to do this is to take the elements to be the left cosets of the basic subgroups (here we take left cosets because we let automorphisms act on the right). But clearly there is nothing unique about this choice of elements; for example if M is any structure, then M is bi-interpretable with another structure M^+ got from M by adding a rigid finite structure at the side of M. Or indeed we can replace M by a structure whose elements are the m-element subsets of M, as in Example 2.8 in Section 2 above.

More important, we can take any $\varnothing$-definable equivalence relation E on the elements of M, and form a structure M^E by adding to M all the equivalence classes of E to M as new elements, with the help of a relation symbol to tie each equivalence class to its elements. The structure M^E is bi-interpretable with M. Structures got from M by iterating this device finitely often are called *finite slices of* M^{eq}. (The reference is to Shelah's M^{eq}, but for our purposes it is best to stick to finite slices—M^{eq} itself is not even ω-categorical.) This construction is extremely useful for the structure classification, because stability theorems often refer to elements of M^{eq}, and it greatly simplifies our work if we can assume that these elements lie already in the structure M.

This is a strong argument for beginning the classification at the level of equivalence (2).

4 Structures and automorphism groups

Henceforth we shall be working with equivalence (2). In other words, we shall be classifying t.c. structures up to their topological automorphism groups. By Theorem 3.1 this is the same thing as working with the abstract automorphism groups, since the topology is easily recoverable.

As noted earlier, we shall make automorphisms act on the right. We shall need some translations between groups and structures.

THEOREM 4.1. The open subgroups of $\mathrm{Aut}(M)$ are exactly the subgroups of the form $\mathrm{Aut}(M)_{(a)}$ with a in M^{eq}.

Proof. We already know that the pointwise stabilizer of an element is open. In the other direction, suppose G is an open subgroup of $\mathrm{Aut}(M)$. Then G contains some basic open subgroup $\mathrm{Aut}(M)_{(\bar{b})}$. Consider the pairs $(\bar{b}, \bar{b}\gamma)$

with $\gamma \in G$. These pairs lie in a finite number of orbits under $\mathrm{Aut}(M)$; choose $(\bar{b}, \bar{b}\gamma_i)$ $(i < m)$ as representatives of the orbits. Note that $G = \bigcup_{i<m} \gamma_i \,\mathrm{Aut}(M)_{(\bar{b})}$.

Let R be the relation $\{(\bar{b}\alpha, \bar{b}\gamma_i\alpha) : i < m, \alpha \in \mathrm{Aut}(M)\}$. Then R is a union of orbits of $\mathrm{Aut}(M)$, so that it is $\varnothing$-definable in M. Also one easily checks that R is an equivalence relation, using the fact that G is a group. Let c in M^{eq} be the R-equivalence class of $\bar{b}$. Then $\mathrm{Aut}(M)_{(c)} = \bigcup_{i<m} \gamma_i \,\mathrm{Aut}(M)_{(\bar{b})} = G$. □

THEOREM 4.2. If M is a countable t.c. structure and $\bar{a}$ is a tuple of elements of M, then the structure $(M, \bar{a})$ got by adding names for the elements of $\bar{a}$ is also a t.c. structure, and its automorphism group is the open subgroup $\mathrm{Aut}(M)_{(\bar{a})}$. Moreover every open subgroup of $\mathrm{Aut}(M)$ is the automorphism group of a structure formed from M in this way (but using a finite slice of M^{eq} if necessary).

Proof. This all follows from the discussion above, except for the statement that $(M, \bar{a})$ is t.c. One can verify the Svenonius property directly, and adding parameters doesn't affect either the non-two-cardinal property or the finite rank property. □

If G is the pointwise stabilizer $\mathrm{Aut}(M)_{(c)}$ of an element c and α is an automorphism of M, then $\alpha^{-1}G\alpha$ is the pointwise stabilizer of $c\alpha$. Hence the pointwise stabilizer of the orbit $c\,\mathrm{Aut}(M)$ of c is the intersection $\bigcap_{\alpha \in \mathrm{Aut}(M)} \alpha^{-1}G\alpha$ of all conjugates of G; group theorists refer to this normal subgroup of $\mathrm{Aut}(M)$ as the *core* of G in $\mathrm{Aut}(M)$.

Suppose now that X is a $\varnothing$-definable subset of M or of some finite slice of M^{eq}. We form a structure $M|X$: the domain of $M|X$ is X, and the relations of $M|X$ are the restrictions to X of the $\varnothing$-definable relations of M. We call $M|X$ the *restriction* of M to X. We refer to structures formed from M in this way as *induced $\varnothing$-definable substructures* of M.

This definition calls for three comments. First, the elements of X may come from M^{eq} rather than M. So strictly we should say 'induced $\varnothing$-definable substructure of a finite slice of M^{eq}'; but I prefer to save ink. Second, even if X is a definable set in M, $M|X$ need not be exactly a substructure of M—for example if M has functions taking elements of X to elements outside X. This will never matter. Third, every induced $\varnothing$-definable substructure of M is a structure interpretable in M (without parameters); but the converse is far from true. For example we can interpret in M a structure N which consists of the domain of M with no relations or functions; N will generally not be an induced $\varnothing$-definable substructure of M. The next theorem fails badly if we only require $M|X$ to be interpretable in M, or even a reduct of M.

THEOREM 4.3. Let M be a countable t.c. structure and let $M|X$ be an induced $\varnothing$-definable substructure of M. Then $M|X$ is a countable t.c.

structure. Moreover there is an induced surjective group homomorphism $\mu\colon \mathrm{Aut}(M) \to \mathrm{Aut}(M|X)$. The map μ is continuous and open, and its kernel is the intersection of the cores of $\mathrm{Aut}(M)_{(c_i)}$ where c_i $(i < m)$ are representatives of the finitely many orbits of $\mathrm{Aut}(M)$ whose union is the set X.

Proof. Replacing M by a finite slice of M^{eq} if necessary, we can assume that X is a $\varnothing$-definable set in M. Then every automorphism of M fixes X setwise and restricts to an automorphism of $M|X$; μ is the restriction map, and it is clearly a homomorphism. Since automorphisms of M restrict to $M|X$, $M|X$ has the Svenonius property and hence is ω-categorical. Since $M|X$ is interpretable in M, it has the finite rank property. The proof that $M|X$ inherits the non-two-cardinal property is more involved, but it rests on the following fact: if $e\colon M|X \to M|X$ is an elementary embedding, then every element of $M|X$ which is algebraic over $\mathrm{Im}(e)$ in the sense of M is already in $\mathrm{Im}(e)$. This fact holds because the relation in M which makes the element algebraic is expressible in $M|X$ too.

To say that μ is surjective is to say that every automorphism of $M|X$ extends to an automorphism of M. When X is finite, this property follows from the ω-categoricity of M by a back-and-forth argument. When X is infinite, we use model theory to show that every automorphism β of $M|X$ can be extended to an elementary embedding $\alpha\colon M \to M'$ where M' is some countable elementary extension of M; but then M' is isomorphic to M, and so the non-two-cardinal property implies that α is an isomorphism, which induces an automorphism of M extending β. Continuity of μ is easy from the definitions. The fact that μ is open is less easy; see Hodges and Pillay (in press, Lemma 5) or the section starting on p. 22 in this volume. The description of the kernel of μ follows from the remarks before the theorem. □

This theorem has a converse. If K is the intersection of the cores of finitely many open subgroups of $\mathrm{Aut}(M)$, then K is a closed normal subgroup of $\mathrm{Aut}(M)$. Thus we can form the group $\mathrm{Aut}(M)/K$ with the induced topology, so that the natural homomorphism μ from $\mathrm{Aut}(M)$ to $\mathrm{Aut}(M)/K$ is continuous and open with K as kernel. It can be verified that $\mathrm{Aut}(M)/K$ is the automorphism group of the induced substructure $M|X$ where X is the $\varnothing$-definable set of all elements (of M^{eq} if necessary) which are fixed by every element of K.

The theorem also has another kind of converse. Instead of inferring that $M|X$ is t.c. from the fact that M is t.c., we can sometimes go in the other direction.

THEOREM 4.4. Suppose M is a countable structure, N is a substructure of M whose domain is $\phi(M)$ for some formula $\phi(x)$ of L, and every automorphism of N extends to an automorphism of M. Suppose also that whenever

M' is a structure elementarily equivalent to M with $\phi(M')$ countable, then M' is also countable. If N is t.c. then M is t.c.

Proof. Hodges and Pillay (in press, Theorem 10). □

Very recently Kikyo and Tsuboi (in press) made a thorough study of this theme.

5 The strictly minimal set

A totally categorical structure M is said to be *strictly minimal* if it is infinite and its automorphism group $\mathrm{Aut}(M)$ has the two properties:

1. $\mathrm{Aut}(M)$ acts transitively and primitively on $\mathrm{Dom}(M)$,
2. if X is a finite set of elements of M then $\mathrm{Aut}(M)_{(X)}$ has one infinite orbit and at most a finite number of finite orbits.

We say that the finite set X is *closed* if $\mathrm{Aut}(M)_{(X)}$ is transitive on the complement of X; the *closure* of a finite set is the smallest closed set containing it. It is not hard to show that every t.c. strictly minimal structure is strongly minimal, and the algebraic closure of a finite set is its closure. In the other direction, if A is a strongly minimal t.c. structure, then we can turn A into a strictly minimal t.c. structure by 'projectivizing', i.e., removing the 'origin' (the elements in the closure of the empty set), and then identifying any two elements if they 'determine the same line through the origin' (i.e., if their singletons have the same closure).

A set X of elements of a strictly minimal structure is *closed* if X contains the closure of every finite subset of X. The closed sets form a lattice known as the *geometry* of the structure.

Example 2.1 in Section 2 above is strictly minimal. All its subsets are closed, so its geometry is a boolean algebra; this case is said to be *degenerate*.

Example 2.4 (the projective geometry) is strictly minimal; the closed sets are exactly the projective flats. The lattice is not boolean, but like a boolean algebra it is modular. Notice that we can recover the field F_q from the geometry: each projective line contains $q+1$ points, so that the sup of any two distinct atoms in the geometry has exactly $q+1$ atoms below it.

Example 2.6 (the affine geometry) is strictly minimal, and its closed sets are the affine flats. The lattice of closed sets is not modular; but we can construct a modular lattice from it by performing the following operation on the affine geometry M: fix a point p, and form the projective geometry whose points are the affine lines of M passing through p, and whose lines are the affine planes of M containing p. This operation is called *localizing at* p. Model-theoretically it interprets a projective geometry in M; its effect on the geometry is to take an atom and remove all closed sets which are not $\geqslant$ that atom. The geometry of M is said to be *locally modular*.

These examples illustrate all the possibilities.

THEOREM 5.1. (Cherlin–Mills–Zil'ber) Let M be a countable strictly minimal t.c. structure. Then its geometry is one of the following:

1. The power-set of $\mathrm{Dom}(M)$.
2. The geometry of a projective space over a finite field, where $\mathrm{Dom}(M)$ is the set of points of the projective space.
3. As the previous, but with an affine space instead of a projective space.

Proof. See Cherlin, Harrington and Lachlan (Cherlin *et al.*, 1985, Theorem 2.1) for a proof by the classification of finite simple groups, or Zil'ber (1993, Theorem 3.0.1) for a direct proof. □

The next theorem also rests on the Cherlin–Mills–Zil'ber Theorem, and is from Cherlin *et al.* (1985, Corollary 2.5 and Lemma 8.1).

THEOREM 5.2. Let M be a countable t.c. structure. Then some induced $\varnothing$-definable substructure of M is strictly minimal. Any two strictly minimal modular (i.e., not affine) $\varnothing$-definable induced substructures of M are bi-definable by a unique $\varnothing$-definable bi-definition.

So we can start to classify countable t.c. structures M by noting the geometries of their induced $\varnothing$-definable strictly minimal substructures.

The first case—let us call it D—is where one of these induced strictly minimal structures A has degenerate geometry, and hence they all do by Theorem 5.2. In fact any induced strongly minimal substructure will be degenerate too, even if its definition uses parameters. This is because we can add the parameters to M, so that this induced substructure becomes $\varnothing$-definable, while A remains degenerate.

Let us write P_q for the case where some induced $\varnothing$-definable strictly minimal substructure has the geometry of a projective space over the field F_q of q elements. This includes the case where there is an induced $\varnothing$-definable affine geometry B over F_q. The reason is that we can form a projective geometry A over the same field by taking as points the parallelism classes of lines in B—this achieves the same effect as localizing but without adding a parameter.

Sometimes a strictly minimal structure carries a little more information than its geometry, and we can use this extra to refine the classification. For example, when the geometry of M is that of a projective space over the field F_q, the automorphism group of M lies between the projective linear group $\mathrm{PGL}(\omega, F_q)$ and the projective semilinear group $\mathrm{P\Gamma L}(\omega, F_q)$. We saw in Examples 2.3 and 2.4 how to get the extreme cases. There can be some groups strictly between $\mathrm{PGL}(\omega, F_q)$ and $\mathrm{P\Gamma L}(\omega, F_q)$, but only finitely many. The group $\mathrm{PGL}(\omega, F_q)$ is a normal subgroup of $\mathrm{P\Gamma L}(\omega, F_q)$, and the factor group is the group $\mathrm{Out}(F_q)$ of outer automorphisms of the field F_q (Artin, 1957, Chapter II.10). So the intermediate groups are the extensions of $\mathrm{PGL}(\omega, F_q)$ by subgroups of $\mathrm{Out}(F_q)$. Exactly the analogous statement holds in the affine case.

It takes some work to translate these descriptions of $\mathrm{Aut}(M)$ into usable model-theoretic descriptions of the relations on M. See for example Zil'ber (1993, Theorem 5.11); he shows that every definable relation of M is a boolean combination of relations induced by linear equations in the underlying vector space. In the disintegrated case there is no chance of any extra structure; the automorphism group is the full symmetric group, and so the strictly minimal set carries only relations definable from equality.

Notice that if M is a t.c. structure of type P_q, we may well be able to interpret in M a strictly minimal structure which is not over the field F_q. For example if M itself is a vector space over F_9, we can interpret in it an affine space A over the field F_3—in fact we can reach A from just the abelian group structure of M. And of course we can interpret in M an infinite set with no structure.

But there are limits. For example one can show:

THEOREM 5.3. No infinite projective or affine geometry over a field F_q is interpretable in a t.c. structure of type P_r or D, where q and r are powers of distinct primes.

6 Groups interpretable

This is a convenient moment to assemble some information that we shall need later, about groups interpretable in t.c. structures. In the following lemmas, G is an abstract group, not a topological or permutation group.

LEMMA 6.1. If G is a group which is interpretable in a t.c. structure, then G has the descending chain condition on definable subgroups. Hence there is a minimum definable subgroup G° of finite index in G; this subgroup is $\varnothing$-definable and hence normal.

Proof. An infinite descending chain of definable subgroups would give an infinite tree of cosets, contradicting the finite rank property. The rest follows from the fact that the intersection of two subgroups of finite index is again of finite index. □

The subgroup G° is called the *connected component* of G. The group G is said to be *connected* if $G = G^\circ$.

LEMMA 6.2. Let G be a connected group which is interpretable in a t.c. structure of type P_{p^i} for some prime p. Then G is abelian; in fact G is the direct sum of a finite abelian group and an abelian p-group of finite exponent.

Proof. By Löwenheim–Skolem we can assume G is at most countable. We know that G is abelian from work by Baur, Cherlin and Macintyre (Baur *et al.*, 1979) (though today there are slicker proofs, as for example Poizat (1987, Theorem 1.14)). Also G has finite exponent by ω-categoricity, and hence is a direct sum of cyclic groups. For each prime q the q-socle is a vector space over the field F_q, and thus is either finite or a strongly minimal set. In the latter case its geometry is projective over F_q, and so q must equal p by Theorem 5.3. □

LEMMA 6.3. Let G be a group which is interpretable in a t.c. structure of type D. Then G is finite.

Proof. It suffices to prove this for connected groups G, by Lemma 6.1. The argument of the previous lemma shows that if G was infinite, there would be a strongly minimal set with geometry of affine type, which is impossible. □

LEMMA 6.4. Let G be an abelian group of Morley rank 1 which is interpretable in a t.c. structure of type P_{p^i} for some prime p. Then G is the direct sum of a finite abelian group and an infinite elementary abelian p-group (i.e., a direct sum of infinitely many cyclic groups of order p).

Proof. By ω-categoricity G has finite exponent and is a direct sum of cyclic groups. Applying Lemma 6.2 to G°, G is the direct sum of a finite group and an infinite abelian p-group of finite exponent. If G has infinitely many summands of order divisible by p^2, then G has Morley rank at least 2. □

7 The Zil'ber ladder

This section introduces our main decomposition theorem, Zil'ber's Ladder Theorem for t.c. structures. (Zil'ber has a more general version for uncountably categorical structures too.)

Suppose B is a countable t.c. structure and A is a $\varnothing$-definable induced substructure of B. By an *orbit of B over A* we mean the orbit under $\mathrm{Aut}(B)_{(A)}$ of some element of B which is not in A.

THEOREM 7.1. Every orbit of B over A is defined by a formula of L with parameters from A.

Proof. Let T be the complete theory of B. Since T is totally transcendental, there is a prime model M of T over A; for every element b of M, the type of b over A is isolated by some formula whose parameters lie in A. Since M is prime over A, we can suppose that it is an elementary substructure of B. But then the non-two-cardinal property of B implies that $M = B$. Suppose the type of b over A is isolated by the formula $\phi(x)$ with parameters from A. Then certainly every automorphism of B fixing A pointwise will take b to an element satisfying ϕ, so that the orbit of b over A lies inside $\phi(B)$. A back-and-forth argument gives the converse inclusion. □

Let us write Gal(X, A) for the 'Galois' group of all permutations of the orbit X which are induced by automorphisms of B fixing A pointwise.

We say that B is an *affine pre-cover* of A if

1. A is an infinite induced $\varnothing$-definable substructure of B.
2. Suppose X is an orbit of B over A. Then the group Gal(X, A) is abelian and is interpretable in A (possibly with parameters) as an abstract group $G(X)$, and the action of $G(X)$ on X is definable in B with parameters from A.

The name 'affine' refers to the fact that in the rank 1 case, the connected component of each group $G(X)$ carries an affine geometry (see Lemma 6.4).

One noteworthy consequence of the definition of affine pre-covers is that for each orbit X of B over A, Gal(X, A) acts regularly on X. For consider any element g of $G(X)$, and suppose $bg = b$ for some element b of X. Since the action of $G(X)$ is definable with parameters in A, we have $(b\alpha)g = b\alpha$ for each $\alpha \in$ Gal(X, A). But Gal(X, A) is transitive on X, so that g fixes all of X.

We say that B is a *finite pre-cover* of A if

1. A is an infinite induced $\varnothing$-definable substructure of B.
2. Each orbit of B over A is finite.

In the case of finite pre-covers, Gal(X, A) need not act regularly—there are easy counterexamples. Hence in general it will not be possible to interpret Gal(X, A) and its action in A and B as with affine pre-covers. However, each finite pre-cover of A is bi-interpretable with a finite pre-cover of A in which all the Galois groups act regularly, as follows (Evans and Hrushovski, 1993, Lemma 1.8). Since each orbit X is finite, we can list its elements as a tuple $\bar{b}$. The induced action of Gal(X, A) on the orbit of $\bar{b}$ is regular.

The group Gal(X, A) can be interpreted as a group $G(X)$ in A^{eq}, since every finite structure is interpretable in any structure with two or more elements; likewise the action of $G(X)$ on X can be defined trivially, using the elements of X as parameters. But in general the action of $G(X)$ is not definable with parameters from A, even when the cover is made regular; otherwise we could use the bi-interpretation to define the action in the original pre-cover.

For both affine and finite pre-covers, we call the group $G(X)$ the *binding group* of the orbit X. A pre-cover is said to be *regular* if all its binding groups act regularly. The *rank* of a pre-cover is the supremum of the Morley ranks of its binding groups.

Now we are equipped to state Zil'ber's Ladder Theorem.

THEOREM 7.2. (Zil'ber, 1993, Theorem 5.0.2) Let M be a countable t.c. structure. Then there is a chain $M_0 \subseteq \ldots \subseteq M_n$ of induced $\varnothing$-definable substructures such that:

1. M_0 is strictly minimal;
2. $M_n = M$; and
3. for each $i < n$, M_{i+1} is either an affine pre-cover or a finite pre-cover of M_i, of rank at most 1.

The PROOF of Zil'ber's result is too elaborate even to sketch. But it uses two main ingredients. The first is a device of stability theory, which tells us roughly that any element of M lies in a set X which is parametrized by elements of some set Y of lower Morley rank than M. The second is a trick for using these parameters to define the Galois group of X over Y as a group interpretable in the structure. The rest is by induction on the Morley rank of the structure. The fullest account is in Zil'ber (1993, Chapter V). □

The length n of the chain in the ladder theorem is not unique, but it is clearly at least $\mu - 1$ where μ is the Morley rank of M. Zil'ber's proof yields a chain of length at most $2\mu + 1$.

COROLLARY 7.3. If M is a countable t.c. structure of type D, then M lies in the algebraic closure of a strictly minimal set of M (in other words, M is almost strongly minimal).

Proof. This follows from the ladder theorem together with Lemma 6.3. See Ahlbrandt (1987) for a discussion. □

Zil'ber's theorem can be refined in several ways. As we saw in the previous section, there is no loss in supposing that all the pre-covers are regular. The next two lemmas say more.

LEMMA 7.4. Suppose A and B are countable t.c. structures, and let B be a pre-cover of A. Then we can choose a single formula of L (with varying parameters from A) to define all the orbits of B over A, and likewise a single formula of L to define all the binding groups $G(X)$. If the pre-cover is affine, so that the actions of the binding groups are definable with parameters from A, then we can choose a single formula of L to define all these actions.

Proof. Either ω-categoricity or compactness will give us a finite number of formulas, and the rest is by an easy coding. □

When we have applied the lemma above, there will be a finite number k such that for each orbit X of B over A there is a k-tuple of parameters in A from which we can define X, $G(X)$ and (in the affine case) the action of $G(X)$ on X. Suppose $k = 1$. Generally there will be some orbit over A which can be defined by any one of several different parameters from

A, using the same formula of L. But replacing B by another structure bi-interpretable with B (where the orbit is copied into several different orbits, one for each choice of parameter), we can suppose that each orbit has a unique defining parameter.

There will then be a map

$$\pi\colon \mathrm{Dom}(B) \setminus \mathrm{Dom}(A) \to \mathrm{Dom}(A)$$

which is $\varnothing$-definable in B and takes each element of $\mathrm{Dom}(B) \setminus \mathrm{Dom}(A)$ to the unique parameter used to define its orbit over A. When this map exists, the pre-cover B of A is said to be a *cover* of A. We call π the *fibre map*, and we call $\pi^{-1}(a)$ the *fibre over a*. We write G_a for the binding group which acts on this fibre.

Not all pre-covers are covers. A trivial counterexample is the structure $A^{\langle k \rangle}$ which comes from A by adding as new elements all the k-tuples of elements of A, together with the projection functions. Then $A^{\langle k \rangle}$ is a finite pre-cover of A, and all its binding groups are trivial.

LEMMA 7.5. In the ladder theorem, we can assume that all pre-covers are covers, except for some pre-covers of the form $A^{\langle k \rangle}$.

Proof. By the discussion above and the previous lemma, each pre-cover B of A in the ladder can be factored into a pre-cover $A^{\langle k \rangle)}$ of A followed by a cover B' of $A^{\langle k \rangle}$ which is bi-interpretable with B. □

8 Analysis of covers

Zil'ber's Ladder Theorem brings the classification tantalizingly close: when we know the possible strictly minimal t.c. structures, all that remains is to classify the possible affine and finite covers of rank 1. This turns out to lead into some very elegant mathematics, but I suspect we have only scratched the surface of the classification. Anyway, let me report what I know.

THEOREM 8.1. Let B be an affine or a finite cover of the countable t.c. structure A. Then B is a countable t.c. structure.

Proof. This follows from Theorem 4.4. □

Hence it is enough to classify covers; there is no further question about which covers yield t.c. structures and which don't.

For the rest of the story, we take the affine covers and the finite covers separately. We assume throughout that A and hence B are countable t.c. structures.

8.1 The affine case

We say that B is a *principal affine cover* of A if A is an induced $\varnothing$-definable substructure of B and the $\varnothing$-definable relations of B are those of A together with the following:

1. A fibre map $\pi\colon \mathrm{Dom}(B) \setminus \mathrm{Dom}(A) \to \mathrm{Dom}(A)$.
2. A formula $\phi(x, y)$ of L which for each element a in the image of π defines in A a set $\phi(A, a)$, and a formula $\chi(x_1, x_2, x_3, y)$ of L such that $\chi(A^3, a)$ is the multiplication relation of an abelian group G_a on the set $\phi(A, a)$.
3. A formula ψ of L which for each element a in the image of π defines a regular action of the group G_a on the set $\pi^{-1}(a)$.

Note that this definition does make B an affine cover of A; the orbits of B over A are exactly the fibres, and the binding groups are the groups G_a.

THEOREM 8.2. If B is an affine cover of A with rank 1, and A is a countable t.c. structure of type P_{p^i} for some prime p, then there is an intermediate structure C such that C is a finite cover of A and B is an affine cover of C in which every binding group is an elementary abelian p-group.

Proof. By Lemma 6.4, each binding group $G(X)$ of B over A is a direct sum of a finite abelian group and an elementary abelian p-group. The p-socle $G(X)[p]$, consisting of the elements of $G(X)$ killed by p, is a definable subgroup of $G(X)$. Call two elements of X *equivalent* if some element of $G(X)[p]$ takes one to the other. The equivalence classes form finitely many sets of imprimitivity, which are permuted by the finite group $G(X)/G(X)[p]$. Form C by adding to A the sets of imprimitivity for each orbit X of B over A. □

Every affine cover B of A is an expansion of a unique principal affine cover B_0 of A. We reach B_0 by throwing away all the relations on $\mathrm{Dom}(B)$ except those of A and the definitions of the binding groups and their actions. So we can classify the affine covers by first classifying the principal affine covers, and then classifying the expansions of each one.

There is not much to say on the principal affine covers. By Theorem 8.2 we can assume that each binding group is an elementary abelian p-group. We need to choose the image of the fibre map π, and for each a in this image we need to find an appropriate elementary abelian p-group G_a.

But as a first step to studying the expansions, we write down a short exact sequence:

$$K_0 \rightarrowtail \mathrm{Aut}(B_0) \xrightarrow{\mu_0} \mathrm{Aut}(A) \qquad (1)$$

Here B_0 is a principal affine cover of A. By Theorem 4.3 the natural map μ_0 from $\mathrm{Aut}(B_0)$ to $\mathrm{Aut}(A)$ is a surjective, continuous and open homomorphism. Its kernel K_0 is $\mathrm{Aut}(B_0)_{(A)}$, and by inspection this is the product

$\prod_a G_a$ of all the binding groups G_a. Since K_0 comes from $\mathrm{Aut}(B_0)$ by adding constants, it is a closed subgroup of $\mathrm{Aut}(B_0)$.

LEMMA 8.3. The short exact sequence (1) splits. The induced action of $\mathrm{Aut}(A)$ on K_0 is exactly that implied by the action of $\mathrm{Aut}(A)$ on the binding groups (which lie in A^{eq}).

Now suppose we try to recover B from B_0 by adding the relations that we removed before. Adding relations means moving to a closed subgroup, and so $\mathrm{Aut}(B)$ is a closed subgroup of $\mathrm{Aut}(B_0)$. The restriction μ of μ_0 to $\mathrm{Aut}(B)$ is still onto $\mathrm{Aut}(A)$, and it is still open and continuous by Theorem 4.3. The kernel of μ is $K = K_0 \cap \mathrm{Aut}(B)$, which is a closed subgroup both of K_0 and of $\mathrm{Aut}(B)$. Thus we have the diagram:

$$
\begin{array}{ccccc}
K_0 & \rightarrowtail & \mathrm{Aut}(B_0) & \xrightarrow{\mu_0} & \mathrm{Aut}(A) \\
\downarrow & & \downarrow & & \| \\
K & \rightarrowtail & G = \mathrm{Aut}(B) & \xrightarrow{\mu} & \mathrm{Aut}(A)
\end{array}
\tag{2}
$$

Following Ahlbrandt and Ziegler (1991*b*), we classify the possible groups $\mathrm{Aut}(B)$ in this diagram by first finding the possible kernels K and then filling in the lower line.

THEOREM 8.4. (Ahlbrandt and Ziegler, 1991*b*) For any subgroup K of K_0, the following are equivalent.

1. There is a cover B of A got by expanding B_0, so that the commutative diagram (2) holds.
2. K is a topologically closed subgroup of K_0 which is also closed under the action of $\mathrm{Aut}(A)$ on K_0.

We refer to the groups which satisfy the conditions of this theorem as the *kernels expanding* B_0 *over* A.

Ahlbrandt and Ziegler (1991*a*) gave a complete list of the possible kernels when A is a vector space over the 2-element field F_2, and A itself as additive group acts as binding group G_a for each element a of A. Evans and Cameron observed (unpublished) that if we replace F_2 here by any finite field F_q, then the problem of finding the possible kernels is closely related to the problem of classifying all Reed–Muller codes over the field F_q. This latter problem is not yet completely solved.

THEOREM 8.5. Given the kernel K, there is a natural 1–1 correspondence between

1. the possible groups G which complete the diagram (2), modulo conjugacy within $\mathrm{Aut}(B_0)$, and

2. the elements of the first cohomology group $H^1(\mathrm{Aut}(A), K_0/K)$.

Moreover, every such group G, as a group acting on $\mathrm{Dom}(B_0)$, is closed and hence is the automorphism group of a cover B of A got by expanding B_0.

Proof. Ahlbrandt and Ziegler (1991*b*) for the first sentence, Hodges and Pillay (in press) for the second. □

Note that G is given as a permutation group, not as a topological group. Hence it determines B more precisely than we need for our classification up to bi-interpretability. (And conjugating G within $\mathrm{Aut}(B_0)$ only applies an automorphism to B.)

Calculations of cohomology groups appear in the literature quite regularly. But Evans, Hodkinson and I found no trace of the calculations we needed for handling some quite simple questions about covers involving abelian groups. So we did the calculations ourselves and published them (Evans *et al.*, 1991). There is evidence that it may sometimes be easiest to compute the relevant cohomological information by model-theoretic arguments, as Pillay does in Section 6 of Hodges and Pillay (in press). In short, this area of work may well end up by feeding more information into group theory than it took out.

8.2 The finite case

One of the hazards of the finite case is that it is not clear whether we should really be dealing with covers rather than pre-covers. Hrushovski (1989*b*) gave a complete analysis of countable t.c. structures of type D, and with Evans (Evans and Hrushovski, 1993) he extended the analysis to almost strongly minimal countable t.c. structures. Although covers appear in the title of the Evans–Hrushovski paper, these authors soon find that they need to work with a more complicated notion which in some sense incorporates a pre-cover of the strictly minimal set.

They also find that they need to study the group of automorphisms of a fibre $\pi^{-1}(a)$ when we hold fixed the element a rather than all the elements of A; this group is the *fibre group* on a, and in general it is larger than the binding group G_a.

For example let A be the structure of Example 2.8 in Section 2 above, with $k = 2$. Above the element representing a pair $\{a, b\}$ put two new elements a', b', and a 3-ary relation S so that $S(a', x, y)$ holds when x, y are elements of A representing pairs whose intersection is a; and likewise for b'. Let B consist of A with all these new elements and relations. Then the fibre groups are cyclic of order 2 but the binding groups all vanish.

This example also has the property that any automorphism of B which holds pointwise fixed all but one of the fibres must pointwise fix that fibre too. Evans and Hrushovski call a finite cover with this property *linked*.

I quote the main result of their paper.

THEOREM 8.6. Let M be a countable t.c. structure of Morley rank n which lies in the algebraic closure of some $\varnothing$-definable modular strictly minimal set D in M^{eq}. Then there exist $\varnothing$-definable subsets

$$M_0, D \subseteq M_{1,0} \subseteq M_1 \subseteq M_{2,0} \subseteq \ldots \subseteq M_{n,0} \subseteq M_n$$

of M^{eq} such that

1. M_0 is finite, $\mathrm{acl}^{\mathrm{eq}}(\varnothing) = \mathrm{dcl}^{\mathrm{eq}}(M_0)$ and M_i has Morley rank i.
2. $\mathrm{Aut}(M_{1,0})_{(M_0 \cup D)}$ is nilpotent by finite-abelian.
3. For $2 \leqslant i \leqslant n$, $\mathrm{Aut}(M_{i,0})_{(M_{i-1})}$ is nilpotent, and for $1 \leqslant i \leqslant n$, $\mathrm{Aut}(M_i)_{(M_{i,0})}$ is a direct product of finite groups.
4. $M \subseteq M_n$.

The reader will have noticed that this theorem never mentions covers. Nevertheless there are various cover-like devices hidden in it. The steps $M_i/M_{i,0}$ correspond to the principal covers of the affine case, except that they are pre-covers rather than covers; the kernel group is the product of the separate binding groups. Hrushovski (1989*b*) gives a concrete description of how M_i is built from $M_{i,0}$. The steps $M_{i,0}/M_{i-1}$ carry the interaction between the fibres. An example which Evans and Hrushovski quote is where V is an infinite vector space over a finite field (with the origin removed), A is the associated projective space and B consists of A, V and the projection map from V to A. Here any automorphism which keeps A fixed must fix each line of V setwise, and hence it must be a dilation. Thus the kernel is finite; Evans and Hrushovski call such covers *superlinked.*

The Evans–Hrushovski Theorem leaves many things undecided. For example it is not yet known whether we can replace 'nilpotent' in clause 3 by 'abelian' in the disintegrated case (on which see Evans (1994)). This reflects the fact that we are still very short of concrete examples. So the work of cataloguing must go on.

Apart from the work of Evans and Hrushovski, we are still nibbling at small special cases to see what kind of pattern is going to emerge.

For example, what are the finite covers of A when A is simply a countable set with no further structure? Hodkinson and I considered this in Spring 1986, as part of quite a different programme. We showed by an Ehrenfeucht–Mostowski argument (unpublished) that all such covers split, in the sense that the short exact sequence

$$\ker \mu \longrightarrow \mathrm{Aut}(B) \stackrel{\mu}{\twoheadrightarrow} \mathrm{Aut}(A) \qquad (3)$$

is split exact. Then all such covers are reducts of the cover got by adding a relation $<$ which linearly orders each orbit separately. This result was rediscovered recently by Ziegler (unpublished), who went on to classify the possible kernel groups.

Ivanov (this volume) considered a number of other natural special cases. For example we can copy the cohomological approach of the affine case, starting with a principal cover B_0 and expanding it to structures B in ways which respect $\mathrm{Aut}(A)$. Ivanov classifies the possibilities (taking A to be a set or a projective geometry) when $\mathrm{Aut}(B)$ is required to contain a particular subgroup of $\mathrm{Aut}(B_0)$, for example the group generated by certain cyclic permutations of the orbits.

Permutations and the axiom of choice

J. K. Truss

School of Mathematics,
The University of Leeds,
Leeds,
LS2 9JT,
U.K.
Electronic mail address: jkt@dcs.leeds.ac.uk

1 Introduction

The purpose of this paper is to explore the connections between permutation groups, definability, and the axiom of choice (AC). The fact that these three topics are closely related is perhaps rather well known, but it is possible to bring out into the open some of the precise connections, which may be spelt out in quite a systematic way. To start from a particular point of view, in the original models for the negation of the axiom of choice there was an explicit use of permutation groups. I refer to the Fraenkel–Mostowski method. Moreover, this was developed to a high degree of elegance, particularly in the work of Mostowski where he was able, by careful choice of the group and the 'support structure', to illustrate non-implications between various weak versions of AC. For instance, he found a model in which every set can be *linearly* ordered (the *ordering principle*) but AC fails (Mostowski, 1939), and (1945) he investigated the connections between versions of AC for families of *finite* sets. More sophisticated constructions were used by other authors, for instance Läuchli (1964), who gave a model in which any family of non-empty finite sets has a choice function, but the ordering principle fails.

The Fraenkel–Mostowski method was makeshift in the sense that at the time when it was devised no-one knew how to construct models of Zermelo–Fraenkel set theory in which AC was false. For this reason, a related system in which models could possess non-trivial $\in$-automorphisms was used. After the advent of forcing, most results obtained by FM methods could be transferred straight to ZF. To begin with, these were all done on an *ad hoc* basis, but then more precise methods of direct transfer were discovered. The first general result of this sort was the Jech–Sochor Theorem (Jech and Sochor, 1966), and the method was considerably extended by Pincus (1972), relying principally on a careful analysis of the Halpern–Levy model (Halpern and Levy, 1971).

Since in ZF there can be no non-trivial $\in$-automorphisms of the universe, if we are to use permutational methods the permutations must lie elsewhere. In Cohen's work he viewed these as acting on the 'label space'. In the more streamlined boolean-valued treatment of forcing due to Scott and Solovay (described by Bell (1985)), the permutations are automorphisms of a complete boolean algebra. Here they extend not to automorphisms of the universe, but rather to automorphisms of the *boolean-valued* universe. If the intuition behind the Fraenkel–Mostowski method was that to negate AC we need to find a symmetrical family of sets such that any choice function for it has to be asymmetrical (and hence excluded), the modified intuition for Cohen models is that we should find a family which with boolean value **1** is sufficiently symmetrical, but such that any choice function is not. This cannot be directly because of lack of symmetry in the true ZF universe, but it can be because of lack of symmetry in its boolean counterpart, or perhaps better, because of lack of *definability*. This is where the other theme, of definability, comes into the picture.

In Section 2 we give a recap of the Fraenkel–Mostowski method, illustrating by giving a standard construction required later. Next in Section 3 we examine some classical independence proofs related to weak versions of the axiom of choice, and show how a uniform treatment can be given by using universal-homogeneous structures. This follows ideas of Pincus (1976). A particularly beautiful illustration of the relationship between $\neg$ AC models and permutation groups comes about in the Mostowski–Gauntt theory of finite versions of the axiom of choice, where the permutation groups are finite. We describe the corresponding 'Galois theory' for this situation in Section 4, emphasizing the different ways in which this works out corresponding to implications between the various modified finite versions of the axiom of choice studied. Then in Section 5 we explore a fascinating area where the connections with model theory seem to be very strong, and describe work still in progress (Creed, to appear; Truss, in press) concerning set-theoretic analogues of the model-theoretic notion of 'strongly minimal set', where we aim to perform a 'classification' of so-called *amorphous* and *o-amorphous* sets.

2 Definability, choice, and Fraenkel–Mostowski models

I shall now recall the classical construction of Fraenkel and Mostowski, which is used to form models in which the axiom of choice is false. The method predates Cohen's techniques by many years, and was at that time the best that could be done in obtaining $\neg$ AC models. It was carried out in the context of a modified set theory which can accommodate the existence of 'atoms' (or *urelemente*). These are objects which are not sets, but which can be members of sets. The point of allowing atoms is the following.

The key idea in obtaining a model for $\neg$ AC is to obtain a set which is reasonably 'symmetrical', but such that any choice function for its members must be rather 'asymmetrical'. Since atoms are by their very nature set-theoretically indistinguishable, the notions of symmetry and asymmetry of sets built up from them make good sense, and if we have so arranged things that our universe contains only symmetrical objects, a set may lie in the model without there being any choice function for it.

The Fraenkel–Mostowski method takes this approach fairly literally; that is, there really *is* a symmetry group acting, and we do apply a criterion of how symmetrical sets should be for them to lie in the model. The group elements are taken to be automorphisms of the set-theoretical universe. This explains why in this approach it is necessary to assume there are atoms. For, in ordinary ZF set theory, one can prove by transfinite induction on rank that the only $\in$-automorphism of the universe is the identity. This is because if g is an $\in$-automorphism then for every set x, $xg = \{yg : y \in x\}$.

We work therefore in the theory FM, which is obtained from ZF by modifying the axiom of extensionality to allow the existence of atoms. It now says that if two sets are non-empty, and have the same members, then they are equal. The empty sets are then either the 'true' empty set, or atoms. To distinguish which are which we adjoin a unary predicate symbol $U(x)$, to express 'x is an atom'. Jech (1970) shows how to find a model for FM built out of certain of the sets of a model for ZF (which we may think of as adjoining the desired atoms 'at the side'), so there is no problem about the existence of such a thing (relative to that of a model of ZF).

Let $\mathfrak{M}$ therefore be a model of FMC (= FM + AC) in which the family U of atoms forms a set (though there are versions of the method where U is a class). Suppose that G is a group of permutations of U, and that $\mathscr{F}$ is a filter of subgroups of G closed under conjugacy. These are then the ingredients necessary for construction of an FM model. Before we can do this we first have to say how G will act on the members of $\mathfrak{M}$. This is uniquely determined by the requirement that it should respect the membership relation, so as above we have to let $xg = \{yg : y \in x\}$ for each x. Once G has been allowed to act on the whole of $\mathfrak{M}$, we at once obtain stabilizers in the usual way. It is convenient to distinguish setwise and pointwise stabilizers, which are written as $G_{\{x\}} =_{\text{def}} \{g \in G : xg = x\}$ (the setwise stabilizer of x) and $G_{(x)} =_{\text{def}} \{g \in G : yg = y \text{ for all } y \in x\}$ (the pointwise stabilizer of x). The resulting FM model $\mathfrak{N}$ then consists of those members of $\mathfrak{M}$ which are hereditarily symmetric with respect to $\mathscr{F}$. Thus we may write $\mathfrak{N} = \{x \in \mathfrak{M} : x \subseteq \mathfrak{N} \text{ and } G_{\{x\}} \in \mathscr{F}\}$ (which is a definition by transfinite induction).

It is necessary to check that this gives a model of FM. That means that all the axioms of FM are to hold when relativized to $\mathfrak{N}$. Now by definition, $\mathfrak{N}$ is certainly *transitive*, meaning that any member of a member of $\mathfrak{N}$ is itself in $\mathfrak{N}$, and from this extensionality (modified of course) and foundation hold. All the 'standard' sets, those whose transitive closures contain no atoms, are automatically in $\mathfrak{N}$, so the axiom of infinity holds, and the axioms of union and power set are straightforward to check (where the power set of $X \in \mathfrak{N}$ in $\mathfrak{N}$ equals the intersection of the power set of X in $\mathfrak{M}$ with $\mathfrak{N}$). It is for the axiom of replacement that the closure of $\mathscr{F}$ under conjugacy is required. Details are given in Jech (1970, Lemma 100).

The idea behind the failure of the axiom of choice in $\mathfrak{N}$ (for suitable choice of group and filter), is that many sets will have been put into $\mathfrak{N}$ on the grounds that they are 'symmetrical enough' (that is, their stabilizer lies in $\mathscr{F}$), but that any well-ordering for them (or choice function) would have to be asymmetrical, and hence will be absent. Let us give a classical case, for constructing a model containing what in Truss (in press) I call a 'strictly amorphous set', due to Fraenkel (1922) (though it was Mostowski (1938) who actually showed that U is amorphous there).

Here we take U to be infinite, and let G be the group of all permutations of U (in $\mathfrak{M}$). For $\mathscr{F}$ we take the filter generated by the stabilizers of the members of U. More precisely

$$\mathscr{F} = \{H \leqslant G : \exists A \subseteq U\,(A \text{ finite and } H \geqslant G_{(A)})\}.$$

Since $g^{-1}G_{(A)}g = G_{(Ag)}$, it is immediate that $\mathscr{F}$ is closed under conjugacy. This defines the model. The idea is that U will be the strictly amorphous set in $\mathfrak{N}$, so first we should check that U *does* lie in $\mathfrak{N}$. Each member of U lies in $\mathfrak{N}$ since it has no members itself, and its stabilizer was explicitly put into the filter. It follows that U is in $\mathfrak{N}$, now that we have shown that all its members are, since its stabilizer equals G. It is at the *next* step that many sets of $\mathfrak{M}$ have been omitted. In fact the only subsets of U which are in $\mathfrak{N}$ are those which are finite or cofinite (= complement of finite). For suppose that $X \subseteq U$ lies in $\mathfrak{N}$ and is infinite. There must therefore be a finite subset A of U such that $G_{(A)} \leqslant G_{\{X\}}$. Since X is infinite there is $x \in X - A$, and as $G_{(A)}$ acts transitively on $U - A$, X must contain the whole of $U - A$, so is cofinite.

Saying that a set is *amorphous* means that it is infinite, but is not the disjoint union of two infinite sets. What we have therefore seen is that U is amorphous in $\mathfrak{N}$. The existence of an amorphous set clearly contradicts the axiom of choice, so that AC is false in $\mathfrak{N}$. One way of viewing amorphous sets is that they are infinite sets which are 'Dedekind finite' in a very strong sense, and this is how they are viewed for instance in Levy (1958) and Truss (1974*a*). Recall that a set is said to be *Dedekind finite* if it has no countably infinite subset. The model just described, and that of Mostowski (1939), are the classical examples of models containing infinite but Dedekind finite sets.

A stronger statement is also true of U, as is expressed by saying that it is strictly amorphous. This just means in addition that in any partition of the set into infinitely many pieces, all except finitely many of the pieces are singletons. Note that this is analogous to the notion of a *strictly minimal set* in model theory, indicating the first of the many connections between axiom of choice properties and ideas in model theory. If the fact that U is amorphous in $\mathfrak{N}$ corresponds to the transitivity of $G_{(A)}$ on $U - A$, the fact that it is strictly amorphous corresponds rather to its *primitivity* there. The easy proof is omitted.

To analyse the properties of a model of this sort in more detail, one really needs a 'support structure'. This consists of a mechanism for relating properties of arbitrary sets in the model to the behaviour of G and $\mathscr{F}$. It is conceptually easiest to handle when there is some global form of AC available, as is the case in L, (the 'constructible universe'), since then one can assert the existence of a (definable) class which supplies the support for all sets simultaneously. This is not really necessary however, and local choice, that is choice for arbitrary families of non-empty sets (rather than choice for the the class of *all* sets) is quite adequate. The idea of a 'support' is defined as follows. We say that one set, x, *supports* another, y, if $G_{(x)} \leqslant G_{\{y\}}$, that is to say, any member of G which fixes every element of x also fixes y. The key point which makes the reduction work is that any family all of whose members are supported by one fixed set can be well-ordered in the model. This is because (as is easy to see) a set can be well-ordered in an FM model if and only if its *pointwise* stabilizer lies in $\mathscr{F}$.

Now by definition, any set in our model has a support which is a finite subset of U. What we want is to be able to make a simultaneous choice of supports for all sets. Since any set will have many supports (for instance, if A supports x and $B \supseteq A$ then B also supports A), this is not always immediate. If x has a *minimal* support, then things generally work out well, and that is indeed what happens here. The relevant lemma is as follows:

LEMMA 2.1. For any finite subsets A and B of U,

$$G_{(A\cap B)} = \langle G_{(A)}, G_{(B)} \rangle.$$

REMARK. To see that this is sufficient to obtain minimal supports, let x be any set in $\mathfrak{N}$. Then there is some finite $A \subseteq U$ such that $G_{\{x\}} \geqslant G_{(A)}$. Let such A be chosen of least possible cardinality. Then it is contained in any other support B for x, for as $G_{(A\cap B)} = \langle G_{(A)}, G_{(B)} \rangle$, $A \cap B$ is also a support for x, so by minimality of $|A|$, $A \subseteq B$. The precise use to which minimal supports are put will be illustrated in Section 3.

Proof of Lemma 2.1. That $G_{(A\cap B)} \geqslant \langle G_{(A)}, G_{(B)} \rangle$ is immediate. Conversely, let $g \in G_{(A\cap B)}$. Let $h \in G_{(B)}$ be such that $(A-B)h \cap (A\cup B)g^{-1} =$

$\varnothing$. Then $hg \in G_{(A\cap B)}$ and $(A \cup B) \cap (A - B)hg = \varnothing$. Define k by

$$xk = \begin{cases} xhg & \text{if } x \in A - B, \\ x(hg)^{-1} & \text{if } x \in (A - B)hg, \\ x & \text{otherwise.} \end{cases}$$

Then $k \in G_{(B)}$ and $hgk \in G_{(A)}$, so that

$$g = h^{-1}(hgk)k^{-1} \in \langle G_{(A)}, G_{(B)} \rangle. \qquad \square$$

Since it is part of our thesis that the notions of definability and choice are closely related, let us remark that there is an alternative description of this model. We may say that it consists precisely of those members of $\mathfrak{M}$ which are hereditarily definable over U, with standard sets also allowed as parameters. Here when we talk about definability *over* U, we mean that members of U may be employed as parameters of the definition. And by a 'standard' set we just mean one whose transitive closure contains no atoms. The relevant material about definability was all discussed in Myhill and Scott (1971) (where it was *ordinals* which were allowed as extra parameters, but the same techniques apply here). The facts that all the members of U, and U itself, are then in this model are immediate. Moreover it is rather clear that it is the same model as was previously defined permutationally. The construction is more direct, one could say; for here we explicitly 'put into' the model exactly the sets we want—U, the members of U, and unavoidably (in view of the axioms of set theory) the finite and cofinite subsets of U. But *no other* subsets of U. It has to be said however that the models are generally easier to work with in terms of the permutational definition.

The great advantage of the Fraenkel–Mostowski method is its simplicity. In a ZF framework one has to work quite a lot harder to achieve the same effect. Either a permutational approach, or one using ideas of definability is possible. Even in the latter case, many of the arguments work out best using permutations—as one can see by looking at some of the key papers (Cohen, 1966; Levy, 1966; Solovay, 1970). In a Fraenkel–Mostowski model one can concentrate on the issues concerned with the axiom of choice which are genuinely features required in the construction, avoiding additional complications needed to make the forcing work. And one can argue, as is done in Truss (in press), that for certain questions, FM is in any case the 'right' theory to be working in when focussing on the model-theoretic rather than the set-theoretic aspects.

3 Classical independence proofs

Levy (1965) summarized several of the independence questions which at that time were still unresolved between weak versions of the axiom of choice. Many of these have since been settled.

The main ones considered are as follows:

C_n: The axiom of choice for families of n-element sets ($1 \leqslant n < \aleph_0$), which says that any family of n-element sets has a choice function.

$C_{<\omega}$: The axiom of choice for families of non-empty finite sets.

OP: The ordering principle: any set can be linearly ordered.

OE: The order-extension principle: any partial ordering can be extended to a linear ordering.

BPIT: The boolean prime ideal theorem: any boolean algebra has a prime ideal.

The principal implications between these are as follows:

$$\mathrm{AC} \Rightarrow \mathrm{BPIT} \Rightarrow \mathrm{OE} \Rightarrow \mathrm{OP} \Rightarrow \mathrm{C}_{<\omega}. \qquad (\star)$$

None of the reverse implications is provable (in ZF) as was shown by Halpern, Felgner, Mathias, and Läuchli, respectively. The proofs used involved a variety of techniques, but an elegant idea of Pincus's means that a quite uniform treatment is possible, involving the model-theoretic idea of 'universal-homogeneous structure'. As illustration, let us consider the question as to whether the axiom of choice for families of finite sets implies the ordering principle. This was settled by Läuchli (1964) using a Fraenkel–Mostowski model. In his key lemma he showed that there is a group G having the following properties:

(i) if $A \subseteq \Omega$ is finite there are $g, h \in G_{(A)}$ and distinct $a, b, c \in \Omega$ such that $ag = b$, $bg = c$, and $bh = c$, $ch = a$;

(ii) if $A \subseteq \Omega$ is finite then $G_{\{A\}} = G_{(A)}$ (pointwise stabilizer = setwise stabilizer);

(iii) if $A, B \subseteq \Omega$ are finite then $G_{(A \cap B)} = \langle G_{(A)}, G_{(B)} \rangle$.

His proof was quite involved, and an alternative method, using a combination of forcing and the Fraenkel–Mostowski construction was given in Truss (1974*a*), using an idea of Gauntt's (Gauntt, 1970). Then in Pincus (1976) it was shown how to do the same thing in a purely FM setting by the use of universal-homogeneous structures. Since this also settles a question raised by P. M. Neumann at one of the Oxford–Queen Mary College series of seminars, let me describe the idea. Neumann's question was the following: note that if G is a permutation group on the set Ω, and G preserves a linear ordering on Ω, then no member of G has a non-trivial finite cycle. Is the converse true? That is to say, if no member of G has a non-trivial finite cycle, is there necessarily a linear ordering on Ω preserved by G? Negative answers to this were rapidly supplied by Cameron and others. In fact Läuchli's group will suffice, (ii) precisely saying that no member of G has non-trivial finite cycles, and (i) telling us that G cannot preserve a linear ordering. The third condition is not needed at all for this particular result.

I shall now present a simplified proof of Läuchli's result based on Pincus's idea, and then describe constructions using various other universal-homogeneous structures (without giving full details), some of which settle the non-implications mentioned in ($\star$).

THEOREM 3.1. There is a permutation group G on a countably infinite set Ω which preserves no linear ordering on Ω and such that no member of G has a non-trivial finite cycle.

Proof. (Pincus) Consider structures of the form $\langle \Omega : f_1, f_2, f_3, \ldots \rangle$, where for each $n \geq 1$, f_n is a choice function for the set $[\Omega]^n$ of n-element subsets of Ω. Moreover suppose that $|\Omega| = \aleph_0$ and that $\langle \Omega : f_1, f_2, f_3, \ldots \rangle$ is universal-homogeneous in the usual sense, that is, it is obtained in a generic fashion as a countable union of finite substructures. Note that strictly speaking these structures are not first-order, but they can easily be replaced by equivalent first-order structures as in Pincus (1976) by instead using for each n an n-ary function f_n which chooses one of its arguments, and is symmetric under interchange of arguments. Let $G = \mathrm{Aut}\, \langle \Omega : f_1, f_2, f_3, \ldots \rangle$. Then it is immediate that no member of G can have a finite non-trivial cycle, since if A is a finite orbit of $g \in G$ with $|A| > 1$, then g cannot preserve $f_{|A|}$ on A.

Now let us check that G satisfies Läuchli's property (i) above, and hence that it can preserve no linear ordering on Ω. Let A be the given finite subset of Ω, and choose distinct a, b, and c not in A. Let $\Omega_1 = A \cup \{a, b, c\}$ and turn Ω_1 into a structure of the same type by defining f'_n for $n \leqslant |\Omega| + 3$ by $f'_{|X|}(X) = f_{|A \cap X|}(A \cap X)$ if $A \cap X \neq \varnothing$, $f'_2\{a, b\} = a$, $f'_2\{b, c\} = b$, $f'_2\{c, a\} = c$, and $f'_3\{a, b, c\} = a$. Note that as we have defined it, $\langle \Omega_1, f'_1, f'_2, f'_3, \ldots \rangle$ need no longer be a substructure of $\langle \Omega, f_1, f_2, f_3, \ldots \rangle$. But as $\langle \Omega, f_1, f_2, f_3, \ldots \rangle$ was chosen to be universal-homogeneous, by changing the choice of a, b, and c, we may suppose that it *is*. It remains to find appropriate g and h. If we define g_1 and h_1 to fix A pointwise, and so that $ag_1 = b, bg_1 = c$ & $bh_1 = c, ch_1 = a$, then g_1 is an isomorphism from $A \cup \{a, b\}$ to $A \cup \{b, c\}$, so by homogeneity extends to the required automorphism g of Ω, and similarly for h. □

Next we discuss the other non-implications mentioned above. In each case the construction involves consideration of a suitable universal-homogeneous structure naturally tailored to the problem in hand. This may be defined by Fraïssé's method using a suitable amalgamation class of structures. In addition there has to be an analogue of Mostowski's 'support lemma', which, rather than appealing to the condition given in Pincus (1976), one can verify separately in each case without too much difficulty. Felgner's proof (Felgner, 1972) of OE $\not\Rightarrow$ BPIT used a Cohen model, but

(iv) describes what may be viewed as the 'natural' corresponding universal-homogeneous structure for providing an FM proof. Most of the proof that this construction works is complete, but it remains to verify all cases of OE in the model. In addition we mention two other universal-homogeneous structures needed for consistencies in Section 5.

(i) U has the trivial structure. Here the automorphism group is just $Sym(U)$.
(ii) U carries the structure of a countable-dimensional projective geometry over a finite field, F_q.
(iii) $U \cong (\mathbb{Q}, <)$. The automorphism group consists of the order-preserving permutations.
(iv) U is a countable atomless boolean algebra with a linear ordering extending the partial ordering of the algebra.
(v) U is a countable universal partial ordering $(U, \leqslant)$ with an (independent) linear ordering $\preceq$.
(vi) U is as in the proof of Theorem 3.1.

As mentioned before, a key point in using these structures to obtain the desired consistency is an appropriate support lemma, which enables us to assign minimal supports to all members of the model, and hence to relate the structure of general sets to that of the family of finite subsets of U.

LEMMA 3.2. For each of the permutation groups just listed, and for any finite subsets A and B of U (except that for (ii), A and B have to be subspaces too, and for (iv) they have to be subalgebras),

$$G_{(A\cap B)} = \langle G_{(A)}, G_{(B)} \rangle.$$

Proof. The result was proved for the first case in Lemma 2.1, and the proof of (iii) is essentially the same. For the other cases some modification is required. For instance for (ii) we take $h \in G_{(B)}$ to be such that $(A-B)h$ is disjoint from the *span* of $(A \cup B)g^{-1}$, and k to be some member of G such that $xk = xhg$ if $x \in A - B$, $xk = x(hg)^{-1}$ if $x \in (A-B)hg$, and $xk = x$ if $x \in B$. But we cannot any longer insist that it is the identity for other points (since it must preserve the geometry). Cases (iv), (v), and (vi) are discussed in Felgner and Truss (to appear), Jech (1973), and Pincus (1976) respectively. □

Now let us see how the models resulting from these structures are used to establish the stated FM consistencies. In each case we suppose that $\mathfrak{M}$ is a model of FMC in which U is the set of all atoms, and U is countable. Since the structures described above all have countable domains, we may suppose that U is indexed by any one of them, and we get a natural action of the automorphism group of the structure on U. Let us denote this by

G. The supports are taken to be finite, that is $\mathscr{F}$ is the filter of subgroups of G generated by $\{G_x : x \in U\}$. Let $\mathfrak{N}$ be the resulting FM model.

THEOREM 3.3. The Fraenkel–Mostowski models defined naturally from the six structures listed have the following properties:

(i) U is strictly amorphous;

(ii) U is amorphous but not strictly amorphous, and carries a non-degenerate modular geometry, (we say that it has *projective type*);

(iii) BPIT holds but AC does not;

(iv) BPIT fails;

(v) OP holds but OE does not;

(vi) $C_{<\omega}$ holds but OP does not.

Proof. (i) has already been shown in Section 2, and that U is amorphous in (ii) follows from the fact that the pointwise stabilizer of any finite subset of U acts transitively on a cofinite subset. The geometry on U is preserved by definition of G. Note that we cannot now however index it by the one-dimensional subspaces of an $\aleph_0$-dimensional vector space over F_q. The indexing has been 'lost' in passing from $\mathfrak{M}$ to $\mathfrak{N}$. But the non-degeneracy and modularity have not.

(iii) The proof that BPIT holds in $\mathfrak{N}$ is beyond the scope of this brief survey and involves a certain amount of combinatorics (Halpern and Levy, 1971; Jech, 1973), though an alternative proof, going by way of the compactness theorem and using ideas of Ehrenfeucht and Mostowski, was given by Pincus (1976). That AC is false in $\mathfrak{N}$ is however easy to see, since any subset of U must be a finite union of intervals and points, having finite support A (and as G acts transitively on the open intervals defined by A). This is the prototype of an *o-amorphous* set (see Section 5), and we deduce easily that U has no countable subset, and in particular, cannot be well-ordered.

(iv) Since U is indexed by a countable boolean algebra (with linear ordering) and G preserves the boolean algebra structure, it is still a boolean algebra in $\mathfrak{N}$. To show that BPIT fails in $\mathfrak{N}$ the algebra U itself is used, and it is shown to have no prime ideal in $\mathfrak{N}$. We remark that the construction is specifically designed so that OE should hold in $\mathfrak{N}$. With this object, an extension of the natural partial ordering (as a boolean algebra) of U was explicitly put into $\mathfrak{N}$. It is anticipated that it will follow from this, and the existence of minimal finite supports, that *all* partial orderings in $\mathfrak{N}$ can be extended to linear orderings.

(v) The fact that any set has a unique finite support in the model means that any set can be put into $1-1$ correspondence with a subset of $\alpha \times e(U)$ for some ordinal α, where $e(U)$ is the family of finite subsets of U. But the linear ordering $\preceq$ of U can be lifted to a linear ordering of $\alpha \times e(U)$, thus verifying OP in $\mathfrak{N}$. To see that OE fails, we show that the partial ordering $\leqslant$ on U (which in $\mathfrak{M}$ is the countable universal partial ordering) has no extension to a linear ordering in $\mathfrak{N}$. For if $\leqslant^*$ were such an extension, it would have to have finite support A, say. By universality and homogeneity there are x, y, z, t exceeding all members of A in both $\leqslant$ and $\preceq$ such that the only relations between them are given by $x \leqslant y$, $z \leqslant t$, $x \preceq t \preceq y \preceq z$. By looking at the isomorphic substructures $A \cup \{x, z\}$, $A \cup \{x, t\}$, $A \cup \{y, z\}$, $A \cup \{t, y\}$, and using homogeneity again, and the fact that $G_{(A)} \leqslant G_{\{\leqslant^*\}}$, we find that either all or none of $x \leqslant^* z$, $x \leqslant^* t$, $y \leqslant^* z$, $t \leqslant^* y$ hold. Since $x \leqslant y$ we cannot have both $t \leqslant^* x$ and $y \leqslant^* t$. Since $z \leqslant t$ we cannot have both $t \leqslant^* y$ and $y \leqslant^* z$. Thus we have the desired contradiction.

(vi) That any family of finite sets has a choice function in $\mathfrak{N}$ follows by using finite supports, and using the choice functions f_n for the set of n-element subsets of U which have been explicitly included. To see that U cannot be ordered in the model, suppose on the contrary that $\leqslant$ is a linear ordering of U supported by $A \subseteq U$. Then by property (i) we know that there are $g, h \in G_{(A)}$ and distinct $a, b, c \in \Omega$ such that $ag = b$, $bg = c$, and $bh = c$, $ch = a$. But as $G_{(A)}$ preserves $\leqslant$, $a \leqslant b \Leftrightarrow b \leqslant c \Leftrightarrow c \leqslant a$, and as $\leqslant$ is meant to be a linear ordering, this gives a contradiction. $\square$

Let us remark on the transfer of these results to ZF. In the early days of forcing, these transfers were done separately for each model. Then in the Jech–Sochor Theorem (Jech and Sochor, 1966) a method for automatically deriving a ZF consistency corresponding to an FM consistency was given, for statements of a certain special form ('boundable' statements). This was subsequently extended by Pincus (1972). The basic idea is that the 'totally indistinguishable' atoms should be replaced by some 'sufficiently indistinguishable' sets. These may be reals, sets of reals, sets of sets of reals,..., depending on the statement to be transferred. For instance, since an amorphous set cannot be linearly ordered, it is hopeless to try to transfer the consistency of the existence of an amorphous set by replacing the atoms by reals, since any set of reals *can* be ordered. The next thing to try is to represent them as a set of sets of reals, and this turns out to be good enough, in each of (i) and (ii).

For (iii), (iv), and (v), it is actually *simpler* in some respects for the ZF case than for FM. The point is here that there is a ('generic') linear ordering on the structure, and if we take the atoms to be represented as reals, it may be taken as the usual ordering. In the case of Mostowski's ordered model (iii), this is all that is needed. In the next two instances we have to put on additional structure, which for (iv) is a boolean algebra structure whose partial ordering relation restricts that on $\mathbb{R}$ (but is otherwise 'generic'), and for (v) is a generic partial ordering of $\mathbb{R}$ (meaning one which is independent of the usual ordering).

As remarked by Mathias (1974) there are a number of non-transferable FM consistencies. These seem mainly to be based on the statement 'the power set of any well-ordered set can be well-ordered', which implies AC in ZF, but not in FM (Rubin and Rubin, 1976). I conclude this section by considering one such statement. The point of the example is not so much that the statement is non-transferable, but that in the FM case, its truth in a model is guaranteed by a simple group-theoretical condition, and there is a corresponding condition in the ZF case which instead guarantees the truth of a slightly weaker statement. Perhaps rather than saying it is non-transferable, we should say that its naturally transferred version is an appropriate weakening of the original. The first statement is as follows.

($*$) For any set X, if the set $[X]^2$ of 2-element subsets of X has a choice function, then X can be well-ordered.

Now this particular statement implies AC in ZF, but not in FM. To derive AC from it in ZF, we may prove from it that the power set of any well-orderable set can be well-ordered, and then appeal to Rubin and Rubin (1976, p. 76) to derive AC. The weaker statement which we claim corresponds to ($*$) in the ZF setting is this:

($**$) For any set X, if $[X]^2$ has a choice function, then X can be mapped 1$-$1 into a set of sets of ordinals.

THEOREM 3.4. In a Fraenkel–Mostowski model $\mathfrak{N}$ defined by U, G, and $\mathscr{F}$, if $\mathscr{F}$ contains a dense set of groups which are generated by involutions, then ($*$) holds.

Proof. Suppose the given statement is true, and that F is a choice function in $\mathfrak{N}$ for $[X]^2$. Let $G_{\{X\}} \cap G_{\{F\}} \geqslant H \in \mathscr{F}$, where H is generated by involutions. We show that $G_{(X)} \geqslant H$. If not, let $g \in H - G_{(X)}$. Write g as a product of involutions. At least one of these must be in $H - G_{(X)}$, so we may assume that $g^2 = 1$. Let $x \in X$, $xg \neq x$. Thus $F\{x, xg\} = y \in \{x, xg\}$, so $(\{x, xg\}, y) \in F$. As $g \in H \leqslant G_{\{F\}}$, $(\{xg, xg^2\}, yg) \in F$. As $xg^2 = x$, $yg = y$, a contradiction. Therefore X is pointwise fixed by H, and we deduce that X can be well-ordered in $\mathfrak{N}$.

As examples of models where this occurs we may take $|U| = \kappa$ for any infinite cardinal κ, $G = Sym(U)$, and let $\mathscr{F}$ be generated by

$$\{G_{(A)} : A \subseteq U, |A| < \kappa\}.$$

This generalizes Fraenkel's model above (which was the case $\kappa = \aleph_0$). □

The analogous result in ZF is obtained by regarding the relevant models of ZF as also being formed by passing to a symmetric submodel. That is, if we work in the boolean-valued universe $V^{\mathbf{B}}$, where $\mathbf{B} \in V$ is a complete boolean algebra, we may suppose that G is a group of automorphisms of

B, and that $\mathscr{F}$ is a filter of subgroups of G closed under conjugacy. The resulting symmetric boolean extension of V consists of all boolean terms which are hereditarily symmetric with respect to $\mathscr{F}$, and the corresponding ZF model is obtained as usual by choice of a generic ultrafilter on **B**. See Jech (1973) for more details.

THEOREM 3.5. In a Cohen model defined by **B**, G, and $\mathscr{F}$, if $\mathscr{F}$ contains a dense set of groups which are generated by involutions, then $(**)$ holds.

The proof, which we omit, is similar to that of the previous theorem. See Truss (1978) for instance.

In this section I have tried to illustrate how certain consistencies can be achieved by examining appropriate universal-homogeneous structures, and building Fraenkel–Mostowski models based on them as in Pincus (1976). Another possible question is to start with the universal-homogeneous structure, and to ask what the properties of the resulting model then turn out to be. Two examples which might be worth investigating are the random graph, and the countable atomless boolean algebra. One could ask then for a closer tie-up between the properties of the structure one started with, and those of the model.

4 Finite versions of the axiom of choice

A beautiful illustration of the relationship between choice and symmetry arises in the consideration of the finite versions of the axiom of choice studied initially by Tarski and Mostowski, and later by others. Here, since it is provable in ZF that any finite family of non-empty sets has a choice function, by a 'finite' version of the axiom of choice we must mean that there is a choice function for a *family* of finite sets. Mostowski realized that in trying to choose 'effectively' a member of a finite set (or any set, come to that), what is essentially involved is achieving a reduction in symmetry. Before the choice is made, there is no particular reason to prefer one member of the set to another, so they all have an equal status, but in making a selection, some reason has to be found for preferring one element to another. As a result of this observation, he formulated various combinatorial conditions on appropriate finite permutation groups which characterize implications between different finite versions of the axiom of choice, very much in the spirit of Galois theory. I briefly recall the main idea of elementary Galois theory, to illustrate the analogies with the present question.

Consider the solution of algebraic equations where 'radicals' are permitted. This means that, as well as performing standard algebraic manipulations such as addition and multiplication, we are allowed to extract roots—square roots, cube roots, and so on. The key observation made by Galois, which made his analysis possible, was that the extraction of a root precisely corresponds to a reduction of symmetry among the roots. To begin with, there is nothing to choose between the three roots α, β, and γ, of a cubic equation, shall we say, and this is more formally expressed by saying that the 'data' consisting as it does of the elementary symmetric functions $\alpha+\beta+\gamma$, $\alpha\beta+\alpha\gamma+\beta\gamma$, $\alpha\beta\gamma$ is unchanged under the action of all permutations of $\{\alpha,\beta,\gamma\}$. Now any other symmetric (rational) function of $\{\alpha,\beta,\gamma\}$ can be expressed in terms of the three particular symmetric functions so given, and the key to solving the equation is finding a (more) asymmetric function whose *square* is symmetric; here $(\alpha-\beta)(\beta-\gamma)(\gamma-\alpha)$ will do; and then a totally asymmetrical function whose cube has the same symmetry as $(\alpha-\beta)(\beta-\gamma)(\gamma-\alpha)$; here $\alpha+\beta\omega+\gamma\omega^2$ will do, where ω is a primitive cube root of unity. This computation corresponds to the choice of subgroups

$$S_3 \rhd A_3 \rhd 1$$

and the type of radical extracted to the index of the relevant subgroup in the next larger one.

We now introduce the principal variants of the finite axiom of choice we study:

C_n: Any family of n-element sets has a choice function.

$\mathrm{C}_n^{\mathrm{o}}$: Any linearly ordered family of n-element sets has a choice function.

C_n^*: Any well-ordered family of n-element sets has a choice function.

C_n^ω: Any countable family of n-element sets has a choice function.

Let us first make some easy remarks about these.

LEMMA 4.1. C_0 is false; C_1 is true; $\mathrm{C}_n \Rightarrow \mathrm{C}_n^{\mathrm{o}} \Rightarrow \mathrm{C}_n^* \Rightarrow \mathrm{C}_n^\omega$; if k divides n then $\mathrm{C}_n \Rightarrow \mathrm{C}_k$, and similarly for $\mathrm{C}_n^{\mathrm{o}}$, C_n^*, and C_n^ω.

Proof. (Last part) We let $mk=n$, and suppose that X is a family of k-element sets and Y is a fixed m-element set. Then $\{x\times Y : x\in X\}$ is a family of n-element sets. By C_n it has a choice function f. Define g by $g(x)=\xi$ if $\exists\eta\, g(x\times Y)=(\xi,\eta)$. Then g is a choice function for X. □

This proof is due to Tarski, who derived various basic results about implications between the C_n, which were greatly extended by Mostowski (1945). Finally Gauntt (1970) proved that one of Mostowski's conditions, $D(n,Z)$ (see below), is necessary and sufficient for $\forall m\in Z\, \mathrm{C}_m \to \mathrm{C}_n$ to be provable. Following a suggestion of A. Levy, I extended these results in various directions (Truss, 1973) (and he independently undertook a generalization, following a slightly different approach (Levy, 1973)). For instance an analysis of the $\mathrm{C}_n^{\mathrm{o}}$, and C_n^*, and the connections between them, was given, and 'mixed' kinds were allowed. Also Z could be infinite, as I now discuss.

If Z is infinite, the natural composite finite choice axiom to take is $\forall n\in Z\, \mathrm{C}_n$. But this is weaker than the principle we actually wish to consider, which is

C_Z: If X is a family of sets such that $\forall x{\in}X\ |x| \in Z$, then X has a choice function

(similarly for $C^{o}_{Z}, C^{*}_{Z}, C^{\omega}_{Z}$). That $C_Z \Rightarrow \forall n{\in}Z\, C_n$ is clear, but the converse is false for infinite Z, since the existence of separate choice functions f_n for $\{x \in X : |x| = n\}$ for each $n \in Z$ does not at all mean that a simultaneous choice for all $n \in Z$ is possible.

So much for the finite versions of the axiom of choice considered. We next introduce the group-theoretical conditions used by Mostowski (1945) and Gauntt (1970) to analyse the interconnections between them.

$D(n, Z)$: For any fixed-point-free subgroup G of S_n there are a subgroup H of G and proper subgroups $K_1, \ldots, K_r$ of H such that $\sum |H : K_i| \in Z$.
$L(n, Z)$: For any fixed-point-free subgroup G of S_n there are proper subgroups $K_1, \ldots, K_r$ of H such that $\sum |G : K_i| \in Z$.
$K(n, Z)$: For any fixed-point-free subgroup G of S_n there are a subgroup H of G^m, some m, and proper subgroups $K_1, \ldots, K_r$ of H such that $\sum |H : K_i| \in Z$.
$M(n, Z)$: If $n = p_1 + \cdots + p_s$ is an expression for n as the sum of (not necessarily distinct) primes, then there are $\alpha_i \geqslant 0$ such that $\sum \alpha_i p_i \in Z$.

In addition, in Truss (1973) I considered versions of these where S_n is allowed to act on the structure built up from an n-element set by taking power sets ω times. The notation is as follows. Let $e(X)$ be the set of finite subsets of the set X, and $e_n(X)$ be given inductively by $e_0(X) = X$, $e_{n+1}(X) = e(e_n(X))$, $e_\omega(X) = \bigcup_{n\in\omega} e_n(X)$. We choose some fixed n-element set X_n such that each member of $e_\omega(X_n)$ 'appears' only once. For instance X_n may consist of any n distinct infinite subsets of ω. There are three further conditions which we wish to consider.

$A(n, Z)$: For any fixed-point-free subgroup G of $\mathrm{Sym}(X_n)$ there is Y in $e_\omega(X_n)$ such that $|Y| \in Z$ and $\forall \eta{\in}Y\ G_{\{Y\}} \cap G \nleqslant G_{\{\eta\}}$.
$B(n, Z)$: For any fixed-point-free subgroup G of $\mathrm{Sym}(X_n)$ there is Y in $e_\omega(X_n)$ such that $|Y| \in Z$, $G_{\{Y\}} = G$ and $\forall \eta{\in}Y\ G \nleqslant G_{\{\eta\}}$.
$C(n, Z)$: For any fixed-point-free subgroup G of $\mathrm{Sym}(X_n)$ there is Y in $e_\omega(X_{n\cdot m})$ for some m such that $|Y| \in Z$ and $\forall \eta{\in}Y\ G^m \nleqslant G^m_{\{\eta\}}$, where G^m acts on $X_{n\cdot m}$ in the natural way.

The main results about these, obtained by combining the results of Mostowski (1945), Gauntt (1970), Truss (1973) are as follows.

THEOREM 4.2. The following are true:

(i) $D(n, Z) \Leftrightarrow A(n, Z) \Leftrightarrow (C_Z \to C_n) \Leftrightarrow (C_Z \to C^{o}_{n})$;
(ii) $L(n, Z) \Leftrightarrow B(n, Z) \Leftrightarrow (C^{o}_{Z} \to C^{o}_{n}) \Leftrightarrow (C^{o}_{Z} \to C^{*}_{n}) \Leftrightarrow$
$(C^{o}_{Z} \to C^{\omega}_{n}) \Leftrightarrow (C^{*}_{Z} \to C^{*}_{n}) \Leftrightarrow (C^{*}_{Z} \to C^{\omega}_{n}) \Leftrightarrow (C^{\omega}_{Z} \to C^{\omega}_{n})$;

(iii) $M(n,Z) \Leftrightarrow K(n,Z) \Leftrightarrow C(n,Z) \Leftrightarrow (\mathrm{C}_Z \rightarrow \mathrm{C}_n^*) \Leftrightarrow$
$(\mathrm{C}_Z \rightarrow \mathrm{C}_n^{\omega})$;

(iv) $\mathrm{C}_Z^{\circ} \nrightarrow \mathrm{C}_n, \quad \mathrm{C}_Z^* \nrightarrow \mathrm{C}_n^{\circ}, \quad \mathrm{C}_Z^{\omega} \nrightarrow \mathrm{C}_n^*$;

where by an implication such as $\mathrm{C}_Z \rightarrow \mathrm{C}_n^{\circ}$ we mean 'is provable in FM' (or ZF actually).

The idea behind the proofs is to show that the effective choice of an element from a set X corresponds to the selection of less and less symmetrical 'objects'. These may be elements of the set, subsets, sets of sequences of elements of the set, and so on, in short, members of $e_{\omega}(X)$. Rather than giving details of this, I shall illustrate how it works out in one or two cases, parallelling the Galois theory example given above. Observe first that $L(4,\{2,3\})$ is true, which means that $\mathrm{C}_{\{2,3\}}^* \rightarrow \mathrm{C}_4^*$ is provable. (On the other hand, $L(n,\{2,3,\ldots,n-1\})$ is false for $n \geqslant 5$. The alternating group A_n provides a counter-example, as it has no subgroup of index less than n). I tabulate below an appropriate chain of subgroups and corresponding choices in $e_{\omega}(X_4)$.

S_4	$\{a,b,c,d\}$	$\{\{\{a,b\},\{c,d\}\},\{\{a,c\},\{b,d\}\},\{\{a,d\},\{b,c\}\}\}$
		applying C_3^* to the set of 3 2+2 partitions
$C_2 \operatorname{Wr} C_2$		$\{\{a,b\},\{c,d\}\}$
		applying C_2^*
$C_2 \times C_2$		$\{a,b\}$
		applying C_2^* again
$\{1\} \times C_2$		a

The key difference here between $D(n,Z)$ and $L(n,Z)$ is that at each stage in applying $L(n,Z)$ ($\Leftrightarrow B(n,Z)$) to a set derived from a well-ordered family of n-element sets, there is only one element corresponding to each member of the family, so that it is still well-ordered. If only the weaker condition $D(n,Z)$ holds ($\Leftrightarrow A(n,Z)$) we have to be satisfied with *several* sets corresponding to each original set, so that the family to which finite choice is to be applied need no longer be well-orderable. Thus the example just given can be modified to show that $\mathrm{C}_2 \rightarrow \mathrm{C}_4$ by choosing a member from each pair of doubletons on the first line, and then one from each doubleton, after which there is sufficient asymmetry to select a single member of $\{a,b,c,d\}$.

The reason Mostowski failed to prove that $\mathrm{C}_Z \rightarrow \mathrm{C}_n \Rightarrow D(n,Z)$ was that he concentrated on the case of X well-ordered, and so the best he could hope for was $\mathrm{C}_Z \rightarrow \mathrm{C}_n \Rightarrow K(n,Z)$. The first value of n for which $K(n,Z) \mathbin{\&} \neg D(n,Z)$ can hold is 15, where $Z = \{3,5,13\}$. Observe that

$K(n,Z) \Leftrightarrow M(n,Z)$, which holds here:

$$\begin{array}{rclclcl} 15 & = & 3+3+3+3+3 & = & 5+5+5 & = & 7+5+3 \\ & & 3.1=3 & & 5.1=5 & & 5.1=5 \\ & = & 7+2+2+2+2 & = & 11+2+2 & = & 13+2 \\ & & 7+3.2=13 & & 11+2=13 & & 13.1=13. \end{array}$$

We now give Gauntt's verification of $\neg D(15,\{3,5,13\})$ (Gauntt, 1970). Let

$$\begin{array}{rcl} a & = & (1\,2\,3\,4\,5\,6\,7)(8\,9\,10\,11\,12\,13\,14), \\ b & = & (8\,10)(9\,12)(11\,15)(13\,14), \\ c & = & (9\,11)(10\,13)(12\,15)(14\,8), \quad \text{and} \\ d & = & (10\,12)(11\,14)(13\,15)(8\,9). \end{array}$$

Let G be generated by a, b, c, and d and let K be generated by b, c and d. Since b, c, and d commute pairwise, K is abelian of order 8. Now

$$a^{-1}ba = c, \quad a^{-1}ca = d, \quad a^{-1}da = (11\,13)(12\,8)(14\,15)(9\,10) = bd.$$

Hence a normalizes K. Therefore $G = \langle a \rangle K$, so $|G| = 56$. Moreover G is clearly a fixed-point-free group of permutations of $\{1,\ldots,15\}$, and $b^{-1}ab = bab = acb \notin \langle a \rangle$. Hence G has at least 7 elements of order 7. By Sylow's Theorem, G has 48 elements of order 7. If H is a proper subgroup of G whose order is divisible by 7, then again by Sylow's Theorem, H has only 6 elements of order 7. Hence G has no subgroups of order 14 or 28. It follows easily that $\sum |H{:}K_i| \neq 3,5$, or 13.

Despite the failure of $D(15,\{3,5,13\})$, $M(n,Z)$ is known to be sufficient for $C_Z \to C_n$ if $n \leqslant 20$ (except $n = 15$), or if n is prime, or if $n = 22, 24, 25, 26, 30, 33, 34, 36, 42, 44, 45$, so the delay in establishing the correct equivalence here is understandable.

Now I consider some slightly different versions of AC, namely of the so-called *axiom of dependent choices*, DC. In this case we make a sequence of choices, each of which may depend on previous ones. The way this works out when the sets are to be finite is as follows.

DC_Z: If T is a finite-branching tree in which every node has exactly n immediate successors for some $n \in Z$, then T has an infinite branch.

Note that in any finitely branching tree, every level is finite, so the axiom of choice for countable families of finite sets implies DC_Z. We may ask however whether, for example, $C_2^\omega \to DC_2$ ($C_2 \to DC_2$ is clear, by the way). In fact $C_2^\omega \nrightarrow DC_2$, and we even have

$$\forall n\, C_n^* \nrightarrow DC_m \quad (m > 1).$$

Notice that DC_Z is *not* the same as $\forall n{\in}Z\,\mathrm{DC}_Z$, even for finite Z. For example, $\mathrm{DC}_{\{2,3\}}$ is not the same as $\mathrm{DC}_2\,\&\,\mathrm{DC}_3$. This is because $\mathrm{DC}_{\{2,3\}}$ applies to trees where the branching may vary between 2 and 3, whereas DC_2 and DC_3 can only be applied to trees with constant branching degrees. Extending Theorem 4.2 in this case we have

THEOREM 4.3. With the notation of Theorem 4.2, we have

(i) $D(n,Z) \Leftrightarrow (\mathrm{C}_Z \to \mathrm{DC}_n)$,

(ii) $L(n,Z) \Leftrightarrow (\mathrm{DC}_Z \to \mathrm{DC}_n) \Leftrightarrow (\mathrm{DC}_Z \to \mathrm{C}_n^\omega) \Leftrightarrow$
$\forall m{\in}Z\,(\mathrm{DC}_m \to \mathrm{C}_n^\omega)$,

(iii) $\mathrm{DC}_Z \nrightarrow \mathrm{C}_n^*, \quad \mathrm{C}_Z^{\mathrm{o}} \nrightarrow \mathrm{DC}_n$,

and, in addition, the following conjecture:

CONJECTURE 4.4.

$$\forall m{\in}Z\,(\mathrm{DC}_m \to \mathrm{DC}_n) \quad \Leftrightarrow \quad \text{for some } m \in Z,\ (\mathrm{DC}_m \to \mathrm{DC}_n).$$

I now sketch the essentials of the construction of a Fraenkel–Mostowski model in which C_2^* holds but DC_2 fails (Truss, 1976). We take for set U of atoms an infinite binary tree, and let G be the group of permutations of U generated by the action of the cyclic group on each set of immediate successors (so that it is a direct limit of finitely iterated wreath products).

If for any group H, we let $\Phi_2(H)$ be the intersection of all subgroups of index 2, then the key point in the construction is to show that there is a proper filter $\mathscr{F}$ of subgroups of G closed under this 'Frattini-like' operation, which may then be used to define an FM model for the desired consistency. In fact, if we let G_n be the subgroup of G consisting of elements which fix pointwise the first n levels of the tree, and act on each subtree starting from the nth level in the same way, then clearly $G \cong G_n$, and we can show that $G_{n+1} \leqslant \Phi_2(G_n)$, for all n, so that $\{G_n : n \in \omega\}$ generates a proper filter $\mathscr{F}$. In the resulting model, U itself then provides a counterexample to DC_2, and the point of the Frattini-like construction of the filter is that this is sufficient to guarantee the truth of C_2^*.

As in the usual Frattini theory, every commutator and every square lies in $\Phi_2(G)$. So we just have to show that every member of G_1 is a product of commutators and squares. Now G_1 is the set of all elements of the form $\langle x,x\rangle$, and G is generated by $G \times G$ (where here the action of the two Gs is taken on the left and right subtrees) and y, where $y^{-1}\langle a,b\rangle y = \langle b,a\rangle$. Thus

$$\begin{aligned}\langle x,x\rangle &= \langle x,x^{-1}\rangle.\langle 1,x^2\rangle\\ &= (y^{-1}\langle x,1\rangle^{-1}y\langle x,1\rangle)\langle 1,x\rangle^2\end{aligned}$$

so $\langle x,x\rangle$ is a product of a commutator and a square, as desired.

5 Set-theoretic analogues of model-theoretic notions

We have seen above how appropriate model-theoretic constructions may correspond to constructions for independences between weak versions of the axiom of choice. I now consider what has for model theory become a very significant notion, since the Baldwin–Lachlan Theorem (Baldwin and Lachlan, 1971), and which has given a lot of impetus to the study of models of small Morley rank, namely the notion of a strongly minimal set. The corresponding notion in set theory already existed, though the name of 'amorphous' did not emerge until later. The definitions are as follows. A model is said to be *minimal* if it is infinite, but is not the disjoint union of two definable infinite sets, and *strongly* minimal means that this persists even under elementary extensions. The corresponding set-theoretic notion, (which is vacuous in the presence of the axiom of choice) is that a set is said to be *amorphous* if it is not the disjoint union of any two infinite sets (definable or not).

In Cherlin *et al.* (1985) an analysis of strongly minimal sets is given which involves consideration also of *strictly* minimal sets, being those which carry no non-trivial 0-definable equivalence relation. The idea is that general strongly minimal sets may be constructed from these by adjoining appropriate 'fibres'. Correspondingly there is a notion of *strictly* amorphous set, being one which carries no non-trivial partition at all.

Let us note that our definition of 'amorphous' is in truth analogous rather to a strengthening of 'strongly minimal' which could be called *higher-order* strongly minimal, since we are working in set theory, and no restriction is placed on the type of variable used in any definition. It is for this reason that it turns out that the only examples of amorphous sets are those which correspond to $\aleph_0$-*categorical* strongly minimal sets.

The aim of Truss (in press), only partially realized, is to give a 'classification' of amorphous sets. Now we have to be clear about what is meant by this. Usually, a classification theorem for a class of structures will isolate invariants ('classifiers') corresponding to the structures so that two structures receive the same classifier if and only if they are isomorphic. This is no good here, since it is easy, essentially by the construction given in Section 2, to construct Fraenkel–Mostowski models containing arbitrarily large sets (or even proper classes) of amorphous sets which look essentially indistinguishable set-theoretically, for example which are strictly amorphous. What is rather wanted is a notion of 'externally' isomorphic, or what comes down to the same thing, elementarily equivalent, inside a suitable structure. With this idea, it is argued in Truss (in press) that there is just a *set* of classifiers, which has cardinality $2^{\aleph_0}$. This is so far only proved with regard to certain special classifiers, sufficient to capture the 'bounded' amorphous sets.

An amorphous set U is said to be *bounded* if it has a strictly amorphous partition into finite sets. We gave an example of a strictly amorphous set in Section 2, and a modified construction produces a variety of bounded amorphous sets. One of the main results of Truss (in press) is that there is only one bounded amorphous set corresponding to each classifier. For the reasons given above, one has to be careful about what exactly is meant. The precise statement is an example of a 'reconstruction' result, which aims to characterize the properties of a model internally. Thus, a certain inner-model construction is performed, giving rise to a submodel $\mathfrak{N}$ of a model $\mathfrak{M}$. The desire is to show that in a sense, $\mathfrak{N}$ has not 'forgotten' where it came from. That is, there is a notion of forcing in $\mathfrak{N}$ such that on adjoining an $\mathfrak{N}$-generic filter $\mathscr{F}$, we return to $\mathfrak{M}$; or less ambitiously, when we perform the same construction by which $\mathfrak{N}$ was formed from $\mathfrak{M}$ inside $\mathfrak{N}[\mathscr{F}]$, we obtain $\mathfrak{N}$ again. A method like this was used in Truss (1974*b*) to obtain (admittedly a very weak) reconstruction result for Solovay's model (Solovay, 1970). In fact it seems true to say that it is generally the case that very strong hypotheses have to be imposed on $\mathfrak{N}$ in order to make the reconstruction work.

The main subdivision of the class of all amorphous sets is as follows.

(i) U is bounded (as just defined).

(ii) U is said to be of *projective type* if there is some non-degenerate pregeometry (as defined by Cameron (1990) for instance) on U satisfying certain conditions (such as the exchange property). Observe that from this it will follow that U is unbounded. This case splits into two, depending on whether there is or is not a bound on the cardinalities of the finite fields associated with geometries on (partitions of) U.

(iii) U may be unbounded but not of projective type.

Theorem 3.3 (i), (ii) illustrated two of these cases. Here is another.

THEOREM 5.1. There is an FM model in which the set of atoms forms an unbounded amorphous set which is not of projective type.

Proof. We give just one possible construction. Let $\mathfrak{M}$ be a model for FMC in which the set U of atoms has cardinality $\aleph_0$ and let $U = \{u_n : n \in \omega\}$. Let G be the group of all permutations of U with finite support, and for each $k \geqslant 0$ let π_k be the partition

$$\{\{u_0, u_1, \ldots, u_{2^k-1}\}, \{u_{2^k}, u_{2^k+1}, \ldots, u_{2.2^k-1}\}, \{u_{2.2^k}, u_{2.2^k+1}, \ldots, u_{3.2^k-1}\}, \ldots\}$$

of U. Let G_k be the setwise stabilizer of π_k in G, and let $\mathscr{F}$ be the filter generated by $\{G_k : k \geqslant 0\} \cup \{G_u : u \in U\}$. Then $\mathscr{F}$ is closed under conjugacy, since G contained only elements of *finite* support (and this was why we had to restrict to that subgroup), so we obtain a corresponding

FM model $\mathfrak{N}$. To see that U is amorphous in $\mathfrak{N}$ note that for any k, $\bigcap\{G_i : i \leqslant k\} \cap \bigcap\{G_{u_j} : j < 2^k\}$ acts transitively on $\{u_i : i \geqslant 2^k\}$, and the arguments previously used apply. Clearly U has partitions into 2^k-element sets for each k, so U is unbounded amorphous in $\mathfrak{N}$. $\square$

The proof of the following result is given in Truss (in press).

THEOREM 5.2. Suppose that the set U of atoms is strictly amorphous, and that $W \subseteq V(U)$ is a transitive model of FM containing U for which the standard sets fulfil AC. Then W is equal to the Fraenkel subuniverse of $W[w]$, where w is a W-generic well-ordering of U in type ω.

Here by a 'W-generic well-ordering in type ω' we mean a well-ordering obtained in the natural way from a W-generic subset of the notion of forcing P in W consisting of all finite sequences of distinct elements of U, partially ordered by end-extension. Similar work is carried out by Creed (to appear) with regard to the notion of an o-minimal set. According to the definition of Pillay and Steinhorn (1986), a structure $\mathfrak{A}$ is *o-minimal* if it is linearly ordered, and the only definable subsets of $\mathfrak{A}$ (with parameters allowed) are finite unions of intervals with endpoints in $A \cup \{\pm\infty\}$. The analogous definition of U being *o-amorphous* is that U is linearly ordered, and the *only* subsets of U are finite unions of intervals with endpoints in $U \cup \{\pm\infty\}$.

What is the correct notion of 'strictly o-amorphous'? Note that any o-amorphous set can automatically be split into arbitrarily many infinite subsets, just by taking intervals. So it may seem as if something a bit more complicated is required, and one possible definition is as follows: $(X, <)$ is said to be *strictly o-amorphous* if it is o-amorphous, and there is no disjoint pair (a, b), (c, d) of isomorphic or anti-isomorphic non-empty intervals. This however is easily seen to be equivalent to saying that there is no partition of U containing infinitely many non-singleton finite sets. One of the main results of Creed (to appear) is then that there is essentially only one strictly o-amorphous set, subject to the same provisos as Theorem 5.2.

What about an analogue of Theorem 5.1? The notion of a *bounded* o-amorphous set may be introduced much as before, but now things work out quite differently.

THEOREM 5.3. (Creed) Every o-amorphous set is bounded.

The reason for this is roughly speaking that any non-trivial partition of U into finite sets provides us with a partition into finitely many intervals, and one of these can be canonically selected. If U were unbounded o-amorphous, we would then be able to select a nested ω-sequence of intervals, which is clearly impossible in an o-amorphous set.

Finally I mention that if in the definition of U being o-amorphous we relax the requirement that the endpoints of the intervals should *lie* in U, then the class of sets thus obtained is greatly enriched, and many interesting configurations are possible. It is unclear at present however whether it will be feasible to carry out any sort of 'classification' in this case too.

Relational structures and dimensions

Frank O. Wagner

Mathematical Institute,
24-29 St. Giles,
Oxford,
OX1 3LB,
U.K.
Electronic mail address: wagner@maths.oxford.ac.uk

1 Introduction

It had been conjectured that, in every countable stable theory which is categorical in some power, one can define finitely many classical geometries (cf. Example 6.2) which determine the structure. More precisely, there were two conjectures.

(1) Lachlan (1974): Any stable $\aleph_0$-categorical theory is totally transcendental.

(2) Zil'ber (1984*b*): Any uncountably categorical theory which is not locally modular is bi-interpretable with an algebraically closed field.

In case (1) the deep analysis of $\aleph_0$-categorical $\aleph_0$-stable structures of Cherlin, Harrington and Lachlan (Cherlin *et al.*, 1985) would yield the geometries. By results of Lachlan (1974) and Zil'ber (1984*b*) (the Trichotomy Theorem), both conjectures are equivalent to conjectures about certain incidence structures on points and lines, called *pseudoplanes* (cf. the definition in Example 5.3), with stability and categoricity properties. Had they turned out to be true, they would have seriously limited the realm of stable structures.

In a series of surprising constructions, Hrushovski used an adaptation of the Ehrenfeucht–Fraïssé homogeneous universal relational structures to refute both conjectures. In particular, he constructed a strongly minimal set with a new geometry (Hrushovski, 1988*b*), thus contradicting (2), a stable $\aleph_0$-categorical pseudoplane (Hrushovski, 1988*b*), refuting (1), and he also used his method to amalgamate two strongly minimal structures (with an additional technical property) to a new strongly minimal structure living on the same domain as the other two (Hrushovski, 1992*a*), disproving a variant of Zil'ber's Conjecture. In Hrushovski (1993*a*) he also begins an analysis of the properties of strongly minimal structures obtained by his method and gives more examples of regular types, thereby answering various questions about symmetric almost-orthogonality (cf. Hrushovski, 1989*a*) and the locally modular vs. strongly regular dichotomy (cf. Hrushovski and Shelah, 1989).

It has thus turned out that among stable structures we encounter much more variety than previously thought, and Hrushovski's techniques are still being applied to construct new ones. In this vein, Baldwin (in press) has found an almost strongly minimal non-Desarguesian projective plane, contradicting another conjecture of Zil'ber (1984*b*) of a more geometric nature; and Baudisch (1992) has built a non-abelian connected uncountably categorical group which does not interpret a field and hence yields a group counterexample to (2). On the $\aleph_0$-categorical side, Herwig (1992) has been analysing weights and preweights in Hrushovski's example, and he has modified the construction to get a small theory with a type of infinite preweight with respect to itself (for definitions cf. the remark on p. 170). This may form the first step towards constructing a stable theory with more than one, but finitely many non-isomorphic countable models. The question of the existence of such a theory is still open; by a result of Lachlan (1973), such a theory cannot be superstable.

We shall give an axiomatic approach to these constructions, prove some of the results in question and analyse the structures thus obtained. While this article is an expansion and correction of Wagner (1988), it was also influenced by papers of Baldwin (in press), Baldwin and Shi (to appear), Herwig (1992), Herwig (in press), Kueker and Laskowski (1992) and Goode (1989). Of course, most of the propositions and methods are due to Hrushovski. Finally, I should like to thank Bernhard Herwig for some helpful discussions on the subject.

2 Closure

$\mathscr{L}$ shall be a fixed countable relational language and $\mathfrak{C}$ a class of finite $\mathscr{L}$-structures closed under isomorphism and substructures, and containing only countably many isomorphism types. Let $A \leqslant B$ (A is *self-sufficient* or *closed* in B) be a reflexive and transitive relation on elements $A \subseteq B$ of $\mathfrak{C}$, which is invariant under isomorphism. We write $A \nleqslant B$, if $A \subseteq B$, but $A \leqslant B$ does not hold. Put $A \nleqslant_{\min} B$, if $A \nleqslant B$, but any B' with $A \subseteq B' \subset B$ satisfies $A \leqslant B'$. Consider the following set of axioms.

C1 $A \subseteq B' \subseteq B \in \mathfrak{C}$ and $A \leqslant B$ implies $A \leqslant B'$.

C2 There are no infinite chains $A_1 \nleqslant_{\min} A_2 \nleqslant_{\min} \dots$ with $A_i \in \mathfrak{C}$ for $i \in \omega$.

C3 $\varnothing \leqslant A$ for all $A \in \mathfrak{C}$.

C4 Suppose $A \in \mathfrak{C}$ and $A_1, A_2 \subseteq A$. Then $A_1 \cap A_2 \nleqslant A_1$ implies $A_2 \nleqslant A$.

Remember that the *age*, $\mathrm{Age}(\mathfrak{M})$, of a structure $\mathfrak{M}$ is the class of finite structures isomorphic to a substructure of $\mathfrak{M}$. For an infinite structure $\mathfrak{M}$

with $\operatorname{Age}(\mathfrak{M}) \subset \mathfrak{C}$ and $A \subset \mathfrak{M}$ finite, put $A \leqslant \mathfrak{M}$ if $A \leqslant B$ for all finite B with $A \subseteq B \subset \mathfrak{M}$.

LEMMA 2.1. If $(\mathfrak{C}, \leqslant)$ satisfies C2, $\operatorname{Age}(\mathfrak{M}) \subseteq \mathfrak{C}$ and $A \subset \mathfrak{M}$ is finite, then there is a finite A' such that $A \subseteq A' \leqslant \mathfrak{M}$.

Proof. By C2 there is a maximal $\nleqslant_{\min}$-chain $A \nleqslant_{\min} A_1 \nleqslant_{\min} \cdots \nleqslant_{\min} A_n = A'$ of substructures of $\mathfrak{M}$. By maximality, there can be no finite B with $A' \nleqslant B \subset \mathfrak{M}$, as such B would have to contain some $A' \nleqslant_{\min} B'$. Hence $A' \leqslant \mathfrak{M}$. □

PROPOSITION 2.2. Suppose $(\mathfrak{C}, \leqslant)$ satisfies C2 and C4, $\operatorname{Age}(\mathfrak{M}) \subseteq \mathfrak{C}$ and $A \subset \mathfrak{M}$ is finite. Then there is a unique smallest closed superset of A in $\mathfrak{M}$.

Proof. By Lemma 2.1, there is at least one finite $A \subseteq A' \leqslant \mathfrak{M}$. Suppose A_1 and A_2 are minimal with $A \subseteq A_i \leqslant \mathfrak{M}$ $(i = 1, 2)$. If $A_1 \cap A_2 \neq A_1$, by minimality of A_1 we have $A_1 \cap A_2 \nleqslant A_1$. So C4 implies $A_2 \nleqslant A_1 \cup A_2 \subset \mathfrak{M}$, contradicting $A_2 \leqslant \mathfrak{M}$. □

We call the set given by Proposition 2.2 the *closure* of A in $\mathfrak{M}$, $\operatorname{cl}_{\mathfrak{M}}(A)$. If the ambient model is clear, we shall omit the subscript. Note that $\operatorname{cl}(\operatorname{cl}(A)) = \operatorname{cl}(A)$, and for $A \subseteq B$, $\operatorname{cl}(A) \leqslant \operatorname{cl}(B)$. In particular,

$$\operatorname{cl}(A \cap B) \leqslant \operatorname{cl}(A) \cap \operatorname{cl}(B) = \operatorname{cl}(\operatorname{cl}(A) \cap \operatorname{cl}(B)).$$

For infinite A, we put $\operatorname{cl}(A) = \bigcup\{\operatorname{cl}(A') : A' \subset A \text{ finite}\}$. Obviously, if the ambient model $\mathfrak{M}$ is saturated, $\operatorname{cl}(A) \subseteq \operatorname{acl}(A)$. Hence if $\mathfrak{M} \prec \mathfrak{N}$ and $\operatorname{Age}(\mathfrak{N}) \subset \mathfrak{C}$, then $A \leqslant \mathfrak{M}$ implies $A \leqslant \mathfrak{N}$.

Suppose $A_0 \leqslant A_i \in \mathfrak{C}$ for $i = 1, 2$. A *good amalgam* of A_1 and A_2 over A_0 is a set $D \in \mathfrak{C}$ such that there are embeddings $f_i \colon A_i \to D$ $(i = 1, 2)$ with $f_i A_i \leqslant D$ for $i = 1, 2$ and $f_1\restriction_{A_0} = f_2\restriction_{A_0}$. From now on, suppose that $\mathfrak{C}$ has the $\leqslant$-amalgamation property, i.e., for any $A_0 \leqslant A_i \in \mathfrak{C}$ $(i = 1, 2)$ there is a good amalgam of A_1 and A_2 over A_0.

PROPOSITION 2.3. If $(\mathfrak{C}, \leqslant)$ satisfies C1–3, then there is a countable structure $\mathfrak{M}$ satisfying:

- F1 $\operatorname{Age}(\mathfrak{M}) \subset \mathfrak{C}$;
- F2 if $A \leqslant \mathfrak{M}$ and $A \leqslant B \in \mathfrak{C}$, then there is an embedding $f \colon B \to \mathfrak{M}$ with $f\restriction_A = \operatorname{id}_A$ and $fB \leqslant \mathfrak{M}$.

Furthermore, any isomorphism between finite closed substructures of two countable structures satisfying F1,2 can be extended to an isomorphism of the whole. In particular, such a structure is uniquely described by F1,2 and is (first-order) homogeneous.

Proof. Enumerate representatives for the isomorphism types in $\mathfrak{C}$ as

$$A_1, A_2, A_3, \ldots.$$

We want to construct a chain $M_0 \leqslant M_1 \leqslant M_2 \leqslant \ldots$ of finite $M_i \in \mathfrak{C}$ for $i < \omega$ such that if $A_i \cong B_1 \leqslant B_2 \cong A_j$ and $B_1 \leqslant M_{n-1}$ for some $i, j \leqslant n-1$, then there is an embedding $f\colon B_2 \to M_n$ with $f\restriction_{B_1} = \mathrm{id}_{B_1}$ and $fB_2 \leqslant M_n$. We can start by taking $M_0 = \varnothing$. So suppose we have found $M_0 \leqslant M_1 \leqslant \cdots \leqslant M_n$. Enumerate representatives for the pairs $B_1 \leqslant B_2$ with $A_i \cong B_1 \leqslant M_n$ and $A_j \cong B_2$ for some $i, j \leqslant n$. As M_n and n are finite, there are only finitely many such pairs, say m. Put $D_0 = M_n$. For $0 \leqslant i < m$, if $D_0 \leqslant D_i$, then the first set of the ith pair is closed in D_0 and hence in D_i, so there is a good amalgam D_{i+1} of D_i and the second set of that pair over the first one. In particular $D_0 \leqslant D_1 \leqslant \ldots \leqslant D_m$. Put $M_{n+1} := D_m$. Then M_{n+1} satisfies the requirements.

Put $\mathfrak{M} = \bigcup_{i<\omega} M_i$. Then $\mathfrak{M}$ is countable, and $\mathrm{Age}(\mathfrak{M}) \subset \mathfrak{C}$, as any finite subset is contained in almost all of the M_i. Suppose $A \leqslant \mathfrak{M}$, $A \leqslant B \in \mathfrak{C}$. Then $A \cong A_i$, $B \cong A_j$, $A \subseteq M_k$ for some $i, j, k < \omega$. If $n = \max\{i, j, k\}$, then $A \leqslant \mathfrak{M}$ implies $A \leqslant M_n$, so there is a copy B' of B over A closed in M_{n+1}. If $B' \subseteq C \subset \mathfrak{M}$ is finite, then $C \subseteq M_m$ for some $n \leqslant m < \omega$, so $B' \leqslant M_n \leqslant M_m$ implies $B' \leqslant C$ by C1. Hence $\mathfrak{M}$ satisfies F1 and F2. (This also shows that a finite subset is closed in $\mathfrak{M}$ iff it is closed in almost all M_i.)

For the furthermore part, suppose f_0 is an isomorphism between $A \leqslant \mathfrak{M}$ and $A' \leqslant \mathfrak{M}'$. By countability we can write $\mathfrak{M} = \bigcup_{i<\omega} A_i$, $\mathfrak{M}' = \bigcup_{i<\omega} A'_i$, for increasing chains of finite $A_i, A'_i \in \mathfrak{C}$, and choose in addition $A = A_0$, $A' = A'_0$. Suppose we have found an isomorphism f_n between some M_n and M'_n with $A_n \subseteq M_n \leqslant \mathfrak{M}$ and $A'_n \subseteq M'_n \leqslant \mathfrak{M}'$ extending f_0. By Lemma 2.1, there is a finite B with $M_n \cup A_{n+1} \subseteq B \leqslant \mathfrak{M}$. Then $M_n \leqslant \mathfrak{M}$ implies $M_n \leqslant B$, so we can use F2 to find a copy $B' \leqslant \mathfrak{M}'$ such that $B'/M'_n \cong B/M_n$. This will extend f_n to $g\colon B \to B'$ with $A_{n+1} \subseteq B \leqslant \mathfrak{M}$. Similarly we can then extend g^{-1} to some $f^{-1}_{n+1}\colon M'_{n+1} \to M_{n+1}$ with $A'_{n+1} \subseteq M'_{n+1} \leqslant \mathfrak{M}'$. Then $f = \bigcup_{i<\omega} f_i$ is the required isomorphism. □

Corollary 2.4. Suppose $(\mathfrak{C}, \leqslant)$ satisfies C1–4 and $\mathfrak{M}$ satisfies F1 and F2. If $A, B \subset \mathfrak{M}$ are closed with the same (atomic) diagram, then they have the same type over $\varnothing$.

Proof. Consider an isomorphism $\sigma\colon A \to B$ and a finite closed subset A_0 of A. Then $B_0 = \sigma A_0$ is closed in B, but as A and B are closed, A_0 and B_0 are closed in $\mathfrak{M}$. So it suffices to prove the corollary for finite sets. By a back-and-forth argument we only have to show that for any isomorphic finite closed A, B in $\mathfrak{M}$ and finite closed $A \leqslant C \leqslant \mathfrak{M}$, there is $B \leqslant D \leqslant \mathfrak{M}$ with $(A, C) \cong (B, D)$. But this is guaranteed by F2. □

We call a countable model satisfying F1 and F2 a *generic model* for $(\mathfrak{C}, \leqslant)$. Henceforth $\mathfrak{M}$ will be a generic model.

REMARK. If $\mathfrak{C}$ contains only finitely many non-isomorphic structures of size n for any $n < \omega$, then F1 is (infinitely) first-order axiomatizable. This holds in particular if $\mathscr{L}$ is finite. More generally, if $\mathfrak{C}$ is a universal elementary class, that is, a theory with a universal set of axioms, then F1 is first-order axiomatizable. If in addition F2 is satisfied by a countable saturated model, then $\mathfrak{M}$ is saturated and $T = \mathrm{Th}(\mathfrak{M})$ is small (i.e., $S(\varnothing)$ is countable); if F2 is first-order axiomatizable, then T is $\aleph_0$-categorical. On the other hand, if $\mathfrak{M}$ is saturated, then any elementarily equivalent model also has its age contained in $\mathfrak{C}$, and any countable saturated elementarily equivalent model also satisfies F2.

Recall that a structure is *weakly saturated* if it realizes all pure types (i.e., all types over $\varnothing$).

PROPOSITION 2.5. Suppose the generic model $\mathfrak{M}$ is weakly saturated and that $A \leqslant \mathfrak{M} \prec \mathfrak{N}$ implies $A \leqslant \mathfrak{N}$. Then $\mathfrak{M}$ is saturated.

Proof. If $\mathfrak{M}$ is weakly saturated then in particular the theory must be small (there are only countably many pure types) and there is a countable saturated model $\mathfrak{N}$. Note that $\mathrm{Age}(\mathfrak{N}) \subset \mathfrak{C}$ by weak saturation. We claim that $\mathfrak{N}$ satisfies F2. So let $A \leqslant \mathfrak{N}$ be finite and $A \leqslant B \in \mathfrak{C}$. By weak saturation of $\mathfrak{M}$ and saturation of $\mathfrak{N}$, we may embed (a copy of) $\mathfrak{M}$ elementarily in $\mathfrak{N}$ over A. Then $A \leqslant \mathfrak{M}$, so by F2 there is a copy $B' \leqslant \mathfrak{M}$ of B over A, and as $\mathfrak{M} \prec \mathfrak{N}$, we have $B' \leqslant \mathfrak{N}$. Hence, by Proposition 2.3, $\mathfrak{M} \simeq \mathfrak{N}$. □

We now consider the following variant of $\leqslant$-amalgamation.

DEFINITION 2.6. We say that A is *n-closed* in B ($A \leqslant^n B$) if for all $A \subseteq B' \subseteq B$ with $|B' - A| \leqslant n$ we have $A \leqslant B'$. For $n = \omega$, this is our ordinary closure. A class $\mathfrak{C}$ has the $\leqslant^*$-amalgamation property if for all $A \leqslant B \in \mathfrak{C}$ and $n < \omega$ there is $m = m(A, B, n) < \omega$ such that $A \leqslant^m C \in \mathfrak{C}$ implies that there is $D \in \mathfrak{C}$ and embeddings $f: B \to D$ and $g: C \to D$ with $f{\restriction_A} = g{\restriction_A}$, $fB \leqslant^n D$ and $gC \leqslant D$.

Note that if C4 is satisfied and $A \leqslant^n B \leqslant C$, then we have $A \leqslant^n C$: if $A \subseteq C' \subseteq C$ with $|C' - A| \leqslant n$, then for $B' = C' \cap B$ we have $A \leqslant B'$ and $B' = C' \cap B \leqslant C'$, as $B \leqslant C' \cup B \subseteq C$ (by C4). Hence $A \leqslant C'$, as required.

There also is a corresponding version of F2.

F2* For any $A \leqslant B$ in $\mathfrak{C}$, any $n < \omega$, and any formula φ in the diagram of B with parameters in A there is $m = m(A, B, n, \varphi) < \omega$ such that if $A \leqslant^m \mathfrak{M}$, then there is $B' \leqslant^n \mathfrak{M}$ satisfying φ.

REMARK. If the language is finite, $A \leqslant^n \mathfrak{M}$ is definable and F2* is first-order (and also the diagram is a single formula!); for infinite languages this requires the definability of $A \leqslant^n \mathfrak{M}$ (i.e., for any $A \leqslant^n \mathfrak{M}$ there is ψ true of A such that if A' satisfies ψ, then $A' \leqslant^n \mathfrak{M}$) and some uniformity of the function m. If the generic model is saturated, by compactness it must satisfy F1 and F2* (if they are first-order axiomatizable) for suitable $m(A, B, n, \varphi)$, so the theory given by F1 and F2* is consistent. It is also complete: any countable saturated model of F1 and F2* satisfies F2, and a back-and-forth argument shows that they are equivalent in $\mathscr{L}_{\infty\omega}$.

PROPOSITION 2.7. Suppose $(\mathfrak{C}, \leqslant)$ satisfies C1–4 and has the $\leqslant^*$-amalgamation property. Then the generic model satisfies F2*.

Proof. We construct a countable model $\mathfrak{N}$ which satisfies F1, F2 and F2*. By Proposition 2.3 it must be isomorphic to the generic model.

We construct $\mathfrak{N}$ as the union of an increasing chain of closed sets M_i. This time, instead of amalgamating merely whenever we have a situation $A \leqslant M_i$, $A \leqslant B$, in order to obtain F2, we also amalgamate situations $A \leqslant B$, $A \leqslant^m M_i$, where $m = m(A, B, n, \varphi)$, and this will yield F2*. By the $\leqslant^*$-amalgamation property we have an amalgam M_{i+1} with $M_i \leqslant M_{i+1}$ and $B \leqslant^n M_{i+1}$, whence $B \leqslant^n \bigcup M_i = \mathfrak{N}$. We do this in such a way that we amalgamate over all possible situations. As we have in particular amalgamated over all $A \leqslant B$, $A \leqslant M_i$, the resulting model must satisfy F2. □

PROPOSITION 2.8. Suppose $(\mathfrak{C}, \leqslant)$ satisfies C1–4 and has the $\leqslant^*$-amalgamation property. Suppose further that F1 and F2* are first-order axiomatizable (this holds in particular for finite languages). Then the generic model is saturated.

Proof. First we prove that the theory is small. So consider a countable saturated model. By the remark on p. 158 it satisfies F2, so by Corollary 2.4 the type of a closed set is determined by its (atomic) diagram. But by our restriction on $\mathfrak{C}$ there are only countably many possible diagrams, hence only countably many types of closed sets over $\varnothing$. But clearly the type of any element is implied by the type of its closure, so there are only countably many types over $\varnothing$.

So there is a countable saturated model, which must satisfy F2*. By saturation it satisfies F2 and therefore is isomorphic to the generic model. □

Note, however, that the theory need not be $\aleph_0$-categorical: a uncountable saturated model need not satisfy F2. The proof also shows that the theory is 1-model-complete, that is, if $\mathfrak{N}_0 \subseteq \mathfrak{N}_1$ are models of the theory satisfying the same universal sentences with parameters from $\mathfrak{N}_0$, then $\mathfrak{N}_0$ is already elementary in $\mathfrak{N}_1$.

DEFINITION 2.9. Let $A_0 \leqslant A_i \in \mathfrak{C}$ $(i = 1, 2)$ and $A_1 \cap A_2 = A_0$. The *free amalgam* of A_1 and A_2 over A_0 is the structure with underlying set $A_1 \cup A_2$, whose only relations are those induced from A_1 and A_2. We denote it by $A_1 \amalg_{A_0} A_2$.

REMARK. Note that if $B \subset A_1 \amalg_{A_0} A_2$, then B is the free amalgam of $B \cap A_1$ and $B \cap A_2$ over $B \cap A_0$. Furthermore, if C4 is satisfied, $A_0 \leqslant A_i$ implies $A_0 \cap B \leqslant A_i \cap B$. Thus, if the original sets are freely amalgamated over a closed subset, so are the small ones. However, in non-relational structures this need no longer be true, and here lies one of the main obstacles to generalizing these methods.

PROPOSITION 2.10. Suppose $\mathfrak{C}$ is closed under free amalgamation and the A_i are closed in $A_1 \amalg_{A_0} A_2$, where the A_i are as in Definition 2.9. Then closure equals algebraic closure in the generic model $\mathfrak{M}$.

Proof. Suppose $a \in \mathrm{acl}(A) - \mathrm{cl}(A)$, of multiplicity n over $\mathrm{cl}(A)$, that is, the type of a over A has n realizations. Consider a closed set B containing a and its conjugates, with $\mathrm{cl}(A) \subset B \leqslant \mathfrak{M}$, and let B_0 be an isomorphic copy of B over $\mathrm{cl}(A)$. Then $\mathrm{cl}(A) \leqslant B$, so $\mathrm{cl}(A) \leqslant B \amalg_{\mathrm{cl(A)}} B_0$ and there is a copy $B' \amalg_{\mathrm{cl(A)}} B'' \leqslant \mathfrak{M}$. But then by Corollary 2.4, as $B, B', B'' \leqslant \mathfrak{M}$, all the elements in B' and B'' corresponding to the conjugates of a have the same type over $\mathrm{cl}(A)$, contradicting that the multiplicity of a over $\mathrm{cl}(A)$ is n. □

Finally we remark that for infinite languages, $A \leqslant B$ need not be definable.

EXAMPLE 2.11. Let $\mathscr{L}$ consist of equivalence relations E_i $(i < \omega)$, and let $\mathfrak{C}$ consist of all finite $\mathscr{L}$-structures such that E_{i+1} refines each E_i-class for all $i < \omega$. Write aE_i^A for the E_i-class containing a in the structure A. Define $A \nleqslant B$ to hold if there is $a \in A$ such that $|\bigcap_i aE_i^A| = 1$ and $|\bigcap_i aE_i^B| \geqslant 2$. This satisfies C1–3. Then $\mathfrak{C}$ is countable, has good amalgams, and the generic model is saturated. However, in the prime model $\mathfrak{M}_0$ the type $\{aE_ix : i \in \omega\}$ is realized only once for all $a \in \mathfrak{M}_0$ (namely by a), so any subset is closed, but it does not remain closed in the saturated model. Note, though, that C4 is not satisfied: the closure of a set is not unique.

REMARK. Clearly, if for any $A \nleqslant B$ there is a (quantifier-free) formula φ_B true for AB such that for any B' with $\models \varphi_B(AB')$ we have $A \nleqslant B'$, then closure is type-definable: A is closed iff A satisfies

$$\{\neg \exists Y \ \varphi_B(X, Y) : A \nleqslant B\}.$$

This holds in particular if $\mathscr{L}$ is finite. If closure is type-definable, then $A \leqslant \mathfrak{M} \prec \mathfrak{N}$ implies $A \leqslant \mathfrak{N}$.

3 Dimension

Suppose $(\mathfrak{C}, \leqslant)$ satisfies C1–4 and $\mathfrak{M}$ is a generic model. Let $\mathfrak{F}$ be the set of closed finite subsets of $\mathfrak{M}$.

DEFINITION 3.1. A *dimension function* for $\mathfrak{M}$ is a map $d: \mathfrak{F} \to \mathbb{R}_0^+$ which satisfies:

D1 for all $A, B \in \mathfrak{F}$, if $A \subseteq B$, then $d(A) \leqslant d(B)$;
D2 for all $A, B \in \mathfrak{F}$, $d(\mathrm{cl}(A \cup B)) + d(A \cap B) \leqslant d(A) + d(B)$;
D3 for all $A, B \in \mathfrak{F}$, if $A \cong B$, then $d(A) = d(B)$.

Put $d(A) = d(\mathrm{cl}(A))$ for any finite $A \subset \mathfrak{M}$, and $d(A/B) = d(A \cup B) - d(B)$. By D1, this is nonnegative. For infinite B, let

$$d(A/B) = \inf\{d(A/B') : B' \subset B \text{ finite}\}.$$

For finite $A_1, A_2 \in \mathfrak{M}$ put $A_1 \perp_B A_2$ (A_1 and A_2 are *independent* over B) iff $d(A_1/A_2B) = d(A_1/B)$ and $\mathrm{cl}(A_1B) \cap \mathrm{cl}(A_2B) = \mathrm{cl}(B)$. For infinite sets A_1, A_2, let $A_1 \perp_B A_2$ iff $A_1' \perp_B A_2'$ for all finite $A_1' \subseteq A_1$, $A_2' \subseteq A_2$.

LEMMA 3.2. For all finite $A, A_1, A_2 \subset \mathfrak{M}$ and all $B \subseteq \mathfrak{M}$:

1. $d(A/B) \geqslant 0$;
2. $B \subseteq B' \subseteq \mathfrak{M}$ implies $d(A/B) \geqslant d(A/B')$;
3. $d(A/B) = d(\mathrm{cl}(A)/B) = d(A/\mathrm{cl}(B)) = d(\mathrm{cl}(A)/\mathrm{cl}(B))$;
4. $d(A_1A_2/B) = d(A_1/A_2B) + d(A_2/B)$.

In particular the definitions of $A_1 \perp_B A_2$ for the finite and infinite case agree for finite A_1, A_2, and $A_1 \perp_B A_2$ iff $A_2 \perp_B A_1$ iff $\mathrm{cl}(A_1B) \cap \mathrm{cl}(A_2B) = \mathrm{cl}(B)$ and $d(A_1A_2/B) = d(A_1/B) + d(A_2/B)$.

Proof. (1) follows from D1. For (2), it is sufficient to consider finite B'. But then

$$d(AB) + d(B') \geqslant d(ABB') + d(\mathrm{cl}(AB) \cap \mathrm{cl}(B')) \geqslant d(AB') + d(B).$$

(3) and (4) follow from the definitions. □

Now suppose that $\mathfrak{M}$ is saturated. Then d can be defined for arbitrary finite closed subsets A of any model $\mathfrak{M}' \models \mathrm{Th}(\mathfrak{M})$ as follows. Let $A' \in \mathfrak{F}$ be isomorphic to A, and put $d(A) = d(A')$. Again we put $d(A) = d(\mathrm{cl}(A))$ for arbitrary finite A. Lemma 3.2 still holds in this context.

We now want to relate independence and free amalgams. There are two possible conditions, depending on the stability class intended.

DW For any a, X there is finite $X_0 \subseteq X$ with $d(a/X) = d(a/X_0)$. For any two finite closed A and B, if $A \perp_{A \cap B} B$ then we have $AB = A \amalg_{A \cap B} B$ and AB is closed.

DS For any closed A and B: if they are not freely amalgamated over $A \cap B$, or if $A \cup B$ is not closed, then there is $\gamma > 0$ and finite $A_0 \subseteq A$, $B_0 \subseteq B$, such that for all finite closed A' and B' with $A_0 \subseteq A' \subseteq A$ and $B_0 \subseteq B' \subseteq B$, we have $d(A') + d(B') \geqslant d(A'B') + d(A' \cap B') + \gamma$.

LEMMA 3.3. Suppose d satisfies D1–3 and DW or DS, A and B are closed in $\mathfrak{M}$, and $A \perp_{A\cap B} B$. Then A and B are freely amalgamated over $A \cap B$ and AB is closed.

Proof. If d satisfies DW and A and B are not freely amalgamated or AB is not closed, let $A_0 \leqslant A$ and $B_0 \leqslant B$ be finite closed subsets witnessing this. Let C be such that $A \cap B \geqslant C \geqslant A_0 \cap B_0$, $d(A_0/B) = d(A_0/C)$ and $d(B_0/A) = d(B_0/C)$, and put $A' = \mathrm{cl}(A_0C)$, $B' = \mathrm{cl}(B_0C)$, $C' = A' \cap B'$. Then

$$\begin{aligned} d(A'/B') &\geqslant d(A'/B) = d(A_0/B) \\ &= d(A_0/C) = d(A'/C) \geqslant d(A'/C') \geqslant d(A'/B'), \end{aligned}$$

so A' and B' are independent and closed over C', contradicting the second assertion of DW.

Suppose now d satisfies DS and A and B are not freely amalgamated over $C = A \cap B$, or they are freely amalgamated over C, but $A \amalg_C B$ is not closed. Let γ, A_0, B_0 be as in DS. Let $C_0 \subseteq C$ be finite and closed, such that $|d(A_0/C) - d(A_0/C_0)| < \gamma/3$, $|d(B_0/C) - d(B_0/C_0)| < \gamma/3$ and $|d(A_0B_0/C) - d(A_0B_0/C_0)| < \gamma/3$. Then

$$\begin{aligned} d(A_0B_0C_0) &= d(A_0B_0/C_0) + d(C_0) \\ &> d(A_0B_0/C) + d(C_0) - \gamma/3 \\ &= d(A_0/B_0C) + d(B_0/C) + d(C_0) - \gamma/3 \\ &= d(A_0/C) + d(B_0/C) + d(C_0) - \gamma/3 \\ &> d(A_0/C_0) + d(B_0/C_0) + d(C_0) - \gamma \\ &= d(A_0C_0) + d(B_0C_0) - d(C_0) - \gamma \\ &\geqslant d(\mathrm{cl}(A_0B_0C_0)) + d(\mathrm{cl}(A_0C_0) \cap \mathrm{cl}(B_0C_0)) - d(C_0) + \gamma - \gamma \\ &\geqslant d(A_0B_0C_0), \end{aligned}$$

a contradiction. □

PROPOSITION 3.4. If d satisfies DS (and D1–3, $(\mathfrak{C}, \leqslant)$ satisfies C1–4 and the generic model is saturated), then $T = \mathrm{Th}(\mathfrak{M})$ is stable.

Proof. Let $\mathfrak{N} \succ \mathfrak{M}$ be countable saturated, $X \subset \mathfrak{N}$ be closed and $a \in \mathfrak{N}$. Then $d(a/X)$ is the infimum of the dimension of a over the finite subsets of X, and countably many of them suffice to approximate $d(a/X)$ arbitrarily well. So there is a countable subset X_0 of X with $d(a/X) = d(a/X_0)$. Put $X_1 = \mathrm{cl}(X_0a) \cap X$.

1. $|X_1| \leqslant \aleph_0$,
2. $\mathrm{cl}(aX_1) \cap X \subseteq X_1$ (so as X is closed, X_1 is closed), and
3. for every $\epsilon > 0$, if $d(a/X) < \epsilon$ then there is finite $X' \subseteq X_1$ with $d(a/X') < \epsilon$.

Let $A = \mathrm{cl}(X_1 a)$, so A is countable. By (3), $d(a/X) = d(a/X_1)$. But

$$\begin{aligned} d(A/X) &= d(AX_1/X) = d(\mathrm{cl}(aX_1)/X) \\ &= d(a/X) = d(a/X_1) = d(A/X_1). \end{aligned}$$

By (2), $A \cap X = X_1$. Hence by Lemma 3.3, A and X are freely amalgamated over X_1 and AX is closed. But $\mathrm{tp}(A/X_1)$ and $\mathrm{tp}(X_1/X)$ (i.e., the choice of X_1) together with $A \amalg_{X_1} X$ determines $\mathrm{diag}(AX)$, and by Corollary 2.4 the diagram of a closed set determines its type, hence also $\mathrm{tp}(a/X)$. So there are at most $2^\omega |X|^\omega$ many types over X. Therefore T is stable. □

PROPOSITION 3.5. If d satisfies DW (and D1–3, $(\mathfrak{C}, \leqslant)$ satisfies C1–4, and the generic model is saturated), then T is ω-stable.

Proof. Let $\mathfrak{N}, X$ be as in the proof of Proposition 3.4 and let $X_0 \subseteq X$ be finite with $d(a/X) = d(a/X_0)$. Put $A = \mathrm{cl}(aX_0)$, a finite set, and $X_1 := A \cap X$. Then

$$d(A/X) \leqslant d(A/X_1) = d(a/X_1) \leqslant d(a/X_0) = d(a/X) = d(A/X),$$

so equality holds all the way through. In particular $A \mathop{\smile\!\!\!\!\!\!\vert\,}_{X_1} X$, so by Lemma 3.3 A and X are freely amalgamated over X_1 and AX is closed. Again $\mathrm{tp}(A/X_0)$ and $\mathrm{tp}(X_0/X)$ determine $\mathrm{tp}(A/X)$, whence there are at most $\omega \cdot |X|^{<\omega} = \omega + |X|$ many types over X (remember that as the generic model is saturated, T is small!). □

REMARK. If the group generated by $d(\mathfrak{F})$ or merely the difference $\{x - y : x, y \in d(\mathfrak{F})\}$ is discrete, then for any a and X there is a finite $X_0 \subseteq X$ with $d(a/X) = d(a/X_0)$, as $\inf\{d(a/X') : X' \subseteq X \text{ finite}\}$ must be attained.

PROPOSITION 3.6. Suppose d satisfies D1–3 and DS or DW, $(\mathfrak{C}, \leqslant)$ satisfies C1–4 and the $\leqslant$-amalgamation property, and the generic model $\mathfrak{M}$ is saturated, then for a model $\mathfrak{N} \models \mathrm{Th}(\mathfrak{M})$ and an algebraically closed subset A of N, any type over A is stationary and and its non-forking extension to $\mathfrak{N}$ is the free amalgam.

Proof. We may assume that $\mathfrak{N}$ is sufficiently saturated and A countable. As A is algebraically closed, it is closed in $\mathfrak{N}$. So suppose $\mathrm{tp}(b/\mathfrak{N})$ does not fork over A and consider the closure $B = \mathrm{acl}(bA)$ (in some big elementary supermodel). There is a countable A_0 with $A \subseteq A_0 \subset \mathfrak{N}$ and $d(b/\mathfrak{N}) = d(b/A_0)$, and by Lemma 3.3 $\mathrm{cl}(BA_0)$ and $\mathfrak{N}$ are freely amalgamated over $\mathrm{cl}(BA_0) \cap \mathfrak{N}$, and the amalgam is closed. Suppose there were $a \in \mathrm{cl}(BA_0) -$

$B\mathfrak{N}$. As B is algebraically closed, $\mathfrak{N}$ sufficiently saturated and $\text{tp}(b/\mathfrak{N})$ does not fork over A, there is a copy A_0' of A_0 in $\mathfrak{N}$, conjugate to A_0 over B, with $a \notin \text{cl}(BA_0')$. But also $d(b/\mathfrak{N}) = d(b/A_0')$ and $\text{cl}(B\mathfrak{N})$ is the free amalgam of $\text{cl}(BA_0')$ and $\mathfrak{N}$ over $\text{cl}(BA_0') \cap \mathfrak{N}$, whence $a \notin \text{cl}(B\mathfrak{N})$, contradiction. Therefore $B\mathfrak{N}$ is closed.

On the other hand, if there is some tuple $n \in \mathfrak{N}$, $n \notin A$, such that a relation $R(b', n)$ holds for some $b' \in B - A$, then we can find such $n \notin \text{cl}(BA_0) \cap \mathfrak{N}$ by saturation of $\mathfrak{N}$, algebraic closure of A and the non-forking of $\text{tp}(b/\mathfrak{N})$ over A. But this contradicts the fact that $\text{cl}(BA_0)$ and $\mathfrak{N}$ are freely amalgamated over $\text{cl}(BA_0) \cap \mathfrak{N}$.

Therefore the non-forking extension of $\text{tp}(B/A)$ over $\mathfrak{N}$ must be the free amalgam of B and $\mathfrak{N}$ over A, which is closed. Hence by Corollary 2.4 its type is uniquely determined, and $\text{tp}(b/A)$ is stationary. □

In particular this shows that finite equivalence relations have representatives for their classes which are algebraic over their set of definition.

COROLLARY 3.7. Under the assumptions of Proposition 3.6, if $A = \text{cl}(a\mathfrak{M})$ and $B = \text{cl}(b\mathfrak{M})$ where $\mathfrak{M}$ is $\aleph_1$-saturated and $\mathfrak{M} = A \cap B$, the following are equivalent:

1. $d(A/B) = d(A/\mathfrak{M})$ (i.e., $A \mathop{\smile\!\!\!\!\!\!|}_{\mathfrak{M}} B$);
2. $AB = A \amalg_{\mathfrak{M}} B$ and is closed; and
3. $\text{tp}(A/B)$ does not fork over $\mathfrak{M}$.

Proof. (1) implies (2) is Lemma 3.3. So assume (2) and consider a non-forking extension A' of $\text{tp}(A/\mathfrak{M})$ to B. By Proposition 3.6, $A'B = A' \amalg_{\mathfrak{M}} B$ and this is closed. Hence there is an automorphism taking $A'B$ to AB, and since $\text{tp}(A'/\mathfrak{M})$ does not fork over B, we have (3). Conversely, (3) implies (2) follows immediately from Proposition 3.6. Note that so far we have not required $\mathfrak{M}$ to be a model or saturated, and $d(B/\mathfrak{M})$ might also be infinite.

Now assume (3) and $d(A/B) < d(A/\mathfrak{M})$. Consider a finite $\bar{b} \in B$ with $d(a/\bar{b}) < d(a/\mathfrak{M})$, a countable algebraically closed $X \subset \mathfrak{M}$ such that $d(a/X) = d(a/\mathfrak{M})$ and $\text{tp}(ab/\mathfrak{M})$ does not fork over X and is stationary, and a realization $\bar{b}' \in \mathfrak{M}$ of $\text{tp}(\bar{b}/X)$. Then a and b are independent over X (as $\text{tp}(a/B)$ does not fork over $\mathfrak{M}$ and $\text{tp}(a/\mathfrak{M})$ does not fork over X) and a and b' are independent over X (as $\text{tp}(a/\mathfrak{M})$ does not fork over X). Since b and b' have the same strong type over X, it follows that $\text{tp}(a\bar{b}X) = \text{tp}(a\bar{b}'X)$. Hence $d(a/\bar{b}) = d(a/\bar{b}') \geqslant d(a/\mathfrak{M})$, which is a contradiction. □

Under additional assumptions, we may do with less saturation or even without assuming $A \cap B$ to be a model.

4 Predimension

Let $\mathfrak{C}$ be a class of finite $\mathscr{L}$-structures, closed under isomorphism and substructures and with only countably many isomorphism types.

DEFINITION 4.1. $\delta\colon \mathfrak{C} \to \mathbb{R}_0^+$ is a *predimension* if

- P1 there is no infinite chain $A_1 \subset A_2 \subset \cdots$ of $A_i \in \mathfrak{C}$ with $\delta(A_i) > \delta(A_{i+1})$ for $i \in \omega$,
- P2 for all $AB \in \mathfrak{C}$, $\delta(AB) + \delta(A \cap B) \leqslant \delta(A) + \delta(B)$,
- P3 if $A \cong B \in \mathfrak{C}$, then $\delta(A) = \delta(B)$, and
- P4 $\delta(\varnothing) = 0$.

If instead of P4 we only have $\delta(\varnothing) = \min \delta(\mathfrak{C})$, we can put $\delta'(A) = \delta(A) - \delta(\varnothing)$ and obtain a predimension δ'. We want to attach a closure relation $\leqslant$ to δ.

EXAMPLE 4.2. Suppose for all $m > 0$ there is $n < \omega$ such that $\delta(A) < m$ implies $|A| < n$, and $\delta(A) = 0$ iff $A = \varnothing$. We can then define $A \leqslant B$ iff $\delta(A) < \delta(A')$ for all $A \subset A' \subseteq B$ (notice the strict inequality). Then $\leqslant$ is reflexive and transitive. Reflexivity is obvious (for $A = B$ the condition is empty), and if $A \leqslant B \leqslant C$ and $A \subset A' \subseteq C$, then $\delta(A'B) \geqslant \delta(B)$ and $\delta(A' \cap B) \geqslant \delta(A)$ and at least one inequality is strict. Thus $\delta(A') \geqslant \delta(A'B) + \delta(A' \cap B) - \delta(B) > \delta(A)$ by P2. Furthermore $(\mathfrak{C}, \leqslant)$ satisfies C1–3. C1 holds trivially, C3 holds because $\varnothing$ is the only element of $\mathfrak{C}$ of predimension 0, and C2 because for any A there is $n < \omega$ such that $\delta(B) \leqslant \delta(A)$ implies $|B| \leqslant n$, so $n - |A|$ is the required bound. If $\mathfrak{M}$ is a generic model, put $d_{\mathfrak{M}}(A) = \min\{\delta(A') : A \subseteq A' \subset \mathfrak{M}\}$. Then $d_{\mathfrak{M}}$ is a dimension. If there are only finitely many relation symbols, then for any $m > 0$ there are only finitely many isomorphism types of A with $\delta(A) < m$. Hence '$A \leqslant \mathfrak{M}$' is first-order definable (even universal), F2 is first-order and $\mathrm{Th}(\mathfrak{M})$ is $\aleph_0$-categorical.

We should retain from this example the method by which P1 is obtained: large sets are forced to have big predimension. Thus any chain $A_1 \subset A_2 \subset \ldots$ of $A_i \in \mathfrak{C}$ with $\delta(A_i) > \delta(A_{i+1})$ must be finite, and we can even give a bound on its length in terms of $\delta(A_1)$, namely the minimal n such that $|B| \geqslant n + |A_1|$ implies $\delta(B) \geqslant \delta(A_1)$ for all B. Another way is to have δ integer-valued; note that in both cases $\mathrm{Im}(\delta)$ is discrete.

The disadvantage of using the strict inequality in the definition of closure is that in general this will not yield C4, i.e., closures will not be unique and the theory will not be stable. The advantage, however, is an easier amalgamation: there are fewer pairs $A \leqslant B$, and the growth of

$$\min\{\delta(A) : |A| = n\}$$

as a function of n is easier to control. Thus, this construction may be helpful in finding new unstable structures.

As we are interested mainly in stable structures, we shall proceed differently. So, for a model $\mathfrak{M}$ with $\mathrm{Age}(\mathfrak{M}) \subset \mathfrak{C}$ we put $d_{\mathfrak{M}}(A) := \min\{\delta(A') : A \subseteq A' \subset \mathfrak{M}, A' \text{ finite}\}$, which exists by P1 and P2, and $A \leqslant B$ iff

$\delta(A) \leqslant \delta(A')$ for all finite $A \subseteq A' \subseteq B$. As with the dimension, we put $\delta(A/B) = \delta(AB) - \delta(B)$.

LEMMA 4.3. 1. $\leqslant$ is reflexive and transitive;

2. $\leqslant$ satisfies C1–4;
3. $d_{\mathfrak{M}}(A) = \delta(\mathrm{cl}_{\mathfrak{M}}(A))$;
4. $d_{\mathfrak{M}}$ satisfies D1–3; and
5. $B \leqslant BA \cap BB'$ implies $\delta(A/B) \geqslant \delta(A/BB')$.

Proof. (1) Reflexivity is trivial. Suppose $A \leqslant B \leqslant C \in \mathfrak{C}$ and $A \subseteq A' \subseteq C$. Then $\delta(A'B) \geqslant \delta(B)$ and $\delta(A' \cap B) \geqslant \delta(A)$. Therefore $\delta(A') \geqslant \delta(A'B) + \delta(A' \cap B) - \delta(B) \geqslant \delta(A)$.

(2) C1 is trivial. For C2, we note that if $A \not\leqslant_{\min} B$, then $\delta(B) < \delta(A)$, and use P1. C3 follows from P4. Finally, if $AB \in \mathfrak{C}$, $B \leqslant AB$ and $A \cap B \subseteq B' \subseteq A$, then $\delta(B'B) \geqslant \delta(B)$. But $\delta(B') \geqslant \delta(B'B) + \delta(B' \cap B) - \delta(B) \geqslant \delta(B' \cap B)$. As $B' \cap B = A \cap B$, C4 follows.

(3) holds by definition of closure and $d_{\mathfrak{M}}$.

(4) D1 is trivial, D2 holds by (3) and P2 (as the sets in question are closed) and D3 holds by P3 and the definition of $d_{\mathfrak{M}}$.

(5) We have $\delta(AB) + \delta(BB') \geqslant \delta(AB \cap BB') + \delta(ABB') \geqslant \delta(B) + \delta(ABB')$. □

Now suppose that the generic model $\mathfrak{M}$ is saturated. Remember that on p. 160 we defined a continuation of the dimension for closed sets of $\mathfrak{M}$ to closed sets of any $\mathfrak{N} \succ \mathfrak{M}$. By P3 and the definition of closure, this continuation equals $d_{\mathfrak{N}}$.

Now we want to relate independence and free amalgamation on this level. The appropriate conditions are:

PS if A and B are closed, but not freely amalgamated over $A \cap B$, then there is $\gamma > 0$ and finite $A_0 \subseteq A$, $B_0 \subseteq B$, such that for all finite closed $A_0 \subseteq A' \subseteq A$ and $B_0 \subseteq B' \subseteq B$ we have $\delta(A') + \delta(B') \geqslant \delta(A'B') + \delta(A' \cap B') + \gamma$;

PW for any a, X there is finite $X_0 \subseteq X$ with $d(a/X) = d(a/X_0)$. Furthermore, if A and B are finite closed and $A \mathop{\smile\!\!\!\!\!\!|}_{A \cap B} B$, then A and B are freely amalgamated over $A \cap B$.

We do not have to require that for $A \mathop{\smile\!\!\!\!\!\!|}_{A \cap B} B$ the union AB is closed, due to the following lemma.

LEMMA 4.4. If $\leqslant$ satisfies P1–4 (and $\mathfrak{C}$ has the $\leqslant$-amalgamation property), A and B are closed in a generic model and $A \mathop{\smile\!\!\!\!\!\!|}_{A \cap B} B$, then AB is closed.

Proof. Suppose otherwise. Then for some finite closed $A_0 \leqslant A$ and $B_0 \leqslant B$ we have $\mathrm{cl}(A_0B_0) \not\subseteq AB$, and we can suppose $\mathrm{cl}(A_0B_0) \cap A = A_0$ and $\mathrm{cl}(A_0B_0) \cap B = B_0$. Let $X = A_0B_0$ and $Y = \mathrm{cl}(A_0B_0) - A_0B_0$. Then

$\mathrm{cl}(A_0B_0) = YA_0B_0 \supset A_0B_0$, and $\delta(Y/A_0B_0) = \delta(YA_0B_0) - \delta(A_0B_0) = d(A_0B_0) - \delta(A_0B_0) = -\gamma < 0$.

Let C be a finite closed subset of $A \cap B$ such that

$$\begin{aligned} |d(A_0/C) - d(A_0/A\cap B)| &< \gamma/3, \\ |d(B_0/C) - d(B_0/A\cap B)| &< \gamma/3, \quad \text{and} \\ |d(A_0B_0/C) - d(A_0B_0/A\cap B)| &< \gamma/3. \end{aligned}$$

Put $A_1 = \mathrm{cl}(A_0C) \leqslant A$ and $B_1 = \mathrm{cl}(B_0C) \leqslant B$. Now

$$\begin{aligned} & d(A_0B_0/C) > d(A_0B_0/A\cap B) - \gamma/3 \\ &= d(A_0/B_0(A\cap B)) + d(B_0/A\cap B) - \gamma/3 \\ &= d(A_0/A\cap B) + d(B_0/A\cap B) - \gamma/3 \\ &> d(A_0/C) + d(B_0/C) - \gamma \\ &= \delta(A_1/C) + \delta(B_1/C) - \gamma \\ &\geqslant \delta(A_1B_1) + \delta(A_1\cap B_1) - 2\delta(C) - \gamma \\ &\geqslant \delta(A_1B_1/C) - \gamma \\ &= \delta(A_1B_1C) - \delta(C) + \delta(XY) - \delta(X) \\ &\geqslant \delta(A_1B_1CXY) + \delta(A_1B_1C\cap XY) - \delta(C) - \delta(X) \\ &\geqslant d(A_1B_1) + \delta(X) - d(C) - \delta(X) \\ &= d(A_0B_0/C), \end{aligned}$$

as $C \leqslant A_1 \cap B_1$, $\mathrm{cl}(A_1B_1) \supseteq A_1B_1CXY$, $Y \cap AB = \varnothing$ and $A_1B_1 \supseteq A_0B_0 = X$. But this gives the required contradiction. □

Hence PW implies DW. To see that also PS implies DS, we have to prove some uniformity, i.e., if $A \amalg_{A\cap B} B$ is not closed, we have to find $\gamma > 0$ and finite subsets A_0 and B_0 as required by the definition (the case where A and B are not freely amalgamated is settled by PS). So let $A, B, A_0, B_0, X, Y, \gamma$ be as above. Then for $A_0 \leqslant A' \leqslant A$ and $B_0 \leqslant B' \leqslant B$ we have $\delta(A'B') + \delta(YX) \geqslant \delta(YA'B') + \delta(X)$, that is, $\delta(Y/A'B') \leqslant \delta(Y/X)$. Hence

$$\begin{aligned} & d(A') + d(B') - d(A'\cap B') \\ &= \delta(A') + \delta(B') - \delta(A'\cap B') \\ &\geqslant \delta(A'B') \\ &= \delta(A'B'Y) - \delta(A'B'Y/A'B') \\ &\geqslant d(A'B') - \delta(Y/A'B') \\ &\geqslant d(A'B') - \delta(Y/X) \\ &= d(A'B') + \gamma. \end{aligned}$$

COROLLARY 4.5. If δ satisfies P1–4 and PS, $\mathfrak{C}$ has the $\leqslant$-amalgamation property and the generic model is saturated, then the theory of the generic model is stable. If δ satisfies PW instead of PS, then the theory is ω-stable.

Again, if the subgroup generated by $\delta(\mathfrak{C})$ or merely $\delta(\mathfrak{C}) - \delta(\mathfrak{C})$ is discrete, then PS implies PW.

An important condition is the *modularity equation* for the free amalgam:

ME $\delta(A_1 \amalg_{A_0} A_2) + \delta(A_0) = \delta(A_1) + \delta(A_2)$.

PROPOSITION 4.6. Suppose δ satisfies ME. Suppose A and B are freely amalgamated over $A \cap B = C$, and $A' \leqslant A$ and $B' \leqslant B$, with $C = A' \cap B'$. Then $A'B'$ is closed in AB.

Proof. We may assume A and B to be finite. Clearly A' and B' are freely amalgamated over C. So consider D with $A'B' \subseteq D \subseteq AB$. We have

$$\begin{aligned}\delta(D) &= \delta(D \cap A) + \delta(D \cap B) - \delta(D \cap C)\\ &\geqslant \delta(A') + \delta(B') - \delta(C) = \delta(A'B').\end{aligned}$$ □

PROPOSITION 4.7. Suppose δ satisfies ME. Then $A \leqslant B$ and $A \leqslant^n C$ implies for $D = B \amalg_A C$ that $C \leqslant D$ and $B \leqslant^n D$.

Proof. For $D' \subseteq B \amalg_A C$, we have $D' = (D' \cap B) \amalg_{D' \cap A} (D' \cap C)$, whence

$$\delta(D') = \delta(D' \cap B) + \delta(D' \cap C) - \delta(D' \cap A).$$

In particular if $D' \supseteq C$, then

$$\delta(D') = \delta(C) + \delta(D' \cap B) - \delta(A) \geqslant \delta(C),$$

as $A \leqslant B \cap D'$; hence $C \leqslant D$. Similarly if $D' \supseteq B$ and $|D' - B| \leqslant n$, then $|(D' \cap C) - A| \leqslant n$ and $A \leqslant C \cap D'$, whence $\delta(D') \geqslant \delta(B)$ and $B \leqslant D'$. □

In particular we may take $m(A, B, n, \varphi) = n$.

PROPOSITION 4.8. If δ satisfies P1–4, PS or PW, and ME, then the following are equivalent for closed A, B:

1. $A \mathop{\smile\!\!\!\!\!\!\mid}_{A \cap B} B$;
2. $AB = A \amalg_{A \cap B} B$ and is closed; and
3. $\mathrm{tp}(A/B)$ does not fork over $A \cap B$.

Proof. By Corollary 3.7 (or rather the proof of it) we only have to show (2) implies (1). Suppose AB is closed and consider finite closed $A' \subseteq A$ and $B' \subseteq B$. We have to show that there is $C \subseteq A \cap B$ with $d(A'/B') \geqslant d(A'/C)$.

There are finite closed A_0 and B_0 with $A' \subseteq A_0 \subseteq A$ and $B' \subseteq B_0 \subseteq B$, such that A_0B_0 is closed and freely amalgamated over $A_0 \cap B_0 = C_0$. Put

$A'' = \mathrm{cl}(A'C_0)$. Then $A'' \subseteq A_0$, as by ME A_0 is closed in the closed set A_0B_0. Thus $A'' \leqslant A_0$. Hence

$$\begin{aligned} d(A'/B') &\geqslant d(A'/B_0) \\ &= d(A''/B_0) = d(A''B_0) - d(B_0) \\ &= \delta(A''B_0) - \delta(B_0) = \delta(A'') - \delta(C_0) \\ &= d(A'') - d(C_0) = d(A''/C_0) = d(A'/C_0). \end{aligned}$$

This implies (1). □

5 Structures

DEFINITION 5.1. Let R_i, $i \in I$, be the relation symbols in $\mathscr{L}$. We define for a finite $\mathscr{L}$-structure A

$$\delta(A) := |A| - \sum_{i \in I} \alpha_i |R_i^A|,$$

for suitable $\alpha_i > 0$.

We shall call a tuple $\bar{x}$ an *edge* if $\models R_i(\bar{x})$ for some $i \in I$.

LEMMA 5.2. This δ satisfies P2–4, PS and ME (for the class of all finite $\mathscr{L}$-structures and hence also for all subclasses).

Proof. Immediate for P2–4 and ME. Let A and B be two closed sets not freely amalgamated over $A \cap B$. Then there holds a relation $R_i(\bar{a}, \bar{b}, \bar{c})$, with $\varnothing \neq \bar{a} \in A - B$, $\varnothing \neq \bar{b} \in B - A$ and $\bar{c} \in A \cap B$. Take $\gamma = \alpha_i > 0$ and $A_0 = \{\bar{a}\bar{c}\}$, $B_0 = \{\bar{b}\bar{c}\}$. This proves PS. □

REMARK. If all α_i are multiples of some rational number (and in particular if I is finite and all α_i rational), then the group generated by $\delta(\mathfrak{C})$ is discrete. In that case, δ also satisfies PW.

In order to obtain a predimension, however, we may have to restrict ourselves to subclasses of

$$\mathfrak{C}_0 = \{A : A \text{ a finite } \mathscr{L}\text{-structure with } \delta(A') \geqslant 0 \text{ for all } A' \subseteq A\}.$$

We may have to restrict our class further to satisfy P1, and finally we have to prove the $\leqslant$-amalgamation property, axiomatizability of F1 and F2*, and saturation of the generic model. Most of the work usually goes into the proof of the amalgamation property and hence into establishing the existence of the generic model.

EXAMPLE 5.3. (Hrushovski) A stable $\aleph_0$-categorical pseudoplane.

A pseudoplane is a structure with two sorts, points and lines, together with an incidence relation satisfying:

1. on any line there are infinitely many points;
2. through any point there are infinitely many lines;
3. any two lines intersect in only finitely many points; and
4. through any two points there are only finitely many lines.

It was conjectured that such a structure cannot be $\aleph_0$-categorical and stable. This was, however, refuted by Hrushovski, 1988*b*. We shall give a simplified treatment of Hrushovski's example, independently found by the author and Herwig (1992).

Let $\mathscr{L}$ consist of a single binary relation R, and $\delta(A) = |A| - \alpha|R^A|$. (We consider R as an edge in an undirected graph and hence count $R(a,b)$ and $R(b,a)$ only once.) First we show that there is an α and an unbounded convex function f such that the class

$$\mathfrak{C}_f = \left\{ A \in \mathfrak{C}_0 : \begin{array}{l} R \text{ is antireflexive, symmetric and} \\ \delta(A') \geqslant f(|A'|) \text{ for any } A' \subseteq A \end{array} \right\}$$

has the $\leqslant^*$-amalgamation property. We should note that since f is unbounded and $\mathscr{L}$ is finite, for any $n < \omega$ there are only finitely many non-isomorphic $A \in \mathfrak{C}_f$ with $\delta(A) \leqslant n$. Hence P1 is satisfied, closure is definable and the generic model is saturated.

If we consider any A as a point with coordinates $(|A|, \delta(A))$, then $A_0 \leqslant A_1$ implies that A_0 lies below and to the left of A_1. For $A_0 \leqslant A_1, A_2$ the free amalgam $A_1 \amalg_{A_0} A_2$ forms the fourth point of a parallelogram. Hence if the slope s of A_0A_1 does not exceed that of A_0A_2, by convexity of f it is sufficient to assure that the (right) derivative of f at $|A_1|$ does not exceed s, for any such parallelogram. (By the remark on p. 159 this will also assure that any substructure of the free amalgam will lie in $\mathfrak{C}_f$.) Thus we require

$$0 \leqslant f'(|A_1|) \leqslant \text{slope of } A_0A_1 = 1 - \alpha\frac{|R^{A_1}| - |R^{A_0}|}{|A_1| - |A_0|},$$

that is, $f' \geqslant 0$ and

$$\frac{1 - f'(|A_1|)}{\alpha} \geqslant \frac{|R^{A_1}| - |R^{A_0}|}{|A_1| - |A_0|}.$$

Note that $|A_1| - |A_0| \leqslant |A_1|$.

Given $1 > \alpha > 0$, let $k_\alpha(n)$ be the best rational approximation of $1/\alpha$ from below with denominator at most n, and let f_α be the linear interpolation of $f_\alpha(0) = 0$, $f_\alpha(n) = 1 + \sum_{i=1}^{n-1}(1 - \alpha k_\alpha(i))$ for $n > 0$. Then $f = f_\alpha$ is obviously convex, and the slope requirement is easily seen to be satisfied.

We now want to choose α such that f_α is unbounded. Put $i(\alpha) = \sum_{i=1}^{\infty}(1 - \alpha k_\alpha(i))$ (the *index* of α). Then for any $n < \omega$ the set of α with

$i(\alpha) > n$ is open (changing α by a tiny amount does not change the initial approximations), but also dense: If p/q is a rational number in some open interval, then for tiny ϵ the approximations at $n \leqslant n_\epsilon$ for $q/(p+\epsilon)$ will satisfy $r_n/s_n < q/p$, that is $r_n/s_n \leqslant q/p - 1/ps_n \leqslant q/p - 1/pn$. Hence

$$i\left(\frac{p+\epsilon}{q}\right) \geqslant \sum_{n=1}^{n_\epsilon}\left(1 - \frac{p+\epsilon}{q}\cdot\frac{qn-1}{pn}\right) = \sum_{n=1}^{n_\epsilon}\frac{1}{qn} - \epsilon\sum_{n=1}^{n_\epsilon}\frac{qn-1}{pqn},$$

which can be arbitrarily large. So by the Baire Category Theorem the set of $0 < \alpha < 1$ with infinite index is dense. Note that if the index of α is infinite, f is unbounded.

In particular we may choose α of infinite index with $3/4 < \alpha < 1$. Then $f(1) = 1$, $f(2) = 2 - \alpha$, $f(3) = 3 - 2\alpha$ and $f(4) = 4 - 3\alpha$. We set $\{\text{points}\} = \{\text{lines}\}$, and a point a lies on a line b iff $R(a, b)$. For any point a the predimension $\delta(a) = 1$, so a point is always closed and there is a unique 1-type. In particular a point a_0 is closed in the set a_0a_1 of two connected points (which lies in $\mathfrak{C}_f$), so the n-fold free amalgam $\coprod_{a_0} a_0a_i$ lies in $\mathfrak{C}_f$ and a point is incident with arbitrarily many lines. On the other hand, the quadrangle a, b, c, d with $R(a, b)$, $R(b, c)$, $R(c, d)$, $R(d, a)$ has predimension $\delta = 4 - 4\alpha < f(4)$, so no two points can have more than one line in common. Hence the generic model is the required pseudoplane; it is even a linear space. (One easily sees that there exist three points not all on the same line.)

Remark. Recall that the *preweight* of a type p is the supremum of the number of independent elements which may fork with some realization of the type, over the given parameters. The *weight* of the type is the maximum of the preweights of its non-forking extensions; if we require the independent elements to also realize p, we talk of (pre-)weight with respect to itself. (See Makkai (1984), part D, for details.)

Herwig (1992) has shown that the (unique) 1-type in Hrushovski's example has weight ω with respect to itself, and preweight ω. However, in a stable $\aleph_0$–categorical theory no type over a finite set can have preweight ω with respect to itself, as there are only finitely many 2-formulas to witness forking, and any one of them may occur only finitely often. It had been an open problem, however, whether such a type may occur in a small stable theory; it is of particular interest in connection with finding a non-$\aleph_0$-categorical theory with only finitely many non-isomorphic models.

Example 5.4. (Herwig, in press) A small stable theory with a unique 1-type of infinite preweight with respect to itself.

This time our language consists of countably many disjoint binary relations R_i with coefficients $\alpha_i = \alpha/n_i$, where $\alpha \in (0, 1)$ is an irrational of infinite index and $n_i | n_{i+1}$. We note that together with α also the α_i have infinite index. We construct our function f step by step, using approximations

$$\delta_i(A) = |A| - \sum_{j=0}^{i} \alpha_j |R_j^A|.$$

So for a start put $n_0 = 1$ and

$$f(n) = f_{\alpha_0}(n) = 1 + \sum_{i=1}^{n-1}(1 - \alpha_0 k_{\alpha_0}(i))$$

for $n \leqslant k_0$, where k_0 is chosen such that $f(k_0) \geqslant 2$. Let ϵ be the minimal vertical distance from a point $(|A|, \delta_0(A))$ above the curve to it $(|A| \leqslant k_0)$, and choose n_1 such that $k_0(2k_0 - 1)\alpha < n_1\epsilon$. Then for any $|A| \leqslant k_0$, if $(|A|, \delta_0(A))$ lies strictly above the curve, so does $(|A|, \delta(A))$.

Now put $f(n) = f(k_0) + \sum_{i=k_0}^{n-1}(1 - \alpha_1 k_{\alpha_1}(i))$ for $k_0 \leqslant n \leqslant k_1$, where k_1 is chosen such that $f(k_1) \geqslant 3$. Choose n_2 such that $k_1(2k_1 - 1)\alpha/n_2$ is less than the minimal vertical distance from a point $(|A|, \delta_1(A))$ above the curve to it $(|A| \leqslant k_1)$, and so on. Consider the class

$$\mathfrak{C} = \{A : A = \varnothing \vee |A| = 1 \vee \delta(A) > f(|A|)\}.$$

We have to check the $\leqslant^*$-amalgamation property. So let $A \leqslant B, C$ be in $\mathfrak{C}$ and consider the free amalgam D. We may assume $|B| \leqslant |C| = n > 1$ and $k_{r-1} \leqslant n < k_r$. By the minimal slope argument, certainly $(|D|, \delta_r(D))$ lies above the curve by at least as much as $(|C|, \delta_r(C))$ does. But $|D| \leqslant 2|C|$, so

$$\delta_r(D) - \delta(D) \leqslant \tfrac{1}{2}|D|(|D| - 1)\alpha_{r+1} \leqslant k_r(2k_r - 1)\alpha/n_{r+1},$$

and by construction this is less than the vertical distance of $(|C|, \delta(C))$ to f. Hence $D \in \mathfrak{C}$. Therefore a generic model $\mathfrak{M}$ exists (note that $\mathfrak{C}$ is countable!).

As f is convex unbounded, P1 is satisfied. Now pruning out edges only increases the predimension and hence preserves membership in $\mathfrak{C}$. Therefore F1 is first-order axiomatizable and any model of $T = \mathrm{Th}(\mathfrak{M})$ has its age contained in $\mathfrak{C}$. Furthermore, for any A and r with $\delta(A) > r$ there is a quantifier-free formula ψ true for A and such that any A' in ψ also satisfies $\delta(A') > r$ (namely the elementary diagram with regard to sufficiently, but finitely many of the relations). So if $A \leqslant \mathfrak{M}$, there is a universal formula ψ true of A such that any $A' \models \psi$ also satisfies $A' \leqslant \mathfrak{M}$ (for this note that any proper superstructure must have greater predimension, as α is irrational, and that f is unbounded, so we have a bound on the size of a possible closure). Hence F2* is first-order axiomatizable. By Proposition 2.8, $\mathrm{Th}(\mathfrak{M})$ is small and $\mathfrak{M}$ is saturated.

Now $\mathfrak{M}$ has a unique 1-type p (realized by some point a), which is closed in any aa_i with $\models R_i(a, a_i)$. By amalgamation the set $\{a, a_i : 0 < i < \omega\}$, where only $R_i(a, a_i)$ holds, is realized closedly in $\mathfrak{M}$. As

$$\delta\left(a/\bigcup a_i\right) = 1 - \sum_{i>0} \alpha/n_i \geqslant 1 - \alpha \sum_{i>0} 1/2^i = 1 - \alpha > 0,$$

the set $\bigcup a_i$ is closed, the a_i form an infinite independent set of points and a forks with every one of them (the unique non-forking extension over a second point has no edges). Thus p has infinite preweight with respect to itself.

6 Geometry

DEFINITION 6.1. A *pregeometry* is set X together with a relation between elements $x \in X$ and finite subsets $X_0 \subset X$, *x is dependent on X_0*, which satisfies:

1. *Reflexivity:* x is dependent on $\{x\}$;
2. *Extension:* x dependent on X_0 and $X_0 \subseteq X_1$ implies x dependent on X_1;
3. *Transitivity:* x dependent on X_0 and every $y \in X_0$ dependent on X_1 implies x dependent on X_1; and
4. *Symmetry:* x dependent on X_0y but not on X_0 implies y dependent on X_0x.

If x is not dependent on X_0, we say that x is *transcendent over* X_0. If $X_0 \subseteq X$ is infinite, we shall say that x is dependent on X_0 if there is finite $X_1 \subset X_0$ such that x is dependent on X_1. The set of x dependent on X_0 will be denoted by $\mathrm{gcl}(X_0)$, the *(geometric) closure* of X_0. Clearly $\mathrm{gcl}(\mathrm{gcl}(X_0)) = \mathrm{gcl}(X_0)$. A subset X_0 is *independent* if any element $x \in X_0$ is transcendent over $X_0 - \{x\}$. A maximal independent subset of some X_0 is called a *basis*, its cardinality the *dimension* of X_0.

EXAMPLE 6.2. The three classical examples are:

1. the *disintegrated* case with x dependent on X_0 iff $x \in X_0$;
2. a vector space with linear dependency; and
3. a field with algebraic dependency.

Cases (1) and (2) are *locally modular*, i.e., closed sets A, B with $\dim(A \cap B) > 0$ satisfy $\dim(A) + \dim(B) = \dim(\langle A, B\rangle) + \dim(A \cap B)$.

DEFINITION 6.3. A *geometry* is a pregeometry such that the closure of a point is just the point itself.

To any pregeometry we can canonically associate a geometry: on $X - \mathrm{gcl}(\varnothing)$, we divide out by the equivalence relation 'x is dependent on y'. (By symmetry, this is equivalent to 'y is dependent on x' and hence an equivalence relation.)

Remember that a strongly minimal set is an infinite structure with a unique non-algebraic 1-type. Equivalently, every definable subset (in every elementary extension) is either finite or cofinite. In such a structure, algebraic dependency defines a pregeometry. It had been conjectured by Zil'ber that the only strongly minimal pregeometries are locally modular, or interpret a field (which must be algebraically closed). However, Hrushovski has constructed continuum many strongly minimal sets with geometry that is not locally modular geometry, and which do not even interpret a group. His construction even yields arbitrarily homogeneous geometries, that is for every $0 < n < \omega$ he constructs a strongly minimal set with a unique n-type over $\varnothing$, but different $(n+1)$-types. We shall give the construction, following Hrushovski (1993*a*) and Goode (1989).

Our language $\mathscr{L}$ will consist of a single symmetric $(n+1)$-ary relation R, which only holds on distinct elements. We put $\delta(A) = |A| - |R^A|$ for any finite $\mathscr{L}$ structure and consider the class

$$\mathfrak{C}_n = \{A : \delta(A') \geqslant n \text{ for all } A' \subseteq A \text{ with } |A'| \geqslant n\}.$$

This is a universal elementary class, so F1 is first-order, and it has free amalgamation. Therefore the $\leqslant^*$-amalgamation property is satisfied, as are P1 and PW; the theory is ω-stable and the generic model is saturated. It has a unique pure n-type, which consists of n unrelated points. Over any closed set A there is a unique type of maximal Morley rank ω, namely that of a completely unrelated point a, with $d(a/A) = 1$. There are also other non-algebraic points x with $d(x/A) = 0$; these may have arbitrarily large finite Morley rank over A. We can associate to this structure a geometry, that of its unique regular type of Morley rank ω: a point x is dependent on A iff $d(x/A) = 0$ iff $\mathrm{RM}(x/A) < \omega$. While closure in the sense of $\leqslant$ equals algebraic closure, it is smaller than this geometric closure.

We now want to collapse the types of finite rank and make them algebraic. This will result in a strongly minimal set; here algebraic closure will equal geometric closure, but strictly contain $\leqslant$-closure.

Definition 6.4. A pair (X, Y) of disjoint sets is called *minimal* if $\delta(XY) = \delta(Y)$ but $\delta(X'Y) > \delta(Y)$ for any $\varnothing \subset X' \subset X$. It is *biminimal* if there is no $Y' \subset Y$ such that (X, Y') is minimal. If (X, Y) is a minimal pair, then its *basis* is the set $Y_0 \subseteq Y$ of elements linked to X by an edge.

Now (X, Y_0) will be biminimal: first note that XY is the free amalgam $XY_0 \amalg_{Y_0} Y$, so $\delta(XY_0) - \delta(Y_0) = \delta(XY) - \delta(Y) = 0$ and for any $Y_0 \subset Z \subset XY_0$ we have $\delta(Z) - \delta(Y_0) = \delta(ZY) - \delta(Y) > 0$, so the pair XY_0 is minimal. And second, if $Z \subset Y_0$, then Y_0 and X are not freely amalgamated over Z, whence $0 = \delta(Y_0X) - \delta(Y_0) < \delta(ZX) - \delta(Z)$ and (X, Z) is not minimal.

If (X, Y) is minimal and $X \not\subseteq \mathrm{cl}(Y)$, then for any $Y' \supseteq Y$ with $X \not\subseteq \mathrm{cl}(Y')$ we have that XY and Y' are freely amalgamated over Y. In particular the type of XYY' is unique, which implies that $\mathrm{tp}(X/Y)$ has rank at most one. Conversely if $\mathrm{tp}(X/Y)$ has rank one in our structure above, then we find a sequence $X_0, \ldots, X_n$ with $\bigcup X_i = X$ and all $(X_i, YX_0 \ldots X_{i-1})$ minimal. There is a unique j such that $\mathrm{tp}(X_j/YX_0 \ldots X_{j-1})$ has rank one, and for $i \neq j$ $X_i \subset \mathrm{acl}(YX_0 \ldots X_{i-1}) = \mathrm{cl}(YX_0 \ldots X_{i-1})$; if Y is closed, this implies $j = 0$ and $X \subset \mathrm{acl}(YX_0)$. (X_0, Y) is a minimal pair associated to our rank one type over Y. So if we can restrict our class $\mathfrak{C}$ such that for every minimal pair (X, Y) (in fact biminimal pairs will suffice) the number of conjugates of X over Y is bounded, then types of rank one and hence types of finite rank will become algebraic and the result will be a structure of rank one.

We choose a function μ from isomorphism types of biminimal pairs to $\mathbb{N}$ with the property $\mu(X, Y) \geqslant \delta(Y)$. Define $\mathfrak{C}_\mu$ to be the class of those elements of $\mathfrak{C}_n$ which do not contain the $(\mu(X, Y)+1)$-fold amalgam of XY over Y, for any biminimal pair (X, Y). This is again a universal elementary class.

LEMMA 6.5. $\mathfrak{C}_\mu$ has the $\leqslant^*$-amalgamation property.

Proof. We take $m(A, B, n) = |B - A| + n$. So let $A = B_0 < B_1 < \ldots < B_n = B$ be a non-refinable tower of closed sets. We may then amalgamate step by step (i.e. assuming we have amalgamated C and B_i to C_i, we next amalgamate C_i and B_{i+1} over B_i) and thus assume $B_1 = B$. Then either $\delta(B) = \delta(A) + 1$ or $(B - A, A)$ is a minimal pair. In the first case B is obtained by adding a single unrelated point b to A and $bA \amalg_A C \in \mathfrak{C}_\mu$, since b cannot be involved in any biminimal pair of the amalgam, so any biminimal pair is contained in C and satisfies the multiplicity requirement.

In the second case, let A_0 be the base of $(B - A, A)$. If the (atomic) diagram of $(B - A/A_0)$ is realized in $C - A$ by some B', then there will not be any additional edges between B' and A, as $\delta(BA) = \delta(A)$ and $A \leqslant^{|B-A|} C$, so we can take $D = C$. Furthermore $A \leqslant^{n+|B-A|} C$ and $\delta(A) = \delta(B'A)$ implies $B'A \leqslant^n C$. Otherwise we consider the free composite $D = B \amalg_A C$. This will automatically satisfy the closure requirement, so we have to check that it is in $\mathfrak{C}_\mu$. Consider a biminimal pair (X, Y) in D, and denote for any $Z \subseteq D$ the intersections by $Z_A = Z \cap A$, $Z_B = Z \cap (B - A)$ and $Z_C = Z \cap (C - A)$.

CASE 1. $Y_B \neq \varnothing = Y_C$. (Similarly for $Y_C \neq \varnothing = Y_B$.) Suppose X' is a realization of $\mathrm{diag}(X/Y)$ with $X'_C \neq \varnothing$. Then $X'_A X'_B \neq \varnothing$ (as Y_B is non-empty) and we get from free amalgamation and minimality that

$$\begin{aligned}
&\delta(X'Y) + \delta(X'_A Y_A) \\
&\quad = \ \delta(X'_A X'_B Y) + \delta(X'_A X'_C Y_A) \\
&\quad > \ \delta(X'Y) + \delta(X'_A X'_C Y_A),
\end{aligned}$$

whence $\delta(X'_C/X'_AY_A) < 0$, contradicting $A \leqslant^{|B-A|} C$. (Note that $|X'_C| \leqslant |X'| \leqslant |B-A|$.) Therefore all such realizations are contained in B and we use the fact that $B \in \mathfrak{C}_\mu$.

CASE 2. $Y_B = Y_C = \varnothing$. Then for any realization X' of $\mathrm{diag}(X/Y)$ we get

$$\delta(Y) + \delta(X'_AY) = \delta(X'Y) + \delta(X'_AY) = \delta(X'_AX'_BY) + \delta(X'_AX'_CY).$$

But $\delta(X'_B/X'_AY) + \delta(X'_AX'_C/Y) \geqslant 0$, and the inequality is strict unless $X'_AX'_C = \varnothing$ or $X'_AX'_C = X'$. Similarly either $X' = X_C$ or $X_C = \varnothing$, whence X' must equal one of X'_A, X'_B or X'_C. If $X' = X'_B$ then $X' = B - A$ by minimality, and in this case, by the assumption that the atomic diagram of $(B - A/A_0)$ is not realized in $C - A$, it follows that no realization of $\mathrm{diag}(X/Y)$ can lie in C; hence all such realizations lie in $B \in \mathfrak{C}_\mu$. Otherwise all realizations lie in $C \in \mathfrak{C}_\mu$.

CASE 3. $Y_B \neq \varnothing \neq Y_C$. Then for any realization X' neither $X'_B = X'$ nor $X'_C = X'$. Suppose X^i is a sequence of disjoint realizations of $\mathrm{diag}(X/Y)$. Then unless $X^i_B = \varnothing$, we have $\delta(X^i_AX^i_C/Y) > 0$ by minimality, so

$$0 = \delta(X^i/Y) = \delta(X^i_B/X^i_AX^i_CY) + \delta(X^i_AX^i_C/Y) > \delta(X^i_B/X^i_AY)$$

(this uses that X^i_B and X^i_C are freely amalgamated over X^i_AY). If X^* is the union of those X^i_B with $X^i_B \neq \varnothing$ (say k_1 many), then

$$\delta(X^*/AY_B) = \delta(X^*/AY) \leqslant \sum \delta(X^i_B/AY) \leqslant \sum \delta(X^i_B/X^i_AY) \leqslant -k_1.$$

Now $A \leqslant B$ implies $\delta(X^*Y_B/A) \geqslant 0$, whence $\delta(Y_B/A) \geqslant k_1$.

By biminimality $\delta(X^i/Y_AY_C) > 0$. So if $X^i_B = \varnothing$, then by free amalgamation

$$\begin{aligned}\delta(Y_B/Y_AX^i_A) &= \delta(Y_B/Y_AY_CX^i)\\ &= \delta(YX^i) - \delta(Y_AY_CX^i)\\ &< \delta(Y) - \delta(Y_AY_C) = \delta(Y_AY_B) - \delta(Y_A)\\ &= \delta(Y_B/Y_A).\end{aligned}$$

Hence $|R^{Y_AY_BX^i_A}| > |R^{Y_AX^i_A}|$. Therefore, if X^* is the union of those X^i_A with $X^i_B = \varnothing$ (say k_2 many), then

$$\begin{aligned}\delta(Y_B/A) - \delta(Y_B/Y_A) &\leqslant \delta(Y_B/Y_AX^*) - \delta(Y_B/Y_A)\\ &\leqslant -\sum(|R^{Y_AY_BX^i_A}| - |R^{Y_AX^i_A}|)\\ &\leqslant -k_2.\end{aligned}$$

So the total number of realizations X^i, $k_1 + k_2$ satisfies

$$k_1 + k_2 \leqslant \delta(Y_B/A) + \delta(Y_B/Y_A) - \delta(Y_B/A)$$

$$\begin{aligned} &= \delta(Y_B/Y_A) = \delta(Y_B/Y_AY_C) \\ &\leqslant \delta(Y) \\ &\leqslant \mu(X,Y). \end{aligned}$$

Hence $D \in \mathfrak{C}_\mu$. □

Therefore the generic model is saturated; its theory is strongly minimal with a unique pure n-type but various $(n+1)$-types $(n > 0)$. The prime model will be algebraic over n transcendental points; all countable models are uniquely given by the cardinality d of a base $(n \leqslant d \leqslant \omega)$ (and algebraic over that basis). For $n = 0$ the theory obtained will be that of a pure set (and no relations satisfied); for $n = 1$ the pregeometry will be disintegrated: after dividing out by algebraic closure, dependence will be trivial. In any case, the restriction $\delta(A) \geqslant n$ for $|A| \geqslant n$ might have been weakened to $\delta(A) \geqslant 0$ for all A. In that case, we would still have constructed a strongly minimal set, but with many 2-types. Note that the choice of μ was arbitrary (subject to $\mu(X,Y) \geqslant \delta(Y)$), so this construction will yield the continuum many distinct theories.

REMARK. A weakening of Zil'ber's Conjecture was the question whether every uncountably categorical affine plane is Desarguesian. This was refuted by Baldwin, varying Hrushovski's construction.

EXAMPLE 6.6. (Baldwin, in press) Let $\mathfrak{C}$ be the collection of all finite symmetric graphs A such that for $\varnothing \subset A' \subseteq A$ we have

$$\delta(A') = |A'| - \frac{1}{2}|R^{A'}| > \frac{1}{2}$$

and A does not contain a cycle of length 4 (i.e., a quadrangle). This is a universal elementary class. Let μ again be a function on biminimal pairs (X,Y) satisfying:

1. if $X = \{x\}$ and $Y = \{y_1, y_2\}$ with $R(x, y_i)$ (for $i = 1, 2$), then $\mu(X,Y) = 1$;
2. otherwise, $\mu(X,Y) \geqslant 2\delta(Y)$ and $\mu(X,Y) > 2$.

Baldwin shows that there are continuum many possible choices for μ. Again define $\mathfrak{C}_\mu$ to be the class of those elements of $\mathfrak{C}_n$ which do not contain the $(\mu(X,Y)+1)$-fold amalgam of XY over Y, for any biminimal pair (X,Y). Then $\mathfrak{C}_\mu$ will satisfy the $\leqslant^*$-amalgamation property. (The transcendental case splits into $\delta(B/A) = 1$ and $\delta(B/A) = 1/2$, but in either case we may take the free amalgam.) The resulting structure will be uncountably categorical, of Morley rank 2. We interpret the points of the graph as both points and lines of a plane, and edges as incidence. Then the axioms of a projective plane are satisfied, but there is no group definable (as will be shown more generally in the next section). Hence in the associated affine plane there is no group definable and that plane cannot be Desarguesian.

7 Flatness

In this section, we assume that δ satisfies P1–4 and that the generic model is saturated. For any family $\{E_i\}_{i\in I}$ of sets and $\varnothing \neq S \subseteq I$, put $E_S = \bigcap_S E_i$. Let $E_\varnothing = \mathrm{cl}(\bigcup_I E_i)$.

DEFINITION 7.1. A generic structure is *flat* if for any finite number of finite closed subsets $\{E_i\}_{i\in I}$ we have $\sum_{S\subseteq I}(-1)^{|S|}d(E_S) \leqslant 0$.

REMARK. Note that for $|I| = 2$, this is just the modular inequality.

PROPOSITION 7.2. The generic structures from the last section are flat.

Proof. Clearly $\sum_{S\subseteq I}(-1)^{|S|}|E_S| = 0$, as we see by counting points in $\bigcup_I E_i$ setwise, then subtract those in two sets, add again those in three, etc. Similarly for every relation R we have

$$\sum_{S\subseteq I}(-1)^{|S|}|R^{E_S}| = |R^{\bigcup_I E_i}| - |\bigcup_I R^{E_i}| \geqslant 0.$$

Furthermore, for $S \neq \varnothing$ we have $d(E_S) = \delta(E_S)$. Hence

$$\sum_{S\subseteq I}(-1)^{|S|}d(E_S) \leqslant d(\bigcup_I E_i) - \delta(\bigcup_I E_i) \leqslant 0. \qquad \square$$

PROPOSITION 7.3. Suppose a structure is flat. Then any finite family $\{E_i\}_{i\in I}$ of closed sets of finite dimension over some closed set B also satisfies the inequality $\sum_{S\subseteq I}(-1)^{|S|}d(E_S/B) \leqslant 0$.

Proof. Suppose $\sum_{S\subseteq I}(-1)^{|S|}d(E_S/B) \geqslant \gamma > 0$. Let $C \subseteq B$ and F_i $(i \in I)$ be finite closed sets such that $C \subseteq F_i \subseteq E_i$ and for any $S \subseteq I$ we have $|d(F_S/C) - d(E_S/B)| < \gamma/2^{|I|}$. (Remember that the dimension is the limit of the dimension of finite subsets over finite subsets. So for $S \subseteq I$ we can first choose finite closed $C^S \subseteq B$ such that

$$|d(E_S/C^S) - d(E_S/B)| < \gamma/2^{|I|+1}$$

and take a closed finite C with $\bigcup_{S\subseteq I} C^S \subseteq C \subseteq B$. Then for $S \subseteq I$ choose F^S with $C \subseteq F^S \subseteq E_S$, such that

$$|d(F^S/C) - d(E_S/C)| < \gamma/2^{|I|+1}$$

and take F_i for $i \in I$ finite, closed, and big enough so that $F^S \subseteq F_S$ for all $S \subseteq I$.) Then by flatness,

$$\begin{aligned} 0 &\geqslant \sum_S (-1)^{|S|}d(F_S) \\ &= \sum_S (-1)^{|S|}d(F_S C) \end{aligned}$$

$$
\begin{aligned}
&= \sum_S (-1)^{|S|}[d(F_SC) - d(C)] \\
&= \sum_S (-1)^{|S|} d(F_S/C) \\
&> \sum_S (-1)^{|S|} d(E_S/B) - \gamma,
\end{aligned}
$$

a contradiction. □

Recall that given a set X, we can canonically construct a set X^{eq}, where we have added new elements in new sorts for every X-definable class of an $\varnothing$-definable equivalence relation. These new elements are called *imaginaries* because they are virtually present in the old structure; the old elements constitute the *home sort.*

PROPOSITION 7.4. Suppose δ satisfies ME and one of PS or PW, and the generic structure $\mathfrak{M}$ is flat. Let $e \in \mathfrak{M}^{\mathrm{eq}}$, $A \leqslant \mathfrak{M}$ be finite with $e \in \mathrm{acl}^{\mathrm{eq}}(A)$ and A' an independent conjugate of A over $\mathrm{acl}^{\mathrm{eq}}(e)$. If $E = \mathrm{acl}(e)$ (in the home sort) then $d(A/A'E) = d(A/E)$.

Proof. Let A_i be a sequence of n conjugates of A independent over e. For $i \neq j$ put $A_{ij} = \mathrm{cl}(A_iA_j)$. Then for $j \neq k$ we have $A_{ij} \cap A_{ik} = A_i$ (as A_j and A_k are independent over A_i), and for distinct i, j, k, l we have $A_{ij} \cap A_{kl} \leqslant E$. We apply the flatness formula to the family $\{A_{ij} : i < j\}$, over the base set E:

$$
\begin{aligned}
0 \geqslant\ & d\Big(\bigcup A_i/E\Big) - \sum_{i<j} d(A_{ij}/E) \\
& + \sum_i d(A_i/E)\left[\binom{n-1}{2} - \binom{n-1}{3} + \cdots\right] \\
=\ & d(A/E) + (n-1)d(A/A'E) \\
& - \tfrac{1}{2}n(n-1)[d(A/A'E) + d(A/E)] + n(n-2)d(A/E) \\
=\ & (\tfrac{1}{2}n^2 - \tfrac{3}{2}n + 1)[d(A/E) - d(A/A'E)].
\end{aligned}
$$

(Here we have used that the A_i are independent over, say, A_0 and hence by Proposition 4.8 $d(A_i/A_0) = d(A_i/A_0 \ldots A_{i-1})$.) For $n > 2$ this forces $d(A/A'E) = d(A/E)$. □

Note that in the last proof the assumption ME and PS or PW was only used to give the equivalence of non-forking and independence. The same is true in the following:

If e is an imaginary element and B a finite set in the home sort, then we may consider any set A such that e is definable over A and $\mathrm{tp}(B/A)$ does not fork over e, in order to define $\delta(B/e) := d(B/A)$.

LEMMA 7.5. If δ satisfies ME and one of PS or PW, this is well-defined.

Proof. Is is enough to show that if $A' \supseteq A$ and $\text{tp}(B/A')$ does not fork over e (and e is definable over A), then $d(B/A') = d(B/A)$. But this is Proposition 4.8, as $\text{tp}(B/A')$ does not fork over A and we may consider B, A and A' to be closed. □

As e is definable over B', this shows that in a flat stable structure, for every imaginary element e there is some tuple $E = \text{acl}(e)$ in the home sort, such that the dimension of any set B over e is the same as the dimension of B over E. Hence e may be identified for geometric purposes with E. (This property might be called geometric elimination of imaginaries.) Note that for $\aleph_0$-categorical theories, E will be finite.

REMARK. If $\text{Th}(\mathfrak{M})$ is superstable, then there is finite $E_0 \subseteq E$ with $d(E/E_0) = 0$: there cannot be an infinite ascending chain $E_0 < E_1 < \cdots < E$ with $d(E/E_i) > d(E/E_{i+1})$, as $\text{tp}(e/E_{i+1})$ would fork over E_i. We may then identify e with E_0. For theories of finite rank, this will actually give us $e \in \text{acl}^{\text{eq}}(E_0)$.

REMARK. In a theory of finite rank (any rank will do) we may, by the same equation, also define flatness using rank instead of dimension.

PROPOSITION 7.6. Suppose δ satisfies ME and one of PS or PW, and $\mathfrak{M}$ is flat. Then there is no interpretable group of non-zero dimension.

Proof. Suppose otherwise, and let a_1, a_2, a_3 be three independent generic elements. Then $a_1^{-1}a_2$ and $a_1^{-1}a_3$ are also independent generics. Put $E_i = \text{acl}(a_j : j \neq i)$ (in the home sort!) and $E_4 = \text{acl}(a_1^{-1}a_2, a_1^{-1}a_3)$. We add the parameters needed for the definition of the group to the language. Let g be the dimension of a generic group element a, i.e., $g = d(\text{acl}(a)) > 0$. By geometric elimination of imaginaries, $d(\bigcup E_i) = 3g$, $d(E_i) = 2g$, $d(E_i \cap E_j) = g$ $(i \neq j)$, and the intersection of any three distinct sets is $\text{cl}(\varnothing)$. Hence $0 \geqslant 3g - 4(2g) + 6g$, whence $g = 0$. □

REMARK. Suppose again ME and one of PS or PW. If we have free amalgamation, then $\text{cl} = \text{acl}$. Hence, if $a_1 \perp_A a_2$, then $\text{acl}(a_1 a_2 A) = \text{cl}(a_1 A) \cup \text{cl}(a_2 A)$. In particular there cannot be any infinite interpretable group of dimension zero. For a different reason this is also true for a strongly minimal set: If the theory is uncountably categorical, then a set of dimension zero is algebraic and cannot support an infinite group. In particular, none of the constructions in Sections 5 and 6 interpret an infinite group.

Two final remarks. In a third construction, Hrushovski (1992*a*) has merged two given strongly minimal sets (subject to some technical property) to one. That is, he has constructed a strongly minimal set which supports both given structures. This uses the geometric methods expounded so far (with Morley rank replacing dimension), but the actual proof is too complicated to be given in this expository article. The principal idea is to first construct a geometry of rank ω, amalgamating the two given ones, and then to collapse the sets of finite rank in a similar fashion to what we have done in the last section.

We may also leave the realm of relational structures. For example, one might amalgamate vector spaces over some finite field. Then the vector space dimension of some subspace A will replace its cardinality $|A|$ and we have to count relations in a suitable manner. There also is a notion of free amalgam: the direct sum of two spaces over a common subspace, with no extra relations. In this fashion, Baudisch (1992) has constructed an uncountably categorical non-abelian pure group (nilpotent of class 2 and exponent $p > 2$). Alas, the technical difficulties are again considerable, and attempts to build a connected non-abelian $\aleph_0$-categorical stable group—the group analogue of Hrushovski (1988*b*)—in a similar fashion have met with serious obstacles.

Bases in permutation groups

Peter J. Cameron

School of Mathematical Sciences,
Queen Mary and Westfield College,
Mile End Road,
London,
E1 4NS,
U.K.
Electronic mail address: P.J.Cameron@qmw.ac.uk

1 Introduction

Let G be a permutation group on a set X. A *base* for G is a sequence $(x_1, x_2, \ldots)$ of elements of X whose pointwise stabilizer in G is the identity. A base is said to be *irredundant* if no point in it is fixed by the pointwise stabilizer of its predecessors.

The prototypical example is the group $\mathrm{GL}(V)$ of invertible linear transformations of a vector space V, acting on the set of vectors in V: an irredundant base for G is the same thing as a basis for V (apart from the fact that a vector space basis is usually unordered).

Bases arose in the work of Sims on computational permutation group theory. It is important to be able to find a base (and a related 'strong generating set') quickly. An outline of this process will be sketched in the next section.

Given a base B for the permutation group G, the images of B under distinct elements of G are themselves distinct. This suggests, first, that group elements g can be uniquely represented by bases Bg, and second, that groups which permute transitively some special class of bases are of interest. Both these points will be considered below. A related question is, given a structure on X with automorphism group G, how to give the points of X unique representations in terms of a base B (as is well-known when $G = \mathrm{GL}(V)$).

If our group G is of infinite degree, the size of a base may be finite or infinite. In this paper, I will consider only groups with finite base size. There is one general observation that can be made about such groups. Let G have a finite base. Then G is closed in the symmetric group (in the topology of pointwise convergence); for, if a sequence of elements of G converges pointwise, then all elements of the sequence from some point on agree on a base, and so are equal. (This argument presupposes that the degree is countable; in general, we must use more elaborate machinery to define 'pointwise convergence'.) It follows from this that G is the full automorphism group of some relational structure. In fact, we can be much more precise:

PROPOSITION 1.1. Let G be a permutation group on the infinite set X, which has a base of size m. Then G is the automorphism group of a $(2m+2)$-ary relation on X.

Proof. We let $\{O_i : i \in I\}$ be the set of orbits of G on X^{m+1}. We use the well-known fact that any set carries a *rigid* binary relation, one whose automorphism group is trivial. Let S be a rigid binary relation on I, and define the $(2m+2)$-ary relation R by the rule

$$(\bar{x}, \bar{y}) \in R \quad \Leftrightarrow \quad (\exists (i,j) \in S)((\bar{x} \in O_i) \wedge (\bar{y} \in O_j)).$$

Clearly G preserves the relation R; we show that G is the full automorphism group.

The partition $X^{m+1} = \bigcup_{i \in I} O_i$ is preserved by $\mathrm{Aut}(R)$, since the sets O_i are equivalence classes of the relation $\equiv$ defined by $\bar{x} \equiv \bar{y}$ if and only if, for all $\bar{z} \in X^{m+1}$,

$$(\bar{x}, \bar{z}) \in R \quad \Leftrightarrow \quad (\bar{y}, \bar{z}) \in R$$

and

$$(\bar{z}, \bar{x}) \in R \quad \Leftrightarrow \quad (\bar{z}, \bar{y}) \in R.$$

Since S is rigid, automorphisms of R fix each set O_i for $i \in I$. Now let $(x_1, \ldots, x_m)$ be a base. Then the $(m+1)$-tuples $(x_1, \ldots, x_m, y)$, for $y \in X$, all lie in different G-orbits; so an automorphism of R which fixes $x_1, \ldots, x_m$ fixes every point. Now, if h is any automorphism of R, then there is an element $g \in G$ such that $x_i h = x_i g$ for $i = 1, \ldots, m$; then hg^{-1} is an automorphism fixing $x_1, \ldots, x_m$, whence $h = g \in G$. □

REMARK. If the group G is t-transitive, we may take the arity of the relation to be $2(m+1)-t$, by taking the first t components of the $(m+1)$-tuples $\bar{x}$ and $\bar{y}$ in the definition of R to be equal. In particular, a regular permutation group is the automorphism group of a ternary relation (though for many groups a binary relation suffices (Godsil, 1979)).

2 Computing a base

A finite permutation group G on X is usually presented computationally by a set of permutations of X which generate G. Given such a generating set, a simple breadth-first search procedure will compute the orbit of any point x. Moreover, for every point y in the orbit xG, a permutation carrying x to y is found. The set of such permutations is a transversal for the stabilizer G_x in G. Each permutation arises as a word in the generators, and these words form a *Schreier transversal*, that is, any initial segment of any word in the set is also in the set. With this information, an argument due to Schreier gives a set of generators for the subgroup G_x. (Simply take all non-identity expressions of the form sgt^{-1}, where s is a coset representative, g a generator, and t the representative of the coset containing sg.)

Calling this procedure recursively until the group is reduced to the identity gives the *Schreier–Sims algorithm*. Its output consists first of a base $B = (x_1, x_2, \ldots)$ for G, and second, for each i, a set S_i of coset representatives for $G(i)$ in $G(i-1)$, where $G(i)$ is the stabilizer of $x_1, \ldots, x_i$ (so that $G(0) = G$). Now it is clear that, if the base is finite, then the union of the sets S_i generates G; and, moreover, $\bigcup_{j>i} S_j$ generates $G(i)$. Such a set is called a *strong generating set* for G.

Given a base and strong generating set as above, many tasks become easy. For example, given an sequence B' with the same length as B, we can decide whether or not $B' = Bg$ for some $g \in G$, and find the unique g if so. For let $B' = (x'_1, x'_2, \ldots)$. If $x'_1 \notin x_1G$, then no such g exists; otherwise, there is a unique $s_1 \in S_1$ with $x_1s_1 = x'_1$, and we can replace B' by $B's_1^{-1}$ and work in G_{x_1}, using induction.

This remark justifies representing elements $g \in G$ by sequences Bg (which will be much shorter than complete permutations if $|B|$ is small). Also, it provides a membership test for G. Let h be any permutation of X. Test whether $Bh = Bg$ for some $g \in G$. If not, then $h \notin G$; if so, then $h \in G$ if and only if $h = g$. (Using the inductive method outlined, this just involves checking whether h is reduced to the identity when the entire base has been processed.) Finally, it allows us to choose a 'random' (uniformly distributed) element of G, by composing random elements of $S_1, S_2, \ldots$ in reverse order.

We see that the size of B is important here. We turn to this question in the next section.

3 Base size in finite groups

It is easy to ensure that our algorithm for finding a base produces an irredundant base—simply require that each chosen point is not fixed by all generators of the stabilizer of its predecessors. (Such a point must exist as long as the stabilizer is non-trivial). However, there is no guarantee that the base found is as small as possible. Indeed, Furst, Hopcroft and Luks (Furst *et al.*, 1980) observed that the Schreier–Sims algorithm runs in polynomial time (with a small modification to ensure that the number of generators doesn't become too large); but Blaha (1992) showed that finding the minimal base size is NP-hard.

However, for primitive groups the situation seems to be better. This is in part because of a connection between the base size and the order of a finite permutation group (of degree n), as follows:

PROPOSITION 3.1. If G has an irredundant base of size b, then $2^b \leqslant |G| \leqslant n^b$.

This is because, in the stabilizer chain, $|G(i-1){:}G(i)|$ is equal to the length of the orbit of x_i under $G(i-1)$, which lies between 2 and n inclusive; and the product of these indices is $|G|$, since $G(0) = G$, $G(b) = 1$. Said otherwise, the size of an irredundant base lies between $\log_2 |G|/\log_2 n$ and $\log_2 |G|$.

The upper bound for base size is attained by groups with all orbits of length 2. There is evidence now that most subgroups of the symmetric group S_n are similar to this (in other words, uninteresting to a group theorist); see Pyber (1993). On the other hand, there is also some evidence that many, perhaps most, primitive groups are closer to the lower bound.

The main tool for studying primitive groups is the *O'Nan–Scott Theorem*, which reduces many questions about them to questions about groups which are close to being simple (especially their maximal subgroups and their projective representations in non-zero characteristic). The statement which follows is accurate and sufficient for some purposes, but not the strongest possible; but there are many other accounts available (Liebeck *et al.*, 1988, for example). First, a few definitions.

A primitive permutation group G is said to be *basic* if it is not contained in a wreath product of symmetric groups with the product action; that is, G doesn't preserve the structure of the set A^B of all functions from B to A, where neither A nor B is a singleton. A transitive group G is *affine* if it has an elementary abelian regular normal subgroup; in other words, G is contained in an affine group over a finite prime field. Also, G is *diagonal* if its socle N is a direct power T^d of a non-abelian simple group T, and the point stabilizer intersects N in a diagonal subgroup. Finally, G is *almost simple* if its socle is non-abelian simple; that is, $T \leqslant G \leqslant \mathrm{Aut}(T)$ for some non-abelian simple group T. (Note that, unlike the other definitions, the way that such a group acts as a permutation group is completely unspecified.)

THEOREM 3.2. (O'Nan–Scott Theorem) Let G be a basic primitive permutation group on a finite set. Then G is affine, diagonal, or almost simple.

REMARK. 1. The extra precision in the Theorem involves more detailed description of non-basic primitive groups. If such a group G has socle N, then $G \leqslant G_0 \text{ wr } S_d$ and $N = N_0^d$, where G_0 is primitive and N_0 is a normal subgroup of G_0; either N_0 is the socle of G_0, or N_0 and N are regular (the latter, if N is non-abelian, is the so-called *twisted wreath product* case).

2. For extensions of the O'Nan–Scott Theorem to some classes of infinite permutation groups, see Macpherson and Pillay (1994) and Macpherson and Praeger (in press *b*).

Now, in order to get upper bounds for base size in primitive groups, we need

(a) results connecting base size in a wreath product to that in the factors;
(b) bounds for base size in affine, diagonal, and almost simple groups.

Some results of both kinds exist, but they are somewhat inconclusive. We summarize the position for almost simple groups. The following result uses the classification of finite simple groups.

THEOREM 3.3. Let G be a primitive, almost simple, permutation group of degree n. Then one of the following holds:

(a) G is a symmetric or alternating group acting on subsets of fixed size, or partitions into parts of constant size, of its domain;
(b) G is a classical group acting on an orbit of subspaces, or pairs of subspaces of complementary dimension, of its natural module;
(c) $|G| \leqslant n^c$, where c is an absolute constant.

CONJECTURE 3.4. In case (c) of the above Theorem, G has a base of size bounded by an absolute constant.

REMARK. It is known that the conjecture is true in the case of groups whose socle is alternating (and, in fact, all but finitely many such groups have bases of size 2).

Another immediate consequence of Proposition 3.1 is that, if the minimum base size of G is b, then *any* irredundant base for G has size at most $b \log_2 n$. Blaha (1992) has shown that this bound cannot be essentially improved in general.

Faced with this, we may try to be cleverer in our choice of base. We want the stabilizer chain to descend from G to 1 as quickly as possible. One heuristic for this is the *greedy algorithm*: choose each base point x_i so that the orbit length of x_i under $G(i-1)$ (the index $|G(i-1){:}G(i)|$) is as large as possible. Blaha (1992) showed that, if the minimum base size is b, then the greedy algorithm will always find a base of size at most $b(\log\log n + c)$; moreover, this is also best possible. However, it is conjectured that, if G is primitive, then the greedy algorithm will over-estimate the base size by a constant factor at worst. The following example illustrates the worst case known (where the constant is 9/8).

EXAMPLE 3.5. Let $G = S_m$, the symmetric group on $\{1, \ldots, m\}$, and let G act on the set of 2-element subsets of $\{1, \ldots, m\}$, so that the degree is $n = m(m-1)/2$. Then the stabilizer of $\{i, j\}$ and $\{i, k\}$ fixes all three points i, j, k; so there is a base of size roughly $2m/3$, and it is easily seen that this is the smallest possible. (There are some end effects: if $m \equiv 2 \pmod 3$, we have to add one more 2-set to the base in order to fix the last two points.) On the other hand, the greedy algorithm chooses disjoint 2-subsets until at most seven points remain uncovered; then, in order to fix the points of $\{i,j\}$ and $\{k,l\}$, it is necessary to add a pair like $\{j,k\}$ to the base. This gives a base whose size is roughly $3m/4$.

Another heuristic proposed by Maund (1989) is to choose the point x_i whose stabilizer (in $G(i-1)$) has the largest number of orbits. Similar results apply in this case.

4 Sharply t-transitive groups

A permutation group is *geometric* if it permutes its irredundant bases transitively. The reason for the name will appear later. First, we consider a special case of great historic importance.

A permutation group G on X is said to be *sharply t-transitive* if, given any two t-tuples $(x_1,\ldots,x_t)$ and $(y_1,\ldots,y_t)$ of distinct points of X, there is a unique element $g \in G$ such that $x_i g = y_i$ for $i = 1,\ldots,t$. (To avoid trivialities, we assume that $t \leqslant |X|$.) In other words, a group is sharply t-transitive if it is t-transitive and some (or equivalently every) t-tuple of distinct points is a basis. Clearly a sharply t-transitive group is geometric. Sharply 1-transitive groups are regular; since every abstract group acts in this way (Cayley's Theorem), no restrictions about the abstract structure are possible. So we assume that $t > 1$.

More generally, we say that a set of permutations is sharply t-transitive if it satisfies the condition of the above definition.

Sharply t-transitive groups or sets are connected with various classical geometric objects. For example, let S be a sharply 2-transitive set of permutations of the set X. We define an incidence geometry on the point set $X \times X$, with three different kinds of lines:

(a) *horizontal* lines $\{(x,a) : x \in X\}$, for $a \in X$;
(b) *vertical* lines $\{(a,y) : y \in X\}$, for $a \in X$;
(c) *oblique* lines $\{(x,xg) : x \in X\}$, for $g \in S$.

It is easily checked that two points lie on exactly one line. If X is finite, a counting argument shows that the Euclidean parallel postulate holds (in the form of *Playfair's Axiom*: given a point p and line L, a unique line parallel to L passes through p, where two lines are parallel if they are equal or disjoint); so the geometry is an affine plane. If X is infinite, the parallel postulate is equivalent to a subtle additional property of the sharply 2-transitive set. Conversely, given an affine plane, if we choose two parallel classes to be the horizontal and vertical lines, and we identify the point set with $X \times X$ for some set X, then the remaining lines are the graphs of permutations of X, and the set of all such permutations is sharply 2-transitive.

The condition that the set S is a group of permutations is equivalent to the plane being coordinatized by a *nearfield.* (This is an algebraic structure satisfying the axioms for a field except for the commutativity of multiplication and the left distributive law.) In this case, S is isomorphic to the one-dimensional affine group

$$\{x \mapsto xa + b : a, b \in F, a \neq 0\}$$

of permutations of the nearfield F.

The symmetric group of degree n is both sharply n-transitive and sharply $(n-1)$-transitive, and the alternating group is sharply $(n-2)$-transitive. In the second half of the nineteenth century, Mathieu (1871) discovered two 'sporadic' simple groups, M_{11} and M_{12}, which are sharply 4-transitive and sharply 5-transitive respectively; and his contemporary Jordan (1871) showed that this list is complete for $t \geqslant 4$:

THEOREM 4.1. The only sharply 4-transitive finite permutation groups are S_4, S_5, A_6, and M_{11}.

(The general case follows from this by an easy induction.)

Zassenhaus (1935*a*, 1935*b*) determined all the finite sharply t-transitive groups for $t = 2$ and $t = 3$. As we've seen, for $t = 2$, this is equivalent to finding all finite nearfields. These are the *Dickson nearfields*, for which the multiplicative group is metacyclic (these are obtained from Galois fields by 'twisting' with an automorphism), and finitely many *exceptional nearfields.* For $t = 3$, there are only the groups $\mathrm{PGL}(2, q)$, and (if q is an odd square) variants obtained by 'twisting' these. Zassenhaus' arguments are much less elementary than Jordan's.

In 1952, Tits (1952) showed:

THEOREM 4.2. There is no infinite sharply 4-transitive permutation group.

The argument is very similar to Jordan's. Two years later, and independently, Hall (1954) showed a stronger result: *there is no infinite 4-transitive group in which the four-point stabilizer is finite of odd order.* In fact, he determined all groups, finite or infinite, with this property: along with the sharply 4-transitive groups, there is just one more, namely the alternating group A_7. This circle of ideas has been completed in two different ways. First, the complete list of finite 2-transitive groups can be deduced from the classification of the finite simple groups, using results of Maillet, Curtis, Kantor, Seitz, Howlett, Hering, and others. Secondly, Yoshizawa (1979) proved:

THEOREM 4.3. There is no infinite 4-transitive group in which the 4-point stabilizer is finite.

Note that it is not reasonable to expect a short proof of Yoshizawa's Theorem including both finite and infinite groups, since this would be a determination of all the finite 4-transitive groups.

5 Geometric groups

We now return to geometric groups (those which permute their ordered bases transitively). An equivalent condition is that the pointwise stabilizer of any collection of points acts transitively on those points (if any) which it doesn't fix. Using this, we can see the reason for the name:

PROPOSITION 5.1. In a geometric group, the irredundant bases are the bases of a matroid (in other words, the *exchange axiom* holds).

Proof. Let $(x_1, \ldots, x_m)$ and $(y_1, \ldots, y_m)$ be bases. Then $y_1, \ldots, y_m$ cannot all be fixed by the stabilizer of $x_1, \ldots, x_{m-1}$; so at least one of them can be adjoined to $(x_1, \ldots, x_{m-1})$ to form a base. □

REMARK. The flats of the matroid are the fixed-point sets of subgroups of G.

This raises a natural QUESTION: For which permutation groups is it true that

(a) the irredundant bases are the bases of a matroid; or

(b) the bases of minimum size are the bases of a matroid; or

(c) all irredundant bases have the same size?

I suspect that the classes defined by these conditions are not too much larger than the class of geometric groups. For example, even those groups which act naturally on a nice matroid (such as $\mathrm{PGL}(n, q)$, where $n, q > 2$) fail to satisfy any of these conditions.

Geometric groups provide good inductive possibilities. The pointwise stabilizer of any collection of points is geometric; also, the group induced on a subspace of the matroid by its setwise stabilizer is geometric. (A subspace of the matroid is just the fixed-point set of the pointwise stabilizer of some set of points.) So the obvious approach to the classification is to find all geometric groups of small rank, and then see how they can be fitted together. If a geometric group has degree n, and the stabilizer of i independent points fixes l_i points altogether for $i = 0, \ldots, r-1$ (where r is the rank), we say that the *type* is $(\{l_0, \ldots, l_{r-1}\}, n)$. Then the stabilizer of a point has rank $r-1$ and type $(\{l_1, \ldots, l_{r-1}\}, n)$, while the group induced on a hyperplane has rank $r-1$ and type $(\{l_0, \ldots, l_{r-2}\}, l_{r-1})$.

First, we give some examples.

EXAMPLE 5.2. Let H be a regular permutation group of degree $k > 1$. Then the wreath product H wr S_m is geometric, of rank m and type $(\{0, k, 2k, \ldots, (m-1)k\}, mk)$.

EXAMPLE 5.3. In the above example, let H be abelian. Let M be the subgroup

$$\{(h_1, h_2, \ldots, h_m) : h_1 h_2 \ldots h_m = 1\}$$

of the base group H^m of the wreath product, and G the group generated by M and the symmetric group S_m. Then G is geometric, with rank $m-1$ and having the type $(\{0, k, \ldots, (m-2)k\}, mk)$. We can allow the possibility $k = 1$ here. Also, if k is even, the construction can be modified to produce non-split extensions of M by S_n.

EXAMPLE 5.4. The alternating group A_m is geometric of rank $m - 2$.

EXAMPLE 5.5. The stabilizer of d independent points in the general linear group $\mathrm{GL}(m, q)$ is a geometric group of rank $m - d$ and type

$$(\{0, q^{d+1} - q^d, \ldots, q^{m-1} - q^d\}, q^m - q^d)$$

(acting on the vectors it doesn't fix).

EXAMPLE 5.6. The group generated by Example 5.5 and the translation group of the vector space has rank $m - d + 1$ and type

$$(\{0, q^d, q^{d+1}, \ldots, q^{m-1}\}, q^m).$$

We call these examples *generic*. Zil'ber (1988) proved that any finite geometric group of sufficiently large rank (at least 7) is generic, by elementary means. But the strongest result was obtained by Maund (1989), who found all finite geometric groups of rank greater than 1. As suggested above, the crucial step is to find those of rank 2; then induction can be applied. Unfortunately, the geometry is trivial in this case (consisting of the points of a line), so group theory is required. Let $\bar{G}$ be the permutation group induced on the points of the geometry. Then $\bar{G}$ is 2-transitive, and the 2-point stabilizer is centre-by-square. (A group H is *centre-by-square* if $H/Z \cong A \times A$ for some group A, where Z is a subgroup of the centre of H.) Maund uses the classification of finite simple groups to determine the possibilities for $\bar{G}$. (It may be that easier methods can be used. Aschbacher (1971) determined the 2-transitive groups in which the 2-point stabilizer is abelian; these form an important special case.) The final result is as follows.

THEOREM 5.7. A finite geometric group of rank 2 is one of the following:

(a) generic;
(b) sharply 2-transitive;
(c) $Z_{(q-1)/2} \times \mathrm{PSL}(2, q)$, where $q \equiv 3 \pmod 4$, with type

$$(\{0, (q-1)/2\}, (q^2-1)/2);$$

(d) $Z_{q-1} \times \mathrm{Sz}(q)$, with type

$$(\{0, q-1\}, (q-1)(q^2+1));$$

(e) PSL$(3, q)$ for $q = 2$ or 3, with type $(\{0, 2\}, 14)$ or $(\{0, 6\}, 78)$ respectively.

As we saw in the last section, the sharply 2-transitive groups correspond to finite nearfields, and were classified by Zassenhaus. Maund's classification for higher rank follows the outline suggested already, though the details are non-trivial. One example: the 'sporadic' geometric group A_7, with type $(\{0, 1, 3\}, 15)$, arises from fitting together PSL$(3, 2)$ (case (e) of Theorem 5.7) with S_3 (generic Example 5.3 with $k = 1$, $m = 3$, or sharply 2-transitive).

We turn now to infinite geometric groups. There are examples of infinite rank (general linear groups of infinite dimension over finite fields), or of finite rank but infinite type (general linear groups of finite dimension over infinite fields); but I will consider here only the case where both the rank and the type are finite (i.e., if the rank is r, then for $i < r$, the stabilizer of i independent points fixes l_i points altogether, where i is finite). Cameron, Deza and Singhi (Cameron *et al.*, 1988) showed that $l_i - l_{i-1}$ is odd for $i = 1, \ldots, r-1$. Using this, Cameron (1992) gave an elementary proof of the following generalization of Tits' Theorem 4.2:

THEOREM 5.8. There is no infinite geometric group of rank at least 4 and finite type.

The infinite geometric groups of smaller rank have not yet been determined. Consider the rank 3 case. One half of the induction works: the stabilizer of a hyperplane induces on it a finite geometric group K of type $(\{0, l_1\}, l_2)$. The possibilities are severely restricted by Maund's Theorem 5.7 and the fact that l_1 is odd and l_2 is even. We find that K is one of the following:

(a) Example 5.2, with $m = 2$, k odd (in this case, G permutes the points of the geometry 3-transitively, and the stabilizer of three points is finite);

(b) sharply 2-transitive of even degree, hence an affine group AGL$(1, q)$, where q is a power of 2;

(c) case (c) of Theorem 5.7.

It can be shown by *ad hoc* arguments that, in (b), q must be an odd power of 2. (This includes the case $q = 2$, where G is sharply 3-transitive.)

Note that the only known examples of infinite geometric groups of finite rank and type are the sharply t-transitive groups for $t \leqslant 3$.

An interesting problem involves finding a common generalization of Theorems 4.3 and 5.8. The formulation of the problem is not obvious, but would go something like this. A permutation group is *weakly geometric* (of finite rank and type) if there is a sequence $(x_1, \ldots, x_r)$ of points such that the stabilizer of $x_1, \ldots, x_{i-1}$ has a unique infinite orbit and x_i is chosen from this orbit for $i < r$, while the stabilizer of $x_1, \ldots, x_r$ has no infinite orbits. (Perhaps we should add that the $(i-1)$-point stabilizer has only finitely many orbits, while the r-point stabilizer is finite.) The problem would be to prove that there is no such permutation group with $r \geqslant 4$.

6 Scales of measurement

Bases and transitivity of permutation groups have arisen naturally in the work of certain philosophers and mathematical psychologists on the *theory of measurement.* I will outline some of this material. For further information, see the three-volume *Fundamentals of Measurement* (Krantz *et al.*, 1971, 1989, 1990).

Measurement of an empirical quantity involves the assignment of a number to that quantity. There are two reasons for using numbers here. Firstly, they form a convenient and familiar fixed reference set (the use of the natural numbers in this role for counting pre-dates history); and secondly, they allow the resources of mathematics to be brought to bear, so that theories of the empirical world can be given mathematical form. The most important property of both the empirical objects and the numbers is *order*; we always assume that the objects are ordered, and that the map from objects to numbers is order-preserving.

Obviously, there are many different ways in which measurement can be made. Stevens (1946) put forward the point of view that a *scale of measurement* consists, in general, not of a single measurement function $f: S \to N$ (where S and N are the sets of objects and numbers respectively), but of all functions obtained from a given one by multiplication by an element of a specified group G of order-preserving permutations of N. Thus, different groups define different kinds of scales.

Normally, we take N to be the real numbers or some subset (perhaps the positive real numbers). Four types of scale are usually recognized. These are:

(a) *nominal* scales. In these, the number assigned to an object is regarded as a fixed 'name' for the object, and no variation is possible; so the group G is trivial. Counting is of this form.

(b) *ratio* scales. The prototype is measuring length. We are free to choose a unit of length; different measurements are related by a multiplicative factor. Thus G is the group $\{x \mapsto ax : a > 0\}$ of multiplications by positive real numbers. Statements like 'the length of this rod is twice the length of that one' are *meaningful* (i.e., independent of the unit of measurement).

(c) *interval* scales. The standard example is temperature, assuming no knowledge of absolute zero. The numbers assigned to two objects fix the measurement; the group is the *affine group*

$$\mathrm{Aff}(\mathbb{R}) = \{x \mapsto ax + b : a, b \in \mathbb{R}, a > 0\}.$$

(d) *ordinal* scales. We assume that only the order of objects is significant, not any numerical properties of the representation. G is the group $A(\mathbb{R})$ of all order-preserving permutations of $\mathbb{R}$.

In case (b), we can convert into an 'additive' representation using the whole of $\mathbb{R}$ by the simple expedient of taking logarithms.

Two important parameters of the scale of measurement associated with the group G are its *degree of homogeneity* and its *degree of uniqueness.* The scale is m-*homogeneous*, respectively m-*unique*, if, given any two m-tuples in increasing order, say $x_1 < \ldots < x_m$ and $y_1 < \ldots < y_m$, there is at least one (resp., at most one) element $g \in G$ such that $x_i g = y_i$ for $i = 1, \ldots, m$. The degree of homogeneity (resp., degree of uniqueness) is the greatest (resp., least) m for which the scale is m-homogeneous (resp. m-unique). (The possibility $m = \infty$ is allowed: infinite homogeneity means m-homogeneity for all finite m, while infinite uniqueness means m-uniqueness for no finite m.) Thus, m-homogeneity means that the value of the measurement can be prescribed arbitrarily at m given points (provided only that order is preserved); m-uniqueness means that the value of the measurement at m points determines its value everywhere. Also, m-homogeneity agrees with the concept of the same name defined by permutation group theorists (in the special case of groups of order-preserving permutations), while m-uniqueness is the assertion that every m-tuple of distinct points is a base.

The *scale type* of the scale associated with a group G is the pair (k, l), where k is the degree of homogeneity and l the degree of uniqueness. Clearly $k \leqslant l$. The scale types for the four standard scales are $(0,0)$, $(1,1)$, $(2,2)$ and (∞, ∞) respectively. What other scale types can exist? We associate the scale type with the group G, and use the terms *homogeneous* and *finitely unique* to mean 1-homogeneous and m-unique for some finite m, respectively. The following important theorem was proved by Alper (1987):

THEOREM 6.1. Let G be a homogeneous and finitely unique subgroup of $A(\mathbb{R})$, with scale type (k, l). Then $(k, l) = (1,1)$, $(1,2)$ or $(2,2)$, and G is conjugate to a subgroup of $\mathrm{Aff}(\mathbb{R})$ containing the additive group of $\mathbb{R}$.

This theorem justifies, to some extent, the use of the 'standard' scales. But what happens if other number systems of measurement are used? Cameron (1989) showed:

THEOREM 6.2. For any pair (k, l) of positive integers with $k < l$, there is a subgroup G of $A(\mathbb{Q})$ with scale type (k, l).

The proof of this theorem uses Fraïssé's method (Fraïssé, 1953) of constructing 'homogeneous' relational structures (a different use of this word!), and a technique due to Tits (1974) for making the free group of countable rank act with prescribed transitivity. Apart from philosophical considerations about the result of a physical measurement, one reason for being interested in $\mathbb{Q}$ is that, if there is a group of order-preserving permutations of any infinite ordered set with any scale type (k, l), where $k \geqslant 2$, then there is such a group of order-preserving permutations of $\mathbb{Q}$. This follows from the downward Löwenheim–Skolem Theorem (p. 7) the ordered set, the group action, and the degrees of homogeneity and uniqueness can be expressed by first order sentences; if the set of such sentences has an infinite model, then it has a countable model; and a countable ordered set admitting a 2-homogeneous group is necessarily dense, and so is isomorphic to $\mathbb{Q}$ by Cantor's Theorem.

This adds interest to the case $k = l$ not covered by Cameron's Theorem (the case which most closely resembles the sharply t-transitive groups of Section 4):

QUESTION 6.3. Is there a subgroup of $A(\mathbb{Q})$ which is *sharply k-homogeneous*, i.e., has scale type (k, k), for any $k > 2$?

Note that it suffices to consider the case $k - 3$ in order to show nonexistence of such groups in general; for the stabilizer of a point in such a group acts with scale type $(k-1, k-1)$ on the points to the left, and to the right, of the given point. Moreover, for $k = 3$, these induced groups of scale type $(2, 2)$ look nothing like our standard example, the affine group $\mathrm{Aff}(\mathbb{Q})$. For, in $\mathrm{Aff}(\mathbb{Q})$, the elements without fixed points, together with the identity, form the unique minimal normal subgroup, and act with scale type $(1, 1)$. But, if G has scale type $(3, 3)$ and $H = G_0$ (acting on the positive rationals), then the stabilizer of any negative number u acts with scale type $(1, 1)$ on the positive rationals, and these subgroups, for distinct u, intersect pairwise only in the identity.

7 Sharply t-transitive semigroups

There is a close analogy between permutations of a set, and arbitrary mappings of the set to itself. A set S of mappings is *sharply t-transitive* if, given any t distinct points $x_1, \ldots, x_t \in X$, and any t points $y_1, \ldots, y_t \in X$ (not necessarily distinct), there is a unique function $f \in S$ with $x_i f = y_i$ for $i = 1, \ldots, t$. Corresponding to a group of permutations, we have the concept of a semigroup of functions (a set closed under composition).

In a remarkable analogue of Jordan's Theorem, Pasini (1991) gave a complete determination of the sharply t-transitive semigroups of functions on finite sets for $t > 1$. I present the argument here because, almost unchanged, it applies to the infinite case as well. First, some examples.

EXAMPLE 7.1. If $|X| = n$, then the set X^X of all functions from X to itself is sharply n-transitive (and is obviously a semigroup).

EXAMPLE 7.2. Let G be a sharply 2-transitive set of permutations of X, and C the set of constant functions on X. Then $G \cup C$ is a sharply 2-transitive set of functions. If G is a group, then $G \cup C$ is a semigroup.

EXAMPLE 7.3. Pasini discovered a sporadic example, which has several equivalent descriptions. It is a sharply 3-transitive set of functions on a 4-set. It can be defined as the set of 'uniform' functions on a 4-set (where a function is uniform if points in its range have constant numbers of pre-images); or as the set of affine functions (polynomial of degree at most 1) on a two-dimensional vector space over GF(2); or as the set of polynomial functions of degree at most 2 on GF(4). (Pasini used the third description.)

THEOREM 7.4. A (finite or infinite) sharply t-transitive semigroup of functions on X, with $2 \leqslant t \leqslant n = |X|$, is one of the above Examples 7.1–7.3.

REMARK. Of course, t is finite. Clearly, if $t = n$, then we have Example 7.1.

Proof. Let S be a sharply t-transitive semigroup on X. For the first step, we assume only that $t > 1$.

We show first that any $f \in S$ satisfying $|Xf| \geqslant t$ is injective. Suppose that such an f is not injective. Choose $x_1, \ldots, x_t \in Xf$ with $|x_1 f^{-1}| > 1$. Let $g, g' \in S$ satisfy $x_i g = x_i g' = x_1$ for $i \neq 2$, $x_2 g = x_1$, $x_2 g' = x_2$. Then $fg \neq fg'$ (since they differ on points of $x_2 f^{-1}$), but fg and fg' agree on the (at least t) points of $\bigcup_{i \neq 2} x_i f^{-1}$. This contradicts the sharp t-transitivity.

In fact, even if X is infinite, the injective elements of S are permutations. For take distinct points $x_1, \ldots, x_t$, and let e be the unique element of S fixing them. Then e^2 also fixes these points, so $e^2 = e$; and now e (being an injective idempotent) is the identity map. Let g be any injection in S, and let $h \in S$ be such that $x_i gh = x_i$ for $i = 1, \ldots, t$. Then $gh = hg = e$, and so g is a permutation.

Now the set of permutations is closed under composition, and so forms a group. By what we have proved, it is a sharply t-transitive group. (The unique element of S mapping t given points to t given distinct points is necessarily injective.) By the results of Jordan, Zassenhaus, and Tits, this places strong restrictions on n, t and S, which we do not use (since the argument below is more elementary). Note, however, that in the case $t = 2$, every element of S is constant or injective, and so S is as in Example 7.2.

Next we show that S contains all the constant functions. Take $a \in X$. There is certainly a function $f \in S$ which is not injective; its range has at most $t - 1$ points, so there is a function g mapping the range to a. Now fg is the constant function with value a.

It follows that, for any $f \in S$ which is not constant, we have $|xf^{-1}| \leqslant t-1$ for all $x \in X$.

From now on, we assume that $2 < t < n$. There is a function in S which is neither constant nor injective. Each of the at most $t-1$ points in its range has at most $t-1$ pre-images; so $|X| \leqslant (t-1)^2$, that is, X is finite: say $|X| = n$.

Let $f \in S$ be neither constant nor injective. Choose $x_1 \in Xf$, and let x_2 be another point. Composing f with a function fixing x_1 and mapping the remainder of Xf to x_2, we obtain a function f' satisfying

$$xf' = \begin{cases} x_1 & \text{if } xf = x_1 \\ x_2 & \text{otherwise.} \end{cases}$$

Thus, $n - |x_1 f^{-1}| \leqslant t-1$, or $|x_1 f^{-1}| \geqslant n-t+1$. This applies to each point of the range of f. We may assume that $|Xf| = t-1$; then

$$n \geqslant (t-1)(n-t+1).$$

From this, we deduce that $(t-2)(n-t) \leqslant 1$. Both factors are positive by assumption; so $t-2 = n-t = 1$, whence $t = 3$, $n = 4$, and it is easy to see that we have Example 7.3. □

If the closure condition (that S is a semigroup) is relaxed, the situation is quite different. Now there is no need for the functions to map the set X to itself. Thus, we say that a set S of functions from X to Y to be *sharply t-transitive* if, given any distinct $x_1, \ldots, x_t \in X$ and any $y_1, \ldots, y_t \in Y$, there is a unique $f \in S$ mapping x_i to y_i for $i = 1, \ldots, t$. These are special cases of orthogonal arrays, and occur in experimental design, etc. They have geometric significance like that of sharply t-transitive sets of permutations. Here are two examples.

Example 7.5. A finite sharply 2-transitive set S of functions from an n-set X to itself exists if and only if there is an affine plane of order n. For, given such a set, we take the point set of the plane to be $X \times X$, and the lines to be the 'vertical' lines $\{(a, x) : x \in X\}$ for each $a \in X$, and the graphs of the functions in S.

Example 7.6. Let $\bar{X}$ be a subset of the point set of the Desarguesian projective plane $\mathrm{PG}(2, F)$, with underlying vector space V. Suppose that $\bar{X}$ is an *arc* (i.e., no three of its points are collinear). Choose a non-zero vector x spanning each point $\bar{x} \in \bar{X}$; let X be the set of such vectors. Let S be the set of functions from X to F induced by elements of the dual space V^* of V. Then S is sharply 3-transitive, since any three elements of X form a basis for V. This example generalizes to higher-dimensional spaces.

8 Linear groups

We turn to another variant of these ideas. We replace sets and mappings by vector spaces and linear maps. Thus, a base for a subgroup of $\mathrm{GL}(V)$ is just a base for the induced permutation group on the vectors of V. The vectors of an irredundant base are linearly independent.

Little work has been done on this in general. I will confine my attention to analogues of sharply t-transitive groups. A *sharply t-transitive linear group* should be defined as a subgroup of $\mathrm{GL}(V)$ which acts sharply transitively on the set of linearly independent t-tuples of vectors of V. A group with this property, for $t = 1$, is just a *Singer group* on the vector space. Examples are given by multiplicative groups of fields (or, more generally, nearfields). For $t > 1$, however, such a group acts 2-transitively on the points of the projective space. In the finite case, it follows from the results of Cameron and Kantor (1979) that the only such groups are $\mathrm{GL}(V)$, with $t = n = \dim(V)$, and one sporadic example, $A_7 \leqslant \mathrm{GL}(4,2)$, which is sharply 3-transitive.

For sets of invertible maps, Cameron, Deza and Frankl (Cameron *et al.*, 1987) proved the following result.

THEOREM 8.1. Let t, n be positive integers with $1 < t \leqslant n$, and let q be a prime power. If a sharply t-transitive subset of $\mathrm{GL}(n,q)$ exists, then

$$n \leqslant \begin{cases} (5t+1)/4, & \text{if } n \text{ is even;} \\ (5t-2)/4, & \text{if } n \text{ is odd.} \end{cases}$$

In other words, if $t > 1$, then t is at least roughly $\frac{4}{5}n$. Note that the example $A_7 \leqslant \mathrm{GL}(4,2)$ attains the bound. No other examples with $t > 1$ are known.

Next we consider the linear analogue of Pasini's Theorem in Section 7; that is, we consider semigroups of linear maps from V to itself. It turns out that the only possibilities are the analogues of Example 7.1, and the proof is even easier. A set S of linear maps from V to itself is *sharply t-transitive* if, given a linearly independent t-tuple $(x_1, \ldots, x_t)$ of vectors of V and an arbitrary t-tuple $(y_1, \ldots, y_t)$, there is a unique $f \in S$ such that $x_i f = y_i$ for $i = 1, \ldots, t$.

PROPOSITION 8.2. Let S be a sharply t-transitive semigroup of linear transformations of the vector space V, where $t > 1$. Then $\dim(V) = t$, and $S = \mathrm{Hom}(V, V)$.

Proof. Step 1: If $f \in S$ with $\mathrm{rank}(f) \geqslant t$, then f is injective.

For, if not, then there exist vectors $v_1, \ldots, v_t$ such that $v_1 f, \ldots, v_t f$ are linearly independent, and a non-zero vector $w \in \mathrm{Ker}(f)$. If g and g' are such that $v_i fg = 0$ $(i = 1, \ldots, t)$, $v_i fg' = 0$ $(i = 1, \ldots, t-1)$, and $v_t fg' \neq 0$, then fg and fg' agree on the t independent vectors $w, v_1, \ldots, v_{t-1}$ but not on v_t.

Step 2: If $f \in S$ with nullity$(f) \geqslant t$, then $f = 0$.

For, if not, let $v_1, \ldots, v_t$ be linearly independent vectors in Ker(f), and w a vector with $wf \neq 0$. If $g, g' \in S$ differ on wf, then fg, fg' agree on $v_1, \ldots, v_t$ but not on w.

Step 3: S contains a map of rank 1.

For we can choose $f \in S$ neither injective nor zero—let $v_1 f = 0$, $v_2 f \neq 0$—and let g map the image of f onto a one-dimensional subspace (possible since $1 \leqslant \text{rank}(f) < t$), and take fg.

Now, if h is as in Step 3, then rank$(h) = 1$ and nullity$(h) \leqslant t - 1$; so $\dim(V) \leqslant t$, whence equality holds. □

Finally what happens if the closure condition is relaxed (so that we are considering linear analogues of orthogonal arrays)? Delsarte (1978) showed that they always exist:

THEOREM 8.3. *Let $t < m, n$ be positive integers and q a prime power. Then there is a sharply t-transitive subset of* Hom(V, W) *where V and W are m- and n-dimensional vector spaces over* GF(q).

9 Independence algebras

Several ideas from the preceding sections come together in the study of independence algebras. These were introduced by Fountain and Lewin (1992), as a setting for generalization of earlier results about the semigroup generated by non-identity idempotent endomorphisms of a set or of a vector space. The results presented here are due to Cameron and Szabó (to appear).

An *independence algebra* is an algebra A with the following properties:

(a) the subalgebras of A have the exchange property;

(b) any map from a basis of A into A extends to an endomorphism of A.

(A *basis* is a minimal generating set. It follows from (a) that any two bases have the same cardinality.) Note that the operations of the algebra are not mentioned, so any classification will be up to *equivalence*, where two algebras are said to be equivalent if there is a bijection between them preserving subalgebras and endomorphisms.

The connection with the earlier material is provided by the following result.

PROPOSITION 9.1. *The automorphism group of an independence algebra is a geometric group, whose bases are precisely the bases of the algebra.*

In any independence algebra A, we let $C = C(A)$ be the subalgebra generated by the empty set, and refer to the elements of C as *constants*. (The prototype is the zero of a vector space.)

PROPOSITION 9.2. A one-dimensional independence algebra determines (and is determined by) a group acting on a set.

For if A is such an algebra, then Aut(A) acts regularly on $A \setminus C$, so we can identify this set with the group $G = \mathrm{Aut}(A)$. There is a left action of G on C defined by $g \cdot f_c = f_{g(c)}$, where f_c is the unique endomorphism mapping $1 \in G$ to $c \in C$. The construction reverses.

More generally, an independence algebra is called *trivial* if its subalgebra lattice is the lattice of subsets of a set. Any trivial independence algebra has the form $(X \times G) \cup C$, where X is a set, G a group, and C a set on which G acts; and any such triple (X, G, C) determines an algebra. Non-trivial independence algebras, however, are more restricted:

THEOREM 9.3. The subalgebra lattice of a non-trivial independence algebra A is isomorphic to the subspace lattice of a projective or affine space, depending on whether $C(A)$ is non-empty or empty.

Examples can be constructed from a vector space or an affine space with a distinguished subspace, and from a sharply 2-transitive group. The determination of finite independence algebras is in progress; details will appear later (Cameron and Szabó, to appear).

Canonical expansions of countably categorical structures

Simon R. Thomas

Department of Mathematics,
Rutgers University,
New Brunswick,
NJ 08903,
U.S.A.
Electronic mail address: sthomas@math.rutgers.edu

1 Introduction

A permutation group acting on a set Ω is said to be highly homogenous if it acts transitively on the set of n-element subsets of Ω for each $n \in \mathbb{N}$. Cameron (1976) has proved the following result.

THEOREM 1.1. Suppose that G is a highly homogeneous group acting on an infinite set Ω. If G is r-transitive but not $(r+1)$-transitive, then $r \leqslant 3$ and there is a linear or circular order on Ω preserved or reversed by all elements of G.

Thomas (1986) considered the analogous problem for groups of semilinear transformations acting on the infinite dimensional projective space $\mathrm{PG}(\infty, q)$ over the finite field $\mathrm{GF}(q)$. The following results were obtained.

THEOREM 1.2. Suppose that $G \leqslant \mathrm{P\Gamma L}(\infty, q)$ acts transitively on the set $\mathrm{PG}^{(n)}(\infty, q)$ of n-dimensional subspaces of $\mathrm{PG}(\infty, q)$ for each $n \in \mathbb{N}$. If G acts primitively on the points of $\mathrm{PG}(\infty, q)$, then $\mathrm{PGL}(X) \leqslant G_{\{X\}}/G_{(X)}$ for each finite dimensional subspace X.

The hypothesis that G acts primitively cannot be omitted. In fact, as the following result shows, it is possible for $G_{\{X\}}$ to act trivially on each finite dimensional subspace X.

THEOREM 1.3. There is a dense linear ordering of the points of $\mathrm{PG}(\infty, q)$ such that the group $G \leqslant \mathrm{P\Gamma L}(\infty, q)$ of order preserving transformations acts transitively on $\mathrm{PG}^{(n)}(\infty, q)$ for each $n \in \mathbb{N}$.

The final result in Thomas (1986) gives a partial description of an arbitrary imprimitive group action.

THEOREM 1.4. Suppose that a group $G \leqslant \mathrm{P\Gamma L}(\infty, q)$ acts transitively on $\mathrm{PG}^{(n)}(\infty, q)$ for each $n \in \mathbb{N}$. If G acts imprimitively on the points on $\mathrm{PG}(\infty, q)$, then there exists a dense linear order without endpoints $\langle D, < \rangle$ and a G-invariant set of subspaces $\{S_a : a \in D\}$ satisfying the following conditions:

(i) $\mathrm{PG}(\infty, q) = \bigcup_{a \in D} S_a$;
(ii) if $a < b$, then $S_a \subsetneq S_b$;
(iii) $\dim(S_a/T_a) = 1$ where $T_a = \bigcup_{b<a} S_b$.

If $\mathrm{PG}(\infty, q)$ has countable dimension, then there is a unique such configuration up to isomorphism, which can be obtained as follows. Let $\{v_q : q \in \mathbb{Q}\}$ be a basis of the underlying vector space $V(\infty, q)$ and define $S_q = \langle v_r : r \leqslant q \rangle$ for each $q \in \mathbb{Q}$.

In this article, I will consider the general model-theoretic problem which underlies the above results. Throughout, all structures $\mathfrak{M}$ will be countable. In particular, an 'infinite dimensional' vector space will always mean a vector space of dimension $\aleph_0$.

DEFINITION 1.5. Let $\mathfrak{M}$ be an ω-categorical structure with underlying set M. The structure $\mathfrak{N}$ for the language L is defined to be a *reduct* of $\mathfrak{M}$ if:

(i) $\mathfrak{N}$ has the same underlying set M;
(ii) for each $R \in L$, $R^{\mathfrak{N}}$ is definable without parameters in $\mathfrak{M}$.

Two reducts $\mathfrak{N}_1$, $\mathfrak{N}_2$ of $\mathfrak{M}$ are said to be *equivalent* iff each is a reduct of the other. This occurs iff $\mathrm{Aut}(\mathfrak{N}_1) = \mathrm{Aut}(\mathfrak{N}_2)$. Thus classifying the reducts of $\mathfrak{M}$ up to equivalence is the same as classifying the closed permutation groups $\mathrm{Aut}(\mathfrak{M}) \leqslant G \leqslant \mathrm{Sym}(M)$. Thomas (1991) classified the reducts of the universal homogenous graph, showing that there are just five reducts up to equivalence. Further work on the problem of classifying the reducts of structures which are homogeneous for a finite relational language can be found in Bennett (1993) and Thomas (to appear).

DEFINITION 1.6. Let $\mathfrak{M}$ be an ω-categorical structure. $\mathfrak{M}$ is an *expansion* of a structure $\mathfrak{N}$ iff $\mathfrak{N}$ is a reduct of $\mathfrak{M}$.

Schmerl (1980) has shown that the problem of classifying the ω-categorical expansions of an ω-categorical structure is completely intractable. For example, he has shown that if $\mathfrak{N}$ is an arbitrary ω-categorical structure and $\mathfrak{N}^*$ is an ω-categorical structure with trivial algebraic closure, then $\mathfrak{N}$ has an ω-categorical expansion $\mathfrak{M}$ such that $\mathfrak{N}^*$ is a reduct of $\mathfrak{M}$. To obtain a more manageable problem, we shall consider an extremely restricted class of expansions.

DEFINITION 1.7. Let $\mathfrak{M}$ be an ω-categorical structure with underlying set M. $\mathfrak{M}$ is a *strongly canonical expansion* of the structure $\mathfrak{N}$ if:

(i) $\mathfrak{M}$ is an expansion of $\mathfrak{N}$;
(ii) $\mathrm{Aut}(\mathfrak{M})$ and $\mathrm{Aut}(\mathfrak{N})$ have the same orbits on the set

$$\{X \subset M : |X| < \omega\}.$$

For example, Cameron's analysis (Cameron, 1976) yields a classification of the strongly canonical expansions of a countably infinite set Ω with no extra structure; and in particular shows that Ω has finitely many strongly canonical expansions up to isomorphism. Here two expansions $\mathfrak{M}_1, \mathfrak{M}_2$ of the structure $\mathfrak{N}$ with underlying set N are said to be isomorphic if there exists $\pi \in \mathrm{Aut}(\mathfrak{N})$ such that

$$(\mathrm{Aut}(\mathfrak{M}_1), N) \overset{\pi}{\cong} (\mathrm{Aut}(\mathfrak{M}_2), N).$$

This simply says that $\mathrm{Aut}(\mathfrak{M}_1)$ and $\mathrm{Aut}(\mathfrak{M}_2)$ are conjugate subgroups of $\mathrm{Aut}(\mathfrak{N})$.

QUESTION 1.8. Does every ω-categorical structure have finitely many strongly canonical expansions up to isomorphism?

The next result shows that the notion of a strongly canonical expansion is usually too restrictive for structures with non-trivial algebraic closure.

THEOREM 1.9. Suppose that the ω-categorical structure $\mathfrak{N}$ with underlying set N satisfies the following conditions:

(i) $\mathrm{Aut}(\mathfrak{N})$ acts transitively on N;
(ii) $\mathrm{acl}(a) = \{a\}$ for all $a \in N$;
(iii) there exists a finite subset $F \subset N$ such that $\mathrm{acl}(F) \neq F$.

Then $\mathfrak{N}$ has *no* non-trivial strongly canonical expansions.

Proof. Let $G = \mathrm{Aut}(\mathfrak{N})$. First we shall show how the result follows easily from the following property.

For every finite subset $X \subset N$, there exists a finite subset $X \subset Y \subset N$ such that $G_{\{Y\}} \leqslant G_{(X)}$. $\quad (*)$

Suppose then that $(*)$ holds, and let $\mathfrak{M}$ be a strongly canonical expansion of $\mathfrak{N}$. Let $\pi \in G$ and let $X \subset N$ be any finite subset. Choose a finite subset $X \subset Y \subset N$ such that $G_{\{Y\}} \leqslant G_{(X)}$. Since $\mathfrak{M}$ is a strongly canonical expansion of $\mathfrak{N}$, there exists $g \in \mathrm{Aut}(\mathfrak{M})$ such that $g[Y] = \pi[Y]$. Then $g^{-1}\pi[Y] = Y$, and so $g\restriction X = \pi\restriction X$. Since X was arbitrary, $\pi \in \mathrm{Aut}(\mathfrak{M})$. Hence $\mathrm{Aut}(\mathfrak{M}) = G$, and $\mathfrak{M}$ is a trivial strongly canonical expansion.

Thus it suffices to prove $(*)$. Let $X = \{a_1, \ldots, a_n\}$ be an arbitrary finite subset of N. Let F be a finite set of size f such that $\mathrm{acl}(F) \neq F$. Choose large integers $m_i > |\mathrm{acl}(F)| + 2f$ for $1 \leqslant i \leqslant n$. Then there exist pairwise disjoint subsets F^i_ℓ for $1 \leqslant i \leqslant n$, $1 \leqslant \ell \leqslant m_i$ satisfying the following conditions:

(a) $\mathrm{tp}(F^i_\ell) = \mathrm{tp}(F)$;

(b) $F^i_\ell \cap X = \varnothing$;

(c) $a_i \in \mathrm{acl}(F^i_\ell)$;

(d) if $k \neq i$, then $a_k \notin \mathrm{acl}\left(\bigcup_{1 \leqslant \ell \leqslant m_i} F^i_\ell\right)$;

(e) if $1 \leqslant \ell_1 < \cdots < \ell_f \leqslant m_i$, $1 \leqslant t_1 < \cdots < t_f \leqslant m_i$ then

$$\mathrm{tp}(F^i_{\ell_1}, \ldots, F^i_{\ell_f}/Z_i) = \mathrm{tp}(F^i_{t_1}, \ldots, F^i_{t_f}/Z_i)$$

where $Z_i = X \cup \bigcup_{\substack{j \neq i \\ 1 \leqslant r \leqslant m_j}} F^j_r$.

To see this, let N be an integer which is much larger than $\max\{m_i : 1 \leqslant i \leqslant n\}$. Then we can clearly find pairwise disjoint subsets F^i_ℓ for $1 \leqslant i \leqslant n$, $1 \leqslant \ell \leqslant N$ satisfying (a), (b), and (c). Using the fact that $\mathrm{acl}(a_i) = \{a_i\}$, we can replace our initial choice of F^i_ℓ by a more suitable one if necessary, so that

$$\mathrm{acl}\Big(\bigcup_{1 \leqslant \ell \leqslant N} F^i_\ell\Big) \cap X = \{a_i\}$$

for each $1 \leqslant i \leqslant n$. Finally, repeated applications of Ramsey's Theorem yield sets F^i_ℓ for $1 \leqslant i \leqslant n$, $1 \leqslant \ell \leqslant m_i$ which also satisfy (e).

Let

$$Y = X \cup \bigcup_{\substack{1 \leqslant \ell \leqslant m_i \\ 1 \leqslant i \leqslant n}} F^i_\ell.$$

Next we shall show that $G_{\{Y\}} \leqslant G_{\{X\}}$. If not, then there exists $b \in Y \smallsetminus X$ and pairwise disjoint $F'_t \subset Y \smallsetminus \{b\}$ for $1 \leqslant t \leqslant m$, where $m = \min\{m_i : 1 \leqslant i \leqslant n\}$ such that $\mathrm{tp}(F'_t) = \mathrm{tp}(F)$ and $b \in \mathrm{acl}(F'_t)$.

Let $b \in F^i_\ell$. Since $m > f$, there exists F'_t such that $F'_t \cap F^i_\ell = \varnothing$. By (e), we may suppose that $\ell = f + 1$ and that

$$F'_t \subset Z_i \cup \bigcup_{1 \leqslant r \leqslant f} F^i_r \cup \bigcup_{m_i - f + 1 \leqslant s \leqslant m_i} F^i_s.$$

Again using (e), this implies that $\mathrm{acl}(F'_t) \cap F^i_p \neq \varnothing$ for all $f + 1 \leqslant p \leqslant m_i - f$, which contradicts the fact that $m_i > |\mathrm{acl}(F)| + 2f$. Thus we have established that $G_{\{Y\}} \leqslant G_{\{X\}}$.

In the rest of the proof, we shall consider what happens when we allow the integers $\{m_i : 1 \leqslant i \leqslant n\}$ to vary. For each $1 \leqslant i \leqslant n$, let c_i be the number of distinct subsets $F' \subset Y \smallsetminus X$ such that $\mathrm{tp}(F') = \mathrm{tp}(F)$ and $a_i \in \mathrm{acl}(F')$. Using (e), it is easily shown that there exist polynomials $p_i \in \mathbb{Q}[x_1, \ldots, x_n]$ such that $c_i = p_i[m_1, \ldots, m_n]$ for $1 \leqslant i \leqslant n$. Suppose that $1 \leqslant i < j \leqslant n$, and define

$$m'_k = \begin{cases} 0 & \text{if } k \neq i \\ m_i & \text{if } k = i. \end{cases}$$

Using (d), we see that $p_j[m'_1, \ldots, m'_n] = 0$. Since $p_i[m'_1, \ldots, m'_n] \geqslant m_i$, it follows that the polynomials $\{p_i : 1 \leqslant i \leqslant n\}$ are pairwise distinct. This implies that there exist natural numbers $d_i > |\operatorname{acl}(F)| + 2f$ for $1 \leqslant i \leqslant n$ such that the values $p_j[d_1, \ldots, d_n]$ for $1 \leqslant j \leqslant n$ are pairwise distinct. Notice that if conditions (a) to (e) hold, then there are only finitely many possibilities for each of the polynomials p_i. Thus we can suppose that $d_i \leqslant m_i$ for $1 \leqslant i \leqslant n$. By shrinking Y if necessary, we can suppose that $m_i = d_i$ for $1 \leqslant i \leqslant n$. This implies that

(i) $m_i > |\operatorname{acl}(F)| + 2f$ for $1 \leqslant i \leqslant n$; and
(ii) $c_i \neq c_j$ for $1 \leqslant i < j \leqslant n$.

It is now clear that Y satisfies our requirements. □

DEFINITION 1.10. Let $\mathfrak{M}$ be an ω-categorical structure with underlying set M. $\mathfrak{M}$ is a *canonical expansion* of the structure $\mathfrak{N}$ if:

(i) $\mathfrak{M}$ is an expansion of $\mathfrak{N}$; and
(ii) $\operatorname{Aut}(\mathfrak{M})$ and $\operatorname{Aut}(\mathfrak{N})$ have the same orbits on the set

$$\{X \subset M : |X| < \omega,\ X \text{ is algebraically closed in } \mathfrak{N}\}.$$

QUESTION 1.11. Does every ω-categorical structure have finitely many canonical expansions up to isomorphism?

It is perhaps worth pointing out that Theorem 1.4 is not enough to settle this question for the structure $\mathrm{PG}(\infty, q)$, and that this case remains open. However, the following easy observation allows us to deal with most structures which are homogeneous for a finite relational language.

THEOREM 1.12. Let $\mathfrak{N}$ be an ω-categorical structure with underlying set N. Suppose that for each finite subset $X \subset N$, there exists a finite algebraically closed subset $X \subset Y \subset N$ such that $G_{\{Y\}} \leqslant G_{(X)}$. Then $\mathfrak{N}$ has *no* non-trivial canonical expansions.

COROLLARY 1.13. The universal homogenous graph Γ has *no* non-trivial canonical expansions.

Proof. Every finite graph embeds in a finite rigid graph. □

The remaining sections of this article will be devoted to the special case of the canonical expansions of infinite dimensional symplectic spaces over finite fields.

2 Symplectic spaces

Let $V(\infty, q)$ be an infinite dimensional vector space over the finite field $\mathrm{GF}(q)$, equipped with a non-degenerate symplectic form σ. (A *symplectic form* is a skew-symmetric bilinear form σ such that $\sigma(v,v) = 0$ for all $v \in V(\infty, q)$. We say that σ is *non-degenerate* if for all $0 \neq v \in V(\infty, q)$ there exists $u \in V(\infty, q)$ such that $\sigma(v, u) \neq 0$.) Let $\mathrm{PG}(\infty, q)$ be the corresponding projective space. If $P = \langle u \rangle$ and $Q = \langle v \rangle$ are two points of $\mathrm{PG}(\infty, q)$, then we define the orthogonality relation $\perp$ by

$$P \perp Q \quad \text{iff} \quad \sigma(u, v) = 0.$$

If S is a subspace of $\mathrm{PG}(\infty, q)$, then

$$S^{\perp} = \{Q \in \mathrm{PG}(\infty, q) : Q \perp P \text{ for all } P \in S\}$$

and $\operatorname{rad} S = S \cap S^{\perp}$. We say that S is *totally isotropic* if $S \subseteq S^{\perp}$, and that S is *non-degenerate* if $S \cap S^{\perp} = \varnothing$. Throughout, we will be using vector space dimensions, so that points P and lines L of $\mathrm{PG}(\infty, q)$ will have dimensions 1 and 2 respectively. A non-degenerate line will often be called a *hyperbolic* line.

We shall classify the canonical expansions of the symplectic space

$$\mathrm{SPG}(\infty, q).$$

This consists of the underlying set $\mathrm{PG}(\infty, q)$ equipped with two relations:

(a) the ternary relation of collinearity of projective points; and
(b) the binary relation of orthogonality $\perp$.

Let $\mathrm{Sp}(\infty, q)$ denote the group of linear transformations of $V(\infty, q)$ which preserve σ, and let $\mathrm{GSp}(\infty, q)$ denote the group of linear transformations g such that there exists $\lambda \in \mathrm{GF}(q)^*$ with

$$\sigma(g(u), g(v)) = \lambda \sigma(u, v)$$

for all $u, v \in V(\infty, q)$. Let

$$\Gamma\mathrm{Sp}(\infty, q) \cong \mathrm{GSp}(\infty, q) \rtimes \mathrm{Aut}(\mathrm{GF}(q))$$

be the corresponding group of semilinear transformations. Then, using the fundamental theorem of projective geometry, it is straightforward to prove that

$$\mathrm{Aut}(\mathrm{SPG}(\infty, q)) = \mathrm{P}\Gamma\mathrm{Sp}(\infty, q)$$

the group of projective transformations induced on $\mathrm{PG}(\infty, q)$ by $\Gamma\mathrm{Sp}(\infty, q)$. The group of projective transformations induced by $\mathrm{Sp}(\infty, q)$ will be denoted by $\mathrm{PSp}(\infty, q)$. Notice that $\mathrm{PSp}(\infty, q)$ is a normal subgroup of finite index in $\mathrm{P}\Gamma\mathrm{Sp}(\infty, q)$.

LEMMA 2.1. (Witt's Lemma) Suppose that (V_1, σ_1) and (V_2, σ_2) are isomorphic finite dimensional (but not necessarily non-degenerate) symplectic spaces, and that W_i is a subspace of V_i for $i = 1, 2$. Further assume that g is an isomorphism from (W_1, σ_1) to (W_2, σ_2). Then g can be extended to an isomorphism from (V_1, σ_1) to (V_2, σ_2).

Suppose that S, T are finite dimensional subspaces of $\mathrm{SPG}(\infty, q)$. Then S and T, equipped with the induced symplectic structure, are isomorphic iff $\dim S = \dim T$ and $\dim \operatorname{rad} S = \dim \operatorname{rad} T$. In this case, an easy back-and-forth argument based on Witt's Lemma shows that there exists $g \in \mathrm{PSp}(\infty, q)$ such that $g[S] = T$. (It is perhaps worth pointing out that this argument shows that $\mathrm{SPG}(\infty, q)$ is ω-categorical. It also shows that the non-degenerate symplectic form over the $\aleph_0$-dimensional space $V(\infty, q)$ is unique up to isomorphism.)

In particular, $\mathrm{P\Gamma Sp}(\infty, q)$ and $\mathrm{PSp}(\infty, q)$ have the same orbits on the set of finite dimensional subspaces of $\mathrm{SPG}(\infty, q)$.

DEFINITION 2.2. A subgroup $G \leqslant \mathrm{P\Gamma Sp}(\infty, q)$ is said to be *Witt homogeneous* if G and $\mathrm{P\Gamma Sp}(\infty, q)$ have the same orbits on the set of finite dimensional subspaces of $\mathrm{SPG}(\infty, q)$.

The proof of the following result will occupy the remaining sections of this article.

THEOREM 2.3. Suppose that $G \leqslant \mathrm{P\Gamma Sp}(\infty, q)$ is Witt homogeneous. Then

$$\mathrm{PSp}(2t, q) \leqslant G_{\{T\}}/G_{(T)}$$

for each non-degenerate $2t$-dimensional subspace T of $\mathrm{SPG}(\infty, q)$.

In particular, if $\mathfrak{M}$ is a canonical expansion of $\mathrm{SPG}(\infty, q)$ and $G = \mathrm{Aut}(\mathfrak{M})$ then

$$\mathrm{PSp}(\infty, q) \leqslant G \leqslant \mathrm{P\Gamma Sp}(\infty, q).$$

Hence $\mathrm{SPG}(\infty, q)$ has finitely many canonical expansions.

3 Incidence matrices

Throughout this section, we will work with a non-degenerate symplectic space V over $\mathrm{GF}(q)$ of dimension $2n$. Let Γ be the automorphism group of V. Let $T < V$ be a non-degenerate subspace of dimension $2f < 2n$, and let $S < T$ be a totally isotropic subspace of dimension e. Let $\mathscr{T} = \{g[T] : g \in \Gamma\}$ and $\mathscr{S} = \{g[S] : g \in \Gamma\}$. Consider the matrix M whose columns are indexed by $\mathscr{S}$ and whose rows are indexed by $\mathscr{T}$, defined by

$$m_{T,S} = \begin{cases} 1 & \text{if } T > S \\ 0 & \text{otherwise.} \end{cases}$$

We shall prove that if n is sufficiently large, then M has rank $|\mathscr{S}|$, i.e., that the columns of M are linearly independent. For each $S \in \mathscr{S}$, let $C(S)$ be the corresponding column. Suppose that there exists a relation of linear dependence

$$C(S) = \sum_{S' \neq S} a_{S'} C(S')$$

where each $a_{S'} \in \mathbb{R}$. Let $g \in \Gamma_{\{S\}}$. Then for all $S_0 \in \mathscr{S}$ and $T_0 \in \mathscr{T}$,

$$m_{g[S_0],g[T_0]} = m_{S_0,T_0}$$

and so

$$\begin{aligned} m_{S,T_0} &= m_{S,g[T_0]} \\ &= \sum_{S \neq S'} a_{g[S']} m_{g[S'],g[T_0]} \\ &= \sum_{S \neq S'} a_{g[S']} m_{S',T_0} \end{aligned}$$

Thus we obtain that

$$C(S) = \sum_{S' \neq S} a_{g[S']} C(S')$$

and hence that

$$C(S) = \sum_{S' \neq S} b_{S'} C(S')$$

where

$$b_{S'} = \frac{1}{|\Gamma_{\{S\}}|} \sum_{g \in \Gamma_{\{S\}}} a_{g[S']}$$

depends only on the $\Gamma_{\{S\}}$-orbit of S'. Now the orbit of $S' \neq S$ under $\Gamma_{\{S\}}$ is determined by the following data:

(a) $\dim(S \cap S') = i < e$;
(b) $\dim(S \cap \operatorname{rad}\langle S, S' \rangle) = j$.

Note that $S \cap S' \leqslant \operatorname{rad}\langle S, S' \rangle$, so that $i \leqslant j$. We define a total ordering on the set

$$\{(i,j) : i \leqslant j \leqslant e,\ i < e\}$$

by setting

$$(i,j) < (i',j')$$

iff either

(c) $i < i'$, or
(d) $i = i'$ and $j > j'$.

From now on, suppose that $n \geqslant e+f$. This means that every pair (i,j) is realized by some $S' \neq S$. Now fix (i,j) and let this pair be realized by S'. We shall define a non-degenerate $2f$-space $T_{ij} \in \mathscr{T}$ such that

(e) $S' < T_{ij}$, and

(f) if $S'' < T_{ij}$ is totally isotropic of dimension e and S'' realizes (i^*, j^*), then $(i^*, j^*) \leqslant (i,j)$.

Let $S \cap \mathrm{rad}\langle S, S'\rangle = R \oplus (S \cap S')$. Then we choose T_{ij} such that

(g) $T_{ij} \cap S = S' \cap S$;

(h) $R \perp T_{ij}$

(i) $S' < T_{ij}$.

Suppose that $S'' < T_{ij}$ is totally isotropic of dimension e, and let S'' realize (i^*, j^*). Then $S'' \cap S \leqslant T_{ij} \cap S = S' \cap S$, so that $i^* \leqslant i$. Thus we can assume that $i^* = i$, i.e., that $S'' \cap S = S' \cap S$. But it is clear that $R, S'' \cap S \leqslant \mathrm{rad}\,\langle S, S''\rangle$ and so $j^* \geqslant j$. Hence $(i^*, j^*) \leqslant (i,j)$, as required.

Now we shall prove by induction on (i,j) that if S' realizes (i,j), then $b_{S'} = 0$. First suppose that $(i,j) = (0,e)$. Every totally isotropic e-space $S'' < T_{0,e}$ realizes $(0,e)$. Let the number of such subspaces be c. Then by considering the $T_{0,e}^{\mathrm{th}}$ row of M, we find that

$$0 = m_{S,T_{0,e}} = \sum_{S'' < T_{0,e}} b_{S''} = c\, b_{S'}$$

and so $b_{S'} = 0$. Now assume that the result holds for all $(i^*, j^*) < (i,j)$. Suppose that there are d totally isotropic e-spaces $S'' < T_{ij}$ which realize (i,j). Then we obtain that

$$0 = m_{S,T_{i,j}} = \sum_{S'' < T_{i,j}} b_{S''} = d\, b_{S'}$$

and so $b_{S'} = 0$. But now $b_{S'} = 0$ for all $S' \neq S$, which is impossible. Hence we have proved the following result.

THEOREM 3.1. Suppose that $e \leqslant f$ and $e + f \leqslant n$. Let V be a non-degenerate symplectic space over $\mathrm{GF}(q)$ of dimension $2n$; and let $\mathscr{S}, \mathscr{T}$ be the sets of totally isotropic e-spaces, non-degenerate $2f$-spaces of V. Then the incidence matrix M of $\mathscr{T}$ vs. $\mathscr{S}$ has rank $|\mathscr{S}|$.

4 Totally isotropic subspaces

From now on, we fix a Witt homogeneous group $G \leqslant \mathrm{P\Gamma Sp}(\infty, q)$. Let R be a finite dimensional totally isotropic subspace. By Witt homogeneity, $G_{\{R\}}$ acts transitively on the set of non-degenerate $2d$-dimensional subspaces of $R^{\perp}/R$ for every $d \geqslant 1$.

LEMMA 4.1. For every $e \geqslant 1$, $G_{\{R\}}$ acts transitively on the set of totally isotropic e-spaces of $R^{\perp}/R$.

Proof. Suppose, not, and let e be a counterexample. Clearly $G_{\{R\}}$ has finitely many orbits on totally isotropic e-spaces of $R^{\perp}/R$; say $\Delta_1, \ldots, \Delta_r$. Let V be a non-degenerate $2n$-dimensional subspace of $R^{\perp}/R$ in which each Δ_i is represented, chosen so that $n \geqslant 2e$. Define an r-colouring of the set $\mathscr{S}$ of totally isotropic e-spaces of V by

$$X(U) = i \quad \text{iff} \quad U \in \Delta_i.$$

Clearly $G_{\{R\}}$ preserves this colouring. Let $\mathscr{T}$ be the set of non-degenerate $2e$-spaces of V. By Theorem 3.1, the incidence matrix M of $\mathscr{T}$ vs. $\mathscr{S}$ has rank $|\mathscr{S}|$. Let M^* be the $|\mathscr{T}| \times r$ matrix whose columns are

$$C_i = \sum_{X(U)=i} C(U), \quad 1 \leqslant i \leqslant r.$$

Then M^* has rank r, and so there exist at least r pairwise unequal rows. But if the T_1^{th} and T_2^{th} rows are distinct, then T_1 and T_2 cannot be $G_{\{R\}}$-conjugate, which is a contradiction. □

LEMMA 4.2. If S is a finite dimensional totally isotropic subspace, then

$$\mathrm{PSL}(S) \leqslant G_{\{S\}}/G_{(S)}.$$

Proof. We can suppose that $\dim S \geqslant 6$. Let $L, M < S$ be a pair of lines. There exists $\pi \in G$ such that $\pi[L] = M$. By Lemma 4.1, there exists $\varphi \in G_{\{M\}}$ such that $\varphi\pi[S] = S$. Thus $G_{\{S\}}$ acts transitively on the lines of S, and so the result follows from Proposition 8.4 of Cameron and Kantor (1979). □

5 Radical action

In this section, we will prove the following result.

LEMMA 5.1. Let T be a finite dimensional subspace and $S = \mathrm{rad}\, T$. Let F be the group induced on S by $G_{\{T\}}$. Then $\mathrm{PSL}(S) \leqslant F$.

We shall make use of the following theorem from Kantor (1979).

THEOREM 5.2. Let $\mathrm{SL}(n,q) \leqslant G \leqslant \Gamma\mathrm{L}(n,q)$ with $n \geqslant 3$. Let $K \leqslant G$ be an irreducible subgroup such that $[G{:}K] \leqslant q^{n(n-1)/2}$. Then $\mathrm{SL}(n,q)$ or $\mathrm{Sp}(n,q) \trianglelefteq K$.

When applying the above result, it is helpful to know that

$$|\,\mathrm{SL}(n,q)| \approx q^{n^2}$$

and that

$$|\,\mathrm{Sp}(n,q)| \approx q^{n^2/2} \approx |\,\mathrm{SL}(n,q)|^{1/2}.$$

Finally, we shall make use of the following 'splitting result', which is also needed in the next section.

LEMMA 5.3. Suppose that T is a finite dimensional vector space over $\mathrm{GF}(q)$, $S < T$ and $\dim T/S = t$. Suppose that $\Gamma \leqslant \mathrm{GL}(T)$ satisfies the following conditions:

(a) $g[S] = S$ for all $g \in \Gamma$; and
(b) $\mathrm{SL}(S) \leqslant H$, where H is the group induced on S by Γ.

Then there is an integer $f(t)$ such that if $n = \dim S \geqslant f(t)$, then there exists a subspace $U < T$ such that

(i) $T = S \oplus U$;
(ii) $\mathrm{SL}(S) \leqslant K$, where K is the group induced on S by $\Gamma_{(U)}$.

Proof. Let $q = p^e$, where p is a prime. A *primitive* prime divisor of $q^n - 1$ is a prime $\ell \mid q^n - 1$ such that $\ell \nmid p^i - 1$ for all $1 < p^i < q^n$. By Zsigmondy (1892), either there exists a primitive prime divisor of $q^n - 1$, or $en = 2$ and $p+1$ is a power of 2, or $en = 6$ and $p = 2$. Thus we can suppose that $q^n - 1$ has a primitive prime divisor ℓ. Let $C < \Gamma$ be a cyclic group of order ℓ. By Maschke's Theorem T, regarded as a C-module, is completely reducible. Let $T = \oplus_{i=1}^{r} V_i$ be a decomposition into irreducible C-modules. Every non-trivial C-module has dimension at least n. Hence we can suppose that $V_1 = S$ and that $\dim V_i = 1$ for $2 \leqslant i \leqslant r = t+1$. We claim that $U = \oplus_{i=2}^{t+1} V_i$ satisfies our requirements.

First note that since $C \leqslant \Gamma_{(U)}$, K is certainly an irreducible subgroup of H. Also we have that

$$\begin{aligned}[\Gamma{:}\Gamma_{(U)}] &< |\,\mathrm{GL}(U)| \prod_{i=1}^{t} \frac{(q^{n+t-i+1} - 1)}{(q^i - 1)} \\ &< cq^{nt+d}\end{aligned}$$

for suitably chosen constants c and d. (This means that c and d depend only on t.) This implies that $[H{:}K] < cq^{nt+d}$. Thus if we chose $f(t)$ sufficiently large, Theorem 5.2 yields that $\mathrm{SL}(S) \leqslant K$. □

Now we turn to the proof of Lemma 5.1. Let $\dim T/S = t$ be fixed for the rest of this section. We will prove the result by studying what happens as $n = \dim S$ tends to infinity. In the following argument, the reader should remember that S and T are subspaces of the ambient infinite dimensional space $\mathrm{SPG}(\infty, q)$.

Suppose that the result holds for $n = \dim S \geqslant f(t)$, where f is the function given by Lemma 5.3. Let L be the group induced on T by $G_{\{T\}}$.

Identify T with its underlying vector space, and let $\tilde{L}$ be the group of semilinear transformations corresponding to L. (This means that $\tilde{L}$ is the full inverse image of L under that homomorphism $\Gamma\mathrm{L}(T) \to \mathrm{P}\Gamma\mathrm{L}(T)$.) Let $\Gamma = \tilde{L} \cap \mathrm{GL}(T)$. Then Γ satisfies the hypotheses of Lemma 5.3. Hence there exists a subspace $U < T$ such that $T = S \oplus U$ and $\mathrm{SL}(S) \leqslant K$, where K is the group induced on S by $\Gamma_{(U)}$. Notice that U is a non-degenerate subspace of dimension t. Now let $m < n$ and let $S_0 < S$ be a subspace of dimension m. By considering the action of the setwise stabilizer of $S_0 \oplus U$ in $\Gamma_{(U)}$, we see that the result also holds for m.

From now on, suppose that the result fails for some n. By the previous paragraph, the result must fail for all sufficiently large values of n. Let S be a totally isotropic subspace of very large dimension n and let $V = S^{\perp}/S$. Thus V is an infinite dimensional non-degenerate symplectic space. By Witt homogeneity, $G_{\{S\}}$ acts transitively on the set Ω of non-degenerate t-spaces of V. Let

$$\Omega = \Omega_1 \cup \ldots \cup \Omega_N$$

be the corresponding $G_{(S)}$-orbit decomposition. Let $\Gamma = G_{\{S\}}/G_{(S)}$, and let $\pi\colon \Gamma \to \mathrm{Sym}(N)$ be the induced homomorphism. By Lemma 4.2,

$$\mathrm{PSL}(S) \leqslant \Gamma \leqslant \mathrm{P}\Gamma\mathrm{L}(S).$$

Case 1. Suppose that $\mathrm{PSL}(S) \leqslant \ker \pi$. Let $S \leqslant T$ with $S = \mathrm{rad}\, T$ and $\dim T/S = t$. Let $\varphi \in \mathrm{PSL}(S)$, and let $\Theta \in G_{\{S\}}$ induce φ on S. Since $\varphi \in \ker \pi$, it follows that T/S and $\Theta[T]/S$ lie in the same $G_{(S)}$-orbit, and so we can assume that $\Theta \in G_{\{T\}}$. But then the result holds for $n = \dim S$.

Case 2. Thus we can assume that for all sufficiently large n,

$$\ker \pi \cap \mathrm{PSL}(S) = 1.$$

Hence $\mathrm{PSL}(S)$ acts faithfully on the set $\{\Omega_1, \ldots, \Omega_N\}$. Let $\Gamma_1 < \Gamma$ be the stabilizer of Ω_1. Then

$$N = [\Gamma{:}\Gamma_1] \geqslant [\mathrm{PSL}(S){:}\,\mathrm{PSL}(S) \cap \Gamma_1].$$

Since $\mathrm{PSL}(S) \cap \Gamma_1 \neq \mathrm{PSL}(S)$, we have that $N \to \infty$ as $n \to \infty$. In particular, we can suppose that N is much larger than t.

CLAIM 5.4. If n is sufficiently large, then $|\Gamma_1| \geqslant |\Gamma|^{2/3}$.

Proof. Let $m > t$ be the least integer such that any four non-degenerate t-spaces of V are contained in a non-degenerate m-space, and let Δ be the set of non-degenerate m-spaces of V. Let

$$\Delta = \Delta_1 \cup \ldots \cup \Delta_M$$

be the $G_{(S)}$-orbit decomposition. Let a be the number of non-degenerate t-spaces in a non-degenerate m-space. Let z be the number of ordered pairs

$$(i, \{j_1, j_2, j_3, j_4\})$$

such that

(a) $1 \leqslant i \leqslant M$ and $1 < j_1 < j_2 < j_3 < j_4 \leqslant N$; and
(b) if $Z \in \Delta_i$, then there exist $T_r \in \Omega_{j_r}$, $1 \leqslant r \leqslant 4$, with $T_r < Z$.

Clearly we have that

$$\binom{N}{4} \leqslant z \leqslant M\binom{a}{4}.$$

Since $N \to \infty$ as $n \to \infty$, if n is sufficiently large then $M \geqslant N^3$. As Γ acts transitively on $\{\Delta_i : 1 \leqslant i \leqslant M\}$, we obtain that

$$|\Gamma| \geqslant M \geqslant N^3 = [\Gamma{:}\Gamma_1]^3$$

and hence $|\Gamma_1| \geqslant |\Gamma|^{2/3}$. □

Claim 5.5. There exist arbitrarily large values of $n = \dim S$ such that Γ_1 acts irreducibly on S.

Proof. Note that, since $\mathrm{PSL}(S) \leqslant \Gamma \leqslant \mathrm{P\Gamma L}(S)$,

$$\frac{1}{(q-1,n)}|\,\mathrm{PGL}(S)| \leqslant |\Gamma| \leqslant e|\,\mathrm{PGL}(S)|$$

where $q = p^e$. Not also that

$$|\,\mathrm{PGL}(S)| = q^{n(n-1)/2}\prod_{i=2}^{n}(q^i - 1).$$

Suppose that p_1 is a prime divisor of $\Pi = \prod_{i=2}^{n}(q^i - 1)$, and let P_1 be the highest power of p_1 which divides Π. By Section 4 of Artin (1955), $P_1 \leqslant 2^n(q+1)^n$. Since $|\Gamma_1| \geqslant |\Gamma|^{2/3}$, it follows that $|\Gamma_1|$ has arbitrary large prime divisors as $n \to \infty$. Let ℓ be an extremely large prime divisor of $|\Gamma_1|$.

Let $S \subsetneq T$ with $\mathrm{rad}\, T = S$ and $T/S \in \Omega_1$. Letting H be the group induced on T by $G_{\{T\}}$, we see that H induces Γ_1 on S. Identify T with its underlying vector space and let $\tilde{H}$ be the group of semilinear transformations corresponding to H. Then there exists a cyclic subgroup C of order ℓ such that $C \leqslant \tilde{H} \cap \mathrm{GL}(T)$. Regarded as a C-module, T is completely reducible. Let

$$T = [\oplus_{i=1}^{r_1} A_i] \oplus [\oplus_{i=1}^{r_2} B_i]$$

be a decomposition into irreducible C-modules, where

$$S = \oplus_{i=1}^{r_1} A_i.$$

We can suppose that every non-trivial C-module has dimension greater than t. Thus $\dim B_i = 1$ for each $1 \leqslant i \leqslant r_2 = t$. We may also suppose that A_1 is a non-trivial C-module, and set $T_0 = \langle A_1, \oplus_{i=1}^t B_i \rangle$. Then $\operatorname{rad} T_0 = A_1$, and $G_{\{T_0\}}$ acts irreducibly on A_1. Since $\dim A_1 \to \infty$ as $\ell \to \infty$, Claim 5.5 follows. □

By Claims 5.4 and 5.5, there exist arbitrarily large values of $n = \dim S$ such that Γ_1 acts irreducibly on S and $|\Gamma_1| \geqslant |\Gamma|^{2/3}$. But if n is sufficiently large, Theorem 5.2 implies that $\mathrm{PSL}(S) \leqslant \Gamma_1$. This contradicts the hypothesis of case 2, and so completes the proof of Lemma 5.1.

6 The punchline

Let $\mathscr{F}_1^{2t}$ denote the set of ordered 4-tuples (α, β, ℓ, T) such that $\ell = \langle \alpha, \beta \rangle \subset T$, where α, β are points, ℓ is a totally isotropic line and T is a non-degenerate $2t$-space. Let $\mathscr{F}_2^{2t}$ denote the corresponding set of ordered 4-tuples, where ℓ is a hyperbolic line. In this final section, we will prove the following result.

Lemma 6.1. For all $t \geqslant 2$, G acts transitively on $\mathscr{F}_i^{2t}$ for $i = 1, 2$.

From this we obtain the following result, which completes the proof of Theorem 2.3.

Corollary 6.2. Let T be a non-degenerate $2t$-space and let Γ be the group induced on T by $G_{\{T\}}$. Then $\mathrm{PSp}(2t, q) \leqslant \Gamma$.

Proof. We can assume that $t \geqslant 3$. Clearly $\Gamma_{\{\ell\}}$ acts 2-transitively on every line $\ell \subset T$. Also, if $\alpha \in T$ is a point then Γ_α has exactly two orbits on $T \setminus \{\alpha\}$. (The orbit of $\beta \in T \setminus \{\alpha\}$ under Γ_α is determined by whether the line $\ell = \langle \alpha, \beta \rangle$ is totally isotropic or hyperbolic.) So the corollary follows from Results 4.1, 5.2, and 6.2 of Cameron and Kantor (1979). □

Fix $t \geqslant 2$. Define a graph structure on $\mathscr{F}_i^{2t}$ by setting

$$(\alpha_1, \beta_1, \ell_1, T_1) \sim (\alpha_2, \beta_2, \ell_2, T_2)$$

iff $Q = \langle T_1, T_2 \rangle$ satisfies the following conditions:

(a) $\dim Q = 4t$ and $\dim \operatorname{rad} Q = 2t$; and

(b) if $\pi_j \colon T_j \to Q/\operatorname{rad} Q$ are the natural maps, then

$$\pi_1(\alpha_1, \beta_1, \ell_1, T_1) = \pi_2(\alpha_2, \beta_2, \ell_2, T_2).$$

Claim 6.3. $(\mathscr{F}_i^{2t}, \sim)$ is a connected graph.

Proof. Let $\Gamma = \mathrm{P\Gamma Sp}(\infty, q)$. Then Γ acts as a group of automorphisms of $(\mathscr{F}_i^{2t}, \sim)$. Let Ω be the set of non-degenerate $2t$-spaces. Define a graph on Ω by

$$T_1 E T_2 \quad \text{iff} \quad T_1 \cap T_2 = \varnothing \quad \text{and} \quad \dim \operatorname{rad}\langle T_1, T_2 \rangle = 2t.$$

Since Γ acts primitively on Ω, the graph (Ω, E) is connected. Suppose that $f_1 = (\alpha_1, \beta_1, \ell_1, T_1) \in \mathscr{F}_i^{2t}$ and that $T_1 E T_2$. Let $Q = \langle T_1, T_2 \rangle$ and, for $j = 1, 2$, let $\pi_j : T_j \to Q/\operatorname{rad} Q$ be the natural map. Define $f_2 = (\alpha_2, \beta_2, \ell_2, T_2) \in \mathscr{F}_i^{2t}$ by the requirement that

$$\pi_1(\alpha_1, \beta_1, \ell_1, T_1) = \pi_2(\alpha_2, \beta_2, \ell_2, T_2).$$

Then $f_1 \sim f_2$. Since (Ω, E) is connected, it follows that whenever $f_1 = (\alpha_1, \beta_1, \ell_1, T_1) \in \mathscr{F}_i^{2t}$ and $T_2 \in \Omega$, then there exists $f_2 = (\alpha_2, \beta_2, \ell_2, T_2) \in \mathscr{F}_i^{2t}$ such that f_1 and f_2 lie in the same connected component of $(\mathscr{F}_i^{2t}, \sim)$. This implies that T_2 contains a representative of every connected component of $(\mathscr{F}_i^{2t}, \sim)$. Now choose T_2 such that $T_1 \perp T_2$ and $T_1 \cap T_2 = \varnothing$. Then $\Gamma_{(T_1)}$ acts transitively on the 4-tuples $(\alpha_2, \beta_2, \ell_2, T_2) \in \mathscr{F}_i^{2t}$. Hence each such 4-tuple lies in the connected component contaning f_1. Thus the graph is connected. □

Thus, by Witt homogeneity, it is enough to find a subspace Q such that:

(a) $\dim Q = 4t$ and $\dim \operatorname{rad} Q = 2t$; and

(b) whenever $f_j = (\alpha_j, \beta_j, \ell_j, T_j) \in \mathscr{F}_i^{2t}$ for $j = 1, 2$ are adjacent 4-tuples with $T_1, T_2 < Q$, then there exists $g \in G$ such that $g(\alpha_1, \beta_1, \ell_1, T_1) = (\alpha_2, \beta_2, \ell_2, T_2)$.

By Lemma 5.1, there exists a finite dimensional subspace U such that:

(i) $\dim U/R = 2t$, where $R = \operatorname{rad} U$;

(ii) $\dim R \geqslant \max\{f(2t), 4t\}$, where f is the function given by Lemma 5.3; and

(iii) if F is the group induced on R by $G_{\{U\}}$, then $\mathrm{PSL}(R) \leqslant F$.

From now on, we shall work with the underlying vector space. Using Lemma 5.3, it follows that there exists a subspace Z satisfying the following conditions.

(1) $\dim Z = 6t$.

(2) $\dim S = 4t$, where $S = \operatorname{rad} Z$.

(3) Let Γ be the group induced on Z by $G_{\{Z\}}$, and let $\tilde{\Gamma}$ be the corresponding group of semilinear transformations. Then there exists a subspace $T < Z$ such that:

 (i) $Z = S \oplus T$;

 (ii) $\tilde{\Gamma}_{(T)}$ induces $\mathrm{SL}(S)$ on S.

Clearly T is a non-degenerate $2t$-space. Let $T = \langle v_i : 1 \leqslant i \leqslant 2t \rangle$. Choose $A = \langle a_i : 1 \leqslant i \leqslant 2t \rangle$, $B = \langle b_i : 1 \leqslant i \leqslant 2t \rangle$ with $S = \langle A, B \rangle$. Define

$$\begin{aligned} T_1 &= \langle v_i + a_i : 1 \leqslant i \leqslant 2t \rangle \\ T_2 &= \langle v_i + b_i : 1 \leqslant i \leqslant 2t \rangle \end{aligned}$$

and $Q = \langle T_1, T_2 \rangle$. We will show that Q satisfies our requirements. Suppose that $T_3 < Q$ is an arbitrary non-degenerate $2t$-space. Then $T_3 = \langle v_i + c_i : 1 \leqslant i \leqslant 2t \rangle$ for some $c_i \in S$. Since $Q \cap T = O$, it follows that $\dim\langle c_i : 1 \leqslant i \leqslant 2t \rangle = 2t$. By (3), there exists $g \in \tilde{\Gamma}$ such that

$$g(v_i + a_i) = v_i + c_i$$

for $1 \leqslant i \leqslant 2t$. So it is clear that Q satisfies (a) and (b). This completes the proof of Lemma 6.1.

Some combinatorial aspects of the cover problem for totally categorical theories

A. A. Ivanov

Institute of Mathematics,
Wrocław University,
pl. Grunwaldzki 2/4,
50 - 384 Wrocław,
Poland.
Electronic mail address: logicuwr@plwruw11.bitnet

A finite cover M of a structure N is obtained in the following way. For a finite set F consider a structure $M = N \cup (N \times F)$ with the natural projection π from $N \times F$ onto N. Some new relations may be added to M but no new structure is induced on N. In this paper we usually assume that N is a strictly minimal countably categorical set. Such covers appear in the investigation of totally categorical theories and the problem of a description of them seems very difficult. We approach this 'cover problem' by analysing arities of covers. Mainly, we are interested in what covers can be obtained by adding binary relations to principal finite covers. We show that these covers split and in some natural situations they can be characterized completely.

The case of ternary covers looks much more complicated and it is unclear how our methods could be generalized here. Very interesting results on covers of arbitrary arities are proved by Ahlbrandt and Ziegler (1991*b*). They show that a cover splits if N is projective over $\mathbf{F}_2$ and F has 2 elements.

Our approach provides us with many combinatorial questions concerning arities of covers. Some of them are discussed in Sections 1 and 5.

I would like to thank Dugald Macpherson, Ludomir Newelski, Martin Ziegler and the referee for helpful remarks.

1 Preliminaries

From now on we assume that all our structures are countably categorical. We will mainly consider finite covers of strictly minimal sets. Some results of the paper are true in more general cases. So our definition of a cover differs slightly from the usual one. We do not require the presence of structure groups on each fibre. It does not preserve total categoricity (see Ahlbrandt and Ziegler (1991*b*); Hodges and Pillay (in press)).

DEFINITION 1.1. A countable structure M is a *cover* of N if

(i) N is a 0-definable *fully embedded* subset of M, i.e., every relation on N which is 0-definable in M is 0-definable in N, and every relation on N which is definable with parameters in M is definable with parameters from N;

(ii) there is a 0-definable surjection $\pi\colon M \setminus N \to N$ whose *fibres* $F_a = \pi^{-1}(a)$ have the same cardinality.

A cover is *finite* if its fibres are finite.

DEFINITION 1.2. The *kernel* of a cover M of a structure N is the set of all automorphisms of M which leave N pointwise fixed.

In the paper we always assume that the fibres of M are given as pairwise isomorphic structures

$$(F_a, \rho_{1,a}, \ldots, \rho_{l,a}) \quad (a \in N)$$

where the relations

$$\rho_i = \bigcup\{\rho_{i,a} : a \in N\}$$

are 0-definable in M and no other structure is induced on the fibres in M.

DEFINITION 1.3. A cover M_0 of N is *principal* if it is obtained from N by adding for every $a \in N$ a new set F_a with relations $\rho_{i,a}$ on it and has no other structure. It is considered in the language of N extended by $\pi, \rho_1, \ldots, \rho_l$.

Note that the point here is that there is no interaction *between* the fibres other than that determined by the structure of N.

All the covers of N which have a given structure for the fibres are (up to interdefinability) expansions of the corresponding principal cover M_0 determined by that fibre structure.

The *cover problem* is:

> Determine up to interdefinability all expansions of a principal cover M_0 which induce no new structures on N.

It is known (Ahlbrandt and Ziegler, 1991*b*; Hodges and Pillay, in press) that an expansion M of some principal cover M_0 is a cover of N iff every automorphism of N lifts to an automorphism of M. Thus the cover problem can be restated as follows:

> Determine all expansions M of M_0 for which the projection map $\mathrm{Aut}(M) \to \mathrm{Aut}(N)$ is surjective.

Assume that the theory of M admits a finite language (for example, this is true in the totally categorical case (Hrushovski, 1989*b*)). Then there is a natural number k such that the group $G = \mathrm{Aut}(M)$ is a group of automorphisms of a structure with k-ary relations only. In this case we will say that the group G is *k-closed.* The *arity* of G, $\mathrm{ar}(G)$, is the least k for which G is k-closed; the arity of M is $\mathrm{ar}(\mathrm{Aut}(M))$.

It is natural to investigate covers assuming some restrictions on their arities. Some new questions arise here. In general we do not understand the connection between arities of covers and the arity of the corresponding principal cover M_0 very well. In fact, our knowledge consists of the following observations.

PROPOSITION 1.4. The arity of a principal cover M_0 is

$$\max(\mathrm{ar}(N), \mathrm{ar}(F_a), 2).$$

Proof. Indeed, from the definition of a principal cover we have the inequality $\leqslant$. For the other inequality, we need to analyse the k-orbits of $\mathrm{Aut}(M_0)$ as in Kaluzhnin and Klin (1976). Theorem 2.7 there asserts that the arity of the wreath product of permutation groups (G, J) and (H, L) is $\max(\mathrm{ar}(G), \mathrm{ar}(H), 2)$. Of course, $\mathrm{Aut}(M_0)$ is the wreath product of the permutation groups $(\mathrm{Aut}(F_a), F_a)$ and $(\mathrm{Aut}(N), N)$. This proves the proposition. □

We say that a finite cover $M = N \cup (N \times F)$ is *splitting* if it has a covering expansion by unary predicates P_f $(f \in F)$, such that each unary predicate P_f intersects each fibre in a singleton (Ziegler, 1992). Of course, such an expansion has trivial kernel.

PROPOSITION 1.5. Assume M is a finite cover of N such that $\mathrm{ar}(M) < \mathrm{ar}(N)$. Then M is not splitting.

Proof. Let $M = N \cup (N \times F)$ and $k = \mathrm{ar}(M)$. We may suppose that all new relations of M are defined on $N \times F$. Assume that the expansion of M by unary predicates $P_f = \{(a, f) : a \in N\}$ $(f \in F)$ is a cover of N. We must show that there is a set of 0-definable (in M) k-ary relations on N such that every permutation of N which preserves these relations is an automorphism of N. For every k-ary relation $R(\vec{x})$ of M on $N \times F$ and every k-tuple $\vec{f} \subset F$ consider a relation on N,

$$R((y_1, f_1), (y_2, f_2), \ldots, (y_k, f_k)).$$

Let α be a permutation of N which preserves all these relations together with all k-ary relations of N. Clearly, the natural expansion of α on M preserves all k-ary relations of M. Hence α is an automorphism of N. This contradicts $k < \mathrm{ar}(N)$. □

We do not know if there exists an expansion M of a finite principal cover M_0 of a strictly minimal set N such that M is a cover of N and $\mathrm{ar}(M) < \mathrm{ar}(M_0)$. One of the corollaries of the results of this paper asserts that this is impossible if N is of disintegrated or projective type and there is no structure on fibres.

In Section 2 we characterize finite covers of disintegrated strictly minimal sets. This allows us to find some lower and upper bounds of their arities. In Section 3 we prove that a finite binary cover of a strictly minimal set is splitting. Some natural binary covers are described in Section 4. In Section 5 we consider ternary covers.

Now we recall some other definitions (see Cherlin *et al.* (1985)). In both of them we require M to be ω-stable. Note that any finite cover of a strictly minimal set is totally categorical, and hence ω-stable.

DEFINITION 1.6. Let A be an algebraically closed subset of a strictly minimal set $D \subseteq M$. A *D-envelope of A in M* is a maximal set B which is independent from D over A.

DEFINITION 1.7. A subset A of M is called *ω-homogeneous* if for all finite tuples $\vec{b}, \vec{b'} \in A$ such that $\mathrm{tp}(\vec{b}) = \mathrm{tp}(\vec{b'})$ there exists an automorphism α of M such that $\alpha(\vec{b}) = \vec{b'}$ and $\alpha(A) = A$.

By Theorem 7.3 of Cherlin *et al.* (1985) any non-empty D-envelope of A is ω-homogeneous and uniquely determined by A up to an A-automorphism (i.e., an automorphism that fixes A pointwise).

2 Covers of disintegrated sets

The results of this section will be applied for the proofs of our main theorems in Sections 3 and 4. Here we investigate arities of finite covers of disintegrated strictly minimal sets. We use Ziegler's description of these covers (Ziegler, 1992). Then we introduce simple covers and completely describe their arities in the disintegrated case.

If N is a disintegrated strictly minimal set then $\mathrm{Aut}(N) = \mathrm{Sym}(N)$. Consider a finite principal cover $M_0 = N \cup (N \times F)$ of N. Assume that a fibre F_a is $\{a\} \times F$ ($a \in N$), $\mathrm{card}(F) = n$ and there is no structure on a fibre in M_0. It is easy to see that every covering expansion of M_0 is totally categorical. The group $\mathrm{Aut}(F)$ has the natural action on $N \times F$ by automorphisms of $\mathrm{Aut}(M_0)$ in the following way: for $\alpha \in \mathrm{Aut}(F)$ we put $\alpha(a, f) = (a, \alpha(f))$ and $\alpha(a) = a$, where $a \in N, f \in F$. Let G_0 be the group of these automorphisms in $\mathrm{Aut}(M_0)$.

PROPOSITION 2.1. Let M be an expansion of M_0 inducing no new structure on the disintegrated strictly minimal set N. If G is a subgroup of $\mathrm{Aut}(M) \cap G_0$ then there is an action of $\mathrm{Sym}(N) \times G$ on $N \times F$ by automorphisms of $\mathrm{Aut}(M)$. Automorphisms of M which correspond to elements of $\mathrm{Sym}(N)$ have the corresponding action on N and non-trivial automorphisms for G act trivially on N and have non-trivial projections on every fibre.

Proof. We start with the investigation of N-envelopes of finite subsets of N. It is easy to see that in our case the envelope of C is the set $C \cup (C \times F)$. Since every envelope is ω-homogeneous, for a finite $A \subset N$ and each $a \in N \setminus A$ every permutation α of $A \cup (A \times F)$ preserving types can be extended to an automorphism of M fixing a. This allows us to check that for each $b \in N \setminus A$ there is an automorphism α of M fixing all elements of $A \times F$ such that $\alpha(b) = a$. Thus $\mathrm{tp}(a/A \times F) = \mathrm{tp}(b/A \times F)$ (also this is a consequence of weak elimination of imaginaries for N, see Ziegler (1992) and Evans and Hrushovski (1993)).

If $\{a, b\} \subset N$ then by triviality of N the permutation interchanging a and b can be extended to an automorphism α of M such that for every $c \in N \setminus \{a, b\}$, $\alpha(c) = c$. As above, by ω-homogeneity we may suppose that for each $c \in (N \setminus \{a, b\}) \times F$, $\alpha(c) = c$. So for every $a_0 \in N$ we can find an infinite set $\{b_0, b_1, \ldots\} \subseteq N$ and an infinite set of automorphisms $\{\alpha_0, \alpha_1, \ldots\} \subset \mathrm{Aut}(M)$ such that for each $c_i = (b_i, f) \in \{b_i\} \times F$ and $c_j = (b_j, f) \in \{b_j\} \times F$ we have $\alpha_i(c_i) = \alpha_j(c_j) \in \{a_0\} \times F$ (and the same for α_i^{-1}, α_j^{-1}) and for every $c \in (N \setminus \{a_0, b_i\}) \times F$, $\alpha_i(c) = c$. Thus for every set $\{b_0, \ldots, b_m\}$ every permutation ν of its envelope with the condition $\nu((b_i, g)) = (\nu(b_i), g)$ can be extended to an automorphism of M. Since G acts naturally on $\{b_0, \ldots, b_m\} \times F$ we have the natural action of $\mathrm{Sym}(m) \times G$ on the $\{b_0, \ldots, b_m\} \times F$ by automorphisms of the corresponding envelope.

If A and A' from N have the same finite cardinality then their envelopes are isomorphic. So for every $A' \subset N$ of cardinality m there is an action of $\mathrm{Sym}(m) \times G$ on its envelope by automorphisms.

Now by induction we can define an action of $\mathrm{Sym}_f(N) \times G$ on M by automorphisms of M, where $\mathrm{Sym}_f(N)$ is the set of permutations of N with finite support. By approximation we can define our action of $\mathrm{Sym}(N) \times G$. □

This proof with $\mathbf{Z}_n$ instead of G was presented in the first draft of this paper. Independently, Ziegler proved a similar result (1992). In fact, he assumed the triviality of G. He notes that Proposition 2.1 implies that covers are determined by their kernels. Indeed, in this case every cover is a reduct of some cover with trivial kernel. This cover is defined by unary predicates which are 1-orbits of the action of $\mathrm{Sym}(N)$ in our proposition. Thus, M splits, and $\mathrm{Aut}(M) = \mathrm{Sym}(N) \cdot K$, where K is the kernel of M.

Moreover, Ziegler described kernels of finite covers of disintegrated sets. His description is as follows.

The subgroups K of $\prod_{x\in N} \mathrm{Sym}(F)$ which are kernels of finite covers of N are exactly the groups of the form

$$K_H^L = \{\alpha \in \prod_{x\in N} L : \forall x,y\, (\alpha_x H = \alpha_y H)\},$$

where L is a subgroup of $\mathrm{Sym}(F)$, H is a normal subgroup of L and α_a is the action on F which corresponds to the= $\{\alpha$

$$\begin{aligned} & {}_a : \alpha \in K\}, \\ H(K) \quad &= \quad \{\alpha_a : \alpha \in K \text{ and } \forall x{\neq}a\, (\alpha_x = id)\}. \end{aligned}$$

The groups H and L are permutation groups on F. Let us fix the numbers $\mathrm{ar}(H), \mathrm{ar}(L)$. What can we say now about the arity of the corresponding cover? The upper bound is given by the following proposition.

PROPOSITION 2.2. Let L be a permutation group on a finite set F and H be a normal subgroup. Assume N is a disintegrated set and the group K_H^L defined as above is the kernel of a covering expansion M of a principal cover $M_0 = N \cup (N{\times}F)$. Then $\mathrm{ar}(M) \leqslant \max(2\cdot \mathrm{ar}(H), \mathrm{ar}(L))$.

Proof. Let us put $k = \mathrm{ar}(H)$ and consider k-orbits of H on F. If R is a k-orbit of H then define a relation $S(R)$ on M_0 as follows: $S(R)$ consists of the set of all tuples

$$((a, f_1), \ldots, (a, f_k), (b, f_1'), \ldots, (b, f_k'))$$

such that

$$a, b \in N,\ a \neq b,\ (f_1, \ldots, f_k) \in R, \ \text{ and } (f_1', \ldots, f_k') \in R.$$

Put $S(R, L) = \bigcup\{\gamma(S(R)) : \gamma \in L\}$, where the action of L on $N{\times}F$ is defined by the action of G_0 as in Proposition 2.1. We have to prove that the structure M_0 with all relations $S(R, L)$ and $\mathrm{ar}(L)$-orbits on fibres is a cover with the kernel K_H^L. Since such a cover is unique, then it follows that $\mathrm{ar}(M) \leqslant \max(2k, \mathrm{ar}(L))$.

The structure

$$\begin{aligned} (M_0,\ & \{S(R, L) : R \text{ is a } k\text{-orbit of } H \text{ on } F\}, \\ & \{T(Q) : Q \text{ is an } \mathrm{ar}(L)\text{-orbit of } L \text{ on } F\}) \end{aligned}$$

(where $T(Q)$ is the union of the $\mathrm{ar}(L)$-orbits on fibres which correspond to the $\mathrm{ar}(L)$-orbit Q) is a cover of N because the natural action of every permutation $\rho \in \mathrm{Sym}(N)$ on $N{\times}F$ preserves all relations $S(R, L)$ and $T(Q)$.

If an automorphism α of this structure fixes N pointwise, then for every $x \in N$ the permutation α_x fixes $\mathrm{ar}(L)$-orbits of L on F, and so $\alpha_x \in L$. Since H is a normal subgroup of L, the permutation α_x has a correct action on the set of k-orbits of H. Moreover, the corresponding actions of α_x and α_y are the same because for every k-orbit R of H the set of all

$$((x, \alpha_x(f_1)), \ldots, (x, \alpha_x(f_k)), (y, \alpha_y(f'_1)), \ldots, (y, \alpha_y(f'_k)))$$

such that $(f_1, \ldots, f_k), (f'_1, \ldots, f'_k) \in R$ is in $S(R, L)$. So $\alpha_x^{-1}\alpha_y$ preserves all k-orbits of H on F. Since H is k-closed, $\alpha_x^{-1}\alpha_y \in H$ and $\alpha \in K_H^L$.

By a similar argument we can prove that every $\alpha \in K_H^L$ preserves all $S(R, L)$. So the kernel of our structure is K_H^L. □

REMARK. If we assume in the proof of Proposition 2.2 that

$$k = \max(\mathrm{ar}(H), \mathrm{ar}(L))$$

then the relations $T(Q)$ are not needed. Indeed, to prove $\alpha_x \in L$ it is enough to note that α_x fixes all k-orbits of L. But each of these k-orbits is $\bigcup\{\gamma(R) : \gamma \in L\}$ for some k-orbit R of H.

We will use this remark later for ease of notation.

Simple covers. We will say that a cover M of a structure N is *simple* if it has no non-trivial covering expansion M' such that $\mathrm{Aut}(M')$ is normal in $\mathrm{Aut}(M)$. It seems that simple covers are important. The motivation for this notion is the following.

Consider a finite cover M of a strictly minimal set N. Every proper increasing chain of its covering expansions is finite (Hrushovski, 1989*c*, 1989*b*). Thus there is a covering expansion M' of M such that $\mathrm{Aut}(M')$ is a minimal normal covering subgroup of $\mathrm{Aut}(M)$. By the following proposition this subgroup is unique and simple as a cover.

PROPOSITION 2.3. Suppose that N is ω-categorical (but not necessarily ω-stable) and that $[\mathrm{Aut}(N), \mathrm{Aut}(N)] = \mathrm{Aut}(N)$. Let M be a finite cover of N and M' be an expansion of M such that M' is a cover of N and $\mathrm{Aut}(M')$ is a minimal normal subgroup of $\mathrm{Aut}(M)$ with this condition. Then M' is uniquely determined by this and is simple.

Proof. If M'' is another covering expansion such that $\mathrm{Aut}(M'')$ is a minimal normal subgroup then $\mathrm{Aut}(M') \cap \mathrm{Aut}(M'')$ is a group of automorphisms of some cover. For the condition

$$[\mathrm{Aut}(N), \mathrm{Aut}(N)] = \mathrm{Aut}(N)$$

allows us to extend automorphisms of $\mathrm{Aut}(N)$ to elements of

$$[\mathrm{Aut}(M'), \mathrm{Aut}(M'')] \subseteq \mathrm{Aut}(M') \cap \mathrm{Aut}(M'').$$

Since $\mathrm{Aut}(M')$ and $\mathrm{Aut}(M'')$ are minimal and normal, they are equal to $\mathrm{Aut}(M') \cap \mathrm{Aut}(M'')$. If M' is not simple then there exists a unique expansion of M' covering N such that its group of automorphisms H is a minimal normal subgroup. So for every $g \in \mathrm{Aut}(M)$, $g^{-1}Hg = H$. This contradicts the minimality of $\mathrm{Aut}(M')$. □

For an infinite set N the condition $[\mathrm{Sym}(N), \mathrm{Sym}(N)] = \mathrm{Sym}(N)$ is true. How do simple covers of N look in Ziegler's notation?

PROPOSITION 2.4. Let K be a kernel of a finite cover M of a disintegrated set N. Then M is simple iff $H(K) = L(K)$.

Proof. It is easy to check that

$$\mathrm{Sym}(N) \cdot (\prod_{x \in N} H(K))$$

is normal in $\mathrm{Sym}(N) \cdot K$, because, for every $\alpha \in K$,

$$\alpha^{-1} \mathrm{Sym}(N) \alpha \subseteq \mathrm{Sym}(N) \cdot (\prod_{x \in N} H(K)).$$

So the condition $H(K) = L(K)$ is necessary.

For the other direction let us assume that $H(K) = L(K)$. Let M' be the simple expansion of M having a minimal normal subgroup $\mathrm{Aut}(M')$ in $\mathrm{Aut}(M)$. We have to prove that the kernel K' of M' is equal to K. It is enough to prove $H(K) = L(K')$ because $H(K') = L(K')$.

Let $\sigma \in H(K)$, $\alpha \in \prod_{x \in N} H(K)$, $\alpha_a = \sigma$ and $\forall x {\neq} a\, (\alpha_x = \mathrm{id})$. Then for every transposition (a, b) on N the automorphism $k = \alpha(a, b)\alpha^{-1}(a, b)^{-1}$ is σ at a. Since $k \in K'$ we have $\sigma \in L(K')$. □

REMARK. The arity of M in Proposition 2.4 is $\max(\mathrm{ar}(N), \mathrm{ar}(H(K)), 2)$.

3 Finite covers

Throughout this section we assume that $M_0 = N \cup (N \times F)$ is a finite principal cover of a strictly minimal structure N and all fibres have cardinality n (thus all covering expansions of M_0 are totally categorical). Our main results here (Theorem 3.2 and Corollary 3.3) describe all covering expansions of M_0 obtained by adding binary relations. Note that when we say that a cover of N is *binary* we mean that it is obtained from M_0 by adding only binary (and unary) relations.

We will say that an expansion M of M_0 is a *minimal cover* if M is a cover of N and every proper expansion of M is not a cover. Hrushovski (1989*c*) calls minimal covers *maximal expansions.* The following statement is a partial case of a more general result of Hrushovski (1989*c*, 1989*b*).

THEOREM 3.1. Every covering expansion of M_0 is a reduct of some minimal cover.

This theorem asserts that every cover is defined by its kernel and some minimal cover. Indeed, if M' is a reduct of a minimal cover M and K is the kernel of M' then $\mathrm{Aut}(M') = \mathrm{Aut}(M)\cdot K$. So we can restrict the cover problem to the problem of description of kernels and minimal covers.

If a cover is splitting then it is an expansion of the corresponding minimal cover (*trivial cover*) defined as in Section 1. Thus a splitting cover is determined by its kernel only.

In this section we prove that every covering expansion of M_0 obtained by adding binary relations only (*binary covering expansion*) is splitting. Also we characterize the kernels of these expansions.

THEOREM 3.2. Every binary covering expansion M of M_0 is splitting.

Proof. Consider all 0-definable relations $\mathbf{r}$ in M such that

$$\begin{array}{c}\text{if } \vec{a} \in \mathbf{r} \text{ then there are } b, b' \in N \text{ such that} \\ \text{all elements of } \vec{a} \text{ are in } F_b \cup F_{b'}.\end{array} \qquad (*)$$

We may assume that these relations, π, and the relations of N are all the basic relations of M.

Let $B = \{b_1, b_2, \ldots\} \subseteq N$ be an infinite independent set. Consider a model M_B which is obtained from B by adding all corresponding B-fibres of M such that all relations of M_B are restrictions of the 0-definable relations of M. It is not difficult to check that M_B is a cover of the disintegrated set B and every automorphism of M_B can be extended to an automorphism of M. By Proposition 2.1 the structure M_B can be considered as $B \cup (B \times F)$ where every fibre F_b is $\{b\} \times F$ and the natural action of $\mathrm{Sym}(B)$ on $B \times F$ defines a subgroup of $\mathrm{Aut}(M_B)$. By Ziegler's description of kernels over disintegrated sets, the kernel K of M_B is K^L_H (see Section 2), where L is the projection of K to $\mathrm{Aut}(F)$ and

$$H = \{\alpha_x : \alpha \in K \text{ and } \forall y \neq x\, (\alpha_y = \mathrm{id})\}.$$

If $b, b' \in B$ then the envelope of $\{b, b'\}$ in M_B is

$$E(b, b') = \{b, b'\} \cup F_b \cup F_{b'}.$$

Thus every elementary map from the kernel of $E(b, b')$ extends to an automorphism in the kernel of M_B. Since the kernel of M_B is K^L_H, the projection of the kernel of $E(b, b')$ on $\mathrm{Aut}(F)$ is L and

$$H = \{\alpha_b : \alpha \in \mathrm{Ker}(E(b, b')), \alpha_{b'} = \mathrm{id}\}.$$

Of course, the kernel of $E(b, b')$ is

$$\{\alpha : \alpha_b, \alpha_{b'} \in L \text{ and } \alpha_b^{-1}\alpha_{b'} \in H\}.$$

As in the proof of Proposition 2.2 (see the remark following the proof) the structure $E(b, b')$ can be considered under relations

$$S(R, L) = \bigcup\{\gamma(S(R)) : \gamma \in L\},$$

where the R's are k-orbits of H for some natural number k and $S(R)$ is the set of all

$$((a, f_1), \ldots, (a, f_k), (a', f'_1), \ldots, (a', f'_k))$$

such that

$$\{a, a'\} = \{b, b'\},\ (f_1, \ldots, f_k) \in R, \text{ and } (f'_1, \ldots, f'_k) \in R.$$

The relations $S(R, L)$ are definable in $E(b, b')$ and the corresponding formulas give a uniform definition of these relations in every pair of fibres of M_B.

The relations $S(R, L)$ on the whole of M_B are the restrictions of 0-definable relations of M which satisfy the condition (*). So these relations are basic in the language of M. Moreover, since every pair of elements of N is independent we can consider every pair of fibres of M with these relations only.

Since every binary relation of M is the union of its restrictions to pairs of fibres, we can assume that M is obtained from the principal cover by adding the relations $S(R, L)$.

In the structure M_B, every relation $S(R, L)$ induces a transitive binary relation on the set of k-orbits of H. This follows from the fact that the natural action of every permutation of B on $B \times F$ is an automorphism. We note next that this is true for M as well as for M_B.

If $a, b, c \in N$ are independent, then this is obvious for $\{a, b, c\} \times F$ because this substructure is isomorphic to a substructure of M_B. In the other case let us choose an independent element $d \in N$. Of course, $S(R, L)$ induces transitive binary relations on k-orbits of H in each of the following substructures:

$$F_a \cup F_b \cup F_d, \quad F_a \cup F_c \cup F_d, \quad F_b \cup F_c \cup F_d.$$

So it is true in $F_a \cup F_b \cup F_c$.

Thus we can consider M as $N \cup N \times F$ with the relations $S(R, L)$ defined as in the proof of Proposition 2.2. In this case every automorphism of N has a natural extension on $N \times F$ which preserves the predicates

$$P_f = \{(x, f) : x \in N\},$$

where $f \in F$. So M is splitting. □

Remark. The statement of this theorem is not true for ternary expansions of M_0. A vector space of countably infinite dimension over a finite field with more than 2 elements is the natural cover of its projective space. This cover does not split.

COROLLARY 3.3. The kernels K of the binary covering expansions of M_0 are characterized by their groups $L(K)$ and $H(K)$ as in Ziegler's description of covers of disintegrated sets.

Proof. Indeed, by the proof of Theorem 3.2 we can assume that a binary cover M is obtained from M_0 by adding the relations $S(R, L)$. Now it is enough to apply the proof of Proposition 2.2. □

COROLLARY 3.4. Let M_0 be a finite principal cover of a strictly minimal set N of disintegrated or projective type and assume that there is no structure on fibres. If M is a cover of N which is an expansion of M_0, then $\mathrm{ar}(M_0) \leqslant \mathrm{ar}(M)$.

Proof. Indeed, from the definition of a cover it is easy to see that $\mathrm{ar}(M) \geqslant 2$. If N is a disintegrated set then $\mathrm{ar}(M_0) = 2$ and there is nothing to prove. In the other case $\mathrm{ar}(M_0) = 3$. If $\mathrm{ar}(M) = 2$ then M is a binary cover. By the proof of Theorem 3.2 (see the remark after Proposition 2.2) we can suppose that all relations of M are $S(R, L)$'s. But in this case every permutation of N can be extended to an automorphism of M. So M_0 is not a reduct of M. □

4 Some binary covering expansions of 2-transitive sets

Theorem 3.2 describes all binary finite covers of strictly minimal sets by their kernels. In this section we show that a certain natural assumption on kernels yields a very simple characterization of binary covers of some 2-transitive sets.

Finite covers of strictly minimal sets. Let M_0 be a principal cover of a strictly minimal set N. Throughout this subsection we assume that all fibres $F_a = \{a\} \times F$ $(a \in N)$ have cardinality p, where p is prime and there is no structure on fibres. Identifying F and $\mathbf{Z}_p$, we can describe some subgroups of $\mathrm{Sym}(F)$ as follows. The multiplicative group $\mathbf{P}(p)$ of $\mathbf{Z}_p$ acts on F by right multiplication. So the group $\mathbf{P}(p) \cdot \mathbf{Z}_p$ may be considered as the group of all linear permutations: $x \to xa + b$ $(a \in \mathbf{P}(p),\ b \in \mathbf{Z}_p)$.

We also assume that for every $z \in \mathbf{Z}_p$ the permutation

$$(a, f) \mapsto (a, f + z)$$

fixing elements of N is an automorphism of $M_0 = N \cup (N \times F)$. Let G_0 be the group of all such automorphisms of the form $(a, f) \mapsto (a, f + z)$.

In the next theorem we will use the following well-known definitions.

The direct product $(G_1, X) \times (G_2, Y)$ of permutation groups consists of all permutations of $X \times Y$ of the form $(g_1, g_2) : (x, y) \mapsto (g_1 x, g_2 y)$.

The wreath product $(G_1, X)\,\mathrm{Wr}\,(G_2, Y)$ is the set of all permutations on $X\times Y$ of the form $\nu_{g,\alpha} : (x, y) \mapsto (\alpha(y)x, gy)$, where $g \in G_2$ and $\alpha : Y \to G_1$ is an arbitrary mapping.

THEOREM 4.1. Let N be strictly minimal, let M be a binary covering expansion of M_0 and assume $G_0 \subseteq \mathrm{Aut}(M)$. Then there is an identification of F_a's with $\{a\}\times F$'s ($a \in N$) such that the action of $\mathrm{Aut}(M)$ on $N\times F$ has one of the following forms:

1. $(J, F)\,\mathrm{Wr}\,(\mathrm{Aut}(N), N)$;
2. $(\mathrm{Aut}(N), N)\times(J, F)$;

where $J = \mathrm{Sym}(F)$ or $J = A \cdot \mathbf{Z}_p \subseteq \mathbf{P}(p) \cdot \mathbf{Z}_p$ for a proper subgroup A of $\mathbf{P}(p)$.

Proof. We may suppose that new binary relations are defined on $N\times F$ only. Indeed, since $G_0 \subseteq \mathrm{Aut}(M)$ and $\mathrm{Aut}(M)$ has a 2-transitive action on N, every automorphism of M_0 preserves every binary relation R of M which is a subset of $N^2 \cup N\times(N\times F) \cup (N\times F)\times N$.

Now let us start with the case of disintegrated N. Let q be a prime number ($p \neq q$). If the cardinality of $A \subset N$ is q then, by Proposition 2.1, there is a regular transitive action of $\mathbf{Z}_q\times\mathbf{Z}_p$ on $A\times F$, as in Proposition 2.1, by automorphisms of our envelope $A \cup A\times F$.

Klin and Poschel (1981, Theorems 3.3 and 3.4) prove that after the corresponding identification $\mathbf{Z}_q\times\mathbf{Z}_p$ with $A\times F$, any 2-closed group on $A\times F$, which includes $\mathbf{Z}_q\times\mathbf{Z}_p$, has one of the following forms:

1. $\mathrm{Sym}(\mathbf{Z}_q\times\mathbf{Z}_p)$;
2. $(J', \mathbf{Z}_q)\,\mathrm{Wr}\,(J, \mathbf{Z}_p)$;
3. $(J, \mathbf{Z}_p)\,\mathrm{Wr}\,(J', \mathbf{Z}_q)$;
4. $(J', \mathbf{Z}_q)\times(J, \mathbf{Z}_p)$;

or it is a subgroup of the holomorph of $\mathbf{Z}_q\times\mathbf{Z}_p$ (the group $(J, \mathbf{Z}_p)$ is defined as above and $(J', \mathbf{Z}_q)$ is defined in a similar way).

Since the group of automorphisms of $A\times F$ with respect to the basic relations of M is 2-closed, we can apply this result. If q is big enough then the last case is impossible because each element of $\mathrm{Sym}(\mathbf{Z}_q)$ extends to an automorphism of the envelope $A \cup A\times F$ (also see Theorem 3.4 in Klin and Poschel (1981)). Since an automorphism of an envelope preserves the projection π, it is easy to see that the group of automorphisms of the binary structure $A\times F$ has one of the following forms:

1. $(J, \mathbf{Z}_p)\,\mathrm{Wr}\,(\mathrm{Sym}(\mathbf{Z}_q), \mathbf{Z}_q)$;
2. $(\mathrm{Sym}(\mathbf{Z}_q), \mathbf{Z}_q)\times(J, \mathbf{Z}_p)$;

where J is defined as above.

Now, as in the proof of Proposition 2.1, we can use induction.

Let us consider the case of a general strictly minimal N. Let $\{b_1,\ldots\} \subseteq N$ be an infinite set of independent elements. Fixing some enumeration of F we can assume that the set of the corresponding tuples $\{b_i\}\times F$ is indiscernible. Thus we can define an action of $\mathrm{Sym}(\{b_1,\ldots\})\times\mathbf{Z}_p$ on $\bigcup\{F_a : a \in \{b_1,\ldots\}\}$ by automorphisms of M. Consider this set with the binary relations of M only. As in the disintegrated case, we can prove that after some identification of $\{b_1,\ldots\}\times\mathbf{Z}_p$ with $\bigcup\{F_a : a \in \{b_1,\ldots\}\}$ the group of automorphisms of the obtained model has one of the following forms:

1. $(J, \mathbf{Z}_p)\,\mathrm{Wr}\,(\mathrm{Sym}(\{b_1,\ldots\}), \{b_1,\ldots\})$;
2. $(\mathrm{Sym}(\{b_1,\ldots\}), \{b_1,\ldots\})\times(J, \mathbf{Z}_p)$.

If in the first case, for some $g, g', i, j (i \neq j)$ and a relation R of M we have $R(g, g')$, $g \in \{b_i\}\times\mathbf{Z}_p$, and $g' \in \{b_j\}\times\mathbf{Z}_p$, then it is true for all elements of $\{b_k\}\times\mathbf{Z}_p$ and $\{b_l\}\times\mathbf{Z}_p$ $(k \neq l)$. So we may suppose that there is no binary relation between different fibres of our set. Since any two elements of N are independent, every pair of fibres of M is an image of some pair of fibres of $\{b_1,\ldots\}$ under an automorphism of M. So there is no binary relation between different fibres of M. Now by an easy verification we obtain the first case of our theorem.

Assume that the group of automorphisms of the binary structure $\bigcup\{F_a : a \in \{b_1,\ldots\}\}$ has the second form. In this case each automorphism fixing fibres is determined by its restriction on some fibre. Now it is easy to prove that there is a 0-definable relation P on $\bigcup\{F_a : a \in \{b_1,\ldots\}\}$ such that P is an 1–1 relation between every two fibres (the corresponding formula defines P in any substructure consisting of two fibres). This 1–1 relation P is definable on all fibres of M because every two elements of M are independent. Since every envelope is ω-homogeneous and an envelope of three independent elements is unique up to an isomorphism, we can define a 1–1 relation

$$E = \left\{ (a,b) : \begin{array}{l} \pi(a) \neq \pi(b)\ \&\ \exists c\,(\pi(a), \pi(b), \pi(c) \text{ are} \\ \text{independent and } P(a,c) \& P(b,c)) \\ \text{or } a = b \end{array} \right\}.$$

Certainly, E induces the equivalence relation

$$\{\langle (b_i, z), (b_j, z)\rangle : i, j \in \omega, z \in \mathbf{Z}_p\}$$

on $\{b_1,\ldots\}\times\mathbf{Z}_p$ (the natural action of $\mathrm{Sym}(\{b_1,\ldots\})$ preserves E). Using E and all binary relations on $\{b_i\}\times\mathbf{Z}_p$ which are 0-definable in the substructure $\{b_i\}\times\mathbf{Z}_p \cup \{b_j\}\times\mathbf{Z}_p$, we can define all 0-definable binary relations of this structure (by coordinatization). The corresponding definitions do not depend on i, j. So we may assume that the basic relations on $\{b_1,\ldots\}\times\mathbf{Z}_p$ are π, E and the binary relations on each fibre. Since every two elements of N are independent, we may also assume that the basic relations of M are π, E, the relations of N and the relations of fibres.

Now let us prove that E is an equivalence relation on $N \times F$. As in the proof of Theorem 3.2, for every $a, b, c \in N$ we can choose an independent element $d \in N$. So the relation E is transitive on each of the following substructures:

$$F_a \cup F_b \cup F_d; \quad F_a \cup F_c \cup F_d; \quad F_b \cup F_c \cup F_d.$$

(They are isomorphic to substructures of $\{b_1, \ldots\} \times \mathbf{Z}_p$.) Thus E is transitive on $F_a \cup F_b \cup F_c$. Since E is an isomorphism of any two fibres from $\{b_1, \ldots\} \times \mathbf{Z}_p$, every two fibres of M are isomorphic by E. So each automorphism of N can be extended by E to an automorphism of M. It is easy to see that every automorphism of some fibre has a unique extension to an automorphism of M fixing N. This yields the second case of our theorem. □

REMARK. Using our method and results due to Klin and Poschel (1981) one can prove that every binary expansion from Theorem 4.1 can be obtained by adding one binary relation.

Infinite covers. In this subsection we consider the case of a principal cover M_0 with infinite fibres. We assume that M_0 is a cover of a 2-transitive structure N (not necessarily strictly minimal) ($M_0 = N \cup (N \times F)$) and the group of automorphisms of a fibre is also 2-transitive. As above we assume that for every $\alpha \in \mathrm{Aut}(F)$ the permutation

$$(a, f) \mapsto (a, \alpha(f))$$

fixing N pointwise is an automorphism of M_0. Let G_0 be the group of these automorphisms.

PROPOSITION 4.2. Let M be a binary covering expansion of M_0 and $G_0 \subset \mathrm{Aut}(M)$. Then there is an identification of F_a's with $\{a\} \times F$'s ($a \in N$) such that the action of $\mathrm{Aut}(M)$ on $N \times F$ has one of the following forms:

1. $(\mathrm{Aut}(F), F)\ \mathrm{Wr}\ (\mathrm{Aut}(N), N)$;
2. $(\mathrm{Aut}(N), N) \times (\mathrm{Aut}(F), F)$.

Proof. As in the proof of Theorem 4.1 we may suppose that new binary relations are defined on $N \times F$ only. Since $\mathrm{Aut}(F)$ is 2-transitive, every 0-definable binary relation on a pair of fibres $E = \{a, b\} \times F$ is one of

$$\begin{gathered} 0, \\ E^2, \\ \mathrm{id} = \{((a, f), (b, f)) : f \in F\}, \\ \mathrm{id}^{-1}, \\ \mathrm{ID} = \{((a, f), (b, f')) : f \neq f'\}, \\ \mathrm{ID}^{-1} \end{gathered}$$

or a boolean combination of these. If only the first two cases are possible then we have the first conclusion of our proposition. Since Aut(N) is 2-transitive we have the second conclusion in the other case. □

5 The cover problem and two-graphs

In this section we investigate ternary expansions of a finite principal cover of a 2-transitive structure. It was noted above that such covers may not be splitting. We hope that the case of an expansion by a ternary relation defining a two-graph is more understandable. It will be shown that sometimes the results of Sections 2–4 can be applied for such an expansion.

We recall that a set τ of 3-subsets of a set X is called a *two-graph* on X if for any 4-subset of X, the number of its 3-subsets belonging to τ is even.

Each two-graph can be defined in the following way. Let (X, R) be a graph. Put

$$R^* = \left\{ (x, y, z) : \begin{array}{l} x, y, z \in X, \text{ and the set } \{x, y, z\} \text{ has} \\ \text{an odd number of edges of the graph } R \end{array} \right\}.$$

Then R^* is a two-graph, and for every two-graph T there is a graph R such that $T = R^*$. As in Cameron (1977) we say that symmetric binary relations R_1 and R_2 are in the same *switching class* if $R_1^* = R_2^*$. (The reason for the term is that R_2 may be obtained from R_1 by choosing a set $A \subseteq X$, and, for those 2-sets which meet A in a singleton, requiring them to satisfy R_2 if and only if they do not satisfy R_1.)

From now on let M_0 be a finite principal cover of a 2-transitive structure N and $M_0 = N \cup (N \times F)$. Let F_N be the union of all fibres. We assume that there is no structure on a fibre. We will investigate covering expansions of M_0 by two-graph relations. This has the following motivation also.

Let R be a binary symmetric irreflexive relation on F_N. Does the structure (M_0, R) determine a binary cover in some indirect way? It is natural to formalize this question in two parts as follows.

> Is there a binary relation R' with the switching class of R such that (M_0, R') is a cover of N? Is it possible to find R' such that $\mathrm{Aut}(M_0, R') = \mathrm{Aut}(M_0, R^*)$?

In the case of the existence of R', as in the first part of this question, the structure (M_0, R^*) is a cover because R^* is definable from R'. If N is a strictly minimal set then this two-graph cover splits (apply our result from Section 3). Thus the existence of a finite non-splitting two-graph cover of a strictly minimal set implies the existence of an 'essentially non-covering' graph on F_N which defines a two-graph cover.

The second part of our formalized question is connected with some ideas of Cameron. Following Cameron (1977), we identify graphs on F_N with the corresponding formal linear combinations over $\mathbf{F}_2$ of the 2-subsets of F_N. The set of graphs on F_N becomes a vector space V, and the operation of symmetric difference defines the addition of the vector space. Here the operation of switching corresponds to adding a complete bipartite graph. Let C be the subspace consisting of all complete bipartite graphs, together with the null graph. A switching class is a coset of C in V. Using results of Cameron (1977) and König's Lemma one can check that this definition of switching class and the above definition are the same.

Let $R + C$ be a switching class. There is a natural action of $G = \mathrm{Aut}(M_0, R^*)$ on the set $R + C$. This action naturally defines a derivation $d: G \to C$ as follows: $d(g) = R^g - R$. Let

$$B^1(G, C) = \{d_P: g \mapsto (P^g - P) \ : \ P \in C\}$$

be the group of inner derivations. Then the element $\gamma = d + B^1(G, C) \in H^1(G, C)$ will be called the first invariant of G and R (Cameron, 1977). Repeating the proof of Theorem 3.1 of Cameron (1977) we have the following statement.

PROPOSITION 5.1. The vanishing of the first invariant of G and R is equivalent to the existence of a graph R_0 such that R and R_0 are in the same switching class and $\mathrm{Aut}(M_0, R_0) = G$.

We do not know an example of a two-graph expansion of M_0 induced by a covering graph expansion such that the corresponding first invariant is not 0. The results below show that in the case of a strictly minimal N we should study finite graphs with some additional binary relations.

We will characterize R^*'s from the first part of our question assuming that N is strictly minimal. Let $M = (M_0, R')$ be a cover of N and R' a binary symmetric irreflexive relation. By Corollary 3.3 we can suppose that there are actions of some groups L and H on F such that the kernel of M is described by L and K as in Section 2 (Ziegler's description). Using this description and the 'trick of transitivity' from the proof of Theorem 3.2, the following proposition can be proved.

PROPOSITION 5.2. Suppose that M satisfies the assumptions above. Then there is an L-invariant binary symmetric relation δ on the set of 1-orbits of H on F such that for every $x, y \in N$ ($x \neq y$) the pair $((x, f), (y, f'))$ is in R' iff the pair of the 1-orbits with representatives f and f' is in δ. Moreover, H is 1-closed in $\mathrm{Aut}(F, R')$.

The corresponding two-graph expansions can be described in the following way:

> Let Q be a graph on F and let $K \leqslant \mathrm{Aut}(F, Q)$ be 1-closed in $\mathrm{Aut}(F, Q)$. For a binary symmetric relation δ on the set of all

1-orbits of K, put

$$\Delta = \left\{ ((x, f_1), (y, f_2)) : \begin{array}{l} x \neq y \text{ and the pair of the} \\ \text{1-orbits of } K \text{ with} \\ \text{representatives } f_1, f_2 \text{ is in } \delta \end{array} \right\}.$$

On F_N, define $T_{Q,\delta} = (Q \cup \Delta)^*$. Then the two-graph expansions of M_0 determined by covering graph expansions of M_0 are the structures of the form $(M_0, T_{Q,\delta})$.

In fact, we do not know other examples of covering two-graph expansions of M_0. We suspect that these new examples must be non-splitting. The following proposition allows us to confirm this under some additional restrictions.

PROPOSITION 5.3. Let N be an arbitrary structure. Assume that a two-graph covering expansion $M = (M_0, T)$ admits elimination of quantifiers in the language $\{P^N, \pi, T\}$, where P^N is a predicate for N. Then every 3-orbit of distinct elements of N defines a two-graph.

Proof. Consider $p(\vec{x}) = \mathrm{tp}(a, b, c)$ for some distinct elements of N. Let a', b', c' be some elements of the corresponding fibres. Since, by quantifier elimination, every permutation of $\{a', b', c'\}$ extends to an automorphism of M, $p(\vec{x})$ is symmetric.

Let $\{a, b, c, d\}$ be a 4-element subset of N and d' corresponding to d as above. If (a, b, d) does not satisfy $p(\vec{x})$ then

$$\{a', b', d'\} \in T \quad \text{iff} \quad \{a', b', c'\} \text{ is not in } T.$$

Thus one of the triples (a', c', d'), (b', c', d') satisfies $\mathrm{tp}(a', b', c')$ and the other satisfies $\mathrm{tp}(a', b', d')$, hence the number of p-subsets in $\{a, b, c, d\}$ is even. □

Thus if N in this proposition does not have a non-trivial 3-orbit defining a two-graph then N is 3-transitive, and if also N is not disintegrated, then M does not split by Proposition 1.5.

QUESTION 5.4. Does a finite two-graph cover of a strictly minimal set split?

QUESTION 5.5. Does a finite splitting two-graph cover of a strictly minimal set contain a graph cover in the corresponding switching class?

A generalization of Jordan groups

S. A. Adeleke

Mathematics Department,
Western Illinois University,
Macomb,
IL61455,
U.S.A.

1 Introduction

For Γ to be a Jordan set in a group (G, Ω) in the usual definition (see p. 73 in the article by Macpherson in this volume) the *pointwise* stabilizer, $G_{(\Omega \setminus \Gamma)}$, must be transitive on Γ. In this paper we relax this condition and demand that the *setwise* stabilizer, $G_{\{\Gamma\}}$, be transitive on Γ. But we also add as assumptions two properties which hold for Jordan sets in the usual definition; namely that the union of any chain of the (new) Jordan sets be a (new) Jordan set, and that the union of any two non-disjoint (new) Jordan sets be a (new) Jordan set. Although infinite simply primitive Jordan groups in the new sense lead to the same G-invariant structures as the old ones of Adeleke and Neumann (in press$b$), there are simply primitive groups which are Jordan in the new sense but not in the usual sense. Immediate examples are the *pathological* groups on linear orders described by Glass (1981, Chapter 6, Section 1.10). We make remarks in Section 4 below about the construction of such examples, and it follows easily that there are more examples of the same type built from semilinear orders and C-relations. Apart from including more groups, the result below should be useful in any eventual classification of *general* infinite simply primitive groups according to their invariant structures. The new Jordan sets and Jordan groups shall be called *c-Jordan sets* and *c-Jordan groups* respectively ('c-Jordan' as in 'Camille Jordan'). The proofs of Adeleke and Neumann (in press$b$, to appear) need several adjustments to handle the present context.

2 Definitions and notation

DEFINITION 2.1. (Usual definition of Jordan sets) In a transitive permutation group (G, Ω), a subset Γ of Ω is a *Jordan set* if $|\Gamma| > 1$ and $G_{(\Omega \setminus \Gamma)}$ is transitive on Γ. It is a *proper* Jordan set if either $|\Omega \setminus \Gamma|$ is infinite, or

$|\Omega \setminus \Gamma|$ is finite and (G, Ω) is not $(|\Omega \setminus \Gamma|+1)$-transitive. The group (G, Ω) is a *Jordan group* if it has a proper Jordan set.

DEFINITION 2.2. Let (G, Ω) be an infinite transitive permutation group. Let $\mathscr{S} \subseteq \mathscr{P}(\Omega)$. Call $\mathscr{S}$ a *c-Jordan system* for (G, Ω), and call elements of $\mathscr{S}$ *c-Jordan sets* for (G, Ω) if the following hold:

(a) $|\Gamma| > 1$ for every $\Gamma \in \mathscr{S}$;
(b) $\mathscr{S}g = \mathscr{S}$ for all $g \in G$;
(c) $G^{\Gamma}_{\{\Gamma\}}$ is transitive on Γ;
(d) if $\Gamma_1, \Gamma_2 \in \mathscr{S}$ with $\Gamma_1 \cap \Gamma_2 \neq \varnothing$ then $\Gamma_1 \cup \Gamma_2 \in \mathscr{S}$;
(e) if $(\Gamma_i : i \in I)$, ordered by inclusion, is a chain of elements of $\mathscr{S}$, then $\bigcup(\Gamma_i : i \in I) \in \mathscr{S}$.

DEFINITION 2.3. Let $\mathscr{S}$ be a c-Jordan system for (G, Ω). If $\Gamma \in \mathscr{S}$, call Γ an *improper* c-Jordan set if $|\Omega \setminus \Gamma|$ is finite and G is $(|\Omega \setminus \Gamma| + 1)$-transitive on Ω; otherwise, call Γ a *proper* c-Jordan set. A *c-Jordan group* is a permutation group having a c-Jordan system with a proper c-Jordan set.

In Section 4, we remark on why the pathological groups in the work of Glass (1981) are c-Jordan but not Jordan.

We recall from p. 78 the definitions of an *upper semilinear order* and a *C-relation.* If $\Gamma, \Delta \subseteq \Omega$ we say that (Γ, Δ) is *typical*, or Γ, Δ form a *typical pair*, if the sets $\Gamma \setminus \Delta$, $\Delta \setminus \Gamma$, $\Gamma \cap \Delta$ are all non-empty. We shall write $A \subset B$ to mean that $A \subseteq B$ and $A \neq B$.

3 Results and proofs

THEOREM 3.1. Let (G, Ω) be an infinite simply primitive permutation group which is c-Jordan. Then G preserves on Ω a dense linear order, a dense upper semilinear order, or a C-relation.

The proof of Theorem 3.1 follows from a sequence of lemmas. The first two lemmas have already occurred in the work of Adeleke and Neumann.

LEMMA 3.2. Let (G, Ω) be an infinite simply primitive group (not necessarily c-Jordan). Suppose that there is a proper subset Σ of Ω satisfying

$$\exists \alpha, \beta \in \Omega \, (\alpha \neq \beta \wedge \forall g \in G \, (\beta \in \Sigma g \longrightarrow \alpha \in \Sigma g)). \tag{$*$}$$

Then G preserves a partial order on Ω.

LEMMA 3.3. Let (G, Ω) be an infinite simply primitive permutation group. Suppose that there is $\Sigma \subset \Omega$ such that $|\Sigma| > 1$, condition $(*)$ does not hold for Σ, and

$$\forall g \in G \, (\Sigma, \Sigma g \text{ are not typical}). \tag{$\dagger$}$$

Then there is a C-relation on Ω invariant under G.

Note. In Lemmas 3.4 to 3.7, (G, Ω) is assumed to be an infinite simply primitive group which is a c-Jordan group with a c-Jordan system $\mathscr{S}$.

LEMMA 3.4. Let Γ be a c-Jordan set for (G, Ω) with $\alpha \in \Gamma$ and $\beta \in \Omega \setminus \Gamma$. Then there is a unique maximal c-Jordan set for (G, Ω) containing α and excluding β.

Notation. Denote by $M\mathscr{S}(\beta|\alpha)$ the maximal c-Jordan set in $\mathscr{S}$, if it exists, containing α and excluding β.

LEMMA 3.5. Let α and β be distinct elements of Ω and suppose that $M\mathscr{S}(\alpha|\beta)$ and $M\mathscr{S}(\beta|\alpha)$ exist and are typical. Then for any $g \in G$, the pair $M\mathscr{S}(\alpha|\beta) \cup M\mathscr{S}(\beta|\alpha), (M\mathscr{S}(\alpha|\beta) \cup M\mathscr{S}(\beta|\alpha))g$ is not typical.

LEMMA 3.6. If there exist distinct elements α and β in Ω such that every c-Jordan set containing β must contain α, then G preserves on Ω a dense upper semilinear order or a dense linear order.

LEMMA 3.7. If (G, Ω) satisfies condition $(*)$ on p. 234 above and the set Σ in $(*)$ is a proper c-Jordan set in Ω, then G preserves on Ω a dense upper semilinear order or a dense linear order.

Proof of Lemma 3.2. We follow the proof of Theorem 5.2 in Adeleke and Neumann (in press*b*). For $\mu, \nu \in \Omega$, define $\mu \leqslant \nu$ to hold if and only if

$$\forall g {\in} G\, (\nu \in \Sigma g \longrightarrow \mu \in \Sigma g).$$

Also define $\mu \equiv \nu$ if and only if $\mu \leqslant \nu$ and $\nu \leqslant \mu$. Then $\equiv$ is a G-invariant equivalence relation which is not universal. Hence it is equality, so $\leqslant$ is a G-invariant partial order on Ω. □

Proof of Lemma 3.3. The proof of this result is in Adeleke and Neumann (to appear). Briefly, it goes as follows. Define

$$C(\alpha; \beta, \gamma) \quad \text{if and only if} \quad \exists g {\in} G\, (\beta, \gamma \in \Sigma g \wedge \alpha \notin \Sigma g).$$

We check that the axioms C1–C5 listed on p. 80 hold. Axioms C1, C3, and the invariance of C are immediate consequences of the definition. Axiom C2 holds because of (†) on p. 234, and C4 holds because Lemma 3.2 fails for Σ. If C5 fails to hold for some α, β, then the union of all Σg containing α (which form a chain by (†) on p. 234) will be a block of (G, Ω) excluding β, contrary to the primitivity of (G, Ω). □

Proof of Lemma 3.4. Let $\mathscr{C} := \{\Sigma_i : i \in I\}$ be a maximal chain of c-Jordan sets containing α and excluding β. Because of the existence of Γ, we know that $\mathscr{C} \neq \varnothing$. By condition (e) of Definition 2.2, $\bigcup \mathscr{C}$ is a c-Jordan set containing α and excluding β. If there is a c-Jordan set $\Lambda \supset \bigcup \mathscr{C}$ with $\beta \notin \Lambda$, then $\mathscr{C} \cup \{\Lambda\}$ will be a chain of c-Jordan sets containing α and excluding β, contrary to the maximality of $\mathscr{C}$. Hence no such Λ exists. Therefore, $\bigcup\mathscr{C}$ is a maximal c-Jordan set containing α and excluding β. From this and Definition 2.2(d), $\bigcup\mathscr{C}$ is unique as claimed in the lemma. □

Proof of Lemma 3.5. Let $\Gamma_1 := M\mathscr{S}(\alpha|\beta)$ and $\Gamma_2 := M\mathscr{S}(\beta|\alpha)$. By Definition 2.2(d) applied to $\mathscr{S}$, $\Gamma_1 \subset \Gamma_1 \cup \Gamma_2 \in \mathscr{S}$. Therefore $\alpha \in \Gamma_1 \cup \Gamma_2$. Similarly $\beta \in \Gamma_1 \cup \Gamma_2$.

Next, we have that for every $\gamma \in \Gamma_1$, $M\mathscr{S}(\alpha|\gamma) = \Gamma_1 \subset \Gamma_1 \cup \Gamma_2$. Also, for every $\delta \in \Gamma_2 \setminus (\Gamma_1 \cup \{\alpha\})$, if $M\mathscr{S}(\alpha|\delta)$ exists, then $M\mathscr{S}(\alpha|\delta) \cap \Gamma_2 \neq \varnothing$ but $M\mathscr{S}(\alpha|\delta) \cap \Gamma_1 = \varnothing$ by Definition 2.2(d) and the definition of Γ_1. We deduce from 2.2(d) that in this case $M\mathscr{S}(\alpha|\delta) \cup \Gamma_2 \in \mathscr{S}$ with $\beta \notin M\mathscr{S}(\alpha|\delta) \cup \Gamma_2$. So, $M\mathscr{S}(\alpha|\delta) \cup \Gamma_2 \subseteq M\mathscr{S}(\beta|\delta)$ which equals Γ_2 by the choice of δ. Thus $M\mathscr{S}(\alpha|\delta) \subseteq \Gamma_2$ whenever $\delta \in \Gamma_2 \setminus (\Gamma_1 \cup \{\alpha\})$ and $M\mathscr{S}(\alpha|\delta)$ exists. Hence,

$$\forall \xi \in (\Gamma_1 \cup \Gamma_2) \setminus \{\alpha\}\, (M\mathscr{S}(\alpha|\xi) \text{ exists}) \rightarrow M\mathscr{S}(\alpha|\xi) \subseteq \Gamma_1 \cup \Gamma_2). \qquad (\S)$$

By Definition 2.2(d) we can translate α to any element of $\Gamma_1 \cup \Gamma_2$ using elements in $G_{\{\Gamma_1 \cup \Gamma_2\}}$. Applying such a translation to (§) we obtain

$$\forall \mu, \rho \in \Gamma_1 \cup \Gamma_2 \text{ with } \mu \neq \rho\, ((M\mathscr{S}(\mu|\rho) \text{ exists}) \rightarrow M\mathscr{S}(\mu|\rho) \subseteq \Gamma_1 \cup \Gamma_2).$$

Consequently,

$$\forall \mu \in \Gamma_1 \cup \Gamma_2\, \forall \nu \in \Omega \setminus (\Gamma_1 \cup \Gamma_2) \begin{pmatrix} M\mathscr{S}(\mu|\nu) \text{ exists} \rightarrow \\ \quad M\mathscr{S}(\mu|\nu) \cap (\Gamma_1 \cup \Gamma_2) = \varnothing \end{pmatrix}.$$

Thus, if there are $\mu \in (\Gamma_1 \cup \Gamma_2) \setminus (\Gamma_1 \cup \Gamma_2)g$ and $\nu \in (\Gamma_1 \cup \Gamma_2)g \setminus (\Gamma_1 \cup \Gamma_2)$, then $M\mathscr{S}(\nu|\mu)$ contains $\Gamma_1 \cup \Gamma_2$ and is disjoint from $(\Gamma_1 \cup \Gamma_2)g$; also, $M\mathscr{S}(\mu|\nu)$ contains $(\Gamma_1 \cup \Gamma_2)g$ and is disjoint from $\Gamma_1 \cup \Gamma_2$. Thus $(\Gamma_1 \cup \Gamma_2) \cap (\Gamma_1 \cup \Gamma_2)g = \varnothing$. □

Proof of Lemma 3.6. This is a generalization of the proof of Theorem 5.2 in Adeleke and Neumann (in press*b*).

Define $\alpha \leqslant \beta$ to hold if and only if every c-Jordan set containing β also contains α. Then, as in the proof of Lemma 3.2, $\leqslant$ is a G-invariant partial order on Ω.

Let $\lambda \geqslant \alpha$ and $\mu \geqslant \alpha$. Suppose $\mu \ngeqslant \lambda$. Then there exists a c-Jordan set Γ containing μ and excluding λ. Also $\alpha \in \Gamma$, since $\alpha \leqslant \mu$. By Lemma 3.4, there exists a maximal c-Jordan set Λ containing μ and excluding λ. Define $\Delta(\Lambda)$ to be the set of all λ' such that Λ is a maximal c-Jordan set containing μ and excluding λ'. Note that $\Delta(\Lambda)$ can be equivalently defined as the set of all λ' not in Λ but present in every c-Jordan set larger than Λ, so in particular the definition is independent of the choice of $\mu \in \Lambda$. So, $\lambda \in \Delta(\Lambda)$. Pick $x \in G_{\{\Lambda\}}$ satisfying $\alpha x = \mu$. Then $\lambda x \geqslant \alpha x = \mu$, and $\lambda x \in \Delta(\Lambda x) = \Delta(\Lambda)$. Now consider λx and λ. Let Θ be any c-Jordan set containing λ. Then it follows progressively that $\alpha \in \Theta$ (since $\lambda \geqslant \alpha$), $\alpha \in \Theta \cap \Lambda$, $\Theta \cup \Lambda$ is a c-Jordan set properly containing Λ, $\lambda x \in \Theta \cup \Lambda$, and finally that $\lambda x \in \Theta$. Hence $\lambda \geqslant \lambda x$. Since $\lambda x \geqslant \mu$, we have $\lambda \geqslant \mu$. Thus, if $\lambda \geqslant \alpha$ and $\mu \geqslant \alpha$ then $\lambda \geqslant \mu$ or $\mu \geqslant \lambda$. We conclude that the set of upper bounds of any α is linearly ordered.

To complete the proof that $\leqslant$ is an upper-semilinear order, we merely show that every pair of elements has an upper bound. Define $\delta \sim \xi$ to mean that there is ν satisfying $\nu \geqslant \delta$ and $\nu \geqslant \xi$. Then, by the conclusion of the last paragraph, $\sim$ is a G-congruence. By the hypothesis of this lemma, the congruence is not equality. So, by the primitivity of G, the congruence is universal, and thus every pair of elements has an upper bound. This shows that $\leqslant$ is an upper semilinear order.

Next, we prove density. Let $\beta \geqslant \alpha$. By the transitivity of G, there is $\gamma \in \Omega$ with $\alpha > \gamma$. Since $\alpha \not\geqslant \beta$, there is a c-Jordan set Γ containing α and excluding β. Now $\gamma \in \Gamma$ since $\alpha > \gamma$. Pick $x \in G_{\{\Gamma\}}$ with $\gamma x = \alpha$. Then $\alpha x > \gamma x = \alpha$ and $\alpha x \in \Gamma$. So, αx and β are upper bounds for α. Since $\alpha x \in \Gamma$ and $\beta \notin \Gamma$, we have $\alpha x \not\geqslant \beta$. Since the upper bounds of α are linearly ordered, we deduce that $\beta > \alpha x > \alpha$, as required.

Finally, we remark that if, for every distinct $\alpha, \beta \in \Omega$ either $\alpha < \beta$ or $\beta < \alpha$, then the upper semilinear order above reduces to a dense linear order. □

Proof of Lemma 3.7. Use the definition of $\leqslant$ in the proof of Lemma 3.2, and, in the proof of Lemma 3.6 (second, third and fourth paragraphs), replace Γ by Σg_1 for some $g_1 \in G$ and Θ by Σg_2 for some $g_2 \in G$. □

Proof of Theorem 3.1. Let Γ be a c-Jordan set for (G, Ω). Suppose $\alpha \in \Gamma$ and $\beta \notin \Gamma$. Let $\Sigma := M\mathscr{S}(\beta|\alpha)$. This exists by Lemma 3.4.

Case 1. If Σ satisfies $(*)$ on p. 234, then by Lemma 3.7, G preserves on Ω a dense upper semilinear order or a dense linear order.

Case 2. If Σ violates $(*)$ but satisfies $(\dagger)$, then by Lemma 3.3, there is a G-invariant C-relation on Ω.

Case 3. Suppose that Σ violates $(*)$ and $(\dagger)$, and that there is $g \in G$ such that $(\Sigma, \Sigma g)$ is typical and $\Sigma \cup \Sigma g \neq \Omega$. Put $\Lambda := \Sigma \cup \Sigma g$. Let $\Delta(\Sigma)$ again denote the set of all β' such that $\Sigma = M\mathscr{S}(\beta'|\alpha)$. Then by the definition of Σ and Definition 2.2(d), $\Delta(\Sigma) \subseteq \Sigma g \setminus \Sigma$ and $\Delta(\Sigma g) \subseteq \Sigma \setminus \Sigma g$. Now, by Lemma 3.5, Λ satisfies $(\dagger)$. If in addition Λ violates $(*)$, then, with Λ in place of Σ in Lemma 3.3, a C-relation is invariant. But, if, in addition to satisfying $(\dagger)$, Λ satisfies $(*)$, then, again with Λ in place of Σ in Lemma 3.7, G preserves on Ω a dense upper semilinear order or a dense linear order.

Case 4. Suppose that Σ violates $(*)$ and $(\dagger)$, and that $\Sigma \cup \Sigma g = \Omega$ whenever Σ and Σg are typical.

If $|\Delta(\Sigma)| > 1$ and $\gamma \in \Delta(\Sigma) \setminus \{\beta\}$, then from the first hypothesis of this case,

$$\exists h \in G\, (\beta \in \Sigma h \wedge \gamma \notin \Sigma h)$$

and from the second hypothesis,

$$\exists x\in G\,(\Sigma h, \Sigma x \text{ are typical}).$$

Since $\gamma \in \Delta(\Sigma) \setminus \Sigma h$, then

$$\Sigma \cap \Sigma h = \varnothing. \tag{‡}$$

From the definition of Σ, $\Sigma x = M\mathscr{S}(\xi_1|\xi_2)$ for some $\xi_1 \in \Sigma h \setminus \Sigma x$ and any $\xi_2 \in \Sigma x \setminus \Sigma h$. Pick $y \in G_{\{\Sigma h\}}$ satisfying $\xi_1 y = \beta$. Then Σh, Σxy are typical and $\Sigma xy = M\mathscr{S}(\beta|\xi_3)$ for some ξ_3. Since $\Sigma\cap\Sigma h = \varnothing$ and $\Sigma xy\cap\Sigma h \neq \varnothing$, we have $\Sigma \neq \Sigma xy$. Since furthermore $\Sigma = M\mathscr{S}(\beta|\alpha)$ and $\Sigma xy = M\mathscr{S}(\beta|\xi_3)$, it follows that $\Sigma \cap \Sigma xy = \varnothing$. So, by (‡), $\Sigma \cap (\Sigma h \cup \Sigma xy) = \varnothing$. Thus $\Sigma h\cup\Sigma xy \neq \Omega$ and $\Sigma h, \Sigma xy$ are typical. This clearly violates the hypothesis of this case. Hence, our former assumption that $|\Delta(\Sigma)| > 1$ does not hold. We must have $|\Delta(\Sigma)| = 1$.

We now show that this makes (G,Ω) 2-transitive, contrary to the assumption that G is simply primitive on Ω. Again, assume that Σ and Σg are typical with $\Sigma\cup\Sigma g = \Omega$. Since $|\Delta(\Sigma)| = 1$, we deduce that $\Delta(\Sigma) = \{\beta\}$ and so $\beta x = \beta$ for all $x \in G_{\{\Sigma\}}$. Let $\omega \in \Sigma$. Then the orbit ωG_β contains Σ. Let $\lambda \in \Sigma g \setminus (\Sigma \cup \{\beta\})$. Then $\lambda g^{-1} \in \Sigma$. Since $\lambda \neq \beta$, we have $\lambda \notin \Delta(\Sigma)$, and so $M\mathscr{S}(\lambda|\omega) \supset \Sigma$. Put $\Theta := M\mathscr{S}(\lambda|\omega)$. Pick $z \in G_{\{\Sigma g\}}$ with $\beta z = \lambda$. Then, as $\Delta(\Sigma g) = \{\beta g\}$, we have $\beta gz = \beta g$ and so $\Sigma z = M\mathscr{S}(\lambda|\beta g)$ which contains Σ and so equals Θ. Hence $\Theta = \Sigma z$ and so $|\Delta(\Theta)| = 1$ with $\Delta(\Theta) = \{\lambda\}$. Thus, $G_{\{\Theta\}}$ fixes λ. In particular there is $g_1 \in G_{\{\Theta\}}$ with $(\beta g)g_1 = \beta$ and $\lambda g_1 = \lambda$. In summary, $(\lambda g^{-1})(gg_1) = \lambda g_1 = \lambda$, and $\beta(gg_1) = \beta$. Consequently, $\lambda \in \lambda g^{-1} G_\beta$ which equals ωG_β since both contain Σ. Since λ is arbitrary in $\Sigma g \setminus (\Sigma \cup \{\beta\})$, we conclude that $\omega G_\beta = (\Sigma \cup \Sigma g) \setminus \{\beta\} = \Omega \setminus \{\beta\}$. It follows that G is 2-transitive on Ω, which is a contradiction. □

4 Some c-Jordan groups which are not Jordan

DEFINITION 4.1. (Glass (1981), Section 1.10 & Chapter 6) A permutation group (G,Ω) is *pathological* if the following hold:

(a) G preserves a dense linear order on Ω;

(b) G is order 2-transitive on Ω; that is

$$\forall\alpha{<}\beta{\in}\Omega\,\forall\gamma{<}\delta{\in}\Omega\,\exists g{\in}G\,(\alpha g = \gamma \wedge \beta g = \delta);$$

(c) for each $g \in G \setminus \{1\}$, the support of g is not bounded below or above;

(d) for every $g_1, g_2 \in G$ we have $g_1 \vee g_2, g_1 \wedge g_2 \in G$, where for all $\alpha \in \Omega$,

$$\alpha(g_1 \vee g_2) := \max\{\alpha g_1, \alpha g_2\},$$

and

$$\alpha(g_1 \wedge g_2) := \max\{\alpha g_1, \alpha g_2\}.$$

It is clear that pathological groups are infinite simply primitive c-Jordan groups with c-Jordan sets of the form $(-\infty, q)$. But they are not Jordan by condition 4.1(c).

INDUCTIVE CONSTRUCTION 4.2. We now describe a direct construction of groups which are c-Jordan but not Jordan and which have invariant linear orders. The groups will satisfy 4.1(a)–(c). The method of the construction is easily extendible to yield groups with invariant semilinear orders or C-relations that are c-Jordan but not Jordan. In brackets below are some remarks which should clarify the idea behind the construction.

4.2.1. Let $x_1, x_2, x_3, \ldots$ be a list of symbols, and let F be the free group $\langle x_1, x_2, x_3, \ldots \rangle$. (Each x_i will be constructed to be a permutation preserving the linear order on $\mathbb{Q}$.)

4.2.2. Suppose that $w_1, w_2, w_3, \ldots$ is a list of all words in $F \setminus \{1\}$, with each $u \in F \setminus \{1\}$ occurring *infinitely often* in the list. (The infinite occurrence of each word in the list allows us to adjust the construction so that each w_i satisfies 4.1(c).)

4.2.3. Let $q_1, q_2, q_3, \ldots$ be an enumeration of elements of $\mathbb{Q}$ and let (a_0, b_0), (a_1, b_1), $(a_2, b_2), \ldots$ be an enumeration of all pairs of elements of $\mathbb{Q}$ with $a_i < b_i$ for each i. (Pairs appear here because we want subsequent maps to be order 2-transitive. The map x_i will be defined so as to take the fixed pair (a_0, b_0) to the pair (a_i, b_i).)

4.2.4. As an initial step, define a map x_i partially as $(a_0, b_0)x_i = (a_i, b_i)$ for $i = 1, 2, 3, \ldots$.

4.2.5. *Step* s_i. If q is the first element of $\mathbb{Q}$ in 4.2.3 satisfying $q \notin (\mathrm{Dom}(x_i) \cap \mathrm{Ran}(x_i))$, then extend the action of x_i in an order-preserving way so that q is in the domain and the range of x_i. (This step ensures that in the end each x_i is a bijection of $\mathbb{Q}$, and is order-preserving.)

Step t_i. Ensure that the word w_i has an element in its support greater than the integer i by extending the action of all the x_j in w_i (one may work with elements larger than all those that have appeared in the construction before this step of the domain and range of all x_j in w_i).

Steps s_i and t_i are all achievable since $(\mathbb{Q}, \leqslant)$ is dense with no upper or lower bound. It is clear that after performing all the steps in (4.2.5), the resulting group satisfies 4.1(a)–(c). It is also clear that the method of construction is easily extendible to produce groups with invariant semilinear orders or C-relations that are c-Jordan but not Jordan.

PART II

Recursively saturated models

Recursive saturation

1 Introduction

The group $\mathrm{Aut}\,\mathfrak{M}$ of automorphisms of a countable structure $\mathfrak{M}$ (whether considered as a permutation group acting on the domain M of $\mathfrak{M}$, as a topological group, or as an abstract group) only describes the first-order structure of $\mathfrak{M}$ accurately if $\mathfrak{M}$ is a suffiently 'rich' or sufficiently 'saturated' structure. Considering just one end of the spectrum for the moment, by Scott's Theorem (p. 20) a countable $\mathscr{L}$-structure $\mathfrak{M}$ is rigid if and only if every element of $\mathfrak{M}$ is $\mathscr{L}_{\omega_1\omega}$-definable, and there are many rather dissimilar such structures. So one obviously necessary condition for $\mathrm{Aut}\,\mathfrak{M}$ to reflect only first-order information is that $\mathfrak{M}$ be homogeneous (or rather, in the terminology given on p. 20, strongly $\aleph_0$-homogeneous), that is that the orbit of any tuple $\bar{a} \in M$ be determined by $\mathrm{tp}_{\mathfrak{M}}(\bar{a})$. But we might also be interested in the first-order properties of tuples of infinite length, subsets of M, equivalence relations on M, etc., and this would lead us to formulate stronger saturation assumptions concerning $\mathfrak{M}$.

The notion of $\aleph_0$-categoricity works very well in this context, as witnessed by Proposition 5.5 on p. 22 and Corollary 7.7 on p. 31. The notion of a countable saturated structure is a natural generalization, but it turns out that not all first-order theories have countable saturated models. In fact there is an even more general notion, that of *recursive saturation*, which ensures sufficient 'richness' of a structure and for which many of the results from the theory of $\aleph_0$-categorical structures turn out to be true. In fact, there are many methods from the theory of recursive saturation that have no counterpart for countable saturated models, and it seems that in many cases it may be easier to apply these methods to give a proof of a result for recursively saturated models than it would be to find, 'bare-handed', a proof of the same result for countable saturated models. Thus this is a promising area for interaction between model theory and permutation group theory, and the papers presented in this section give a good idea of the flavour of the subject.

To prepare the reader for the definitions and arguments to follow, we should say that, despite its name and its definition, the key ideas in recursive saturation have very little to do with 'recursion theory', the theory of computability. The reason computability enters the discussion at all is that many of the arguments (in particular, the proof of the theorem on resplendency) rest on the fact that first-order logic has a formal method of proof which is partially computable. This should be understood in the sense that, given a computable set of axioms, there is an algorithm such that on input σ, a first-order statement in the language, if σ is a consequence of the axioms then the algorithm will return a proof of σ, and if not, then the algorithm will never halt. Essentially, the only recursion theory the reader need know is an intuitive notion of what it means for a function or set to be 'computable' (we shall use the word 'recursive') in the ideal sense in which the machine carrying out the computation may be as large as is needed and may take as much time as needed, provided an answer is eventually given. (Later on, when we talk about Scott sets, we will need a notion of 'relative recursion', that is, computability given some oracle for a possibly non-computable set. All the notions we need here can be found in any standard text on computability, recursion theory, or mathematical logic. Cutland (1980) for instance contains much more than is needed.)

It turns out that recursive saturation has much closer connections with nonstandard models of number theory, or of set theory. These are models which contain finite integers which, seen from 'outside' the model, are infinite. (Strictly speaking, one should talk about 'ω-nonstandard models of set theory', since it is the ω in the model that is nonstandard.) Often arguments go by 'attaching' to a countable recursively saturated model $\mathfrak{M}$ such a model of set theory or arithmetic which describes the notion of truth in $\mathfrak{M}$, and then by formalizing some standard argument inside the model of set theory or arithmetic. Thus it is no accident that two of the papers below concern recursively saturated models of Peano arithmetic, since these provide one of the key ways of describing recursively saturated models. (In fact, it turns out that the automorphism group of a countable recursively saturated model of PA is a particularly interesting group in its own right.)

The other key tool for describing countable recursively saturated models is *resplendency*. Roughly, resplendency says that provided it is consistent to expand a structure to another one, satisfying a given theory in a richer language, then there is indeed such an expansion. In fact the expansion may also be resplendent, so an inductive construction may be given. This is the main tool used in the paper 'Indiscernibles' by Kaye below, where he shows how to use it to obtain indiscernible sequences or sets I in a countable recursively saturated model $\mathfrak{M}$, such that $\operatorname{Aut}\mathfrak{M}$ induces large permutation groups on I.

2 Realizing types

The rest of this chapter will be devoted to presenting the basic definitions and results concerning recursive saturation. All proofs of the results cited here can be found in Kaye (1991). We will try to attribute results when mentioned, but often (especially for this section and the next one) this is rather difficult. The basic definitions and the key result (resplendency) were given by Barwise and Schlipf, and (independently) Ressayre. Independently, and at about the same time, Wilmers (1975) was working on ω-nonstandard models of set theory, and he devised the related notion of $\mathscr{X}$-saturation below, where $\mathscr{X}$ is the set of 'reals' coded in such a model. His work contains many variations on the phenomenon that Barwise and Schlipf were calling resplendency, but the connections between his work, that of Barwise, Schlipf, Ressayre, and with the much earlier work of Scott on what are here termed 'Scott sets' were not worked out until a little later. H. M. Friedman's work (Friedman, 1973) also helped clarify these connections and gave new insights into nonstandard models.

We shall work with a *recursive* first-order language $\mathscr{L}$ throughout this chapter. This is a language $\mathscr{L}$ with a suitable recursive Gödel-numbering $\ulcorner\urcorner: \sigma \mapsto \ulcorner\sigma\urcorner \in \mathbb{N}$ mapping formulas and terms σ injectively into $\mathbb{N}$. 'Suitable' here means that the image of the sets of constants, variables, terms, formulas under $\ulcorner\urcorner$ should all be recursive, and that all the operations on $\mathbb{N}$ corresponding to syntactic operations of $\mathscr{L}$ (composing terms, forming atomic formulas from terms and relations, forming negations, conjunctions, disjunctions, and quantifying formulas) are all recursive. All of the results here remain true for an arbitrary countable language $\mathscr{L}$, provided 'recursive' is reinterpreted to mean 'recursive in some oracle for $\mathscr{L}$'.

We will also need the notion of a language $\mathscr{L}'$ being a *recursive extension* of $\mathscr{L}$. This means that $\mathscr{L}'$ is an extension of $\mathscr{L}$, and that there is a new Gödel-numbering for $\mathscr{L}'$ as above, with the additional property that the set of Gödel-numbers (in the new sense) of all $\mathscr{L}$-terms is recursive, and similarly for $\mathscr{L}$-formulas. Any extension of a recursive language by only finitely many new symbols will always be a recursive extension. However the Gödel-numbering of $\mathscr{L}'$ does not have to be an extension of that for $\mathscr{L}$, so there may be recursive languages $\mathscr{L}$, $\mathscr{L}'$ such that $\mathscr{L}'$ is an extension of $\mathscr{L}$, but $\mathscr{L}'$ is not a *recursive* extension of $\mathscr{L}$. For example, if $A \subset \mathbb{N}$ is non-recursive, $\mathscr{L}$ contains $\aleph_0$ n-ary relations for each $n \in \mathbb{N}$, but no function or constant symbols, and $\mathscr{L}'$ extends $\mathscr{L}$ by having a single new n-ary relation for each $n \in A$, but no other new symbols, then both $\mathscr{L}$ and $\mathscr{L}'$ are recursive, but $\mathscr{L}'$ is not a recursive extension of $\mathscr{L}$.

For the rest of this chapter, fix a recursive language $\mathscr{L}$, with Gödel-numbering $\ulcorner\urcorner$.

We now give the basic definition of a recursively saturated model, together with some related definitions.

DEFINITION 2.1. Let $\mathfrak{M}$ be an $\mathscr{L}$-structure and let M denote the domain of $\mathfrak{M}$. We say that a *semitype over* $\mathfrak{M}$ is a set $p(\bar{v}, \bar{a})$ of $\mathscr{L}(\bar{a})$-formulas in finitely many free variables $\bar{v}$ and finitely many parameters $\bar{a} \in M$ such that

$$\mathfrak{M} \models \exists \bar{v} \bigwedge_{\theta \in S} \theta(\bar{v}, \bar{a})$$

for all finite $S \subseteq p(\bar{v}, \bar{a})$. ('Semi' here is intended to remind the reader that it need not be complete and only has finitely many parameters $\bar{a}$.) We say $p(\bar{v}, \bar{a})$ is *recursive* if $\{\ulcorner\theta(\bar{v}, \bar{w})\urcorner : \theta \in p\}$ is recursive (where $\bar{w}$ is some other tuple of variables). $\mathfrak{M}$ *realizes* $p(\bar{v}, \bar{a})$ if there is some $\bar{b} \in M$ such that $\mathfrak{M} \models \theta(\bar{b}, \bar{a})$ for all $\theta \in p$. Finally, $\mathfrak{M}$ is *recursively saturated* if every recursive semitype over $\mathfrak{M}$ is realized in $\mathfrak{M}$.

A simple union-of-chains argument gives

PROPOSITION 2.2. For all $\mathscr{L}$-structures $\mathfrak{M}$ there is $\mathfrak{N} \succ \mathfrak{M}$ which is recursively saturated and of the same cardinality.

Many simple properties of recursively saturated models follow from this definition just given. For example, every recursively saturated model $\mathfrak{M}$ is $\aleph_0$-homogeneous, that is if $\bar{a}, \bar{b}, c \in M$ with $\mathrm{len}(\bar{a}) = \mathrm{len}(\bar{b}) < \omega$ and $\mathrm{tp}(\bar{a}) = \mathrm{tp}(\bar{b})$ there is $d \in M$ with $\mathrm{tp}(\bar{a}, c) = \mathrm{tp}(\bar{b}, d)$ (the proof of this is sketched in Theorem 2.12 below). Hence every countable recursively saturated model is strongly $\aleph_0$-homogeneous in the model-theoretic sense that types correspond to orbits. The next proposition is another sample, and (together with homogeneity) shows that every countable recursively saturated model has continuum-many automorphisms.

PROPOSITION 2.3. Let $\mathfrak{M}$ be a recursively saturated $\mathscr{L}$-structure with domain M. Then for any $\bar{a} \in M^{<\omega}$ and any $\mathscr{L}$-formula $\theta(x, \bar{a})$, if

$$A = \{b \in M : \mathfrak{M} \models \theta(b, \bar{a})\}$$

is infinite then there are distinct $b, c \in A$ with $\mathrm{tp}(b, \bar{a}) = \mathrm{tp}(c, \bar{a})$.

As a more sophisticated application of these rather simple ideas, we quote

THEOREM 2.4. Let $\mathfrak{M}$ be a countable recursively saturated $\mathscr{L}$-structure. Then $F(2^{\aleph_0})$, the free group of rank $2^{\aleph_0}$, embeds densely into $\mathrm{Aut}\,\mathfrak{M}$.

This is an example of a result that was first proved for $\aleph_0$-categorical structures (Macpherson, 1986*a*, for $F(\aleph_0)$) and then for recursively saturated structures (Kaye, 1992). (Hodges pointed out that a simple modification of the proof yields $F(2^{\aleph_0})$ in place of $F(\aleph_0)$; the theorem just quoted gives the same result for $F(\aleph_0)$ immediately, since M^n has only countably many $\mathrm{Aut}\,\mathfrak{M}$-orbits for each n.)

Of course, to show that a non-recursive semitype p is realized in $\mathfrak{M}$ it suffices that it is logically equivalent to a recursive set, and sometimes the following observation is useful.

PROPOSITION 2.5. (Craig's trick) For each recursively enumerable (r.e.) set of $\mathscr{L}$-formulas there is a recursive (in fact primitive recursive) set of $\mathscr{L}$-formulas logically equivalent to it.

Proof. If p is an r.e. set of formulas, consider

$$\left\{ \overbrace{\theta \wedge \theta \wedge \ldots \wedge \theta}^{n_\theta} : \theta \in p \right\}$$

for some suitable sequence n_θ. □

Here, a *recursively enumerable* set is the range of a recursive function. The most important examples are the sets of (Gödel numbers of) sentences σ that are provable from recursive sets of formulas, the set of consequences of PA, for example.

Occasionally this trick is not enough, and one needs to realize a semitype that is not even r.e. In such cases, Theorem 2.10 below sometimes suffices. We shall work towards this result, giving some other important definitions and results on the way.

Let $2^{<\omega}$ denote all finite sequences of 0s and 1s. We denote the empty sequence by ϵ. For sequences σ, τ, let $\sigma \subseteq_{\mathrm{e}} \tau$ denote 'σ is an initial subsequence of τ.' We say a set $A \subseteq 2^{<\omega}$ is a *tree* if

$$\sigma \subseteq_{\mathrm{e}} \tau \in A \text{ implies } \sigma \in A$$

and that A is a *path* if in addition, for all $\alpha, \beta \in A$, $\alpha \subseteq_{\mathrm{e}} \beta$ or $\beta \subseteq_{\mathrm{e}} \alpha$.

DEFINITION 2.6. (Scott, 1962) Fix a recursive bijection $c: 2^{<\omega} \to \mathbb{N}$, and identify $2^{<\omega}$ and $\mathbb{N}$ via c. A set $\mathscr{X} \subseteq \mathscr{P}(\mathbb{N})$ is a *Scott set* if it is non-empty and:

1. if A, B are in $\mathscr{X}$ then so are $A \cap B$, $A \cup B$;
2. if $A \in \mathscr{X}$ and B is recursive in A (i.e., using an oracle for A) then $B \in \mathscr{X}$;
3. $\mathscr{X}$ is closed under König's Lemma, in the sense that if $A \in \mathscr{X}$ is an infinite tree, then there is an infinite path $B \in \mathscr{X}$ such that $B \subseteq A$.

Since Scott sets are closed under relative recursion (condition 2 above) it is easy to check that this definition does not depend on the choice of the bijection $c: 2^{<\omega} \to \mathbb{N}$, provided this is recursive. The most familiar example of a Scott set is the set $\mathscr{A}$ of all sets $A \subseteq \mathbb{N}$ which are definable in $(\mathbb{N}, +, \cdot, 0, 1, <)$ by a first-order formula. However, proper subsets of $\mathscr{A}$ can be Scott sets too.

For a semitype $p(\bar{v}, \bar{a})$ over a model $\mathfrak{M}$, we write $p \in \mathscr{X}$ to mean $\{\ulcorner \theta(\bar{v}, \bar{w}) \urcorner : \theta \in p\} \in \mathscr{X}$. We use the notation $T \in \mathscr{X}$ similarly, when T is a set of $\mathscr{L}$-sentences. (Many authors simply identify a formula with its Gödel number.) A semitype $p(\bar{v}, \bar{a})$ is *complete* if $p(\bar{c}, \bar{a})$ is complete as a $\mathscr{L}(\bar{a}, \bar{c})$-theory, where $\bar{c}$ is a tuple of new constants not occurring in $\mathscr{L}(\bar{a})$.

DEFINITION 2.7. (Wilmers, 1975) A countable $\mathscr{L}$-structure is *$\mathscr{X}$-saturated* if for all complete semitypes $p(\bar{v},\bar{a})$ over $\mathfrak{M}$ we have

$$p(\bar{v},\bar{a}) \in \mathscr{X} \quad \Longleftrightarrow \quad p(\bar{v},\bar{a}) \text{ is realized in } \mathfrak{M}.$$

Notice that this definition, unlike most 'saturation' definitions, says *exactly* which complete semitypes are realized.

One obvious necessary condition for the existence of a $\mathscr{X}$-saturated model of a $\mathscr{L}$-theory T is that T or some completion of it must be in $\mathscr{X}$. (This is just the right-to-left direction in the definition, where p is the trivial type with no variables and no parameters.) In fact, for countable $\mathscr{X}$, this necessary condition suffices.

THEOREM 2.8. Let $\mathscr{X}$ be a countable Scott set, and let T be a $\mathscr{L}$-theory, with $T \in \mathscr{X}$. Then there is a countable $\mathscr{X}$-saturated model $\mathfrak{M}$ of T.

It turns out that, for an $\mathscr{X}$-saturated structure $\mathfrak{M}$, the left-to-right direction of the bi-implication in the last definition is true even if p is not complete. This follows from

PROPOSITION 2.9. (Scott, 1962) If $\mathscr{L}$ is a recursive language, $\mathscr{X}$ is a Scott set, and T is a consistent $\mathscr{L}$-theory such that $T \in \mathscr{X}$, then there is a complete $\mathscr{L}$-theory $T' \supseteq T$ with $T' \in \mathscr{X}$.

To apply this to $\mathscr{X}$-saturated stuctures, note that if $\mathfrak{M}$ is $\mathscr{X}$-saturated, $\bar{a} \in M$, and $p(\bar{v},\bar{a}) \in \mathscr{X}$ is a semitype over $\mathfrak{M}$ then $p(\bar{v},\bar{a}) \cup \text{tp}(\bar{a}) \in \mathscr{X}$. Applying the last proposition to this, there is a complete semitype $p'(\bar{v},\bar{a})$ in $\mathscr{X}$ extending p, and p' must be realized by $\mathfrak{M}$ by $\mathscr{X}$-saturation. Proposition 2.9 is also the key lemma for proving Theorem 2.8. In rough terms, these last two results say that the proof of the Completeness Theorem in first-order logic 'relativizes' to a Scott set.

We have just seen that an $\mathscr{X}$-saturated structure is recursively saturated. For most structures there is a converse to this.

THEOREM 2.10. For all countable recursively saturated $\mathscr{L}$-structures $\mathfrak{M}$ there is a Scott set $\mathscr{X}$ such that $\mathfrak{M}$ is $\mathscr{X}$-saturated.

It is unknown if the hypothesis 'countable' is needed here. The same theorem can be proved, however, for recursively saturated models of cardinality $\aleph_1$, and for all recursively saturated models of rich theories (see below).

There is one point that deserves further discussion here. In certain cases the Scott set $\mathscr{X}$ for which $\mathfrak{M}$ is $\mathscr{X}$-saturated is uniquely determined by $\mathfrak{M}$. When this happens we call $\mathscr{X}$ the *standard system* of $\mathfrak{M}$ and write $\mathscr{X} = \text{SSy}\,\mathfrak{M}$.

DEFINITION 2.11. (Jensen and Ehrenfeucht, 1976) An $\mathscr{L}$-theory T is *rich* if there is a recursive sequence $\{\phi_n(\bar{v}) : n \in \mathbb{N}\}$ of $\mathscr{L}$-formulas such that, for all disjoint finite sets $X, Y \subset \mathbb{N}$

$$T \vdash \exists \bar{v} \left(\bigwedge_{n \in X} \phi_n(\bar{v}) \wedge \bigwedge_{n \in Y} \neg \phi_n(\bar{v}) \right).$$

A set $A \subseteq \mathbb{N}$ is *coded* in $\mathfrak{M}$ if there is some $\bar{a} \in M$ such that

$$A = \{n : \mathfrak{M} \models \phi_n(\bar{a})\}.$$

The theories PA of arithmetic and ZF of set theory are rich; on the other hand any $\aleph_0$-categorical theory (indeed, any theory with a countably saturated model) fails to be rich, since rich theories have continuum-many n-types for some n.

It turns out that, for a recursively saturated model $\mathfrak{M}$ of a rich theory T, there is a unique $\mathscr{X}$ such that $\mathfrak{M}$ is $\mathscr{X}$-saturated, and that this $\mathscr{X}$, the standard system of $\mathfrak{M}$, is exactly the set of $A \subseteq \mathbb{N}$ that are coded in $\mathfrak{M}$.

Given a countable recursively saturated structure $\mathfrak{M}$, its language $\mathscr{L}$, its theory $\mathrm{Th}\,\mathfrak{M}$, and the set of complete types

$$\mathrm{Tp}(\mathfrak{M}) =_{\mathrm{def}} \{\mathrm{tp}(\bar{a}) : \bar{a} \in M^{<\omega}\}$$

realized in $\mathfrak{M}$ suffice to determine the isomorphism type of $\mathfrak{M}$ exactly. This is the content of the next result, which should probably be attributed to Wilmers and H. M. Friedman. Its proof is typical of the sort of back-and-forth argument for recursively saturated models.

THEOREM 2.12. If $\mathfrak{M}$ and $\mathfrak{N}$ are countable recursively saturated models of the same complete $\mathscr{L}$-theory T, then $\mathfrak{M} \simeq \mathfrak{N}$ if and only if $\mathrm{Tp}(\mathfrak{M}) = \mathrm{Tp}(\mathfrak{N})$.

Proof. By back-and-forth. Let $\bar{a} \in M$, $\bar{b} \in N$ and suppose $\mathrm{tp}(\bar{a}) = \mathrm{tp}(\bar{b})$. Let $c \in M$. Since $\mathrm{Tp}(\mathfrak{M}) = \mathrm{Tp}(\mathfrak{N})$, there are $\bar{b}', d' \in N$ with $\mathrm{tp}(\bar{a}, c) = \mathrm{tp}(\bar{b}', d')$. Now let $d \in N$ realize the following recursive type $p(v, \bar{b}, \bar{b}', d')$.

$$\{\theta(\bar{b}, v) \leftrightarrow \theta(\bar{b}', d') : \theta \in \mathscr{L}\}$$

The 'back' direction is identical. □

In the case of rich theories, we have the following particularly pleasant classification.

THEOREM 2.13. Let T be a rich $\mathscr{L}$-theory, where $\mathscr{L}$ is recursive.

1. For all countable recursively saturated models $\mathfrak{M}, \mathfrak{N} \models T$, we have $\mathfrak{M} \cong \mathfrak{N}$ if and only if $\mathrm{Th}\,\mathfrak{M} = \mathrm{Th}\,\mathfrak{N}$ and $\mathrm{SSy}\,\mathfrak{M} = \mathrm{SSy}\,\mathfrak{N}$.
2. For all Scott sets $\mathscr{X}$ and all complete extensions T' of T, there is a countable recursively saturated model $\mathfrak{M} \models T'$ with $\mathrm{SSy}\,\mathfrak{M} = \mathscr{X}$ if and only if $T' \in \mathscr{X}$.

In the case of models of first-order Peano arithmetic, this yields exactly $2^{\aleph_0}$ countable recursively saturated models for each of the $2^{\aleph_0}$ complete extensions of PA.

3 Resplendency

The definition, and some of the results of the last section concern types that can be realized in recursively saturated models. The other main tool in the theory of recursive saturation is *resplendency* and concerns expansions of the structure.

The main theorem (that countable recursively saturated models are resplendent) is due to Ressayre and (independently) Barwise and Schlipf. (Versions of this result for structures realized in ω-nonstandard models of set theory were also proved independently by Wilmers.) In fact an improved version of this result follows immediately from results already stated, and so we shall give it first, and define resplendency later.

THEOREM 3.1. (Resplendency) Let $\mathfrak{M}$ be a countable recursively saturated $\mathscr{L}$-structure, let T be a set of $\mathscr{L}'$-sentences, where $\mathscr{L}'$ is a recursive extension of $\mathscr{L}$, and suppose that $T + \mathrm{Th}(\mathfrak{M})$ is consistent.

1. If T is recursive, then $\mathfrak{M}$ has an expansion $\mathfrak{M}'$ to a $\mathscr{L}'$-structure which is a model of T.
2. If $\mathfrak{M}$ is $\mathscr{X}$-saturated, where $\mathscr{X}$ is a countable Scott set, and $T \in \mathscr{X}$, then $\mathfrak{M}$ has an $\mathscr{X}$-saturated expansion $\mathfrak{M}' \models T$.

Proof. We prove the stronger statement 2.

By definition of $\mathscr{X}$-saturation, $T + \mathrm{Th}\,\mathfrak{M} \in \mathscr{X}$, so by Theorem 2.8, there is a countable $\mathscr{X}$-saturated model $\mathfrak{N}$ of $T + \mathrm{Th}\,\mathfrak{M} \in \mathscr{X}$. But by Theorem 2.12 $\mathfrak{M} \cong \mathfrak{N}{\restriction}\mathscr{L}$, and we are finished. □

DEFINITION 3.2. A model $\mathfrak{M}$ for a recursive language $\mathscr{L}$ is *resplendent* if and only if for every $\bar{a} \in M^{<\omega}$ and every recursive set of $\mathscr{L}'$-sentences T, where $\mathscr{L}'$ is a recursive extension of $\mathscr{L}(\bar{a})$, if $\mathrm{Th}(\mathfrak{M}, \bar{a}) + T$ is consistent then $(\mathfrak{M}, \bar{a})$ has an expansion $(\mathfrak{M}', \bar{a})$ satisfying T.

This definition isn't quite the received one from the work of Barwise, Schlipf, Ressayre and others, but it perhaps should be. The official definition of resplendency is the one just given, except that $\mathscr{L}'$ may only have finitely many nonlogical symbols not in $\mathscr{L}$, and T must be finite. We have seen that countable recursively saturated structures are resplendent in the stronger sense, and we will see later that a resplendent structure (in the weak sense) for a finite language is recursively saturated, so it perhaps makes little difference, but in practice the stronger version of resplendency is often useful, as are infinite languages.

Incidentally, the weaker version of resplendency is often defined using a notion of a Σ_1^1 formula over a language $\mathscr{L}$. These are simply formulas $\Theta(\bar{v})$

of the form $\exists \bar{f}, \bar{R}\,\theta(\bar{v}, \bar{f}, \bar{R})$, where $\theta(\bar{v}, \bar{f}, \bar{R})$ is a formula of a language $\mathscr{L}'$ extending $\mathscr{L}$ and $\bar{f}, \bar{R}$ are the new functions and relations in $\mathscr{L}'$ but not in $\mathscr{L}$. Thus a $\mathscr{L}$-structure is resplendent (in the weaker sense) iff for every Σ^1_1 sentence $\Theta(\bar{a})$ of $\mathscr{L}(\bar{a})$, if $\Theta(\bar{a}) + \mathrm{Th}(\mathfrak{M}, a)$ is consistent (meaning 'has a model,' in the obvious sense) then $\mathfrak{M} \models \Theta(\bar{a})$.

The connection between recursive sets of sentences and Σ^1_1 sentences is the following theorem.

THEOREM 3.3. (Kleene, 1952) Let $\mathscr{L}$ be a language with only finitely many nonlogical symbols, and let $\phi_i(\bar{v})$ be a recursive set of $\mathscr{L}$-formulas. Then there is a Σ^1_1 formula $\Phi(\bar{v})$ such that, in all infinite $\mathscr{L}$-structures $\mathfrak{M}$,

$$\mathfrak{M} \models \forall \bar{v} \left(\Phi(\bar{v}) \leftrightarrow \bigwedge_i \phi_i(\bar{v}) \right).$$

Proof. See Kaye (1991, pp. 254–57). □

It is easy to see that there is a converse to Theorem 3.1 (even without the cardinality restriction on $\mathfrak{M}$); that is, all resplendent models are recursively saturated. If we take instead the weaker notion of resplendency and we assume the model's language is finite, there there is still a converse, by Kleene's Theorem. The theorem itself, however, is false for uncountable models: there are for example rigid recursively saturated models of PA of cardinality $\aleph_1$.

We should say that, as for recursive saturation, resplendent models exist in all cardinalities: by a union of chains argument any $\mathscr{L}$-structure has an elementary extension that is resplendent and of the same cardinality. This is proved using the Joint Consistency Lemma due to A. Robinson. In fact, the Joint Consistency Lemma can most conveniently be proved using countable resplendent models and recursive saturation—see Kaye (1991) for details.

The chapter by Kaye on indiscernibles in this volume has some examples of how resplendency is typically used. The other main uses are in attaching a nonstandard model of arithmetic or set theory to a given countable recursively saturated model.

Given a countable recursively saturated $\mathscr{L}$-structure $\mathfrak{M}$, we can identify its domain M with the natural numbers $\mathbb{N}$ and obtain an expansion $\mathfrak{M}' = (\mathfrak{M}, 0, 1, +, \cdot, <)$ to the language $\mathscr{L}'$ which has the usual symbols of arithmetic added. This expansion satisfies a *Peano-like theory*, that is it satisfies the axioms of PA* in the language $\mathscr{L}'$, that is it satisfies all the axioms for non-negative parts of discretely ordered rings, and all induction axioms

$$\forall \bar{v}\,((\theta(0, \bar{v}) \wedge \forall x\,(\theta(x, \bar{v}) \rightarrow \theta(x+1, \bar{v}))) \rightarrow \forall x\,\theta(x, \bar{v}))$$

for all first-order formulas θ in $\mathscr{L}'$. (Authors often omit the mention of the language when talking about PA*, using the star as a reminder that there may be functions and relations other than the usual arithmetic ones.) Now we use resplendency, in the form of Theorem 3.1, to give that there are operations $+, \cdot$ on M^2, a relation $< \subset M^2$ and elements $0, 1 \in M$ such that the expansion $\mathfrak{M}'' = (\mathfrak{M}, 0, 1, +, \cdot, <)$ satisfies PA* and is recursively saturated. (Note that these functions and relations can't be the same as before since $(\mathbb{N}, 0, 1, +, \cdot, <)$ is not recursively saturated.) Thus

THEOREM 3.4. Every countable recursively saturated structure is a reduct of a countable recursively saturated model $\mathfrak{M}'$ of PA*.

In his paper below (p. 281), Lascar uses the induction axioms to show that a large class of recursively saturated models of PA* has the small index property (see p. 25), and Kaye (from pp. 293 on) examines the closed normal subgroups of a countable recursively saturated model of PA*.

There is a further refinement of the last theorem that is often useful. In the above argument, once we have identified M with $\mathbb{N}$, and expanded $\mathfrak{M}$ to a structure $\mathfrak{M}'$ for the language $\mathscr{L}' = \mathscr{L} \cup \{0, 1, +, \cdot, <\}$ (so that its reduct to the usual language of arithmetic is the standard structure $(\mathbb{N}, 0, 1, +, \cdot, <)$), we can add further relations to the structure. In particular, we can add a binary relation $S(n, m)$ which is true just in case the formula of $\mathscr{L}'$ with Gödel-number n is true when its free variables $v_0, v_1, \ldots, v_k$ are replaced by $m_0, m_1, \ldots, m_k$, this being the finite sequence coded (in some predetermined and arithmetically definable way) by m. This gives a structure for another language $\mathscr{L}'(S)$, which satisfies PA* for $\mathscr{L}'(S)$, as well as Tarski's definition of truth. (Tarski's definition can be axiomatized in $\mathscr{L}'(S)$ by a (finite or infinite) recursive set of sentences; the set of sentences will be finite just in case $\mathscr{L}$ has finitely many non-logical symbols.) So there is an expansion $\mathfrak{M}$ to the same theory which is recursively saturated. Such relations S are called *satisfaction classes*; and are useful because they can describe the definition of truth in the original model, but in such a way that can be used by an argument that is formalized in the theory PA*. See Kaye (1992) for an argument using satisfaction classes that shows that recursively saturated models satisfy a version of the Ryll-Nardzewski Theorem but with nonstandard integers replacing finite integers in counting the number of Aut $\mathfrak{M}$-orbits.

Satisfaction classes are closely related to recursive saturation. See for example Kaye (1991) for a discussion and proofs of the key results in this line. One of the most important general questions here is the following: if $\mathfrak{M}$ is a recursively saturated model of PA, and we wish to add a satisfaction class as in the last paragraph, but using the original $+$ and $\cdot$ rather than new (added) ones, how far can we succeed? It turns out that this can always be done if we relax our condition that the expanded structure satisfies PA* in the new language (adding a so-called *full satisfaction class*) or if we only require our satisfaction class which defines truth for formulas with Gödel-number less than a small nonstandard bound (a *partial inductive satisfaction class*), and that $\mathfrak{M}$ is recursively saturated is also necessary for either of these expansions to exist. Partial inductive satisfaction classes are particularly useful for the theory of recursively saturated models of PA*.

4 Arithmetically saturated models

It is perhaps surprising that very few interesting variations on the definition of recursive saturation have been found. Modifying 'recursive' in the definition to 'primitive recursive' or 'r.e.' gives the same notion, by Craig's trick. However Kaye, Kossak and Kotlarski (Kaye *et al.*, 1991) did find such a variation with interesting consequences for automorphism groups—at least for models of PA. Schmerl and Kossak have suggested it should be called 'arithmetically saturated' (although the reader should be warned that it is not the same thing as what one gets when one replaces 'recursive' by 'arithmetical' in Definition 2.1).

DEFINITION 4.1. An $\mathscr{L}$-structure is *arithmetically saturated* if if is $\mathscr{X}$-saturated for some Scott set $\mathscr{X}$ closed under the jump operator of recursion theory.

For those that know what it means, we should mention that closure under *arithmetical comprehension* or *arithmetical closure* (meaning closure under both jump and relative recursion) is already enough to imply closure under König's Lemma, so an arithmetically saturated model is just one that is $\mathscr{X}$-saturated for some $\mathscr{X} \subseteq \mathscr{P}(\mathbb{N})$ which is closed under arithmetic comprehension, $\cup$, and $\cap$.

Fortunately, the reader doesn't need to know what the jump operator is. It suffices to note that by part (2) of Theorem 3.1 an arithmetically saturated structure is a reduct of an arithmetically saturated model $\mathfrak{M}$ of PA*, and then to use the following theorem which gives a much more useful characterization.

In this theorem, as usual when considering nonstandard models of arithmetic, we identify each $n \in \mathbb{N}$ with the element of $\mathfrak{M}$ realizing the closed term $1 + 1 + \cdots + 1$ (n 1s); and when convenient we consider the model as a model of ZF with the negation of the axiom of infinity, identifying each $x \in M$ with the set of y for which the yth digit in the binary expansion of x is one. This enables elements of M to 'code' subsets of M and partial functions $M \to M$, the latter by defining, for example, $f(x)$ to be the least y such that the $((x+y)(x+y+1)/2+y)$th binary digit of f is one.

THEOREM 4.2. (Kirby and Paris, 1977) A model $\mathfrak{M}$ of PA* is arithmetically saturated iff it is recursively saturated and for any $f \in M$ coding (in the way just described) a function with $\operatorname{Dom} f \supset \mathbb{N}$ there is $c \in M$ such that

$$\forall n{\in}\mathbb{N}\,(f(n) \in \mathbb{N} \Leftrightarrow f(n) < c).$$

The condition on $\mathfrak{M}$ in this last result concerning all coded functions f with domain including $\mathbb{N}$ is often referred to as $\mathbb{N}$ being *strong* in $\mathfrak{M}$.

Let us conclude this chapter with a simple example from Kaye *et al.* (1991) that demonstrates the usefulness of this theorem. We shall show that a countable arithmetically saturated model of PA* has an automorphism that moves all non-definable points. Perhaps more surprisingly, arithmetical saturation is also a *necessary* condition for the conclusion to be true. In this theorem, $p^{\mathfrak{M}}$ denotes the set of elements of M realizing the type p and $\mathrm{fix}(g)$ denotes the fixed points of g. Also, for a formula $\theta(x, y)$, $t_\theta(x)$ is the least y such that $\theta(x, y)$, if this exists, and 0 otherwise.

THEOREM 4.3. (Kaye, Kossak and Kotlarski) Let $\mathfrak{M}$ be a countable recursively saturated model of PA*, let $\mathfrak{M}_0$ be the elementary submodel of $\mathfrak{M}$ consisting of all definable points (i.e., the domain M_0 of $\mathfrak{M}_0$ is $\mathrm{dcl}(\varnothing)$) and let $G = \mathrm{Aut}\,\mathfrak{M}$. Then the following are equivalent:

1. $\exists g{\in}G\ \mathrm{fix}(g) = M_0$;
2. there exists $a \in M \setminus M_0$ and $g \in G$ such that $aG \cap \mathrm{fix}(g) = \varnothing$;
3. $\forall a{\in}M\ \{\ulcorner\theta\urcorner : t_\theta(a) \notin M_0\} \in \mathrm{SSy}\,\mathfrak{M}$;
4. $\mathbb{N}$ is strong in $\mathfrak{M}$;
5. there are $g \in G$ and an open subgroup $H < G$ such that

$$\forall f{\in}G\,(g^{f\cdot} \notin H).$$

REMARK. Notice that the property in 5 above is a property of the topological group G. Lascar's paper below will improve this, showing that it is in fact a property of G as an abstract group.

Proof. We shall prove $1 \Rightarrow 5 \Rightarrow 2 \Rightarrow 4 \Rightarrow 3 \Rightarrow 1$.

For $1 \Rightarrow 5$, given g as in 1, let $a \in M \setminus M_0$, $H = G_a$. Then for $f \in G$, $b = af^{-1} \notin M_0$, so $bg \neq b$ hence $af^{-1}gf = bgf \neq bf = a$ as required.

For $5 \Rightarrow 2$, suppose that H, g are as in 5, and choose a with $G_a \leqslant H$. Then if $bg = b$ and $f \in G$ with $a = bf$ then $af^{-1}gf = a$, i.e., $g^f \in G_a \leqslant H$. This is a contradiction.

For $2 \Rightarrow 4$, suppose $\mathbb{N}$ is not strong in $\mathfrak{M}$, a, g are as given in 2, $p = \mathrm{tp}(a)$, and $b \in p^{\mathfrak{M}}$. By saturation there is $h \in M$ such that for all $n \in \mathbb{N}$, $h(n)$ is defined and is the least $x \leqslant b$ such that

$$\mathfrak{M} \vDash \forall\ulcorner\phi\urcorner{<}n\,(\phi(x) \leftrightarrow \phi(b)).$$

By saturation again there is $\alpha > \mathbb{N}$ in M such that

$$\mathfrak{M} \vDash \bigwedge_{i\in\mathbb{N}} \forall u{<}\alpha\,(h(u) \text{ is defined and } \theta_i(h(u)))$$

where $p(v) = \{\theta_i(v) : i \in \mathbb{N}\}$. (This saturation argument works because the required type is recursive in p and hence is in SSy $\mathfrak{M}$.) Now suppose $f \in M$ witnesses that $\mathbb{N}$ is not strong. That is

$$\forall c{>}\mathbb{N}\, \exists n{\in}\mathbb{N}\;\; \mathbb{N} < f(n) < c.$$

Put $f' = fg$ and $h' = hg$. Then

$$\forall k{\in}\mathbb{N}\;\; \mathfrak{M} \models \forall u{<}k\, (f(u) < k \to h(f(u)) = h'(f'(u)))$$

so by saturation there is $k > \mathbb{N}$ such that $h(f(u))$ and $h'(f'(u))$ are defined for all $u < k$ and

$$\mathfrak{M} \models \forall u{<}k\, (f(u) < k \to h(f(u)) = h'(f'(u))).$$

Now let $n \in \mathbb{N}$ satisfy $\mathbb{N} < f(n) < \min(k, \alpha)$. Then since $ng = n$ for all $n \in \mathbb{N}$ we have

$$(h(f(n)))g = hg(fg(ng)) = h'(f'(n)) = h(f(n)) \in p^{\mathfrak{M}}$$

as required.

For $4 \Rightarrow 3$, suppose $\mathbb{N}$ is strong in $\mathfrak{M}$, and $a \in M$. By saturation there is $F \in M$ and $b > \mathbb{N}$ such that for all $u < b$ $F(u)$ is defined, and for all θ in two free variables x, y, $F(\ulcorner\theta\urcorner) = t_\theta(a)$, and that $F(u) > \mathbb{N}$ for all other $u < \mathbb{N}$. Similarly there is $F_0 \in M$ and $b_0 > \mathbb{N}$ such that $F_0(u)$ is defined for $u < b_0$, and for all θ we have $F_0(\ulcorner\theta\urcorner) = t_\theta(0)$, $F_0(u)$ being nonstandard for all other u. Using induction in $\mathfrak{M}$ (or alternatively another saturation argument) there is $G \in M$ with

$$G(x) = \begin{cases} \text{the least } y < b \text{ such that } x = F_0(y) \\ b \qquad \text{if there is no such } y. \end{cases}$$

Then, by strength, there is $c > \mathbb{N}$ such that, for all $n \in \mathbb{N}$, $G(F(n)) \in \mathbb{N}$ if and only if $G(F(n)) < c$. But it is evident that $G(F(\ulcorner\theta\urcorner)) \in \mathbb{N}$ iff $t_\theta(a) = t_\phi(0)$ for some ϕ iff $t_\theta(a) \in M_0$. It follows that the set

$$\{\ulcorner\theta\urcorner : t_\theta(a) \notin M_0\}$$

is in SSy $\mathfrak{M}$.

The proof of $3 \Rightarrow 1$ is a back-and-forth argument. Inductively we will have $\bar{a}, \bar{b} \in M^n$ for some n such that $\mathrm{tp}(\bar{a}) = \mathrm{tp}(\bar{b})$ and, for all $\theta(x, y)$,

$$t_\theta(\langle\bar{a}\rangle) = t_\theta(\langle\bar{b}\rangle) \Rightarrow t_\theta(\langle\bar{a}\rangle) \in M_0.$$

(Here,

$$\langle x, y\rangle =_{\mathrm{def}} (x+y)(x+y+1)/2 + y$$

and

$$\langle x_1, \ldots x_n, x_{n+1}\rangle =_{\mathrm{def}} \langle x_1, \ldots, x_n, x_{n+1}\rangle$$

are convenient functions coding tuples as single numbers.) We want to extend this partial isomorphism. Given $d \in M$, we need to find e such that

$$\phi(\bar{a}, d) \leftrightarrow \phi(\bar{b}, e) \tag{$\dagger$}$$

for all ϕ, and

$$t_\theta(\langle \bar{a}, d \rangle) \neq t_\theta(\langle \bar{b}, e \rangle) \tag{$\ddagger$}$$

for all θ such that $t_\theta(\langle \bar{a}, d \rangle) \notin M_0$. This set of formulas is in the standard system, by our assumption 3. It suffices to show it is finitely satisfied.

Suppose $t_1, \ldots, t_k$ are finitely many of the terms t_θ ocurring in ($\ddagger$), and let $\delta_i = t_i(\langle \bar{a}, d \rangle)$. Without loss of generality, we may assume that each δ_i is not in $M_{\bar{a}}$, the definable closure of $\{\bar{a}\}$, for if $t_i(\langle \bar{a}, d \rangle) = t_\theta(\langle \bar{a} \rangle)$ then $\mathrm{tp}(\langle \bar{a}, d \rangle) = \mathrm{tp}(\langle \bar{b}, e \rangle)$ would imply $t_i(\langle \bar{b}, e \rangle) = t_\theta(\langle \bar{b} \rangle)$ and we may use the induction hypothesis to deduce $t_\theta(\langle \bar{a} \rangle) = t_\theta(\langle \bar{b} \rangle) \Leftrightarrow t_\theta(\langle \bar{a} \rangle) \in M_0$. Thus $\Delta = \{\delta_1, \ldots, \delta_k\} \subseteq \Omega := M \smallsetminus M_{\bar{a}}$, and $G_{\bar{a}}$ acts naturally on Ω, with no finite orbits (because of the order on $\mathfrak{M}$ and recursive saturation). Let $h: \bar{a} \to \bar{b}$ be any automorphism of $\mathfrak{M}$. Then by the Separation Lemma, (Theorem 3.1 on p. 17) there is $g \in G_{\bar{a}}$ such that $\Delta g \cap \Delta h^{-1} = \varnothing$. Then $gh: \bar{a} \mapsto \bar{b}$ since $\bar{a}g = \bar{a}$, and $\Delta gh \cap \Delta = \varnothing$, as required. $\square$

Indiscernibles

Richard Kaye

School of Mathematics and Statistics,
The University of Birmingham,
Edgbaston,
Birmingham,
B15 2TT,
U.K.
Electronic mail address: R.W.Kaye@bham.ac.uk

1 Introduction

When investigating the structure of $G = \mathrm{Aut}(\mathfrak{M})$ as a permutation group acting on $M = \mathrm{Dom}(\mathfrak{M})$ it is natural to look for subsets I of M for which the induced group

$$G^I_{\{I\}} =_{\mathrm{def}} \{g{\restriction}I : g \in G_{\{I\}}\} \leqslant \mathrm{Sym}(I)$$

is a better-known permutation group acting on I. In particular, it is natural to look for I such that this group is the full symmetric group on I, or the group of order-automorphisms $\mathrm{Aut}(I,<)$ for some linear order $<$ on I; model-theoretically this corresponds to looking for an indiscernible set I or, respectively, indiscernible sequence $(I,<)$ with the property that any permutation of I (order-automorphism of $\mathrm{Aut}(I,<)$) extends to an automorphism of M. This is typically achieved by constructing I in such a way that M is 'generated' by I—although the precise meaning of 'generate' will differ from case to case. This chapter is devoted to the construction of such I in the case when $\mathfrak{M}$ is countable and recursively saturated.

Throughout the chapter, and unless explicitly stated otherwise, $\mathfrak{M}$ is a countable recursively saturated model for a recursive language $\mathscr{L}$, M is the domain of $\mathfrak{M}$, and $G = \mathrm{Aut}(\mathfrak{M})$.

There are two main theorems presented here. The first is due to J. H. Schmerl and shows that any such $\mathfrak{M}$ has an expansion $(\mathfrak{M},\beta)$ such that, for any countable linear ordered set $(J,\prec)$ with no last element, there is $I \subset M$ and $<$ on I such that $(I,<) \cong (J,\prec)$, $(I,<)$ is an indiscernible sequence for $(\mathfrak{M},\beta)$, and M is the closure of I under the function β.

The second theorem concerns indiscernible sets I on which the full symmetric group is induced. Clearly, in this situation, some further hypotheses on the structure of $\mathfrak{M}$ are required, and we shall assume that

Th($\mathfrak{M}$) is stable. Under this hypothesis, there is an infinite I on which Sym(I) is induced, although there will not necessarily be any embedding $\mathrm{Sym}(I) \hookrightarrow \mathrm{Aut}(\mathfrak{M})$. In this case, the set I generates (or, rather, *determines*) M in a sense rather similar to that in Fraïssé's Theorem describing the unique homogeneous structure with a given age. See Theorem 4.2 for a precise statement.

2 Indiscernibles and combinatorics

We start with some basic definitions.

If I is a set and $n \in \mathbb{N}$ then $[I]^n$ denotes the set of subsets S of I with exactly n elements. Usually I will have a distinguished linear order which is clear from context, so it is possible to identify these subsets with strictly increasing sequences $i_0 < i_1 < \cdots < i_{n-1}$ of elements $\bar{\imath}$ of I. The notation $[I]^{\geqslant n}$ is used to denote $\bigcup_{k \geqslant n} [I]^k$ where (unless stated otherwise) the union is over all (standard) integers $k \in \mathbb{N}$ greater than or equal to n.

A linearly ordered set $(I,<)$ is an *n-indiscernible sequence* of an $\mathscr{L}$-structure $\mathfrak{M}$ iff $I \subseteq M$ (we do not necessarily require the order $<$ to be definable in $\mathfrak{M}$) and, for all $\phi(v_1,\ldots,v_n) \in \mathscr{L}$, all $i_1 < i_2 < \cdots < i_n$ and all $j_1 < j_2 < \cdots < j_n$ from I,

$$M \vDash \phi(i_1,\ldots,i_n) \leftrightarrow \phi(j_1,\ldots,j_n).$$

I is an *indiscernible sequence* iff it is an n-indiscernible sequence for all $n \in \mathbb{N}$. The subset I of M is an *indiscernible set* iff $(I,<)$ is an indiscernible sequence of M for *every* possible order $<$ on I. It is clear that this can be possible only if there is no non-trivial 0-definable order on I, so for example models of arithmetic cannot have indiscernible sets—except trivial one-element ones. $(I,<)$ is an indiscernible sequence *over* a set of parameters $A \subseteq M$ iff $(I,<)$ is an indiscernible sequence for the structure $(\mathfrak{M},a)_{a \in A}$, i.e., if $(I,<)$ is an indiscernible sequence according to the above definitions except that parameters from A are allowed to be present in the formula $\phi(v_1,\ldots,v_n)$. We define the notion of an indiscernible set over a set of parameters A analogously.

Usually, the order $<$ on an indiscernible sequence $(I,<)$ is clear from context. For example, if $(M,0,1,<,+,\cdot)$ is a model of arithmetic then the order $<$ on the indiscernible sequence $(I,<)$ must be the restriction of the order $<$ on M to the set I. When the order on I is clear, we shall simply refer to I as being an indiscernible sequence without mentioning $<$.

If $(I,<)$ is an indiscernible sequence for the $\mathscr{L}$-structure M, we denote the *type of* I, i.e., the set of $\mathscr{L}$-formulas

$$\{\phi(v_0,\ldots,v_{n-1}) : \exists n \in \mathbb{N}\, \exists i_0 < \cdots < i_{n-1} \in I\; \mathfrak{M} \vDash \phi(i_0,\ldots,i_{n-1})\},$$

by $\mathrm{tp}_{\mathfrak{M}}(I)$. If the model $\mathfrak{M}$ is clear from context, we write $\mathrm{tp}_{\mathfrak{M}}(I)$ more simply as $\mathrm{tp}(I)$. If $(I,<)$ is indiscernible over parameters A, $\mathrm{tp}_{\mathfrak{M}}(I/A)$ denotes the type of I in the expanded structure $(\mathfrak{M},a)_{a\in A}$.

Indiscernible sequences and sets are usually constructed using Ramsey's Theorem. In most cases, the finite version of this theorem suffices. Let T be a complete $\mathscr{L}$-theory, where $\mathscr{L}$ is countable. A *Skolemization* of T is an $\mathscr{L}'$ theory T' (where $\mathscr{L}'$ is a countable expansion of $\mathscr{L}$ by adding countably many function symbols) such that: (1) the models of T are precisely the reducts to $\mathscr{L}$ of models of T'; and (2) if $\mathfrak{K}$ and $\mathfrak{M}$ are $\mathscr{L}'$-structures and $\mathfrak{K}\subseteq\mathfrak{M}\vDash T'$ then $\mathfrak{K}\vDash T'$. The result that every theory T has a Skolemization is essentally due to Skolem. The basic result on indiscernibles is due to Ehrenfeucht and Mostowski who showed:

THEOREM 2.1. For any $\mathscr{L}$-theory T with infinite models and any linearly ordered set $(I,<)$ there is a model $\mathfrak{M}$ of T such that $I\subseteq M$ and $(I,<)$ is an indiscernible sequence for $\mathfrak{M}$. Moreover, $\mathfrak{M}$ can be chosen so that there is an embedding of groups

$$\mathrm{Aut}(I,<)\to\mathrm{Aut}(\mathfrak{M}).$$

Proof. Given T, we must show that the following set of $\mathscr{L}(I)$-sentences T^* is consistent with T:

$$\{\phi(i_1,\ldots,i_n)\leftrightarrow\phi(j_1,\ldots,j_n):\phi\in\mathscr{L},\ n\in\mathbb{N},\ \bar{\imath},\bar{\jmath}\in[I]^n\}$$

where elements $i\in I$ are considered as constants in the language $\mathscr{L}(I)$.

Let $\mathfrak{M}\vDash T$ be countable, and identify $M=\mathrm{Dom}(\mathfrak{M})$ with $\mathbb{N}$. Consider $\phi_0,\ldots,\phi_{k-1}\in\mathscr{L}$ with at most n free variables each, and consider any sequence $i_1,\ldots,i_r\in I$ of finite length, given in increasing order. Define $f:[\mathbb{N}]^n\to 2^k$ by

$$f(x_1,\ldots,x_n)=\sum_{i=0}^{k-1}\epsilon_i 2^i$$

where

$$\epsilon_i=\begin{cases}0 & \text{if } \mathfrak{M}\vDash\phi_i(x_1,\ldots,x_n);\\ 1 & \text{otherwise.}\end{cases}$$

The finite Ramsey Theorem says that there is $e\in\mathbb{N}$ such that $\{0,1,\ldots,e\}\subseteq M$ contains a set $J=\{j_1,\ldots,j_r\}$ such that f is constant on $[J]^n$. By interpreting i_l by j_l in M we obtain a model of a finite fragment of T^*. Thus every finite subset of T^* is consistent and so T^* has a model, as required.

For the 'moreover' part, simply use the above argument on a Skolemization T' of T, obtaining a model $\mathfrak{N}'$ of T' with indiscernibles I. Let $\mathfrak{M}'$ be the $\mathscr{L}'$-substructure of $\mathfrak{N}'$ which is the closure of I under terms in $\mathscr{L}'$, i.e., M consists of all $t(\bar{\imath})$ for $\mathscr{L}'$-terms t and $\bar{\imath}\in I$, and let $\mathfrak{M}$ be the reduct $\mathfrak{M}'\restriction\mathscr{L}$. Then any order-automorphism π of $(I,<)$ extends uniquely to a automorphism π^* of $\mathfrak{M}$ defined by

$$(t(\bar{\imath}))^{\pi^*} =_{\text{def}} t(\bar{\imath}^{\pi}). \qquad \square$$

Models $\mathfrak{M}$ which are generated by a sequence of indiscernibles are called Ehrenfeucht–Mostowski models. Schmerl's Theorem says that *any* countable recursively saturated model can be regarded as an Ehrenfeucht–Mostowski model in a rather simple way. To prove this we will need a powerful generalization of the finite Ramsey Theorem, due to Nešetřil and Rödl (1977, 1983) and (independently) to Abramson and Harrington (1978). First, we must give some definitions; the terminology we shall use will be slightly different from either of these two papers, due mostly to the fact that the words such as 'homogeneous' are already overused in this setting.

A *palette* (system of colours) is a finite sequence of finite non-empty sets $\bar{\alpha}=(\alpha_0,\alpha_1,\ldots,\alpha_{n-1})$. A *set system* (also called a *coloured set*, or a *generalized graph*) is a quadruple $\mathfrak{A}=(A,<,\bar{\alpha},\bar{f})$ where $(A,<)$ is a linearly ordered set, $\bar{\alpha}$ is a palette, and $\bar{f}$ is a family of functions

$$f_i\colon [A]^i \to \alpha_i \qquad (i<\operatorname{len}(\bar{\alpha})).$$

When we want to make $\bar{\alpha}$ explicit, we talk about an $\bar{\alpha}$-system ($\bar{\alpha}$-coloured set, or $\bar{\alpha}$-graph). It will also be helpful to think of such $\mathfrak{A}$ as structures for the language $\mathscr{L}_{\bar{\alpha}}$ with relations $<$ and R_{ij} ($j\in\alpha_i$, $i<\operatorname{len}(\bar{\alpha})$) defined by

$$R_{ij}(x_1,\ldots,x_i) \quad\Leftrightarrow\quad \{x_1,\ldots,x_i\}\in[A]^i \wedge f_i(x_1,\ldots,x_i)=j.$$

A *pattern* is the isomorphism-type of some finite set-system. It is convenient to think of patterns as set-systems where the underlying set is some $\{0,1,\ldots,r\}$ ($r\in\omega$) and the order $<$ is the usual one on ω. If $\mathfrak{A}=(A,<,\bar{\alpha},\bar{f})$ is a set-system and $\bar{b}\in A$, we denote by $\operatorname{Patt}(\bar{b})$ the pattern of $\{\bar{b}\}$ as a substructure of $\mathfrak{A}$. Similarly $\operatorname{Patt}(\mathfrak{B})$ is the pattern of the set-system $\mathfrak{B}$.

A function $\chi\colon[A]^e\to m$ (where $e,m<\omega$) will be called an m-colouring of $[\mathfrak{A}]^e$. If $B\subseteq A$, B is *relatively monochromatic for* χ (I am avoiding the use of the word 'homogeneous') iff $\chi(\bar{b})$ only depends on $\operatorname{Patt}(\bar{b})$.

For $\bar{\alpha}$-systems $\mathfrak{A}$ and $\mathfrak{B}$ we write $\mathfrak{A}\to(\mathfrak{B})^e_m$ to denote: for all m-colourings of $[\mathfrak{A}]^e$ there is a subset $\{\bar{b}\}$ of A which is relatively monochromatic for χ and $\operatorname{Patt}(\bar{b})=\operatorname{Patt}(\mathfrak{B})$.

The theorem due to Nešetřil and Rödl, and (independently) Abramson and Harrington that we will need is the following.

THEOREM 2.2. For all palettes $\bar{\alpha}$ all $e,m\in\omega$, and all finite $\bar{\alpha}$-systems $\mathfrak{B}$ there is a finite $\bar{\alpha}$-system $\mathfrak{A}$ such that $\mathfrak{A}\to(\mathfrak{B})^e_m$.

This will be used together with resplendency to construct the indiscernible sequence in Schmerl's Theorem. For a straightforward account of Theorem 2.2 I can recommend Prömel and Voigt (1989) (an easy extra induction argument gives Theorem 2.2 from their statement on p. 314 of the Nešetřil–Rödl result). In fact we will only need Theorem 2.2 in the special case when α_i is a singleton set for all $i \leqslant e$ (in which case 'relatively monochromatic' can be replaced by 'monochromatic', i.e., the colour $\chi(\bar{b})$ of a tuple in the system with the same pattern as $\mathfrak{B}$ is constant, since the pattern $\mathrm{Patt}(\bar{b})$ of e-tuples is always trivial), but it appears that the inductive proof of even this special case needs a version of the full result as inductive hypothesis.

3 J. H. Schmerl's Theorem

For an n-tuple of variables $v_0,\ldots,v_{n-1}$ let $\neq(\bar{v})$ denote the formula

$$\bigwedge_{i<j<n} (v_i \neq v_j)$$

expressing that the v_i are all different.

Let $\mathscr{L}_{\mathrm{CFF}}$ be the first-order language with a single binary function symbol β. CFF (for 'Codes Finite Functions') is the $\mathscr{L}_{\mathrm{CFF}}$-theory with axiom 'there are at least two elements' and axiom scheme

$$\forall \bar{x}\, \forall \bar{y}\, \exists z \left(\neq(\bar{x}) \rightarrow \bigwedge_{i<n} \beta(x_i, z) = y_i \right)$$

for each $n = \mathrm{len}(\bar{x}) \in \mathbb{N}$. $\mathscr{L}_{\mathrm{CFF}}(<)$ is the language obtained by adding a binary relation symbol $<$, and $\mathrm{CFF}(<)$ is the $\mathscr{L}_{\mathrm{CFF}}(<)$-theory with axioms stating that $<$ is a linear order and also the axiom scheme

$$\forall \bar{x}\, \forall \bar{y}\, \forall w\, \exists z > w \left(\neq(\bar{x}) \rightarrow \bigwedge_{i<n} \beta(x_i, z) = y_i \right)$$

over all $n \in \mathbb{N}$.

All models of CFF are infinite: since the model contains distinct elements x_1, x_2 there is z with $\beta(x_1,z) \neq \beta(x_1,x_1)$ and $\beta(x_2,z) \neq \beta(x_2,x_2)$, so there are at least three elements, ..., and so on.

Notice that, if $(M,\beta,\ldots)$ is a model of CFF and $\prec$ is any linear-ordering of M with order-type ω, then the axioms of $\mathrm{CFF}(<)$ are satisfied. Hence by Theorem 3.1 on p. 250 the model has another linear order such that $(M,\beta,<,\ldots)$ is a recursively saturated model of $\mathrm{CFF}(<)$.

The typical example of a model of $\mathrm{CFF}(<)$ is a model of first-order arithmetic. If $(M,0,1,+,\cdot,<)$ is such a model and $\beta(x,y) = (y)_x$ is the function whose value is the $(x+1)$st element of the sequence

$$(y)_0, (y)_1, (y)_2, \ldots$$

defined in some standard way (using the Chinese Remainder Theorem perhaps, as in Kaye (1991, Chapter 5)) then $(M,\beta,<) \vDash \mathrm{CFF}(<)$. For this reason Theorem 3.1 below is often stated as a theorem concerning models of arithmetic, whereas in fact the axioms of CFF suffice.

The β function is extended to arguments which are n-tuples for any $n \in \mathbb{N}$ as follows.

$$\beta(x) = x$$

and

$$\beta(x_0,\ldots,x_{n-1},x_n) = \beta(\beta(x_0,\ldots,x_{n-1}),x_n)$$

for all $n > 0$. This is simply for convenience in the proof below: we will be choosing sequences $b_0,\ldots,b_{n-1}$ and considering possible values of $\beta(b_0,\ldots,b_{n-1},x_n)$. These new βs don't really do any more than the original $\beta(x,y)$.

Schmerl's Theorem concerns models M which are expansions of models of CFF(<) (or, more generally, interpret CFF(<)).

THEOREM 3.1. (Schmerl, 1985, 1989) Let

$$\mathfrak{M} = (M,\beta,<,\ldots) \models \mathrm{CFF}(<)$$

be countable and recursively saturated, and let $(J,\prec)$ be a countable linear order with no last element. Then there is an indiscernible sequence $I \subseteq M$ with $(I,< \restriction I) \cong (J,\prec)$ generating M in the sense that, for every $a \in M \setminus I$ there are $n \in \mathbb{N}$, an $\mathscr{L}_{\mathrm{CFF}}$ term $t(v_1,\ldots,v_n)$ and $\bar{\imath} \in I$ such that $a = t(\bar{\imath})$.

Before we prove this note that, in the particular case when the model $(M,\beta,<,\ldots)$ is a model of arithmetic, the set I must be cofinal in M, that is $\forall a \in M\, \exists i \in I\ \mathfrak{M} \models i > a$. This is because, in arithmetic, one can put a small nonstandard bound ν on the number of $\mathscr{L}_{\mathrm{CFF}}$ terms, and so, if $a \in M$ were such that every $i \in I$ is less than a, there would be at most $\nu \cdot a^\nu$ elements of M of the form $t(\bar{\imath})$ ($\bar{\imath} \in I$, $t \in \mathscr{L}_{\mathrm{CFF}}$), a contradiction. Thus, in this case at least, the condition that $(J,\prec)$ has no last element is certainly necessary.

Proof of Theorem 3.1. It will turn out that, to start with at least, it suffices to construct I of order-type ω. My presentation will give sufficient conditions for a tuple $\bar{b}$ to be in such a set I, and it will (I hope) give a fairly good understanding of the ideas involved to readers with only a slight acquaintance with recursive saturation.

The idea is to inductively define two infinite lists of elements of M:

$$b_0,b_1,b_2,\ldots,b_n,\ldots$$

and

$$c_0,c_1,c_2,\ldots,c_n,\ldots$$

At the nth stage of the construction we will have a rather complicated inductive hypothesis on the elements $a_0,a_1,b_0,b_1,\ldots,b_{n-1},c_0,c_1,\ldots,c_{n-1}$ of

M (where a_0,a_1 are fixed parameters from M to be chosen in a moment). This hypothesis will be expressed by saying that a certain formula

$$\Theta_n(a_0,a_1,b_0,\ldots,b_{n-1},c_0,\ldots,c_{n-1},R)$$

is consistent with the theory of $(\mathfrak{M},a_0,a_1,b_1,\ldots,b_n,c_1,\ldots,c_n)$. The formulas Θ_n will be Σ^1_1 over the first-order language of $\mathfrak{M}$ expanded by adding the constants $a_0,a_1,b_1,\ldots,b_n,c_1,\ldots,c_n$ and also a new unary predicate R. By resplendency, the consistency of Θ_n is equivalent to saying that there is a set $J\subseteq M$ such that the expansion $(\mathfrak{M},a_0,a_1,b_0,\ldots,b_{n-1},c_0,\ldots,c_{n-1},J)$ satisfies Θ_n.

During the construction, we will show how to arrange that *every* element of M is b_j or c_j for some $j\in\mathbb{N}$. This, together with some of the other properties expressed by the Θ_n about R will ensure that, at the end of the construction, our set $I =_{\text{def}} \{b_0,b_1,\ldots,b_n,\ldots\}$ is as required.

We shall first show how to use recursive saturation to find a sequence $a_0,a_1,\ldots,a_n,\ldots$ in M such that

$$\bigwedge_k \neq(a_0,\ldots,a_{k-1}) \wedge \bigwedge_k \beta(a_k,a_0)=a_{k+1}.$$

Indeed, let $x_1,\ldots,x_n\in M$ be all different and otherwise arbitrary. Since models of CFF are infinite, by axioms of CFF there are infinitely many $x_0\in M$ such that $\bigwedge_{k<n}\beta(x_k,x_0)=x_{k+1}$. (Indeed, if $z\in M\smallsetminus\{x_1,\ldots,x_n\}$ then for any $y\in M$ there is such an x_0 with $\beta(z,x_0)=y$.) So at least one of these infinitely many x_0 must be different from any of the $x_1,\ldots,x_n$. Repeating this, we deduce that the following type $p(x_0,x_1)$ is finitely satisfied on M:

$$\left\{\exists x_2,\ldots,x_n\left(\neq(x_0,\ldots,x_n) \wedge \bigwedge_{k<n}\beta(x_k,x_0)=x_{k+1}\right):n\in\mathbb{N}\right\}.$$

So by recursive saturation there are a_0,a_1 in M realizing $p(x_0,x_1)$, and defining $a_{k+1}=\beta(a_k,a_0)$ for all $k>1$ we obtain our sequence $a_0,a_1,\ldots,a_n,\ldots$ as required.

In the argument that follows, we will arrange that every element c of $M\smallsetminus I$ will satisfy

$$\beta(a_{n+1},i_0,i_1,\ldots,i_n)=c \tag{1}$$

for some $n\in\mathbb{N}$ and some $i_0<i_1<\cdots<i_n$ from I. We shall also arrange that

$$\beta(i_0,i_1)=a_0 \tag{2}$$

and

$$\beta(a_0,i_0)=a_1 \tag{3}$$

for all $i_0 < i_1$ from I. This will ensure that M is generated by I using the function β. It also suggests the induction hypotheses (that is, the formulas Θ_n) that we will need.

The formulas $\Theta_0(a_0,a_1,R),\ldots,\Theta_n(a_0,a_1,b_0,\ldots,b_{n-1},c_0,\ldots,c_{n-1},R),\ldots$, will be defined as follows: $\Theta_0(a_0,a_1,R)$ is the formula which states that

1. $\forall(i_0,i_1)\in[R]^2\,\big(\beta(i_0,i_1)=a_0 \wedge \beta(a_0,i_0)=a_1\big)$, and
2. R is 0-free

and $\Theta_n(a_0,a_1,b_0,\ldots,b_{n-1},c_0,\ldots,c_{n-1},R)$ is the conjunction of

1. $\forall(i_0,i_1)\in[R]^2\,\big(\beta(i_0,i_1)=a_0 \wedge \beta(a_0,i_0)=a_1\big)$
2. $\bar{b}\in[R]^n$
3. $\bigwedge_{k<n}\beta(a_{k+1},b_0,b_1,\ldots,b_k)=c_k$
4. R is n-free
5. R is n-indiscernible over $\{a_0,a_1\}$

The definition of 'n-free' will be given in a moment. Notice that the first three clauses in Θ_n are first-order, and the last (that R is n-indiscernible over $\{a_0,a_1\}$) is a recursive conjunction of first-order formulas. Thus Θ_n is Σ^1_1 provided 'n-free' is Σ^1_1.

To motivate the definition of n-free, notice that at some stage of the construction we will be given $R\subseteq M$ such that $\mathfrak{M}\models\Theta_n(\bar{a},\bar{b},\bar{c},R)$ and will want to add an arbitrary element $d\in M$ to the list $c_0,\ldots,c_{n-1}$. To do this, we would like to find $i_n>\bar{b}$ in R such that $\beta(a_{n+1},b_0,b_1,\ldots,b_{n-1},i_n)=d$. This is the sort of thing that 'n-free' says, except it is even more complicated than this: 'n-free' must also say that the construction can be carried out indefinitely. Also, since at each stage we must improve our n-indiscernible sequence R to get an $(n+1)$-indiscernible sequence, each stage will involve a 'Ramsey's Theorem' type of argument, and our notion of 'n-free' must be preserved under this part of the argument too.

After thinking about this for a bit, one arrives at the following idea: with a fixed $\mathfrak{M}$, β and a_0,a_1 in mind, say that a set $I\subseteq M$ is n-free iff it has at least n elements and, for all finite $Y\subseteq I$ and all functions $f\colon[Y]^{\geqslant n}\to M$ there is $z\in I$ greater than all $y\in Y$ coding f in the sense that

$$\beta(a_{\mathrm{len}(\bar{y})+1},\bar{y},z)=f(\bar{y})$$

for all $\bar{y}\in[Y]^{\geqslant n}$. This is the definition in Schmerl's second paper (1989) and would be the simplest approach. The statement of the preservation of this notion under the required 'Ramsey argument' is stated there as Lemma 2 on p. 1385 (no proof is given, and the reader is referred to a similar lemma in Schmerl (1985)). Unfortunately, the present author was unable to verify this lemma, and the proof of the analogous lemma in the previous paper seems to contain an error. So instead, I offer the following alternative which (like a different argument in Schmerl (1985)) uses Theorem 2.2 from the previous section, but is closer in style to the presentation in Schmerl's second paper.

DEFINITION 3.2. A non-empty subset $I \subseteq M$ is *n-free* iff it has at least n elements and, for all finite $Y \subseteq I$ and all functions

$$f:[Y]^{\geqslant n} \rightarrow M$$

there is $z \in I$, greater than all $y \in Y$, coding the 'first-order' part of f in the sense that the map

$$\alpha: f(\bar{y}) \mapsto \beta(a_{s+1}, \bar{y}, z)$$

defined on all $\bar{y} \in [Y]^s$, for all $s \geqslant n$, can be extended to an automorphism of $(\mathfrak{M}, a_0, a_1, x)_{x \in Y}$. Note that, by saturation, this is equivalent to

$$(\mathfrak{M}, a_0, a_1, x, f(\bar{y}))_{x \in Y, \bar{y} \in [Y]^{\geqslant n}} \equiv (\mathfrak{M}, a_0, a_1, x, \beta(a_{\mathrm{len}(\bar{y})+1}, \bar{y}, z))_{x \in Y, \bar{y} \in [Y]^{\geqslant n}}.$$

Note that if $I \subseteq M$ is n-free then it must be infinite. (To prove this, first find $u_0, u_1, u_2, \ldots$ in M such that the types $\mathrm{tp}(a_0, a_1, u_i)$ are all different—for example, take u_i so that $\beta(a_i, u_i) = a_1$ and $\beta(a_j, u_i) = a_0$ for all other j—and then consider constant functions f_i with value u_i.) Also, 'R is n-free' is expressed by a Σ^1_1 formula of the language with R and a_0, a_1 added. To see this observe that R is n-free iff

$$\bigwedge_{s \geqslant n} \forall x_1, \ldots, x_s \in R\ \forall f_1, \ldots, f_t$$

$$\exists z \in R \bigwedge_{\theta} \left(\theta(a_0, a_1, \bar{x}, \beta_1, \ldots, \beta_t) \leftrightarrow \theta(a_0, a_1, \bar{x}, f_1, \ldots, f_t)\right)$$

where $\beta_i = \beta(a_{\mathrm{len}(\bar{y})+1}, \bar{y}, z)$ for $\bar{y}$ = the ith element of $[x_1, \ldots, x_s]^{\geqslant n}$, and $t = \sum_{i \geqslant n} \binom{s}{i}$ is the number of such $\bar{y}$. This is equivalent in $\mathfrak{M}$ to a recursive conjunction of first-order formulas, since the '$\exists z$' can be taken inside the '$\bigwedge_\theta$' by recursive saturation as follows:

$$\bigwedge_{s \geqslant n} \bigwedge_{r \in \mathbb{N}} \forall y_1, \ldots, y_s \in R\ \forall f_1, \ldots, f_t$$

$$\exists z \in R \bigwedge_{\ulcorner \theta \urcorner < r} \left(\theta(a_0, a_1, \bar{y}, \beta_1, \ldots, \beta_t) \leftrightarrow \theta(a_0, a_1, \bar{y}, f_1, \ldots, f_t)\right).$$

Thus the notion 'n-free' is Σ^1_1 by Kleene's Theorem (Theorem 3.3, on p. 251).

LEMMA 3.3. $\mathrm{Th}(\mathfrak{M}, a_0, a_1) + \Theta_0(a_0, a_1, R)$ is consistent.

Proof. We must find a 0-free set $R \subseteq M$ such that

$$\beta(i, j) = a_0 \wedge \beta(a_0, i) = a_1$$

for all $i<j$ in R. It suffices of course to choose R so that the map α in the definition of 0-free is the identity. We shall construct such an R inductively as the set $\{i_0,i_1,\ldots\}$, the sequence $i_0,i_1,\ldots,i_n,\ldots\in M$ being an increasing sequence.

Fix some enumeration of all functions g in

$$\{g:[S]^{\geqslant 0}\to M : S\subseteq\mathbb{N} \text{ is finite}\}$$

as $g_r:[S_r]^{\geqslant 0}\to M$, where $S_r\subseteq\{0,1,\ldots,r-1\}$. At a given stage $k\geqslant 0$ of the construction, the set $R_k =_{\text{def}} \{i_0,i_1,\ldots,i_{k-1}\}$ will have been defined and satisfies:

1. $\forall i,j\in R_k\ (i<j\to\beta(i,j)=a_0)$;
2. $\forall i\in R_k\ \beta(a_0,i)=a_1$;
3. for all $0\leqslant r<r'$, $0\leqslant s<s'$, for all $\bar{\imath}\in[R_k]^r$ and for all $\bar{\jmath}\in[R_k]^s$,
$$\big((a_{r'+1},\bar{\imath})\neq(a_{s'+1},\bar{\jmath})\to\beta(a_{r'+1},\bar{\imath})\neq\beta(a_{s'+1},\bar{\jmath})\big);$$
4. for all $0\leqslant r<r'$, for all $\bar{\imath}\in[R_k]^r$ and for all $j\in R_k$
$$\big(\neq(\beta(a_{r'+1},\bar{\imath}),j,a_0,a_1)\big).$$

(This is trivially true for $k=0$ and $R_k=\varnothing$.) Given $R_k=\{i_0,i_1,\ldots,i_{k-1}\}$, we shall now show how to find i_k.

Let $Y_k\subseteq R_k$ be $\{i_j : j\in S_k\}$, and define $f_k:[Y_k]^{\geqslant 0}\to M$ by

$$f_k(i_{j_0},\ldots,i_{j_s})=g_k(j_0,\ldots,j_s).$$

It is necessary to find i_k satisfying:

1. $i_k>i_{k-1}$;
2. $\beta(a_0,i_k)=a_1$;
3. $\forall j\in R_k\ (\beta(j,i_k)=a_0)$;
4. for all $s\geqslant 0$ and for all $\bar{y}\in[Y_k]^s$, $\big(\beta(a_{s+1},\bar{y},i_k)=f_k(\bar{y})\big)$;
5. $(i_k\neq a_0\wedge i_k\neq a_1)$;
6. for all $0\leqslant r<r'$ and for all $\bar{\imath}\in[R_k]^r$, $\big(\beta(a_{r'+1},\bar{\imath})\neq i_k\big)$;
7. for all $0\leqslant r<r'$, for all $\bar{\imath}\in[R_k]^{r-1}$ and for all $j\in R_k$,
$$\big(\beta(a_{r'+1},\bar{\imath},i_k)\neq j\big);$$
8. for all $0\leqslant r<r'$, $0\leqslant s<s'$, for all $\bar{\imath}\in[R_k]^r$, and for all $\bar{\jmath}\in[R_k]^{s-1}$,
$$\big(\beta(a_{r'+1},\bar{\imath})\neq\beta(a_{s'+1},\bar{\jmath},i_k)\big);$$
9. for all $0\leqslant r<r'$, $0\leqslant s<s'$, for all $\bar{\imath}\in[R_k]^{r-1}$, and for all $\bar{\jmath}\in[R_k]^{s-1}$,
$$\big((a_{r'+1},\bar{\imath})\neq(a_{s'+1},\bar{\jmath})\to\beta(a_{r'+1},\bar{\imath},i_k)\neq\beta(a_{s'+1},\bar{\jmath},i_k)\big);$$
10. for all $0\leqslant r<r'$ and for all $\bar{\imath}\in[R_k]^{r-1}$, $\beta(a_{r'+1},\bar{\imath},i_k)\neq i_k$.

That is, we require i_k to satisfy a certain type in M over the parameters $a_0,a_1,i_0,\ldots,i_{k-1}$. Thus it suffices to check that this type is finitely satisfied (since M is recursively saturated). But any finite set of formulas from this type is $i_k > i_{k-1}$ together with a finite set of equalities and inequalities of one of the forms

1. $i_k \neq u$
2. $\beta(t_1(a_0,a_1,i_0,\ldots,i_{k-1}),i_k) = s$
3. $\beta(t_2(a_0,a_1,i_0,\ldots,i_{k-1}),i_k) \neq r$
4. $\beta(t_3(a_0,a_1,i_0,\ldots,i_{k-1}),i_k) \neq \beta(t_4(a_0,a_1,i_0,\ldots,i_{k-1}),i_k)$
5. $\beta(t_5(a_0,a_1,i_0,\ldots,i_{k-1}),i_k) \neq i_k$

for certain terms $t_1,\ldots,t_5$ involving the function β only, and various elements u,r,s of M. These sets of formulas are satisfied, for since M is infinite, we can find distinct values $\gamma(t_l(a_0,a_1,i_0,\ldots,i_{k-1}))$ in M for each $\beta(t_l(a_0,a_1,i_0,\ldots,i_{k-1}),i_k)$ mentioned above so that all the inequalities and equalities of the first four kinds above are satisfied when these values are substituted in. But then, by axioms of CFF($<$), the induction hypothesis 3 on p. 266, and the form that the terms $t_l()$ take, there is $i_k \in M$ greater than i_{k-1} and all the $\gamma(t_l(a_0,a_1,i_0,\ldots,i_{k-1}))$ such that

$$\beta(t_l(a_0,a_1,i_0,\ldots,i_{k-1}),i_k) = \gamma(t_l(a_0,a_1,i_0,\ldots,i_{k-1}))$$

for all $\gamma(t_l(a_0,a_1,i_0,\ldots,i_{k-1}))$. Since i_k is greater than all these γ, the last set of inequalities is satisfied too. □

We now do the induction step of the proof of the theorem. Fix an enumeration

$$\{c_0,c_1,\ldots\}$$

of the domain of $\mathfrak{M}$. At a given stage of the construction we will have

$$\Theta_n(a_0,a_1,b_0,b_1,\ldots,b_{n-1},c_0,c_1,\ldots,c_{n-1},R)$$

as above, consistent with $\mathrm{Th}(\mathfrak{M},\bar{a},\bar{b},\bar{c})$. By Theorem 3.1 on p. 250 there is $R \subseteq M$ satisfying $\Theta_n(\bar{a},\bar{b},\bar{c},R)$ such that $(\mathfrak{M},R)$ is recursively saturated.

The first step is to find an $(n+1)$-free $R' \subseteq R$ such that for all $i_0,\ldots,i_n \in [R']^{n+1}$

$$\mathrm{tp}(a_0,a_1,i_0,\ldots,i_{n-1},\beta(a_{n+1},i_0,\ldots,i_n)) = \mathrm{tp}(a_0,a_1,b_0,\ldots,b_{n-1},c_n).$$

Given such an R', we will then find an $(n+1)$-free $R'' \subseteq R'$ which is also $(n+1)$-indiscernible over $\{a_0,a_1\}$. This suffices for, taking any $b''_0,\ldots,b''_n \in [R'']^{n+1}$, we have:

$$\mathrm{tp}(a_0,a_1,b''_0,\ldots,b''_{n-1},\beta(a_{n+1},b''_0,\ldots,b''_n)) = \mathrm{tp}(a_0,a_1,b_0,\ldots,b_{n-1},c_n)$$

since $\bar{b}'' \in [R']^{n+1}$; so there is $g \in \mathrm{Aut}(\mathfrak{M}, a_0, a_1)$ with

$$g: b''_0, \ldots, b''_{n-1} \mapsto b_0, \ldots, b_{n-1}$$

and

$$g: \beta(a_{n+1}, b''_0, \ldots, b''_n) \mapsto c_n.$$

Thus the set $S = R''^g$ is $(n+1)$-free, $(n+1)$-indiscernible over a_0, a_1, and contains $b_0, \ldots, b_{n-1}$. By putting $b_n = b''^g_n$, S satisfies $\Theta_{n+1}(\bar{a}, \bar{b}, \bar{c}, S)$ as required.

So we must show how to construct R' and R''.

Construction of R'. We construct R' as $\{i_0, i_1, \ldots\}$ in increasing order. Enumerate all functions $f: [Y]^{\geqslant n+1} \to M$ with $Y \subseteq \mathbb{N}$ finite as $f_k: [Y_k]^{\geqslant n+1} \to M$, and assume without loss that $Y_k \subseteq \{0, 1, \ldots, k-1\}$. We will identify each $j \in \mathbb{N}$ with i_j as soon as i_j is defined.

Given $i_0 < i_1 < \cdots < i_{k-1}$ from R' we wish to find $i_k > i_{k-1}$ in R' such that

$$\forall \bar{x} \in [i_0, \ldots, i_{k-1}]^n \ \mathrm{tp}(a_0, a_1, \bar{x}, \beta(a_{n+1}, \bar{x}, i_k)) = \mathrm{tp}(a_0, a_1, \bar{b}, c_n)$$

and

$$\begin{aligned} &(\mathfrak{M}, a_0, a_1, i_0, \ldots, i_{k-1}, \beta(a_{\mathrm{len}(\bar{y})+1}, \bar{y}, i_k))_{\bar{y} \in [i_0, \ldots, i_{k-1}]^{\geqslant n+1}} \\ \equiv \ &(\mathfrak{M}, a_0, a_1, i_0, \ldots, i_{k-1}, f_k(\bar{y}))_{\bar{y} \in [i_0, \ldots, i_{k-1}]^{\geqslant n+1}}. \end{aligned}$$

Now, for each $\bar{x} \in [i_0, \ldots, i_{k-1}]^n$, $\mathrm{tp}(a_0, a_1, \bar{b}) = \mathrm{tp}(a_0, a_1, \bar{x})$, since R is n-indiscernible over $\{a_0, a_1\}$. Hence by homogeneity there is $c_{\bar{x}} \in M$ such that

$$\mathrm{tp}(a_0, a_1, \bar{b}, c_n) = \mathrm{tp}(a_0, a_1, \bar{x}, c_{\bar{x}}).$$

By applying the n-freeness of R to the function

$$g(\bar{x}) = \begin{cases} c_{\bar{x}} & \text{if } \mathrm{len}(\bar{x}) = n \\ f_k(\bar{x}) & \text{if } \mathrm{len}(\bar{x}) > n. \end{cases}$$

we can satisfy the two requirements above.

Construction of R''. Take R' with the properties just described such that $(\mathfrak{M}, R')$ is recursively saturated. Such R' exist, by chronic resplendency, since $(\mathfrak{M}, R)$ is recursively saturated. We simplify the problem by asking for R'' so that

$$\bigwedge_{r \in \mathbb{N}} R'' \text{ is } (n+1, r)\text{-free}$$

and

$$\bigwedge_{r \in \mathbb{N}} \bigwedge_{\ulcorner \theta \urcorner < r} \forall \bar{\imath}, \bar{\jmath} \in [R'']^{n+1} \ (\theta(a_0, a_1, \bar{\imath}) \leftrightarrow \theta(a_0, a_1, \bar{\jmath}))$$

where 'R'' is $(n+1,r)$-free' is defined precisely as for $(n+1)$-free except that we use

$$\bigwedge_{r\leqslant s\geqslant n+1} \forall y_1,\ldots,y_s\in R''(r)\ \forall f_1,\ldots f_t$$

$$\exists z\in R'' \bigwedge_{\ulcorner\theta\urcorner<r} \left(\theta(a_0,a_1,\bar{y},\beta_1,\ldots,\beta_t) \leftrightarrow \theta(a_0,a_1,\bar{y},f_1,\ldots,f_t)\right).$$

Here, $R''(r)$ is the set consisting of the r smallest elements of R'', and the β_i are as in the remarks before Lemma 3.3. Note in particular that an $(n+1,r)$-free set may be finite.

Let us show how to obtain an $(n+1,r)$-free $R''\subseteq R'$ such that

$$\bigwedge_{\ulcorner\theta\urcorner<r} \forall\bar{\imath},\bar{\jmath}\in[R'']^{n+1}\ \left(\theta(a_0,a_1,\bar{\imath}) \leftrightarrow \theta(a_0,a_1,\bar{\jmath})\right).$$

Since R' is $(n+1)$-free there are $s\in\mathbb{N}$ and $x_0,\ldots,x_{s-1}\in R'$ (listed in ascending order) such that $X=\{x_0,\ldots,x_{s-1}\}$ is $(n+1,r)$-free. X is made into a set-system by defining $f_i{:}[X]^i\to\alpha_i$ by $\alpha_i=\{0\}$ and $f_i(\bar{w})=0$ for $i\leqslant n+1$ and $f_{i+1}(w_0,\ldots,w_i)=$

$$\left\{\theta(u_0,u_1,v_0,\ldots,v_{i-1},v_i)\ \middle|\ \begin{array}{l}\ulcorner\theta\urcorner<r \text{ and } \mathfrak{M} \text{ satisfies}\\ \theta(a_0,a_1,w_0,\ldots,w_{i-1},\beta(a_{i+1},w_0,\ldots,w_{i-1},w_i))\end{array}\right\}$$

for $i\geqslant n+1$, noting that this last set is always finite. Now let $\mathfrak{A}=\mathrm{Patt}(X)$, and note that every subset of R' with pattern $\mathfrak{A}$ is $(n+1,r)$-free. Then Theorem 2.2 gives us $\mathfrak{B}$ with domain $\{\hat{y}_0,\hat{y}_1,\ldots,\hat{y}_{t-1}\}$ (listed in increasing order) such that

$$\mathfrak{B} \to (\mathfrak{A})^{n+1}_{2^r}.$$

We now attempt to use the $(n+1)$-freeness of R' to show that there is $Y\subseteq R'$ with $\mathfrak{B}=\mathrm{Patt}(Y)$. Suppose we have $y_0,\ldots,y_{j-1}\in R'$ with

$$\mathrm{Patt}(y_0,\ldots,y_{j-1})=\mathrm{Patt}(\hat{y}_0,\ldots,\hat{y}_{j-1}),$$

i.e.,

$$\hat{}:\{y_0,\ldots,y_{j-1}\}\to\{\hat{y}_0,\ldots,\hat{y}_{j-1}\}\subseteq\mathfrak{B}$$

is an order-isomorphism, and for all $i\geqslant n+1$ and all $\bar{w}\in[Y]^{i+1}$ $f_{i+1}(\hat{\bar{w}})=$

$$\left\{\theta(u_0,u_1,v_0,\ldots,v_{i-1},v_i)\ \middle|\ \begin{array}{l}\ulcorner\theta\urcorner<r \text{ and } \mathfrak{M} \text{ satisfies}\\ \theta(a_0,a_1,w_0,\ldots,w_{i-1},\beta(a_{i+1},w_0,\ldots,w_{i-1},w_i))\end{array}\right\}.$$

Give $\hat{y}_j$, the next element of the domain of $frakB$, we require $y_j\in R'$ so that, for all $\ulcorner\theta\urcorner<r$ and for all tuples $\bar{w}$ from $\{y_0,\ldots,y_{j-1}\}$ listed in increasing order,

$$\theta(a_0,a_1,\bar{w},\beta(a_{\mathrm{len}(\bar{w})+1},\bar{w},y_j)) \leftrightarrow \theta\in f_{len(\bar{w})+1}(\hat{\bar{w}},\hat{y}_j). \qquad (*)$$

Now, this is a finite collection of formulas in a free-variable y_j and parameters $a_0,a_1,y_0,\ldots,y_{j-1}$ that may or may not be satisfied by some $y_j \in R'$. If there is such a y_j we are finished (by induction), and if otherwise, then the axioms of CFF($<$) says there is a *unique maximal* set of formulas θ satisfying $(*)$ as above. In other words, if Ψ_j is the set of all formulas θ such that $(*)$ is true for *some* y_j, then there is a single y_j for which $(*)$ is true for all θ in Ψ_j. We take such a y_j. In general, we find $y_0,\ldots,y_{t-1} \in R'$ so that y_j satisfies the maximum number of formulas

$$\theta(a_0,a_1,\bar{w},\beta(a_{\mathrm{len}(\bar{w})+1},\bar{w},y_j))$$

for $\theta \in f_{\mathrm{len}(\bar{w})+1}(\hat{\bar{w}},\hat{y}_j)$ and $\bar{w} \in [y_0,\ldots,y_{j-1}]^{\geqslant n+1}$ given our choice of the tuple $y_0,\ldots,y_{j-1}$. At the end of this construction we will have found $\mathfrak{B}' = \{y_0,y_1,\ldots,y_{s-1}\}$ in R'. It follows from by the construction and the maximality of each set Ψ_j, and the fact that $\mathfrak{A}$ is the pattern of a finite subset of R', that $\mathfrak{B}' \to (\mathfrak{A})^{n+1}_{2^r}$. To see this, let $f\colon [y_0,y_1,\ldots,y_{s-1}]^{n+1} \to 2^r$, and let $\hat{f}\colon [\hat{y}_0,\hat{y}_1,\ldots,\hat{y}_{s-1}]^{n+1} \to 2^r$ be the map induced from f by the one-to-one correspondence $\hat{\ }$. Suppose that $\{\hat{y}_{i_1},\ldots,\hat{y}_{i_m}\}$ is the domain of a subsystem of $\mathfrak{B}$ which is relatively monochromatic for $\hat{f}$ and of pattern $\mathfrak{A}$. Then the corresponding $\{y_{i_1},\ldots,y_{i_m}\}$ is the domain of a subsystem which is obviously relatively monochromatic for f, but it is easy to see from the construction and an induction on j that

$$\mathrm{Patt}(y_{i_1},\ldots,y_{i_j}) = \mathrm{Patt}(\hat{y}_{i_1},\ldots,\hat{y}_{i_j})$$

for all $j \leqslant m$ as required.

The rest of the argument is standard. For Y as just constructed and $\bar{y} \in [Y]^{n+1}$ let $\chi(\bar{y}) = \sum_{i=0}^{r-1} \epsilon_i 2^i$ where

$$\epsilon_i = \begin{cases} 0 & \text{if } \theta_i(a_0,a_1,\bar{y}) \\ 1 & \text{otherwise} \end{cases}$$

and θ_i is the ith formula. Then there is $Z = \{z_0,\ldots z_{s-1}\} \subseteq Y$ with pattern $\mathfrak{A}$, hence $(n+1,r)$-free, and also relatively monochromatic for χ. But in this case it is actually monochromatic for χ since in the set-system Z, f_i is trivial for $i \leqslant n+1$.

It follows from these considerations that for all $r \in \mathbb{N}$ there is $R'' \subseteq R$ such that

1. for all θ with Gödel number less than r and for all $\bar{\imath},\bar{\jmath} \in [R'']^{n+1}$, $\theta(a_0,a_1,\bar{\imath}) \leftrightarrow \theta(a_0,a_1,\bar{\jmath})$, and
2. for all $y_1,\ldots,y_s \in R''(r)$, for all $f_1,\ldots f_t$, there is $z \in R''(F(r))$ such that

$$\bigwedge_{\ulcorner\theta\urcorner<r} (\theta(a_0,a_1,\bar{y},\beta_1,\ldots,\beta_t) \leftrightarrow \theta(a_0,a_1,\bar{y},f_1,\ldots,f_t)),$$

where $F(r)$ is a suitable upper bound for the cardinality of a $(n+1,r)$-free subset of R'. Since $F()$ is clearly primitive recursive, the family of all such formulas is a recursive one. Let R'' satisfy it, and let $R''(\omega)=\{x_0,x_1,\ldots\}$ be the first ω elements of R''. Then $R''(\omega)$ is $(n+1)$-free with the required properties.

The indiscernible sequence I. This completes the argument showing that if $\Theta_n(\bar{a},\bar{b},\bar{c},R)$ is consistent with the theory $\mathrm{Th}(\mathfrak{M},\bar{a},\bar{b},\bar{c})$, then for all c_n we can find $b_n\in M$ such that $\Theta_{n+1}(\bar{a},\bar{b},b_n,\bar{c},c_n,R)$ is consistent with $\mathrm{Th}(\mathfrak{M},\bar{a},\bar{b},b_n,\bar{c},c_n)$. The verification that $I=_{\mathrm{def}}\{b_0,b_1,\ldots\}$ has the required properties is now straightforward, since $c_0,c_1,\ldots$ exhausts the whole of M, and, by the construction, there are terms t_n involving β only such that $c_n=t_n(b_0,b_1,\ldots,b_n)$ for all $n\in\mathbb{N}$. The sequence I is clearly indiscernible since any finite subsequence of it is contained in some n-indiscernible sequence R, for all n.

The standard system. For an integer $n\in\mathbb{N}$ we let $\eta_n(x_0,x_1,x_2)$ be the $\mathscr{L}_{\mathrm{CFF}}$-formula that says that there are $y_0,y_1,\ldots,y_n$ such that

1. $y_0=x_0\wedge y_1=x_1$
2. $\bigwedge_{1\leqslant j<n}\beta(y_j,y_0)=y_{j+1}$
3. $\neq(\bar{y})$
4. $\beta(y_n,x_2)\neq y_0$

(Compare this with the construction of the sequence $a_0,a_1,\ldots$ given on p. 263.) It follows that CFF is a rich theory (in the sense of Jensen and Ehrenfeucht (1976), see p. 248), since for any finite disjoint sets $R,S\subseteq\mathbb{N}$

$$\mathrm{CFF}\vdash\exists x_0,x_1,x_2\left(\bigwedge_{n\in R}\eta_n(x_0,x_1,x_2)\wedge\bigwedge_{m\in S}\neg\,\eta_m(x_0,x_1,x_2)\right).$$

It follows that, for a recursively saturated model $\mathfrak{M}$ of CFF we may define its *standard system*, $\mathrm{SSy}(\mathfrak{M})$ to be

$$\mathrm{SSy}(\mathfrak{M})=_{\mathrm{def}}\left\{A\subseteq\mathbb{N}:\exists\bar{b}\in M\;\;A=\{n:M\vDash\eta_n(\bar{b})\}\right\}.$$

This allows us to apply general results from the theory of recursive saturation, and we use this now to find indescernible sequences I generating $\mathfrak{M}$ of almost any countable order type.

Other order types. The indiscernible sequence $(I,<)$ constructed so far has order-type ω. Suppose now that $(J,\prec)$ is any countable linearly ordered set with no greatest element. By compactness there is a model $\mathfrak{N}'$ for the language of $\mathfrak{M}$ such that $(J,\prec)$ is an indiscernible sequence for $\mathfrak{N}'$ and

$$\mathrm{tp}_{\mathfrak{N}'}(J)=\mathrm{tp}_{\mathfrak{M}}(I).$$

Let $N \subseteq N' = \mathrm{Dom}(\mathfrak{N}')$ be the closure of J under the function β. We claim that:

1. N is the domain of an elementary substructure $\mathfrak{N}$ of $\mathfrak{N}'$;
2. $\mathfrak{N}$ is recursively saturated; and
3. $\mathrm{SSy}(\mathfrak{M}) = \mathrm{SSy}(\mathfrak{N})$.

It follows from this and standard results (see Theorem 2.13 on p. 249) that $\mathfrak{N}$ and $\mathfrak{M}$ are back-and-forth equivalent, and in particular if J is countable (and so therefore is $\mathfrak{N}$) then $\mathfrak{N}$ and $\mathfrak{M}$ are isomorphic.

The claims are verified easily. For the first, consider typical elements $t_1(\bar{\jmath}),\ldots,t_r(\bar{\jmath})$ from N (where $\bar{\jmath} \in [J]^n$), and suppose

$$\mathfrak{N}' \models \exists x\, \theta(x, t_1(\bar{\jmath}),\ldots,t_r(\bar{\jmath})).$$

Then, for $b_0,\ldots,b_{n-1} \in [I]^n$,

$$\mathfrak{M} \models \exists x\, \theta(x, t_1(b_0,\ldots,b_{n-1}),\ldots,t_r(b_0,\ldots,b_{n-1}))$$

so for some $m \geqslant n$ and some term t_0 of $\mathscr{L}_{\mathrm{CFF}}$

$$\mathfrak{M} \models \theta(t_0(b_0,\ldots,b_{m-1}), t_1(b_0,\ldots,b_{n-1}),\ldots,t_r(b_0,\ldots,b_{n-1})).$$

Hence

$$\mathfrak{N}' \models \theta(t_0(\bar{\jmath},\bar{\jmath}'), t_1(\bar{\jmath}),\ldots,t_r(\bar{\jmath}))$$

for any suitably large $\bar{\jmath}' \in [J]^{m-n}$, which exist by the fact that $(J,\prec)$ has no largest element.

Now suppose $\bar{a} \in N$, and $\Phi(\bar{x},\bar{y})$ is a recursive set of $\mathscr{L}_{\mathfrak{M}}$-formulas such that $\Phi(\bar{x},\bar{a})$ is finitely satisfied in $\mathfrak{N}$ and $\Phi(\bar{x},\bar{y})$ is coded in $\mathfrak{M}$. Since $\bar{a} = \bar{f}(\bar{\jmath})$ for some $\mathscr{L}_{\mathrm{CFF}}$-terms $\bar{f}$ and some $\bar{\jmath} = j_0,\ldots,j_{n-1} \in [J]^n$,

$$\mathfrak{N} \models \exists \bar{x} \bigwedge_{\phi \in S} \phi(\bar{x}, \bar{f}(\bar{\jmath}))$$

for each finite subset $S \subseteq \Phi(\bar{x},\bar{y})$, so

$$\mathfrak{M} \models \exists \bar{x} \bigwedge_{\phi \in S} \phi(\bar{x}, \bar{f}(b_0,\ldots,b_{n-1}))$$

for all finite subsets $S \subseteq \Phi(\bar{x},\bar{y})$. But $\mathfrak{M}$ is recursively saturated and generated by I, so there are $m > n$ and $\mathscr{L}_{\mathrm{CFF}}$-terms $\bar{g}$ such that

$$\mathfrak{M} \models \bigwedge_{\phi \in \Phi} \phi(\bar{g}(b_0,\ldots,b_{m-1}), \bar{f}(b_0,\ldots,b_{n-1})).$$

Since J has no greatest element there are $\bar{\jmath}' = (j_n,\ldots,j_{m-1}) \in J$ such that

$$\mathfrak{N} \models \phi(\bar{g}(\bar{\jmath},\bar{\jmath}'), \bar{f}(\bar{\jmath}))$$

for each $\phi \in \Phi$ by virtue of $\mathrm{tp}_{\mathfrak{N}}(J) = \mathrm{tp}_{\mathfrak{M}}(I)$. Thus $\bar{x} = \bar{g}(\bar{\jmath}, \bar{\jmath}')$ realizes $\Phi(\bar{x}, \bar{a})$ in $\mathfrak{N}$. This shows that $\mathfrak{N}$ is recursively saturated. Finally, notice that for any $\bar{a} = \bar{f}(j_1, \ldots, j_n) \in N$ and any $i_1, \ldots, i_n \in I$ such that $i_k \mapsto j_k$ is an order-isomorphism from $\{\bar{\imath}\}$ to $\{\bar{\jmath}\}$,

$$\forall m \in \mathbb{N}\ (\mathfrak{N} \models \eta_m(\bar{f}(\bar{\jmath})) \Leftrightarrow \mathfrak{M} \models \eta_m(\bar{f}(\bar{\jmath}))),$$

so $\mathrm{SSy}(\mathfrak{N}) = \mathrm{SSy}(\mathfrak{M})$. □

The proof of the theorem is complete. Note that the last part of the proof together with one direction of Karp's theorem immediately yields the following corollary.

COROLLARY 3.4. Let $\mathfrak{M} = (M, \beta, <, \ldots) \models \mathrm{CFF}(<)$ be recursively saturated and countable. Then, for any linear order $(J, \prec)$ with no last element, there is a model $\mathfrak{N} \equiv_{\infty\omega} \mathfrak{M}$ with domain N, having $(J, \prec)$ as an indiscernible sequence $(J \subseteq N)$ such that $\mathfrak{N}$ is generated by J via the function β only.

Schmerl has extended this last result somewhat, building into the construction of $\mathfrak{N}$ further saturation properties (resplendency, and indeed also something apparently stronger still, which he calls 'total resplendency'). See Theorem 1 of Schmerl (1989). The argument there uses the doubtful Ramsey-style lemma (Lemma 2 on p. 1385), but the method presented above will however modify directly.

As already indicated, Schmerl's theorem is not just of interest for models which are already equipped with a beta function and a suitable linear order. Resplendency allows us to pass to an expansion of the original structure, adding an order relation, and possible also the beta function itself. (Of course, the automorphism group of the expansion will typically be a proper closed subgroup of the original one.)

Firstly, recall that any countable recursively saturated model of CFF can be expanded to a recursively saturated model of CFF($<$), so

COROLLARY 3.5. Let $(M, \beta, \ldots) \models \mathrm{CFF}$ be countable and recursively saturated. Then there is an order relation $<$ on M and an infinite sequence of indiscernibles $I \subseteq M$ such that M is generated by I via the function β. The order-type of I can be any countable linear order with no last element.

Secondly, if the model does not even have a β function we may add both the beta function and the order getting a recursively saturated expansion.

COROLLARY 3.6. For any countable recursively saturated model $\mathfrak{M}$ there is an embedding of groups $\mathrm{Aut}(\mathbb{Q}, <) \hookrightarrow \mathrm{Aut}(\mathfrak{M})$.

I would like to close this section mentioning a problem in this area which would seem to be worth investigation. In the last theorem no mention was made as to how 'large' the image of the embedding $\mathrm{Aut}(\mathbb{Q}, <) \hookrightarrow \mathrm{Aut}(\mathfrak{M})$ can be. In particular,

PROBLEM 3.7. If $\mathfrak{M}$ is a countable recursively saturated model of PA, is there an indiscernible sequence I of the same order type as $(\mathbb{Q},<)$ for which the natural embedding $\mathrm{Aut}(\mathbb{Q},<)\hookrightarrow\mathrm{Aut}(\mathfrak{M})$ has dense image in $\mathrm{Aut}(\mathfrak{M})$?

There may be other notions of 'large' that are interesting in this context too. Also, one can ask similar questions about other groups of the form $\mathrm{Aut}(I,<)$ which embed into $\mathrm{Aut}(\mathfrak{M})$ by the same argument.

4 Indiscernible sets

Under the additional assumption that $\mathrm{Th}(\mathfrak{M})$ is *stable* we now proceed to construct infinite indiscernible sets $I\subseteq M$ on which the full symmetric group $\mathrm{Sym}(I)$ is induced by $G=\mathrm{Aut}(\mathfrak{M})$.

But first, however, we mention a counterexample that shows that we cannot expect in general to have $\mathrm{Sym}(I)$ as a subgroup of G. Consider the case when $\mathfrak{M}$ is an algebraically closed field. (In the special case when $\mathfrak{M}$ is countable of transendence degree $\aleph_0$ over its prime subfield, it will be countably saturated and hence recursively saturated, and also ω-stable.) Suppose $g\in\mathrm{Aut}(\mathfrak{M})$ has order 3 for example. Then it is straightforward to check that the fixed field of g,

$$K=\{x\in M : xg=x\}$$

has codimension $[M{:}K]=3$, that is, M has dimension 3 as a K-vector space. This contradicts a result of Artin and Schreier (Jacobson, 1951, Vol III, p. 316) that states that the only possible finite codimension of such K is 2, and this only when K is real-closed with $M=K(\sqrt{-1})$.

We will give a proof of

THEOREM 4.1. (Kaye, 1992) Let $\mathfrak{M}$ be countably infinite, stable and recursively saturated. Then there is an infinite set of indiscernibles $I\subseteq M$ such that any permutation of I extends to an automorphism of $\mathfrak{M}$.

As already indicated, the set I so constructed 'generates' M—but in what sense? Lachlan originally proved the above result for $\aleph_0$-categorical stable $\mathfrak{M}$ using a modification of Fraïssé's notion of *age*. Since our language $\mathscr{L}$ may contain function symbols, let us agree to consider instead the language with relations $R_\phi(x_1,\ldots,x_n)$ for each $\mathscr{L}$-formula $\phi(x_1,\ldots,x_n)$ (and $\mathscr{L}$-structures are made into structures for the new language in the obvious way). We shall continue to refer to these structures as $\mathscr{L}$-structures, even though we now think of them in the new language. (In other words, we have reduced the problem to the case when the structure we consider is relational and has elimination of quantifiers.)

Given an $\mathscr{L}$-structure $\mathfrak{M}$ and an indiscernible set I of $\mathfrak{M}$, we say that I is *locally finite* iff for all finite $A\subseteq M$ there is a finite $B\subseteq I$ such that $I\smallsetminus B$ is indiscernible over $A\cup B$. Given such I, the *I-age of* $\mathfrak{M}$ is the set of all types

$$\mathrm{tp}_{\mathfrak{M}}(I\smallsetminus\{\bar b\}/\{\bar a,\bar b\})$$

where $\bar a\in M$, $\bar b\in I$ are both tuples of finite length, and $I\smallsetminus\{\bar b\}$ is indiscernible over $\{\bar a,\bar b\}$. (More precisely, the I-age of $\mathfrak{M}$ is the collection of all the sets of formulas

$$\theta(u_0,\dots,u_{k-1};v_0,\dots,v_{m-1};w_0,\dots,w_{n-1})$$

for which

$$\mathfrak{M}\models\theta(i_0,\dots,i_{k-1};\bar a;\bar b),\quad \bar\imath\in I\smallsetminus\{\bar b\},\quad k\in\omega,$$

one such set for each $\bar a$, $\bar b$ as above, $\bar u$, $\bar v$, $\bar w$ being fixed ω-sequences of variables.)

The following replaces 'homogeneity' in Fraïssé's Theorem: we say $\mathfrak{M}$ is *homogeneous over* I iff whenever $\bar b,\bar b'\in I$, $\bar a,\bar a'\in M\smallsetminus I$ are of finite length with $I\smallsetminus\{\bar b\}$ indiscernible over $\{\bar a,\bar b\}$ and $I\smallsetminus\{\bar b'\}$ indiscernible over $\{\bar a',\bar b'\}$, and also

$$\forall\bar\imath\in I\smallsetminus\{\bar b,\bar b'\}\ \mathrm{tp}_{\mathfrak{M}}(\bar a,\bar b,\bar\imath)=\mathrm{tp}_{\mathfrak{M}}(\bar a',\bar b',\bar\imath)$$

then there is $g\in\mathrm{Aut}(\mathfrak{M},I)$ (that is, g fixes I setwise) such that $g{:}\,\bar a,\bar b\mapsto\bar a',\bar b'$.

THEOREM 4.2. For all countably infinite recursively saturated stable $\mathfrak{M}$ there is an infinite locally finite indiscernible set I such that $\mathfrak{M}$ is homogeneous over I.

This proves Theorem 4.1, for we have

CLAIM. If $\mathfrak{M},\mathfrak{M}'$ are both countable, $I\subseteq M$ and $I'\subseteq M'$ are infinite locally finite indiscernible sets, $\pi{:}\,I\to I'$ is a bijection, $\mathfrak{M}$ is homogeneous over I and $\mathfrak{M}'$ is homogeneous over I', and I-age($\mathfrak{M}$) equals I'-age($\mathfrak{M}'$) then π extends to an isomorphism $\mathfrak{M}\to\mathfrak{M}'$. In particular this is true when $\mathfrak{M}=\mathfrak{M}'$ and π is an arbitrary permutation of I.

Proof. By back-and-forth: at any stage we have finite tuples $\bar a\in M,\bar a'\in M'$, $\bar b\in I$, $\bar b'\in I'$ such that

1. $\bar b'=\bar b^{\pi}$,
2. $I\smallsetminus\{\bar b\}$ is indiscernible over $\{\bar a,\bar b\}$,
3. $I'\smallsetminus\{\bar b'\}$ indiscernible over $\{\bar a',\bar b'\}$, and
4. $\forall\bar\imath\in I\smallsetminus\{\bar b\}\ \mathrm{tp}_{\mathfrak{M}}(\bar a,\bar b,\bar\imath)=\mathrm{tp}_{\mathfrak{M}'}(\bar a',\bar b',\bar\imath^{\pi})$.

Using the properties, we show how to extend this finite map $\bar a,\bar b\mapsto\bar a',\bar b'$. For example, if $\bar c\in M$ is given, let $\bar d\in I$ be such that $I\smallsetminus\{\bar b,\bar d\}$ is indiscernible over $\{\bar a,\bar b,\bar c,\bar d\}$. Since the I'-age of $\mathfrak{M}'$ contains the I-age of $\mathfrak{M}$ there are $\hat a,\hat b,\hat c,\hat d$ in M' (with $\hat b,\hat d\in I'$) such that

$$\forall\bar\imath\in I\smallsetminus\{\bar b,\bar d\}\ \mathrm{tp}(\hat a,\hat b,\hat c,\hat d,\bar\imath^{\pi})=\mathrm{tp}(\bar a,\bar b,\bar c,\bar d,\bar\imath).$$

By homogeneity there is $\rho \in \mathrm{Aut}(\mathfrak{M}', I')$ with $\bar{a}'^{\rho} = \hat{a}$, $\bar{b}'^{\rho} = \hat{b}$. Put $\bar{c}' = \hat{c}^{\rho^{-1}}$ and $\bar{d}' = \hat{d}^{\rho^{-1}}$ so $\bar{c}' \in M'$ and $\bar{d}' \in I'$ and

$$\forall \bar{\imath} \in I \setminus \{\bar{b}, \bar{d}\}\ \mathrm{tp}_{\mathfrak{M}}(\bar{a}, \bar{b}, \bar{c}, \bar{d}, \bar{\imath}) = \mathrm{tp}_{\mathfrak{M}'}(\bar{a}', \bar{b}', \bar{c}', \bar{d}', \bar{\imath}^{\pi}).$$

The map $\bar{d} \mapsto \bar{d}'$ might not be compatable with π, however, but homogeneity of $\mathfrak{M}'$ guarentees the existence of $\bar{c}'' \in M' \setminus I'$, $\bar{d}'' = \bar{d}^{\pi}$, and $g \in \mathrm{Aut}(\mathfrak{M}', I')$ with g fixing $\bar{a}', \bar{b}'$ pointwise and sending $\bar{c}', \bar{d}'$ to $\bar{c}'', \bar{d}''$. This (together with the fact that $I' \setminus \{\bar{b}', \bar{d}''\}$ is indiscernible over $\{\bar{a}', \bar{b}', \bar{c}'', \bar{d}''\}$) is enough to keep the back-and-forth argument going, and the claim is proved. □

Proof of Theorem 4.2. The idea, clearly, is to carry out the Fraïssé-style argument within the recursively saturated model $\mathfrak{M}$. (Compare the argument in Macpherson (1986*a*) where the Fraïssé argument is done simply in the 'monster model'; resplendency will not immediately yield the case for a countable recursively saturated model from this one since the property being considered is not Σ_1^1.) We will argue in a similar way to that in the previous section: at any given stage n of the constuction we will have chosen $\bar{b}_n$ to lie in our indiscernible set and we will also have considered some other tuple $\bar{c}_n$. The inductive assumption at this stage is that a finite conjunction of Σ_1^1 sentences $\Theta_n(\bar{b}_n, \bar{c}_n, R)$ is consistent with $\mathrm{Th}(\mathfrak{M}, \bar{b}_n, \bar{c}_n)$. We will have to ensure that the $\bar{b}$'s and $\bar{c}$'s eventually exhaust the whole of M, and that the $\bar{b}$'s form an indiscernible set with the required properties. These required properties will be 'forced' by dovetailing in several requirements in such a way as to ensure that each potential requirement is considered at some time in the (infinite) construction. As is typical with this sort of model-theoretic forcing, at stage i, the new Θ_{i+1} will be the old Θ_i with a finite number of new conjuncts that are added that stage.

To be precise, each $\Theta_n(\bar{b}_n, \bar{c}_n, R)$ will be the conjuction of 'R is infinite, each element of $\bar{b}_n$ is in R, and each element of $\bar{c}_n$ is not in R' with finitely many statements of one of the two forms:

1. $R \setminus \{\bar{u}\}$ is indiscernible over $\{\bar{u}, \bar{v}\}$;
2. $\forall \bar{\imath} \in R \setminus \{\bar{u}\}\ \mathrm{tp}_{\mathfrak{M}}(\bar{r}, \bar{s}, \bar{\imath}) = \mathrm{tp}_{\mathfrak{M}}(\bar{r}', \bar{s}', \bar{\imath})$

for suitable $\bar{u}, \bar{v}, \bar{r}, \bar{s}, \bar{r}', \bar{s}'$ from $\{\bar{b}_n, \bar{c}_n\}$.

The facts we will use from stability theory are the following ones due to Shelah:

RESULT 4.3. (1) If $\mathfrak{M}$ is an infinite model of a stable theory and I is an infinite indiscernible sequence (for some linear order) in $\mathfrak{M}$, then I is in fact an indiscernible set.

(2) If $\mathfrak{M}$ is a model of a stable theory and I is an infinite indiscernible set in $\mathfrak{M}$ then, for all finite $A \subset M$ there is a countable set $E \subset I \setminus A$ such that $I \setminus (E \cup A)$ is indiscernible over $E \cup A$.

For proofs of these facts see Shelah (1978, p. 111) or Pillay (1983, pp. 87–88). They are used in the following technical lemma:

LEMMA 4.4. Let $\mathfrak{M}$ be stable, countably infinite and recursively saturated. Let $\bar{b},\bar{c},\bar{d}\in M$ and let $\Theta(\bar{b},\bar{c},R)$ be Σ_1^1 over the indicated language such that

$$\mathfrak{M}\models \exists R\,\Theta(\bar{b},\bar{c},R).$$

Suppose also that, for any $\mathfrak{N},J$ with $\{\bar{b},\bar{c}\}\cup J\subseteq N$ and

$$(\mathfrak{N},J)\models \mathrm{Th}(\mathfrak{M},\bar{b},\bar{c})+\Theta(\bar{b},\bar{c},J),$$

we have

1. J is an infinite indiscernible set, $\bar{b}\in J$; and $\{\bar{c}\}\cap J=\varnothing$, and $J\smallsetminus\{\bar{b}\}$ is indiscernible over $\{\bar{b},\bar{c}\}$.
2. $\mathfrak{N}^*\succ\mathfrak{N}$ implies that $(\mathfrak{N}^*,J)\models\Theta(\bar{b},\bar{c},J)$.

Then $\mathfrak{M}\models\exists R\,(\Theta(\bar{b},\bar{c},R)$ and $R\smallsetminus\{\bar{b},\bar{d}\}$ is indiscernible over $\{\bar{b},\bar{c},\bar{d}\})$.

Proof. Let $(\mathfrak{M},J)\models\Theta(\bar{b},\bar{c},J)$ and let $(\mathfrak{M},J)\prec(\mathfrak{M}^*,J^*)$, where

$$(\mathfrak{M}^*,J^*)\models\Theta(\bar{b},\bar{c},J^*)$$

and J^* is uncountable. Such extensions exist by the compactness theorem, using the fact that J is infinite. By the results from stability just quoted, there is a countable $E\subseteq J^*\smallsetminus\{\bar{b}\}$ such that, in $\mathfrak{M}^*$,

$$J^*\smallsetminus(E\cup\{\bar{b}\})\text{ is indiscernible over }E\cup\{\bar{b},\bar{c},\bar{d}\}.$$

Since J is countable, $J^*\smallsetminus J$ is uncountable. Also, since $J^*\smallsetminus\{\bar{b}\}$ is indiscernible over $\{\bar{b},\bar{c}\}$, there are $\mathfrak{N}\succ\mathfrak{M}^*$ and $\pi\in\mathrm{Aut}(\mathfrak{N},J^*)$ such that π fixes $\{\bar{b},\bar{c}\}$ pointwise, and π sends E into $J^*\smallsetminus J$. It follows that $J\smallsetminus\{\bar{b},\bar{d}^\pi\}$ is indiscernible over $\{\bar{b},\bar{c},\bar{d}^\pi\}$, for if $\bar{\imath}\in J\smallsetminus\{\bar{b}\}$ then $\bar{\imath}^{\pi^{-1}}\notin E$ so, for all $\bar{\imath},\bar{\jmath}\in J\smallsetminus\{\bar{b},\bar{d}^\pi\}$, $\mathrm{tp}(\bar{\imath},\bar{b},\bar{c},\bar{d}^\pi)=\mathrm{tp}(\bar{\imath}^{\pi^{-1}},\bar{b},\bar{c},\bar{d})=\mathrm{tp}(\bar{\jmath}^{\pi^{-1}},\bar{b},\bar{c},\bar{d})=\mathrm{tp}(\bar{\jmath},\bar{b},\bar{c},\bar{d}^\pi)$. But

$$\mathrm{Th}(\mathfrak{M},\bar{b},\bar{c},\bar{d})=\mathrm{Th}(\mathfrak{N},\bar{b},\bar{c},\bar{d}^\pi)$$

and $(\mathfrak{N},J)\models\Theta(\bar{b},\bar{c},J)$ by the assumptions, so

$$\exists R\,(\Theta(\bar{b},\bar{c},R)\text{ and }J\smallsetminus\{\bar{b},\bar{d}\}\text{ is indiscernible over }\{\bar{b},\bar{c},\bar{d}\})$$

is consistent with $\mathrm{Th}(\mathfrak{M},\bar{b},\bar{c},\bar{d})$, hence true in $\mathfrak{M}$ by resplendency. □

REMARK. It is easy to see that, provided each Θ_n is of the form indicated just before Result 4.3, it satisfies the technical conditions of the lemma.

The proof of the theorem is carried out by dovetailing the following three steps, I, II and III.

I. *Adding a new element to the sequence* $\bar{b}$. Unless we force it so, the indiscernible set we will construct may be finite. To ensure this is not the case, add more elements to the sequence $\bar{b}$ as follows. Given $\Theta_i(\bar{b}_i,\bar{c}_i,R)$ at stage i, use resplendency to find $J \subseteq M$ satisfying $\Theta_i(\bar{b}_i,\bar{c}_i,J)$, and take any $d \in J \setminus \{\bar{b}_i\}$. Then add d to $\bar{b}_i$ (i.e., the new $\bar{b}_{i+1}$ is the old $\bar{b}_i$ concatenated with d) and let Θ_{i+1} be the conjunction of Θ_i with the statement that $R \setminus \{\bar{b}_i, d\}$ is indiscernible over $\{\bar{b}_i, d, \bar{c}_i\}$. Clearly this new Θ is consistent and satisfies the conditions in the lemma.

II. *Adding an arbitrary element d to $\{\bar{b}_i,\bar{c}_i\}$.* The $\bar{b}$'s and $\bar{c}$'s must exhaust the whole of M. Given $d \in M$ at stage i, assume that '$d \in R$' is not consistent with $\Theta_i(\bar{b}_i,\bar{c}_i,R)$ and the theory of $(\mathfrak{M},\bar{b}_i,\bar{c}_i)$, so that we cannot add d in the trivial way given in I. Thus Lemma 4.4 applies, and so

$$d \notin R \wedge \Theta_i(\bar{b}_i,\bar{c}_i,R) \wedge$$
$$J \setminus \{\bar{b}\} \text{ is indiscernible over } \{\bar{b}_i,\bar{c}_i,d\}$$

is consistent with the theory of $(\mathfrak{M},\bar{b}_i,\bar{c}_i,d)$. So we let Θ_{i+1} be the above statement, this time adding d to the tuple $\bar{c}_i$, giving $c_{i+1} = (c_i, d)$.

III. *Extending a finite map.* Throughout the construction we have to consider any potential finite maps of the form given by part (2) of the definition of 'homogeneous over I'. Since we do not know at any given stage what the outcome of the construction will be, it is important that we deal with all such *potential* maps, even though further conditions we may make later might prevent them having the properties indicated.

A 'potential map' is given as $\bar{d},\bar{e} \mapsto \bar{d}',\bar{e}'$, where $\{\bar{d},\bar{d}'\} \subseteq \{\bar{b}_i\}$, $\{\bar{e},\bar{e}'\} \subseteq \{\bar{c}_i\}$, and the following is consistent with the current $\Theta_i(\bar{b}_i,\bar{c}_i,R)$ and $\mathrm{Th}(\mathfrak{M},\bar{b}_i,\bar{c}_i)$:

$$R \setminus \{\bar{d}\} \text{ is indiscernible over } \{\bar{d},\bar{e}\}$$
$$R \setminus \{\bar{d}'\} \text{ is indiscernible over } \{\bar{d}',\bar{e}'\}$$
$$\bigwedge_{\theta} \forall \bar{\jmath} \in R \setminus \{\bar{d},\bar{d}'\}\ (\theta(\bar{d},\bar{e},\bar{\jmath}) \leftrightarrow \theta(\bar{d}',\bar{e}',\bar{\jmath}))$$

Let this statement be $\Psi(\bar{d},\bar{e},\bar{d}',\bar{e}',R)$, and let $f \in M$. Just as in the back-and-forth construction, we must extend $\bar{d},\bar{e} \mapsto \bar{d}',\bar{e}'$ to include f in its domain and range. We shall do the first case (that of the domain) here. Note also that by I and II we can assume that $f \in \{\bar{b}_i,\bar{c}_i\}$. For convenience, we reorder and relabel the tuples, writing $\bar{d}$ as $\bar{a},\bar{d}''$, and $\bar{d}'$ as $\bar{a}',\bar{d}''$ where $\{\bar{a}\} \cap \{\bar{a}'\} = \varnothing$, the map $\bar{d} \mapsto \bar{d}'$ sending $\bar{a} \mapsto \bar{a}'$ bijectively, and $\bar{d} \mapsto \bar{d}'$ restricted to $\{\bar{d}''\}$ being a permutation π in $\mathrm{Sym}\{\bar{d}''\}$. By resplendency, let $J \subseteq M$ satisfy Θ_i together with $\Psi(\bar{d},\bar{e},\bar{d}',\bar{e}',J)$. By compactness, there are $\mathfrak{N} \succ \mathfrak{M}$, and $\alpha \in \mathrm{Aut}(\mathfrak{N})$ such that

$$\alpha: \bar{a},\bar{a}',\bar{d}'',\bar{e} \mapsto \bar{a}',\bar{a},\bar{d}''^{\pi},\bar{e}'$$

and α fixes every other point j of J. This is because, for all $\bar{\jmath} \in J \setminus \{\bar{a}, \bar{a}', \bar{d}''\}$ and all formulas θ,

$$\begin{array}{ll} & M \models \theta(\bar{a},\bar{a}',\bar{d}'',\bar{e},\bar{\jmath}) \\ \text{iff} & M \models \theta(\bar{a},\bar{k},\bar{d}'',\bar{e},\bar{\jmath}) \text{ for all } \bar{k} \in J \setminus \{\bar{a},\bar{a}',\bar{d}'',\bar{\jmath}\} \\ \text{iff} & M \models \theta(\bar{a}',\bar{k},\bar{d}''^{\pi},\bar{e}',\bar{\jmath}) \text{ for all } \bar{k} \in J \setminus \{\bar{a},\bar{a}',\bar{d}'',\bar{\jmath}\} \\ \text{iff} & M \models \theta(\bar{a}',\bar{a},\bar{d}''^{\pi},\bar{e}',\bar{\jmath}) \end{array}$$

using our assumption Ψ on J. By inspection, $(\mathfrak{N}, J)$ satisfies Θ_i and Ψ.

Now consider $f \in \{\bar{b}_i, \bar{c}_i\} \setminus \{\bar{d}, \bar{e}\}$. If $f \in J$ then $f^\alpha \in J$ and we let Θ_{i+1} to be Θ_i together with

$$\begin{array}{l} R \setminus \{\bar{d}, f\} \text{ is indiscernible over } \{\bar{d}, f, \bar{e}\} \\ R \setminus \{\bar{d}', f^\alpha\} \text{ is indiscernible over } \{\bar{d}', f^\alpha, \bar{e}'\} \\ \bigwedge_\theta \forall \bar{\jmath} \in R \setminus \{\bar{d},\bar{d}',f,f^\alpha\}\, (\theta(\bar{d},f,\bar{e},\bar{\jmath}) \leftrightarrow \theta(\bar{d}',f^\alpha,\bar{e}',\bar{\jmath})) \end{array}$$

and this is consistent with the theory of $\mathfrak{M}$ as required.

Otherwise, assume that $f \notin J$. Since $f \in \{\bar{b}_i, \bar{c}_i\}$ it suffices to show how the map $\bar{d}, \bar{e} \mapsto \bar{d}', \bar{e}'$ extends to $\bar{b}_i, \bar{c}_i$. By our constuction, α is a permutation of $\bar{b}_i$. To simplify notation, reorder the tuple $\bar{c}_i$, writing it in such way that $\bar{c}_i^\alpha = (g_0, \ldots, g_r, h_0, \ldots, h_s)$ where $g \in M$ and $\bar{h} \in N \setminus M$. Then, in $\mathfrak{N}$, we have

$$\begin{array}{l} \bigwedge_\theta \forall \bar{\jmath} \in J \setminus \{\bar{b}_i\}\, (\theta(\bar{b}_i,\bar{c}_i,\bar{\jmath}) \leftrightarrow \theta(\bar{b}_i^\alpha,\bar{g}\bar{h},\bar{\jmath})) \land \\ J \setminus \{\bar{b}_i\} \text{ is indiscernible over } \{\bar{b}_i, \bar{g}\bar{h}\}. \end{array}$$

Call this formula $\Phi(\bar{b}_i, \bar{c}_i, \bar{g}, \bar{h}, J)$. It follows, by resplendency, that there are $J' \subseteq M$ and $\bar{h}' \in M$ satisfying $\Phi(\bar{b}_i, \bar{c}_i, \bar{g}, \bar{h}, J)$ for J' and $\bar{h}'$ in place of $J, \bar{h}$, together with Θ_i and Ψ. By Lemma 4.4 we can add $\bar{g}, \bar{h}'$ to $\bar{c}_i$, and let Θ_{i+1} be the conjunction of Θ_i, Ψ, Φ and the statement that

$$R \setminus \{\bar{b}_i\} \text{ is indiscernible over } \{\bar{b}_i, \bar{c}_i, \bar{g}, \bar{h}'\}.$$

This concludes the proof of the theorem. $\square$

The small index property and recursively saturated models of Peano arithmetic

Daniel Lascar

C.N.R.S.,
Université Paris 7,
URA 753, Tour 45-55,
2 Place Jussieu,
75251 Paris cédex 05,
France.
Electronic mail address: lascar@logique.jussieu.fr

1 Preliminaries

In this paper $\mathfrak{M}$, $\mathfrak{M}'$, $\mathfrak{M}_0$, etc., will denote countable structures whose universe is M, M' or M_0, etc. We will denote by $\mathrm{Aut}(\mathfrak{M})$ the group of automorphisms of $\mathfrak{M}$ and, if A is a subset of M, $\mathrm{Aut}_A(\mathfrak{M})$ the group of automorphisms of $\mathfrak{M}$ which leave A pointwise fixed. As usual, we may consider $\mathrm{Aut}(\mathfrak{M})$ as a topological group by deciding that a base for open subsets is the set

$$\{\alpha\cdot \mathrm{Aut}_A(\mathfrak{M}) : \alpha \in \mathrm{Aut}(\mathfrak{M}), A \subset M, A \text{ finite}\}.$$

We will make heavy use of the fact that $\mathrm{Aut}(\mathfrak{M})$ is a Polish group. As on p. 25 we will say that $\mathrm{Aut}(\mathfrak{M})$ (or $\mathfrak{M}$) *has the small index property* if every subgroup of $\mathrm{Aut}(\mathfrak{M})$ whose index is strictly less than $2^{\aleph_0}$ is open.

The small index property has been proved for several countable structures. See Hodges (1989) and Hodges *et al.* (1993) for more details. There exist countable $\aleph_0$-categorical structures which fail to have it (Lascar, 1991), see also p. 52 in this volume. Surprisingly, the case of uncountable saturated structures is easier, and it has been proved by Lascar and Shelah (1993) with the help of GCH. The question of the small index property is closely related to the possibility of recovering a structure from its automorphism group, and in the case of the recursively saturated model of Peano arithmetic, it has been raised by Kaye, Kossak and Kotlarski (Kaye *et al.*, 1991).

We will prove here the small index property for some recursively saturated models of Peano arithmetic. The technique of the proof, using a tree of generic automorphisms was invented by Shelah and developed by Hodges, Hodkinson, Lascar and Shelah (Hodges *et al.*, 1993) to prove the small index property for certain countable $\aleph_0$-categorical structures, in particular the random graph (the notion of existentially closed automorphism was not needed there), and also developed by Lascar and Shelah (1993) in the uncountable case. For the sake of self-completeness we will repeat the construction of the tree proving that a non-open subgroup of $\mathrm{Aut}(\mathfrak{M})$ has index $2^{\aleph_0}$. The only original part of this paper is the preparatory part (Section 2).

We begin with some notation. We fix a countable recursive language $\mathscr{L}$ containing the language of arithmetic. We denote by PA the theory consisting of axioms of Peano arithmetic, including the induction axioms for all formulas of $\mathscr{L}$. If $\mathscr{L}'$ is an extension of the language $\mathscr{L}$, $\mathrm{PA}(\mathscr{L}')$ will be the theory PA together with the induction axioms for formulas in $\mathscr{L}'$.

Let $\mathfrak{M}$ be a model of PA. If a and i are two points of M, then $a^{(i)}$ will denote the exponent of the $(i+1)^{\mathrm{st}}$ prime number in the prime decomposition of a. If f is a function which is definable (with parameters) in $\mathfrak{M}$ or in any expansion of $\mathfrak{M}$ satisfying all the Peano axioms, and c is an element of M such that the interval $[0, c[$ is included in the domain of f, then we will denote

$$\langle f(0), f(1), \ldots, f(c-1)\rangle = \prod_{i=0}^{c-1} p_i^{f(i)}$$

where p_i denotes the $(i+1)^{\mathrm{st}}$ prime number. Notice that this product is an element of M.

We will say that a function *belongs to* $\mathfrak{M}$ if its domain is coded by an element of M and it is definable in $\mathfrak{M}$ (equivalently, if its graph can be encoded by an element of M).

DEFINITION 1.1. (Kirby and Paris, 1977) Let $\mathfrak{M}$ be a model of Peano arithmetic. We say that $\mathbb{N}$ is *strong* in $\mathfrak{M}$ if, for any function f which belongs to $\mathfrak{M}$ and whose domain contains $\mathbb{N}$, there exists $d \in \mathfrak{M}$ such that, for all $i \in \mathbb{N}$, $f(i) > \mathbb{N}$ if and only if $f(i) > d$. We say that $\mathfrak{M}$ is *arithmetically saturated* if it is recursively saturated and $\mathbb{N}$ is strong in $\mathfrak{M}$.

It has been proved that for any consistent theory T in $\mathscr{L}$ containing PA, there exist recursively saturated models of T in which $\mathbb{N}$ is strong, and also recursively saturated models in which $\mathbb{N}$ is not strong (Kaye *et al.*, 1991; Kirby and Paris, 1977). Kaye (1991) gives background material on recursive saturation and standard systems (see also the chapter starting on p. 243 in this volume) and the other references already cited give further details.

The aim of this paper is to prove to following:

THEOREM 1.2. Any countable arithmetically saturated model of PA has the small index property.

Let us just give a corollary of Theorem 1.2.

COROLLARY 1.3. For any complete theory T containing PA in $\mathscr{L}$, there exist two recursively saturated models of T whose automorphisms groups are not algebraically isomorphic.

Proof. Let $\mathfrak{M}$ be an arithmetically saturated model of T and $\mathfrak{M}'$ a recursively saturated model of T in which $\mathbb{N}$ is not strong. We know from Corollary 5.5 of Kaye *et al.* (1991; see also Theorem 4.3 on p. 254) that there is no isomorphism from $\mathrm{Aut}(\mathfrak{M})$ onto $\mathrm{Aut}(\mathfrak{M}')$ which is bicontinuous. But, since $\mathfrak{M}$ has the small index property, every isomorphism from $\mathrm{Aut}(\mathfrak{M})$ onto $\mathrm{Aut}(\mathfrak{M}')$ is bicontinuous (Lascar, 1991, Corollary 2.8; see also Theorem 6.3 on p. 24). □

I wish to thank the referee for the detection of numerous mistakes in the previous version of this paper, and also for the improvements he has suggested.

2 The main proposition

The proof will follow the same lines as in Lascar and Shelah (1993). There, we took advantage of the fact that, in an uncoutable saturated structure, there are plenty of submodels of smaller cardinality. Here, we have to find the right notion of 'small submodels'.

DEFINITION 2.1. Let $\mathfrak{M}$ be a recursively saturated model of PA and A a subset of M. We say that A is *small* in $\mathfrak{M}$ if there exists an element a in M such that

$$A = \{a^{(i)} : i \in \omega\}.$$

We remark that, if A is small in $\mathfrak{M}$, then $\mathrm{Aut}_A(\mathfrak{M})$ is an open subgroup of $\mathrm{Aut}(\mathfrak{M})$, since it includes $\mathrm{Aut}_a(\mathfrak{M})$ for some a in M. We remark also that there are only countably many small subsets in $\mathfrak{M}$.

DEFINITION 2.2. Let $\mathfrak{M}_0$ be a small elementary submodel of $\mathfrak{M}$. $\mathrm{Aut}^*(\mathfrak{M}_0)$ denotes the set of automorphisms of $\mathfrak{M}_0$ which can be extended to automorphisms of $\mathfrak{M}$.

Although $\mathrm{Aut}(\mathfrak{M}_0)$ may have cardinality $2^{\aleph_0}$, we see than $\mathrm{Aut}^*(\mathfrak{M}_0)$ is always countable: if c is a point in M such that $M_0 = \{c^{(i)} : i \in \mathbb{N}\}$, and if α is an automorphism of $\mathfrak{M}$, then the restriction of α to M_0 is entirely determined by $\alpha(c)$.

DEFINITION 2.3. Let $\mathfrak{M}_0$ be a small elementary submodel of $\mathfrak{M}$ and let $(\alpha_1, \alpha_2, \ldots, \alpha_k)$ be a sequence of automorphisms of $\mathfrak{M}_0$. We say that $(\alpha_1, \alpha_2, \ldots, \alpha_k)$ is *existentially closed* if the two following conditions are fulfilled:

1. for all ℓ with $1 \leqslant \ell \leqslant k$, $\alpha_\ell \in \mathrm{Aut}^*(\mathfrak{M}_0)$;

2. for all formulas $\varphi(v_0, v_1, \ldots, v_k)$ with parameters in M_0, if there exist $\beta_1, \beta_2, \ldots, \beta_k$ in $\mathrm{Aut}(\mathfrak{M})$ extending $\alpha_1, \alpha_2, \ldots, \alpha_k$ and b in M such that

$$\mathfrak{M} \models \varphi(b, \beta_1(b), \beta_2(b), \ldots, \beta_k(b)),$$

then there exists c in M_0 such that

$$\mathfrak{M} \models \varphi(c, \alpha_1(c), \alpha_2(c), \ldots, \alpha_k(c)).$$

We will need the following lemma.

LEMMA 2.4. Let $\mathscr{L}'$ be a recursive language extending $\mathscr{L}$, $\mathfrak{M}'$ a recursively saturated model of $\mathrm{PA}(\mathscr{L}')$ and a an element of M'. Then there exists a small elementary submodel of $\mathfrak{M}'$ containing a.

Of course, we could drop the assumption '$\mathscr{L}'$ recursive' if we relativize the notion of recursivity.

Proof. This is just another application of the method of Henkin. We consider the set $\mathscr{A}$ of formulas with one free variable v_0 and one constant symbol x to denote the variable of a type to be constructed. Moreover, we demand that the formulas in $\mathscr{A}$ contain only occurrences of $x^{(i)}$ for integers i, and no occurrence of x itself. More precisely:

$$\mathscr{A} = \left\{ \varphi(v_0, x^{(i_1)}, \ldots, x^{(i_k)}) : \begin{array}{l} k \in \mathbb{N}, (i_1, i_2, \ldots, i_k) \in \mathbb{N}^k, \\ \varphi(v_0, v_1, \ldots, v_k) \text{ a formula of } \mathscr{L}' \end{array} \right\}.$$

Let

$$\{\psi_n(v_0) : n \in \omega\}$$

be a recursive enumeration of all formulas of $\mathscr{A}$. For all n in $\mathbb{N}$, define by induction a finite type p_n over a by:

$$p_0 = \{x^{(0)} = a\} \qquad \text{and} \qquad p_n = p_{n-1} \cup \{\exists v_0 \psi_n(v_0) \rightarrow \psi_n(x^{(i)})\}$$

where i is the smallest integer j such that $x^{(j)}$ does not occur in p_{n-1} nor in ψ_n. Then

$$p = \bigcup_{n \in \omega} p_n$$

is a consistent recursively enumerable type, and if b is a realization of it, $\{b^{(i)} : i \in \omega\}$ is a small elementary submodel of $\mathfrak{M}'$ containing a. □

We can now state the main proposition.

PROPOSITION 2.5. Let $\mathfrak{M}$ be an arithmetically saturated model of PA, $a_0, a_1, \ldots, a_k$ be elements in M, and assume that $\mathrm{t}(a_0) = \mathrm{t}(a_1) = \cdots = \mathrm{t}(a_k)$. Then there is a small elementary submodel $\mathfrak{M}_0$ of $\mathfrak{M}$ containing the a_i's, and an existentially closed sequence of automorphisms of $\mathfrak{M}_0$, $(\alpha_1, \alpha_2, \ldots, \alpha_k)$, such that $\alpha_1(a_0) = a_1, \alpha_2(a_0) = a_2, \ldots$ and $\alpha_k(a_0) = a_k$.

Proof. Let

$$\{\varphi_n : n \in \omega\}$$

be a recursive enumeration of all formula in $\mathscr{L}$ with one free variable. We first enlarge $\mathscr{L}$ to $\mathscr{L}'$ by adding a new two-place predicate symbol V; the theory T' will be the theory PA($\mathscr{L}'$) plus the following set of axioms:

$$\{\forall v_0(V(n, v_0) \leftrightarrow \varphi_n(v_0)) : n \in \mathbb{N}\}.$$

Obviously, the theory of $\mathfrak{M}$ is consistent with T'. So, by persistence, there is an expansion $\mathfrak{M}'$ of $\mathfrak{M}$ which is a model of T' and which is recursively saturated (see p. 250). Applying Lemma 2.4, we see that there exists a small elementary substructure $\mathfrak{M}'_0$ of $\mathfrak{M}'$ containing $a_0, a_1, \ldots, a_n$. We may as well assume that M_0, the base set of $\mathfrak{M}'_0$, contains a nonstandard element f.

Let c be an element of M such that $M_0 = \{c^{(i)} : i \in \mathbb{N}\}$. For b_0 and b_1 in M, define $I(b_0, b_1)$ to be the least element of the set

$$\{i : \mathfrak{M}' \models V(i, b_0) \not\leftrightarrow V(i, b_1)\} \cup \{f\}.$$

Now, if $(b_0, b_1, \ldots, b_k)$ is a sequence of elements in M, we define

$$I_k(b_0, b_1, \ldots, b_k) = \min\{I(b_p, b_q) : 0 < p \leqslant q \leqslant k\}.$$

We first remark that

$$I_k(b_0, b_1, \ldots, b_k) > \mathbb{N} \Longleftrightarrow \mathrm{t}(b_0) = \mathrm{t}(b_1) = \cdots = \mathrm{t}(b_k).$$

Secondly, the function I_k is definable and bounded in $\mathfrak{M}'_0$. We can now apply the strength of $\mathbb{N}$ in $\mathfrak{M}$; there exists an element d of M such that for all $i_0, i_1, \ldots, i_k \in \mathbb{N}$,

$$I_k(c^{(i_0)}, c^{(i_1)}, \ldots, c^{(i_k)}) > \mathbb{N} \Longleftrightarrow I_k(c^{(i_0)}, c^{(i_1)}, \ldots, c^{(i_k)}) > d;$$

that is, for all $b_0, b_1, \ldots, b_k \in M_0$,

$$I_k(b_0, b_1, \ldots, b_k) > \mathbb{N} \Longleftrightarrow I_k(b_0, b_1, \ldots, b_k) > d.$$

For $i \in M$, $i < c$, we define by induction inside $\mathfrak{M}'$ a sequence

$$(a_0(i), a_1(i), \ldots, a_k(i)).$$

First, let p and q be two definable functions from M to M and π a definable function from M^2 to M which are such that PA proves

$$n \longmapsto (p(n), q(n)) \text{ is a bijection from } \mathbb{N} \text{ to } \mathbb{N}^2 \text{ and } \pi \text{ is its inverse.}$$

We set $a_\ell(0) = a_\ell$ for $0 \leqslant \ell \leqslant k$. Assume that the $a_\ell(j)$'s have been defined for $0 \leqslant \ell \leqslant k$ and $j < i$, and denote $\langle a_\ell(0), a_\ell(1), \ldots, a_\ell(i-1)\rangle$ by $\overline{a_\ell}(i)$. If

b is an element of M, we will write $\langle \overline{a_\ell}(i)\hat{\ }b\rangle$ instead of

$$\langle a_\ell(0), a_\ell(1), \ldots, a_\ell(i-1), b\rangle.$$

Then

1. if $I_k(\langle \overline{a_0}(i)\hat{\ }c^{(x_0)}\rangle, \ldots, \langle \overline{a_k}(i)\hat{\ }c^{(x_k)}\rangle) > d$ for some $x_0, \ldots, x_k \in M$ such that
$$\mathfrak{M}' \models V(p(i), \langle c^{(q(i))}, c^{(x_0)}, \ldots, c^{(x_k)}\rangle),$$
set $(a_0(i), \ldots, a_k(i))$ equal to $(c^{(x_0)}, \ldots, c^{(x_k)})$ for the minimum such $(x_0, \ldots, x_k)$ under the lexicographic ordering of
$$(\max(x_0, x_1, \ldots, x_k), x_0, x_1, \ldots, x_k).$$
2. if not, then set $a_\ell(i) = 0$, for $0 \leqslant \ell \leqslant k$.

We first remark that, by induction on $i \in \mathbb{N}$, for all ℓ with $0 \leqslant \ell \leqslant k$, $a_\ell(i) \in M_0$. Assume this is true for all $j < i$. For definable subsets S of M or M_0, and working in $\mathfrak{M}'$ or $\mathfrak{M}'_0$ respectively, we define $\nu_S(t_0, t_1, \ldots, t_k) =$

$$\max\left\{ I_k(\langle t_0\hat{\ }y_0\rangle, \ldots, \langle t_k\hat{\ }y_k\rangle) : \begin{array}{l} y_0, \ldots, y_k \in S \text{ and} \\ \mathfrak{M}' \models V(p(i), \langle c^{(q(i))}, y_0, \ldots, y_k\rangle) \end{array} \right\}.$$

Since $\mathfrak{M}'_0 \prec \mathfrak{M}'$ and since the function $\nu_M(t_0, t_1, \ldots, t_k)$ is definable in $\mathfrak{M}'$ without parameters, we see that

$$\begin{aligned} \nu_M(\overline{a_0}(i), \ldots, \overline{a_k}(i)) &= \nu_{M_0}(\overline{a_0}(i), \ldots, \overline{a_k}(i)) \\ &= \nu_C(\overline{a_0}(i), \ldots, \overline{a_k}(i)) \end{aligned}$$

where $C = \{c^{(i)} : i \in M\}$, this last equality is because $M_0 \subset C \subset M$. (Note that ν_C is definable in $\mathfrak{M}'$.) Thus, if we are defining $(a_0(i), \ldots, a_k(i))$ and are in the first case (the other one is obvious), the sequence $x_0, x_1, \ldots, x_k$ is in $\mathbb{N}$ and $(a_0(i), \ldots a_k(i))$ is in M_0 (because of the minimality clause).

Next, observe that it follows from the remark in the last paragraph that for all $i \in \mathbb{N}$,

$$I_k(\overline{a_0}(i), \overline{a_1}(i), \ldots, \overline{a_k}(i)) > d.$$

since each $\overline{a_j}(i)$ is in M_0. Thus $\mathrm{t}(\overline{a_0}(i)) = \mathrm{t}(\overline{a_1}(i)) = \cdots = \mathrm{t}(\overline{a_k}(i))$ and, by resplendency, there exist $\beta_1, \beta_2, \ldots, \beta_k$ in $\mathrm{Aut}(\mathfrak{M})$ such that, for all $i \in \mathbb{N}$, $\beta_1(a_0(i)) = a_1(i)$, $\beta_2(a_0(i)) = a_2(i)$, $\ldots$, and $\beta_k(a_0(i)) = a_k(i)$.

For all ℓ with $0 \leqslant \ell \leqslant k$, let α_ℓ be the partial map from M to itself defined by $\alpha_\ell(a_0(i)) = a_\ell(i)$ for all $i \in \mathbb{N}$. Their images and domains are included in M_0, and they are elementary. Before proving that they are automorphisms of $\mathfrak{M}_0$, let us check that they satisfy condition 2 of Definition 2.3.

CLAIM. Let $\psi(v_0, v_1, \ldots, v_k)$ be a formula of $\mathscr{L}$ with parameters in M_0. Assume that there exist $\gamma_1, \gamma_2, \ldots, \gamma_k$ in $\mathrm{Aut}(\mathfrak{M})$ extending $\alpha_1, \alpha_2, \ldots, \alpha_k$

and $e \in M$ such that $\mathfrak{M} \models \psi(e, \gamma_1(e), \gamma_2(e), \ldots, \gamma_k(e))$. Then there exists $i \in \mathbb{N}$ such that

$$\mathfrak{M} \models \psi(a_0(i), a_1(i), \ldots, a_k(i)).$$

Proof. We may as well assume that there is only one parameter of M_0 in ψ. Say it is $c^{(n)}$, for some $n \in \mathbb{N}$. There exists a formula $\theta(v_0)$ in $\mathscr{L}$ such that

$$\mathrm{PA} \vdash \theta(\langle c^{(n)}, v_0, v_1, \ldots, v_k \rangle) \leftrightarrow \psi(v_0, v_1, \ldots, v_k)$$

and there exists $m \in \mathbb{N}$ such that $\theta(v_0) = \varphi_m(v_0)$. Consider the step $i = \pi(m, n)$ in the construction of the a_ℓ's. Certainly

$$I_k(\langle \overline{a_0}(i)\hat{\ } e \rangle, \langle \overline{a_1}(i)\hat{\ }\gamma_1(e) \rangle, \ldots, \langle \overline{a_k}(i)\hat{\ }\gamma_k(e) \rangle) > \mathbb{N}.$$

Since $\mathfrak{M}'_0$ is an elementary substructure of $\mathfrak{M}'$,

$$\nu_M(\overline{a_0}(i), \ldots, \overline{a_k}(i)) = \nu_{M_0}(\overline{a_0}(i), \ldots, \overline{a_k}(i))$$

so there exists $j_0, j_1, \ldots, j_k \in \mathbb{N}$ such that

$$I_k(\langle \overline{a_0}(i)\hat{\ } c^{(j_0)} \rangle, \langle \overline{a_1}(i)\hat{\ } c^{(j_1)} \rangle, \ldots, \langle \overline{a_k}(i)\hat{\ } c^{(j_k)} \rangle) > \mathbb{N}$$

and

$$\mathfrak{M}' \models V(p(i), \langle c^{(q(i))}, c^{(x_0)}, \ldots, c^{(x_k)} \rangle);$$

equivalently, by the strength of $\mathbb{N}$ in $\mathfrak{M}$ and the choice of d,

$$I_k(\langle \overline{a_0}(i), c^{(j_0)} \rangle, \langle \overline{a_1}(i), c^{(j_1)} \rangle, \ldots, \langle a_k(i), c^{(j_k)} \rangle) > d$$

and

$$\mathfrak{M} \models \varphi_m(\langle c^{(n)}, c^{(j_0)}, \ldots, c^{(j_k)} \rangle);$$

or, equivalently,

$$\mathfrak{M}' \models V(p(i), \langle c^{(q(i))}, c^{(j_0)}, \ldots, c^{(j_k)} \rangle).$$

Thus, we are in case 1 and the conclusion of the claim follows. □

To prove that the α_ℓ's are automorphisms, it suffices to prove that, for all $b \in M_0$, there exists an integer $i \in \mathbb{N}$ such that $b = a_\ell(i)$. We apply the claim to the formula $\psi =$ '$v_\ell = b$', $\gamma_1, \gamma_2, \ldots, \gamma_k = \beta_1, \beta_2, \ldots, \beta_k$ and e equal to b if $\ell = 0$ and to $\beta_\ell^{-1}(b)$ if not.

From the claim, it follows also that the sequence $(\alpha_1, \alpha_2, \ldots, \alpha_k)$ is existentially closed. □

3 Generic automorphisms

We introduce the following notation: if α and β are two automorphisms of $\mathfrak{M}$, or more generally two partial maps whose domains contain a subset A of M, we write '$\alpha \equiv \beta \bmod A$' in place of 'for all $a \in A, \alpha(a) = \beta(a)$'.

DEFINITION 3.1. Let $k \in \mathbb{N}$ and $(\alpha_1, \alpha_2, \ldots, \alpha_k)$ be a sequence of automorphisms of $\mathfrak{M}$. We say that $(\alpha_1, \alpha_2, \ldots, \alpha_k)$ is *generic* if the two following conditions are satisfied.

1. For each finite subset A of M there exists a small elementary submodel $\mathfrak{M}_0$ of $\mathfrak{M}$ such that $A \subset M_0$, $M_0 = \alpha_1[M_0] = \cdots = \alpha_k[M_0]$, and the sequence
$$(\alpha_1 \restriction M_0, \alpha_2 \restriction M_0, \ldots, \alpha_k \restriction M_0)$$
is existentially closed.
2. If $\mathfrak{M}_0$ is a small elementary submodel of $\mathfrak{M}$ such that
$$M_0 = \alpha_1[M_0] = \cdots = \alpha_k[M_0]$$
and $(\alpha_1 \restriction M_0, \alpha_2 \restriction M_0, \ldots, \alpha_k \restriction M_0)$ is existentially closed, $\mathfrak{M}_1$ is a small elementary submodel of $\mathfrak{M}$ containing M_0 and $(\beta_1, \beta_2, \ldots, \beta_k)$ is a sequence of automorphisms in $\mathrm{Aut}^*(\mathfrak{M}_1)$ such that, for all $1 \leqslant \ell \leqslant k$, β_ℓ extends α_ℓ, then there exists $\gamma \in \mathrm{Aut}_{M_0}(\mathfrak{M})$ such that, for $1 \leqslant \ell \leqslant k$, $\gamma \cdot \alpha_\ell \cdot \gamma^{-1}$ extends β_ℓ.

The following proposition is proved by back and forth.

PROPOSITION 3.2. Let $(\alpha_1, \alpha_2, \ldots, \alpha_k)$ and $(\beta_1, \beta_2, \ldots, \beta_k)$ be two generic sequences of automorphisms and $\mathfrak{M}_0$ a small elementary submodel of $\mathfrak{M}$. Assume that for $1 \leqslant \ell \leqslant k$, $\alpha_\ell[M_0] = \beta_\ell[M_0] = M_0$, $\alpha_\ell \equiv \beta_\ell \bmod M_0$ and that the sequence $(\alpha_1 \restriction M_0, \alpha_2 \restriction M_0, \ldots \alpha_k)$ is existentially closed. Then there exists $\gamma \in \mathrm{Aut}(\mathfrak{M})$ such that, for $1 \leqslant \ell \leqslant k$, $\alpha_\ell = \gamma \cdot \beta_\ell \cdot \gamma^{-1}$.

We will give more details for the next proposition.

PROPOSITION 3.3. Let k be a positive integer. Then the set
$$\{(\alpha_1, \alpha_2, \ldots, \alpha_k) \in \mathrm{Aut}^k(\mathfrak{M}) : (\alpha_1, \alpha_2, \ldots, \alpha_k) \text{ is a generic sequence } \}$$
is comeagre in $\mathrm{Aut}^k(\mathfrak{M})$.

Proof. Fix a finite subset A of M. Then the set of all $(\alpha_1, \alpha_2, \ldots, \alpha_k) \in \mathrm{Aut}^k(\mathfrak{M})$ such that there exists $\mathfrak{M}_0 \prec \mathfrak{M}$, $A \subset M_0$, M_0 small, $M_0 = \alpha_1[M_0] = \cdots = \alpha_k[M_0]$ and the sequence $(\alpha_1 \restriction M_0, \alpha_2 \restriction M_0 \ldots \alpha_k \restriction M_0)$ existentially closed is clearly open. Proposition 2.5 shows that it is everywhere dense. So the set of sequences of automorphisms satisfying condition 1 of Definition 3.1 is comeagre.

For the second condition, suppose that the triple
$$(\mathfrak{M}_0, \mathfrak{M}_1, (\alpha_1, \alpha_2, \ldots, \alpha_k))$$
satisfies: (1) $\mathfrak{M}_0$ and $\mathfrak{M}_1$ are small elementary submodels of $\mathfrak{M}$; (2) $\mathfrak{M}_0 \prec \mathfrak{M}_1$; (3) for $1 \leqslant \ell \leqslant k$, $\alpha_\ell \in \mathrm{Aut}^*(\mathfrak{M}_1)$ and $\alpha_\ell[M_0] = M_0$; and (4) the

sequence $(\alpha_1{\restriction}M_0, \alpha_2{\restriction}M_0, \ldots \alpha_k{\restriction}M_0)$ is existentially closed. Define

$$O_k(\mathfrak{M}_0, \mathfrak{M}_1, (\alpha_1, \alpha_2, \ldots, \alpha_k))$$

to be the set of all $(\beta_1, \beta_2, \ldots, \beta_k) \in \mathrm{Aut}^k(\mathfrak{M})$ for which there exists $\gamma \in \mathrm{Aut}_{M_0}(\mathfrak{M})$ such that, for $1 \leqslant \ell \leqslant k$, $\alpha_\ell \equiv \gamma{\cdot}\beta_\ell{\cdot}\gamma^{-1} \bmod M_1$, together with all $(\beta_1, \beta_2, \ldots, \beta_k) \in \mathrm{Aut}^k(\mathfrak{M})$ such that for some ℓ, we have $1 \leqslant \ell \leqslant k$ and $\alpha_\ell \not\equiv \beta_\ell \bmod M_0$. These sets are clearly open, and there is a countable number of them. It suffices to show that they are everywhere dense. This will follow easily from the next claim.

Claim. Let $(\mathfrak{M}_0, \mathfrak{M}_1, (\alpha_1, \alpha_2, \ldots, \alpha_k))$ be as above. Assume that the elements $c_0, c_1, \ldots, c_k$ are such that for $1 \leqslant \ell \leqslant k$ the partial map β_ℓ whose domain is $M_0 \cup \{c_0\}$ defined by $\beta_\ell \equiv \alpha_\ell \bmod M_0$ and $\beta_\ell(c_0) = c_\ell$ is elementary. Then, there exist $\gamma \in \mathrm{Aut}_{M_0}(\mathfrak{M})$ and $(\delta_1, \delta_2, \ldots, \delta_k) \in \mathrm{Aut}^k(\mathfrak{M})$ such that, for $1 \leqslant \ell \leqslant k$, δ_ℓ extends α_ℓ and $\gamma{\cdot}\delta_\ell{\cdot}\gamma^{-1}(c_0) = c_\ell$.

Proof. Let e and f be points of M such that $M_0 = \{e^{(i)} : i \in \mathbb{N}\}$ and $M_1 = \{f^{(i)} : i \in \mathbb{N}\}$. We know that for all ℓ with $1 \leqslant \ell \leqslant k$, α_ℓ can be extended to an automorphism of $\mathfrak{M}$; let f_ℓ be the image of f under such an automorphism.

What we have to do, is to find a sequence $(d_0, d_1, \ldots, d_k)$ such that

$$\mathrm{t}(d_0, d_1, \ldots, d_k/M_0) = \mathrm{t}(c_0, c_1, \ldots, c_k/M_0)$$

and, for all ℓ with $1 \leqslant \ell \leqslant k$, an automorphism $\delta_\ell \in \mathrm{Aut}(\mathfrak{M})$ extending α_ℓ and such that $\delta_\ell(d_0) = d_\ell$. We use the resplendency of $\mathfrak{M}$. It is sufficient to find, for any formula $\varphi(v_0, v_1, \ldots, v_k)$ with parameters in M_0 such that $\mathfrak{M} \models \varphi(c_0, c_1, \ldots, c_k)$, an element b in M and automorphisms β_ℓ of $\mathfrak{M}$ for $1 \leqslant \ell \leqslant k$ such that $\mathfrak{M} \models \varphi(b, \beta_1(b), \ldots, \beta_k(b))$ and, for all $i \in \mathbb{N}$, $\beta_\ell(f^{(i)}) = f_\ell^{(i)}$. Since the sequence $(\alpha_1{\restriction}M_0, \alpha_2{\restriction}M_0, \ldots \alpha_k{\restriction}M_0)$ is existentially closed, we may choose this element b in M_0 and β_ℓ to be any extension of α_ℓ to M. □

This finishes the proof of Proposition 3.3. □

4 The end of the proof

From now on, we will assume that $\mathfrak{M}$ is a strong, countable model of PA. Let $G = \mathrm{Aut}(\mathfrak{M})$. We will need the following lemmas.

Lemma 4.1. Assume that H is a subgroup of G, and that $[G{:}H] < 2^{\aleph_0}$. Then H is not meagre.

Proof. This is Theorem 4.1 of Hodges *et al.* (1993). □

Lemma 4.2. Let $(\alpha_1, \alpha_2, \ldots, \alpha_k)$ be a generic sequence of automorphisms. Then the set

$$X = \{\beta : (\alpha_1, \alpha_2, \ldots, \alpha_k, \beta) \text{ is generic}\}$$

is comeagre in G.

Proof. If we refer to Definition 3.1, we see that X is the intersection of countably many open sets. So, it suffices to prove that X is everywhere dense. Let a and b be two finite sequences in M such that $\mathrm{t}(a) = \mathrm{t}(b)$. We show that there is an automorphism $\beta \in X$ such that $\beta(a) = b$. Let $\mathfrak{M}_0$ be a small elementary submodel of $\mathfrak{M}$ such that $a \in M_0$, $b \in M_0$, and, for all ℓ with $1 \leqslant \ell \leqslant k$, $\alpha_\ell[M_0] = M_0$ and $(\alpha_1{\restriction}M_0, \alpha_2{\restriction}M_0, \ldots, \alpha_k{\restriction}M_0)$ is existentially closed. By Proposition 3.3, there exists a generic sequence $(\alpha'_1, \alpha'_2, \ldots, \alpha'_k, \beta')$ such that $\beta'(a) = b$ and for all ℓ with $1 \leqslant \ell \leqslant k$, $\alpha'_\ell \equiv \alpha_\ell \bmod M_0$. It is clear that $(\alpha'_1, \alpha'_2, \ldots, \alpha'_k)$ is also generic, so by Proposition 3.2 there exists an automorphism $\gamma \in \mathrm{Aut}_{M_0}(\mathfrak{M})$ such that, for all ℓ with $1 \leqslant \ell \leqslant k$, $\alpha'_\ell = \gamma{\cdot}\alpha_\ell{\cdot}\gamma^{-1}$. Set $\beta = \gamma^{-1}{\cdot}\beta'{\cdot}\gamma$. Then the sequence $(\alpha_1, \alpha_2, \ldots, \alpha_k, \beta)$ is generic and $\beta(a) = b$. □

LEMMA 4.3. Let $(\alpha_1, \alpha_2, \ldots, \alpha_k)$ be a generic sequence of automorphisms, and H a subgroup of G such that $[G{:}H] < 2^{\aleph_0}$. Then there exists $\beta \in H$ such that the sequence $(\alpha_1, \alpha_2, \ldots, \alpha_k, \beta)$ is generic.

Proof. Otherwise, by Lemma 4.2, H will be meagre, and by Lemma 4.1, have index $2^{\aleph_0}$ in G. □

LEMMA 4.4. Let $(\alpha_1, \alpha_2, \ldots, \alpha_k)$ be a generic sequence of automorphisms, H a non-open subgroup of G and O a non-empty open subset of G. Then there exists $\beta \in O \setminus H$ such that $(\alpha_1, \alpha_2, \ldots, \alpha_k, \beta)$ is generic.

Proof. Assume not. Then $H \cap O$ is a comeagre subset of O, and this implies that H is open (Lascar, 1991, proof of Lemma 2.6; see Theorem 6.2 on p. 24). □

We may now start our construction. We shall denote by $S = {}^{<\omega}2$ the set of finite sequences of 0s and 1s and by $\overline{S} = {}^{\omega}2$ the set of ω-sequences of 0 and 1. If s and t are two elements of $S \cup \overline{S}$, then $t \leqslant s$ and $t < s$ mean respectively that s is a initial segment and a proper initial segment of t.

We assume, working toward a contradiction, that $[G{:}H] < 2^{\aleph_0}$ and H is a non-open subgroup of G. We shall construct by induction on the length of $s \in S$:

- a family $(\mathfrak{M}_s; s \in S)$ of small elementary submodels of $\mathfrak{M}$;
- two families $(\alpha_s; s \in S)$ and $(\beta_s; s \in S \setminus \{\varnothing\})$ of automorphisms of $\mathfrak{M}$;

in such a way that

1. for all $s \in S$, $t \in S$, if $s < t$ then $\mathfrak{M}_s \prec \mathfrak{M}_t$;
2. for all $\sigma \in \overline{S}$, $\bigcup_{s<\sigma} M_s = M$;
3. for all $s \in S$, the sequence $(\alpha_{s\restriction 0}, \alpha_{s\restriction 1}, \ldots, \alpha_s)$ is generic;

4. for all s, s', t in S, if $t \leqslant s$ and $t \leqslant s'$, then $\beta_s \equiv \beta_{s'} \bmod M_t$;
5. for all s, s', t in S, if $t < s \leqslant s'$, then $\beta_s \cdot \alpha_t \cdot \beta_s^{-1} = \beta_{s'} \cdot \alpha_t \cdot \beta_{s'}^{-1}$;
6. for all $s \in S$, $\beta_{s\hat{}0} \cdot \alpha_s \cdot \beta_{s\hat{}0}^{-1} \in H$ and $\beta_{s\hat{}1} \cdot \alpha_s \cdot \beta_{s\hat{}1}^{-1} \notin H$.

Let $(a_n; n \in \omega)$ be an enumeration of M. We start by constructing $\alpha_\varnothing, \mathfrak{M}_\varnothing, \beta_0$ and β_1: $\alpha_\varnothing$ is a generic automorphism which belongs to H (Lemma 4.3); we choose $\mathfrak{M}_\varnothing$ such that $\mathfrak{M}_\varnothing$ is a small elementary submodel of $\mathfrak{M}$, $\alpha_\varnothing[M_\varnothing] = M_\varnothing$ and $\alpha_\varnothing \restriction M_\varnothing$ is existentially closed. We set $\beta_0 = \mathrm{id}_M$. Now, by Lemma 4.4 there exists a generic automorphism $\gamma \in \mathrm{Aut}_{M_0}(\mathfrak{M})$ such that $\gamma \notin H$, and by Proposition 3.2, there exists β_1 such that $\beta_1 \cdot \alpha_\varnothing \cdot \beta_1^{-1} = \gamma$.

Assume now that $\alpha_s, M_s, \beta_{s\hat{}0}$ and $\beta_{s\hat{}1}$ have been defined. Let t be either $s\hat{}0$ or $s\hat{}1$. We are going to define $\alpha_t, \mathfrak{M}_t, \beta_{t\hat{}0}$ and $\beta_{t\hat{}1}$.

We know that the index of $\beta_t^{-1} \cdot H \cdot \beta_t$ in G is less than $2^{\aleph_0}$. By Lemma 4.3 we can choose $\alpha_t \in \beta_t^{-1} \cdot H \cdot \beta_t$ in such a way that the sequence

$$(\alpha_\varnothing, \alpha_{s\restriction 1}, \ldots, \alpha_s, \alpha_t)$$

is generic. Let $\beta_{t\hat{}0}$ be equal to β_t, so that we know that $\beta_{t\hat{}0} \cdot \alpha_t \cdot \beta_{t\hat{}0}^{-1}$ belongs to H. Now the sequence

$$(\beta_t \cdot \alpha_\varnothing \cdot \beta_t^{-1}, \beta_t \cdot \alpha_{s\restriction 1} \cdot \beta_t^{-1}, \ldots, \beta_t \cdot \alpha_s \cdot \beta_t^{-1}, \beta_t \cdot \alpha_t \cdot \beta_t^{-1})$$

is also generic. Thus there exists a small elementary submodel $\mathfrak{M}'$ of $\mathfrak{M}$ such that

$$\beta_t[M_s \cup \{a_{\mathrm{len}(s)}\}] \subset M'$$

and

$$(\beta_t \cdot \alpha_\varnothing \cdot \beta_t^{-1} \restriction M', \beta_t \cdot \alpha_{s\restriction 1} \cdot \beta_t^{-1} \restriction M', \ldots, \beta_t \cdot \alpha_s \cdot \beta_t^{-1} \restriction M', \beta_t \cdot \alpha_t \cdot \beta_t^{-1} \restriction M')$$

is existentially closed. Let M_t be equal to $\beta_t^{-1}[M']$. By Lemma 4.4 there exists $\gamma \notin H$ such that $\gamma \equiv \beta_t \cdot \alpha_t \cdot \beta_t^{-1} \bmod M'$ and the sequence

$$(\beta_t \cdot \alpha_\varnothing \cdot \beta_t^{-1}, \beta_t \cdot \alpha_{s\restriction 1} \cdot \beta_t^{-1}, \ldots, \beta_t \cdot \alpha_s \cdot \beta_t^{-1}, \gamma)$$

is generic. By Proposition 3.2, there exists $\delta \in \mathrm{Aut}_{M'}(\mathfrak{M})$ such that $\delta \cdot \beta_t \cdot \alpha_t \cdot \beta_t^{-1} \cdot \delta^{-1} = \gamma$ and

$$\delta \cdot \beta_t \cdot \alpha_u \cdot \beta_t^{-1} \cdot \delta^{-1} = \delta \cdot \beta_t \cdot \alpha_u \cdot \beta_t^{-1} \cdot \delta^{-1} \text{ for all } u \leqslant s.$$

We set $\beta_{t\hat{}1} = \delta \cdot \beta_t$. Conditions 1 to 6 are satisfied, and this finishes the construction.

Let $\sigma \in \overline{S}$. Then the infinite sequence $(\beta_{\sigma\restriction n}; n \in \omega)$ is a Cauchy sequence (by conditions 2 and 4). Let β_σ be its limit. If s is an initial segment of σ the sequence $(\beta_{\sigma\restriction n} \cdot \alpha_s \cdot \beta_{\sigma\restriction n}^{-1}; n \in \omega)$ is eventually constant (condition 5) and its limit, who, by continuity, is equal to $\beta_\sigma \cdot \alpha_s \cdot \beta_\sigma^{-1}$ is

$\beta_{s\hat{}0}\cdot\alpha_s\cdot\beta_{s\hat{}0}^{-1}$ if $s\hat{}0 < \sigma$ and $\beta_{s\hat{}1}\cdot\alpha_s\cdot\beta_{s\hat{}1}^{-1}$ if $s\hat{}1 < \sigma$. Now let τ be any other element of $\overline{S}$, and let s be the largest common initial segment of σ and τ. Assume, for example, that $s\hat{}0 < \sigma$ and $s\hat{}1 < \tau$. Then

$$\beta_\sigma\cdot\alpha_s\cdot\beta_\sigma^{-1} \in H \text{ and } \beta_\tau\cdot\alpha_s\cdot\beta_\tau^{-1} \notin H.$$

It follows that $\beta_\sigma\cdot\beta_\tau^{-1}$ does not belong to H. Thus we have found $2^{\aleph_0}$ elements of G which are pairwise incongruent modulo H, and this contradicts the fact that $[G{:}H] < 2^{\aleph_0}$.

A Galois correspondence for countable recursively saturated models of Peano arithmetic

Richard Kaye

School of Mathematics and Statistics,
The University of Birmingham,
Edgbaston,
Birmingham,
B15 2TT,
U.K.
Electronic mail address: R.W.Kaye@bham.ac.uk

1 Introduction

In this paper I will present some new results classifying the normal subgroups of the automorphism groups of countable recursively saturated models of PA. These results can be considered as a Galois correspondence between closed normal subgroups and initial segments of the model.

PA is the first-order theory in the language $\mathscr{L}_{\mathrm{A}} = \{0,1,+,\cdot,<\}$ with the induction scheme for all first-order formulas and also certain basic axioms consisting of those for the theory Q, or (equivalently, modulo induction) PA^-, the axioms for the non-negative parts of discretely ordered rings (Kaye, 1991). In practice, we will work with the Skolemization of $\mathscr{L}_{\mathrm{A}}$ with extra function symbols

$$f_\theta(\bar{y}) = \begin{cases} (\mu x)\, \theta(x,\bar{y}) & \text{if such exists} \\ 0 & \text{otherwise} \end{cases}$$

and refer to 'terms', etc., for this expanded language. Given a model M of PA, the Skolem hull (or definable closure) over a set of parameters A will be denoted $\mathrm{cl}_M(A)$, i.e.,

$$\mathrm{cl}_M(A) =_{\mathrm{def}} \{t(\bar{a})^M : t \text{ a (Skolem-)term}, \bar{a} \in A\}.$$

For $A = \{\bar{a}\}$ we will use $\mathrm{cl}_M(\bar{a})$ instead of $\mathrm{cl}_M(A)$. Let M_0 be $\mathrm{cl}_M(\varnothing)$, the smallest elementary submodel of M. We use I_0 to denote the initial segment of M defined by M_0, i.e., $I_0 =_{\mathrm{def}} \{x \in M : \exists k \in M_0\, x \leqslant k\}$. Occasionally, it will be useful to use the notations $\leqslant a$ and $< a$ for $\{x \in M : M \vDash x \leqslant a\}$ and $\{x \in M : M \vDash x < a\}$ respectively.

In this paper, we shall consider a countable recursively saturated model M of PA and its automorphism group G. The group G is considered as a topological group in the usual way (see the section starting on p. 22 in this volume) and results are presented connecting the normal subgroup structure of G with the initial segments of M. In fact, all the results presented here will also be true in the more general case when M is a countable recursively saturated model of a *Peano-like* theory in a recursive language—i.e., M is the expansion of a model of PA and satisfies induction for all formulas of the language of M.

The overall plan of the paper is as follows. The rest of this section contains introductory remarks and definitions. Section 2 provides a complete account (with proofs) of the basic results connecting initial segments of M and normal subgroups of G. Section 3 presents a proof of the result of Kaye, Kossak and Kotlarski (Kaye *et al.*, 1991) classifying normal subgroups of the form $G_{(A)}$ for $A \subseteq M$, and also contains several new remarks concerning this classification. This material will also be found useful as motivation and as an introduction to the proof of the main theorem, which appears in Section 4. Section 5 sums up the main theorem and discusses further directions for research.

Since M is countable and recursively saturated, G is a group of cardinality $2^{\aleph_0}$, and because of the pairing function $\langle x,y\rangle = (x+y)(x+y+1)/2+y$ in PA it turns out that the usual topology on G can be given by taking as basic open subgroups the groups $G_a = \{g \in G : g(a) = a\}$ over $a \in M$. As usual, the subgroups $G_{(A)}$ and $G_{\{A\}}$ are, respectively, $\{g \in G : \forall a \in A\, g(a) = a\}$ and $\{g \in G : gA = A\}$ where $gA = \{g(a) : a \in A\}$. The set A is G-*invariant* (or simply *invariant* if G is clear) if $G_{\{A\}} = G$.

All the material needed on models of arithmetic and recursive saturation is standard, and can be found in Kaye (1991) for example. In particular, we shall need exponentiation $x^y = z$ (which is definable in $\mathscr{L}_{\mathrm{A}}$) and the 'stack of twos' function 2^a_n defined by $2^a_0 = a$ and $2^a_{n+1} = 2^{2^a_n}$ and also the iterated logarithm $\log^n x$ where $\log^0 x = x$ and $\log^{n+1} x = \lfloor \log_2(\log^n x + 1) \rfloor$. For a definable subset A of M, $\operatorname{card} A$ denotes the cardinality of A in the sense of M. If this is undefined (this only happens when A is unbounded in M) we write $\operatorname{card} A = \infty$. This ∞ is understood to satisfy the usual laws, such as $\infty + x = \infty$. Using a symbol for infinity in this way will mean that (with care) the same arguments work in both the 'bounded' and 'unbounded' cases. For a formula θ or type p, θ^M (respectively p^M) denotes the set of elements of M satisfying that formula (type).

A *cut* in M is an initial segment of M closed under the successor operation $x \mapsto x+1$. If S is a subset of M, $\sup S$ denotes $\{x \in M : \exists s \in S\, x \leqslant s\}$ and $\inf S$ denotes $\{x \in M : \forall s \in S\, x < s\}$. Any element a of M can be thought of as coding a sequence $(a)_0, (a)_1, \ldots$. For example, we could set $(a)_i$ to be the greatest y such that p_i^y divides a, where p_i is the ith prime. The following definition is standard in the literature on models of PA.

DEFINITION 1.1. A cut I is *ω-coded from above* (respectively, *from below*) if there is $a\in M$ coding a descending (respectively ascending) sequence such that $I=\inf\{(a)_n : n\in\mathbb{N}\}$ (respectively $I=\sup\{(a)_n : n\in\mathbb{N}\}$).

As examples, consider the cuts $2^a_{\mathbb{N}} =_{\text{def}} \sup\{2^a_n : n\in\mathbb{N}\}$ (coded from below) and $\log^{\mathbb{N}} a =_{\text{def}} \inf\{\log^n a : n\in\mathbb{N}\}$ (coded from above).

2 Initial segments and normal subgroups

We start off with a slightly unpromising looking definition that turns out to be of key importance.

DEFINITION 2.1. If $g\in G$ we define $\mathrm{I}_{\text{fix}}(g)=\{x\in M : \forall y<x\, g(y)=y\}$. For $H<G$, define $\mathrm{I}_{\text{fix}}(H)=\bigcap_{h\in H}\mathrm{I}_{\text{fix}}(h)$.

Thus $\mathrm{I}_{\text{fix}}(h)$ is the largest initial segment pointwise-fixed by h. We then ask: what $I\subseteq_{\text{e}} M$ occur as $\mathrm{I}_{\text{fix}}(h)$ for some $h\in G$? And: if I, J are initial segments of M, when is $G_{(I)}=G_{(J)}$?

The following proposition has been known for some time.

PROPOSITION 2.2. If $g\in G$ and $I=\mathrm{I}_{\text{fix}}(g)$ then I is closed under exponentiation.

Proof. Suppose that $g\in \mathrm{Aut}(M)$ fixes $\{x\in M : M\vDash x<a\}$ and $y<2^a$. Denote by '$u\in v$' the formula expressing 'the uth digit of the binary expansion of v is 1'. Then, for all $x\in M$, if $M\vDash x\in y$, we have $x<a$ so

$$\begin{aligned} & M\vDash x\in y \\ \Leftrightarrow\quad & M\vDash g(x)\in g(y) \\ \Leftrightarrow\quad & M\vDash x\in g(y),\ \text{since } x<a. \end{aligned}$$

This shows that y and $g(y)$ have the same binary expansion, i.e., they are equal. □

The next lemma (which was discovered independently by Kotlarski (1984), Smoryński (1982) and Vencovská (unpublished)) shows that this is essentially the only restriction on the largest initial segment fixed by g.

LEMMA 2.3. (Kotlarski–Smoryński–Vencovská) Let $a,\bar{b},\bar{c}\in M$, and suppose that for all $k\in\mathbb{N}$ and all formulas φ

$$M\vDash\forall x<2^a_k\,(\varphi(x,\bar{b})\leftrightarrow\varphi(x,\bar{c})).$$

Then there is g in G with $a\in\mathrm{I}_{\text{fix}}(g)$ and $g(\bar{b})=\bar{c}$.

Proof. By 'back-and-forth'. We do one direction only, as the other is similar. Let $d\in M$ be arbitrary. We will find $e\in M$ realizing the type

$$p(e)=\{\forall x<2^a_n\,\big(\varphi(x,\bar{b},d)\leftrightarrow\varphi(x,\bar{c},e)\big) : \varphi\in\mathscr{L}_{\mathrm{A}},\ n\in\mathbb{N}\}.$$

Since this type is recursive, it suffices to check that it is finitely satisfied in M. Let $\varphi_1(x,\bar{y},z),\ldots,\varphi_k(x,\bar{y},z)$ be $\mathscr{L}_{\mathrm{A}}$-formulas, and let $r\in\mathbb{N}$. For each $1\leqslant i\leqslant k$ let

$$s_i\in M\vDash\forall x<2^a_r\,\big(x\in s_i\leftrightarrow\varphi_i(x,\bar{b},d)\big).$$

Put $s_0=2^a_r$ and $s=\langle s_0,s_1,\ldots,s_k\rangle$. The coding of finite sequences can be chosen in such a way that $s<2^a_{r+2}$. Then

$$M\vDash\exists y\bigwedge_{i=1}^{k}\forall x<(s)_0\,\big(x\in(s)_i\leftrightarrow\varphi_i(x,\bar{b},y)\big).$$

So, by our assumption,

$$M\vDash\exists y\bigwedge_{i=1}^{k}\forall x<(s)_0\,\big(x\in(s)_i\leftrightarrow\varphi_i(x,\bar{c},y)\big)$$

hence there is $e\in M$ as required. □

LEMMA 2.4. (Smoryński) Let $a,\bar{b},\bar{c}\in M$ satisfy

$$M\vDash\forall x<2^{2^a}\,(\theta(x,\bar{b})\leftrightarrow\theta(x,\bar{c}))\qquad(\dagger)$$

for all $\theta\in\mathscr{L}_{\mathrm{A}}$, and let $d>2^{na}$ for all $n\in\mathbb{N}$. Then there are $b'<d$ and $c'\neq b'$ in M such that, for all $\theta\in\mathscr{L}_{\mathrm{A}}$

$$M\vDash\forall x<a\,(\theta(x,\bar{b},b')\leftrightarrow\theta(x,\bar{c},c')).$$

Proof. We must realize the type

$$p(y,z)=\{y\neq z\wedge y<d\}\cup\{\forall x<a\,(\theta(x,\bar{b},y)\leftrightarrow\theta(x,\bar{c},z)):\theta\in\mathscr{L}_{\mathrm{A}}\}.$$

Since this is recursive it suffices to show that it is finitely satisfied. Consider formulas $\theta_0,\ldots,\theta_{k-1}$ from $\mathscr{L}_{\mathrm{A}}$. There are at most ka formulas

$$\theta_i(x,\bar{u},v)\qquad(*)$$

where $i<k$ and $x<a$ (and $\bar{u}$, v are fixed free-variables). So there are at most 2^{ka} different sets of such formulas $(*)$ satisfied by $\bar{u}=\bar{b}$ and $v=$ some $b'<d$. Since $d>2^{ka}$, some such set is satisfied by at least two different $b'_0,b'_1<d$. Now let $w\in M$ be such that, for all $i<k$ and $x<a$, the (2^kx+2^i)th binary digit of w is one if $M\vDash\theta_i(x,\bar{b},b'_0)$ and zero otherwise. Such w can be chosen so that $w<2^{2^k\cdot a}$. Then applying $(\dagger)$ to the formula $\theta(w,\bar{u})=_{\mathrm{def}}$

$$\exists v\bigwedge_{i<k}\forall x\big((2^i+2^kx)\in w\rightarrow\theta_i(w,\bar{u},v)\big)$$

there is $c'\in M$ such that, for $i<k$,

$$M\vDash\forall x<a\,(\theta(x,\bar{b},b'_0)\leftrightarrow\theta(x,\bar{c},c')\leftrightarrow\theta(x,\bar{b},b'_1)).$$

Clearly $c' \neq b'_0$ or $c' \neq b'_1$. This shows $p(y,z)$ is finitely satisfied. □

EXERCISE 2.5. If $a \in M$, show that there are continuum-many $g \in \mathrm{Aut}(M)$ fixing $2^a_{\mathbb{N}}$ pointwise.

Smoryński's application of the last two lemmas was the following elegant theorem characterizing those initial segments closed under exponentiation.

THEOREM 2.6. (Smoryński) Let $I \subseteq_e M$ be closed under exponentiation. Then there is $g \in G$ with $\mathrm{I_{fix}}(g) = I$.

Proof. By back-and-forth. At any stage we have $a > I$ and $\bar{b}, \bar{c} \in M$ such that

$$\text{for all } \theta \in \mathscr{L}_{\mathrm{A}} \quad M \models \forall x < a\, (\theta(x,\bar{b}) \leftrightarrow \theta(x,\bar{c})). \qquad (*)$$

Using the proof of the Kotlarski–Smoryński–Vencovská Lemma, and noticing that, since $a > I$ and I is closed under exponentiation,

$$\lfloor \log a \rfloor, \lfloor \log^2 a \rfloor, \lfloor \log^3 a \rfloor, \ldots > I,$$

we have: for every $b' \in M$ there is $c' \in M$ and $a' > I$ such that

$$M \models \forall x < a'\, (\theta(x,\bar{b},b') \leftrightarrow \theta(x,\bar{c},c'))$$

and similarly for the 'back' direction. We also add a third part to the construction to ensure $I = \mathrm{I_{fix}}(f)$, where f is the automorphism being constructed. Given $a, \bar{b}, \bar{c}$ as in $(*)$, consider any $d > I$ from M. We show how to arrange that $d \notin \mathrm{I_{fix}}(f)$.

By replacing a with $\min(a,d)$ we may assume $d > a$, and since I is closed under exponentiation, $\log^2 a = a' > I$. Then the previous lemma gives $b' < d$ and $c' \neq b'$ such that for all $\theta \in \mathscr{L}_{\mathrm{A}}$

$$M \models \forall x < a'\, (\theta(x,\bar{b},b') \leftrightarrow \theta(x,\bar{c},c')).$$

Doing this for each $d > I$ in turn gives the required f. □

Denote by $\exp(I)$ the closure of the initial segment I under exponentiation. We shall also write $I \models \exp$ for $I = \exp(I)$. Then it follows from Smoryński's Theorem that, for initial segments I and J of M, $G_{(I)} = G_{(J)}$ iff $\exp(I) = \exp(J)$.

There is another way of getting a subgroup of G from an initial segment I of M. Denote by $G_{(>I)}$ the subgroup $\{g \in G : \mathrm{I_{fix}}(g) \supsetneq I\}$. It is clearly of interest to know how these relate to the subgroups $G_{(I)}$. To do this, we give first a lemma of independent interest.

LEMMA 2.7. Let $a, b \in M$ satisfy the same type, let $g \in G$ satisfy $g(a) = b$, let $I = \mathrm{I_{fix}}(g)$ and let $J = \inf_n j_n$ where

$$j_n = \max \left\{ j \in M \;:\; \forall \bar{x} < j \bigwedge_{\ulcorner \theta \urcorner < n} (\theta(\bar{x},a) \leftrightarrow \theta(\bar{x},b)) \right\}.$$

Then $I \subsetneq J$.

Proof. Clearly $J \supseteq I$. Also, by using a satisfaction class, the sequence j_n $(n \in \mathbb{N})$ is coded in M by j say. We shall show that $g \notin G_{(J)}$. Suppose otherwise. Since J is ω-coded from above in M, $J \neq \mathbb{N}$. Also, since

$$\exists \bar{x} \leqslant j_n+1 \bigvee_{\ulcorner \theta \urcorner < n} \neg(\theta(\bar{x},a) \leftrightarrow \theta(\bar{x},b)),$$

there is a canonical such $\bar{x}=x_{n1}\ldots x_{nn}$. (For example, take x_{n1} to be least such that $x_{n2}\ldots x_{nn}$ exist, x_{n2} least given this choice of x_{n1}, and so on.) Notice that (1) at least one x_{ni} is moved by g, (2) $\max(x_{n1}\ldots x_{nn})=j_n+1$, and (3) the sequence x_{ij} $(1 \leqslant j \leqslant i,\ i \in \mathbb{N})$ is coded in M, by x say. Put $x'=g(x)$. Then

$$\forall n \in \mathbb{N}\, \exists i \leqslant n\, ((x')_{ni} \neq (x)_{ni} \wedge (x)_{ni} \leqslant j_n+1)$$

and so, for some small nonstandard $\nu \in J$,

$$\forall n < \nu\, \exists i \leqslant n\, ((x')_{ni} \neq (x)_{ni} \wedge (x)_{ni} \leqslant j_n+1).$$

But $(x')_{ni}=g((x)_{ni})$ since $i,n<\nu$, and we are finished. □

Lemma 2.7 is essentially due to Kotlarski and a slightly simpler version of it appears in Kotlarski and Kaye (in press) as do parts (i) and (ii) of the following corollary. Part (iii) is a corrected version of a statement that appeared in Kaye *et al.* (1991).

COROLLARY 2.8. (i) For each initial segment I closed under exponentiation, the closure of $G_{(>I)}$ is $G_{(I)}$ in all cases except the anomolous cases when $I=\log^{\mathbb{N}}(a)$ for some $a \in M$.

(ii) If $I=\log^{\mathbb{N}}(a)$ where $a \in M$, then $G_{(>I)}$ is already closed, being $G_{(J)}$ where $J=2^a_{\mathbb{N}}$.

(iii) If I, J are closed under exponentiation and $I<J$ then $G_{(I)}>G_{(J)}$. If in addition we have that for no $a \in M$ is $I=\log^{\mathbb{N}}(a)$ and $J=2^a_{\mathbb{N}}$, then $G_{(I)}>G_{(>I)}>G_{(J)}$.

Proof. (i) Clearly $\overline{G_{(>I)}} \subseteq G_{(I)}$. Let $g \in G_{(I)}$, $a,b \in M$, $ga=b$. We must show that there is $h \in G_{(>I)}$ with $ha=b$. Let J be the cut defined in Lemma 2.7, so $J \supsetneq I$, and let $c \in J \smallsetminus I$. Then

$$M \vDash \forall x < c \bigwedge_{\theta} (\theta(x,a) \leftrightarrow \theta(x,b))$$

so by Lemma 2.3 there is $h \in G$ fixing $\log^{\mathbb{N}}(c)$ pointwise and $ha=b$. Since $I<\log^{\mathbb{N}}(c)$ we are finished.

(ii) If $g \in G_{(>I)}$ then g fixes $\{x : x < \log^n a\}$ for some n, hence by Proposition 2.2, $g \in G_{(J)}$.

(iii) Use Theorem 2.6. Note that in the second case, if $a \in J \smallsetminus I$, then $I < \log^{\mathbb{N}} a < J$. □

If we are interested in *normal subgroups* it quickly becomes clear that, for initial segments $I \models \exp$, $G_{(I)}$ and $G_{(>I)}$ are normal in G iff I is invariant. By saturation, it can be shown (Kossak, 1989, for example) that an initial segment I is invariant iff $I = \sup(I \cap M_0)$ or $I = \inf((M \smallsetminus I) \cap M_0)$. If the first of these holds we shall say that I is invariant *from below*, and if the second holds, I is invariant *from above*. If both happen we say that I is *doubly invariant*. (It is an easy exercise to show that there are continuum many cuts in M_0, hence there are continuum many invariant cuts in M, but clearly there are at most countably many invariant cuts that are not doubly invariant, since all of these latter cuts are of the form

$$\sup(\{x : x < a\} \cap M_0)$$

or

$$\inf(\{x : x > a\} \cap M_0)$$

for certain $a \in M$.)

By a similar argument to that given in Kaye (1991, pp.76–77) there are continuum-many invariant initial segments I closed under exponentiation, and the next proposition describes all possible order types of the set of such I ordered by inclusion.

PROPOSITION 2.9. The ordered set of invariant initial segments I closed under exponentiation is isomorphic to one of the following:

1. the ordinal 2;
2. $\xi + 1$;
3. $1 + \xi + 1$;

where ξ is the order-type of the Cantor set. All three possibilities can occur.

Proof. The order-type is 2 if there are no nonstandard definable elements (i.e., $M_0 = \mathbb{N}$), and in this case the invariant cuts are M and $\mathbb{N}$.

If $M_0 \neq \mathbb{N}$ then for each nonstandard $a \in M_0$, the cuts $\log^{\mathbb{N}}(a)$ and $2^a_{\mathbb{N}}$ are invariant (from above and below respectively), are closed under exponentiation and have no invariant cut closed under exponentiation between them. If $a \in I_0$ and there is no $b > a$ in $2^a_{\mathbb{N}} \cap M_0$ then

$$a^+ = \inf\{b \in M_0 : b > a\}$$

and

$$a^- = \sup\{b \in M_0 : b < a\}$$

are invariant, closed under exp and have no such cut between them. Moreover, these are the only ways that elements of the class $\mathscr{C}$ of invariant cuts closed under exponentiation can have successors and predecessors. It is clear that there are countably many of these exceptions, that these occur densely in $\mathscr{C}$, and that every non-trivial cut in $\mathscr{C}$ is the union or intersection of a collection of invariant cuts $\log^{\mathbb{N}}(a)$ and $2^a_{\mathbb{N}}$. So, for the case $M_0 \neq \mathbb{N}$, the only thing to check is whether the set of these 'exceptions' or 'gaps' has end points or not. (Recall that the Cantor set has endpoints.)

In the case we are considering, $M_0 \neq \mathbb{N}$, M is always a successor in $\mathscr{C}$. In other words, there is always a largest gap corresponding to $a \in M \setminus I_0$; $a^+ = \inf \varnothing =_{\text{def}} M$, and $a^- = \sup M_0 = I_0$. It remains to check that the two cases when $\mathbb{N} = a^-$ for some a smaller than all non-definable elements, and when there is no such a, i.e., when $\mathbb{N} = \inf(M_0 \setminus \mathbb{N})$, can both occur. These correspond to order-types having, or not having, a smallest 'gap'; that is, they correspond to the order-types $1+\xi+1$ and $\xi+1$ respectively. By saturation, it is easy to check that $\mathbb{N} = a^-$ for some a iff $M_0 \setminus \mathbb{N}$ is coded in M, i.e., the set of Gödel-numbers of formulas θ such that the Skolem term $t_\theta(0)$ is in $M_0 \setminus \mathbb{N}$ is in $\mathrm{SSy}(M)$.

To construct M where $M_0 \setminus \mathbb{N}$ is not coded, let L be any countable nonstandard model of PA in which Π_1-Th($\mathbb{N}$) is not coded, so nonstandard Δ_0 definable elements of L are downward-cofinal in $L \setminus \mathbb{N}$. Let T be any complete consistent extension of PA that is coded in L. Then by the arithmetized completeness theorem L has a recursively saturated end-extension $M \vDash T$. See Kaye (1991) for further explanation of notation and results used here.

Any countable recursively saturated model M with $\mathbb{N}$ strong in M (see p. 254) codes $M_0 \setminus \mathbb{N}$ (Kaye *et al.*, 1991), but to find such a model in which $\mathbb{N}$ is not strong argue as follows. Let $T \neq \mathrm{Th}(\mathbb{N})$ be a complete extension of PA, and let

$$A = \{\ulcorner t \urcorner : T \vdash t(0) > n \text{ for all } n \in \mathbb{N}\}.$$

Then T is recursive in A. There is a countable Scott set $\mathscr{X}$ not closed under arithmetical comprehension containing A. Since $T \in \mathscr{X}$, there is a recursively saturated countable model $M \vDash T$ with $\mathrm{SSy}(M) = \mathscr{X}$. A is coded by construction, but $\mathbb{N}$ is not strong in M since $\mathrm{SSy}(M)$ is not closed under arithmetical comprehension. □

3 The setwise-stabilizer of a type

The previous section considered ways of defining normal subgroups from invariant initial segments. Here we shall consider the apparently more general construction of subgroups $G_{(A)}$ where A is an invariant subset of M. In fact, no new subgroups arise in this way, and to show this is the main object of this section.

The material here also turns out to be a good introduction to understanding the main theorem later on. It also shows how to decide when two normal subgroups $G_{(A)}$ and $G_{(B)}$ are equal, or which is included in the other, by describing the initial segments corresponding to A and B. The exposition is an expanded version of that in Section 6 of Kaye *et al.* (1991).

Clearly, an invariant set A is a union of orbits, and (by saturation) each orbit is the set p^M of elements of M realizing a complete coded type $p(x)$. Thus the problem of understanding normal subgroups $G_{(A)}$ reduces to understanding the groups $G_{(p^M)}$ (which, for simplicity, we shall usually denote $G_{(p)}$).

We now give the definitions of two principal cuts associated with such a type $p(x)$. Suppose $p = \mathrm{tp}_M(a)$ where $a \in M \setminus M_0$. (It is not difficult to see that $(p^M, <) \cong (\mathbb{Q}, <)$. This can be proved directly or by using Lemma 4.1 below). Without loss of generality (using standard results on recursive saturation) we may assume p is equivalent to $\{\phi_n(x) : n \in \mathbb{N}\}$ where:

1. the function $n \mapsto \ulcorner \phi_n(x) \urcorner$ is coded in M;
2. $M \models \forall x\, (\phi_{n+1}(x) \rightarrow \phi_n(x))$ for each n.

We let $j_n = \mathrm{card}\, \phi_n$ denote the cardinality of ϕ_n^M in the sense of M (with the convention that $\mathrm{card}\, \phi_n = \infty$ if ϕ_n^M is unbounded in M). Since the sequence ϕ_n is coded, the definition of each j_n does not involve any parameters, and the model is recursively saturated, we may regard j_n as a coded sequence of elements of $M_0 \cup \{\infty\}$. The two main cuts associated with p are

1. $J_p =_{\mathrm{def}} \inf\{j_n : n \in \mathbb{N}\}$
2. $I_p =_{\mathrm{def}} \sup(J_p \cap M_0)$.

In the special case when $j_n = \infty$ for all n, we define $I_p = J_p = M$. Note that, when p^M is bounded, $\mathbb{N} \subseteq_{\mathrm{e}} I_p \subseteq_{\mathrm{e}} J_p$, I_p is invariant from below, J_p is invariant from above (and ω-coded from above). We shall see that I_p, J_p are closed under $+$, but not necessarily under $\cdot$. Also, it is clear that $J_p \neq \mathbb{N}$ (an overspill or saturation argument) and that I_p is ω-coded from below if and only if $I_p \neq J_p$.

Given a formula $\theta(x)$ and an element $a \in M$ satisfying θ, we shall use the term *θ-index of a* to mean the number $\mathrm{card}\{x < a : M \models \theta(x)\}$ (where card is taken in the sense of the model).

PROPOSITION 3.1. Both J_p and I_p are closed under $+$.

Proof. It suffices to show this for J_p. Now the statement 'the ϕ_n-index of x is odd' is a formula $\theta(x)$ with

$$\mathrm{card}(\theta \wedge \phi_n)^M, \mathrm{card}(\neg\theta \wedge \phi_n)^M \leqslant \lceil \mathrm{card}(\phi_n)^M / 2 \rceil.$$

Moreover, if ϕ_n is in p, then either $\theta(x) \wedge \phi_n(x)$ or $\neg\theta(x) \wedge \phi_n(x)$ is in p, so there is $\phi_k \in p$ with

$$\mathrm{card}\, \phi_k^M \leqslant \lceil \mathrm{card}\, \phi_n^M / 2 \rceil$$

and hence the result. □

PROPOSITION 3.2. Suppose J is ω-coded from above, is invariant from above, and is closed under $+$. Then $J=J_p$ for some $p=\mathrm{tp}(b)$, where $b\in I_0\setminus M_0$. In fact, given $v,w\in M_0$ with $w>v+J =_{\mathrm{def}} \{x:\exists y\in J\, x\leqslant v+y\}$, a suitable b can be found in the interval $[v,w]$ and, if in addition J is not doubly invariant, b can be found in the cut $v+J$.

Proof. Suppose $J=\inf\{(a)_n:n\in\mathbb{N}\}$, and assume without loss of generality that $(a)_n$ is a properly descending coded sequence of elements of M_0. We build $p=\{\phi_n(x):n\in\mathbb{N}\}$ so that, for each ϕ_n,

$$\exists l\in\mathbb{N}\ \mathrm{card}\{x:\phi_n(x)\wedge v\leqslant x<w\}>(a)_l$$

as follows.

First, $\phi_0(x)$ is '$v\leqslant x<w$', so

$$\mathrm{card}\{x:\phi_0(x)\}>w-v>J.$$

Next, $\phi_{2n+1}(x)$ is $\phi_n(x)\wedge\theta_n(x)$ or $\phi_n(x)\wedge\neg\theta_n(x)$ where θ_n is the nth formula in some fixed coded enumeration of all formulas with one free variable, and θ_n or $\neg\theta_n$ is chosen to maximize $\mathrm{card}\{x:\phi_{2n+1}(x)\}$. Since

$$\mathrm{card}\{x:\phi_{2n+1}(x)\}\geqslant\lceil\mathrm{card}\{x:\phi_{2n}(x)\}/2\rceil\geqslant\lceil(a)_l/2\rceil$$

for some l, and since J is closed under $+$, there is $l'\geqslant l$ with

$$\mathrm{card}\{x:\phi_{2n+1}(x)\}>(a)_{l'}.$$

Finally, $\phi_{2n+2}(x)$ is $\phi_{2n+1}(x)\wedge\mathrm{card}\{y:\phi_{2n+1}(y)\wedge y<x\}\leqslant(a)_n$. Thus if $\mathrm{card}\{x:\phi_{2n+1}(x)\}>(a)_l$,

$$\mathrm{card}\{x:\phi_{2n+2}(x)\}>\max((a)_l,(a)_n).$$

Note that, since the sequence θ_n is coded, $p(x)=\{\phi_n(x):n\in\mathbb{N}\}$ is also coded. It is consistent and complete by construction, and so is satisfied in M and moreover $J=J_p$ is easily verified.

The case when J is not doubly invariant is similar. Let $b\in J$ be greater than all definable elements of J, and construct p in a similar way:

$$\phi_0(x) =_{\mathrm{def}} v<x,$$

and

$$\phi_{2n+2} =_{\mathrm{def}} \phi_{2n+1}(x)\wedge\mathrm{card}\{y:\phi_{2n+1}(y)\wedge y<x\}\leqslant(a)_n.$$

This time though, we define $\phi_{2n+1}(x)$ to be $\phi_n(x)\wedge\theta_n(x)$ or $\phi_n(x)\wedge\neg\theta_n(x)$ chosen to maximize $\mathrm{card}\{x:\phi_{2n+1}(x)\wedge x<b+v\}$. Let $I=\sup(M_0\cap J)$. We shall verify that

$$\mathrm{card}\{x:\phi_n(x)\wedge x<b+v\}>I$$

for all n; it will follow that

$$\mathrm{card}\{x : \phi_n(x)\} > J$$

since $\mathrm{card}\{x : \phi_n(x)\}$ is in M_0. But

$$\mathrm{card}\{x : \phi_0(x) \wedge x < b+v\} = b > I$$

and

$$\mathrm{card}\{x : \phi_{2n+1}(x) \wedge x < b+v\} \geqslant \lceil \mathrm{card}\{x : \phi_{2n}(x) \wedge b < b+v\}/2 \rceil > I$$

since I is closed under $+$, and finally

$$\mathrm{card}\{x : \phi_{2n+2}(x) \wedge x < b+v\} = \mathrm{card}\{x : \phi_{2n+1}(x) \wedge x < b+v\}$$

since $b \in J$. □

Can the cuts I_p and J_p actually be equal? More generally, are there doubly invariant cuts $I \subseteq_{\mathrm{e}} M$ that are ω-coded from above (or below)? The answer is provided by the next two results.

RESULT 3.3. Suppose $\mathbb{N}$ is strong in M, and let $I \subseteq_{\mathrm{e}} M$ be doubly invariant. Then I is not ω-coded in M.

RESULT 3.4. Suppose $\mathbb{N}$ is not strong in M. Suppose also $u, v \in M_0$ are such that $[u, v] \cap M_0$ is infinite. Then there are doubly invariant $I, J \subseteq_{\mathrm{e}} M$ with $u \in I, J < v$ such that I is ω-coded from below and J is ω-coded from above.

Under the additional assumption that there is an invariant cut between u, v and closed under $+$, ($\cdot$, exp, etc.) we can easily modify this construction to give I, J that are closed under $+$ (respectively $\cdot$, exp, etc.). Results very similar to the last two were also observed (independently) by Kossak, Kotlarski and Schmerl (Kossak *et al.*, 1993) and the proofs are omitted here.

We now get back on track, and look at normal subgroups again.

DEFINITION 3.5. We define $G_{(p)}$ to be the subgroup consisting of all $g \in G$ that fixes the set p^M pointwise. $G_{(>p)}$ is the set of all $g \in G$ that fixes ϕ_n^M pointwise, for some $n \in \mathbb{N}$, i.e.,

$$G_{(>p)} =_{\mathrm{def}} \bigcup_{\phi(x) \in p(x)} G_{(\phi^M)}$$

where ϕ^M is the set of elements of M satisfying ϕ.

The following is from Kaye *et al.* (1991).

THEOREM 3.6. (Kaye, Kossak and Kotlarski) If $a \in M \setminus M_0$ and $p(x) = \mathrm{tp}(a)$ then

$$G_{(p)} = G_{(J_p)} \qquad \text{and} \qquad G_{(>p)} = G_{(>J_p)}.$$

Proof. To show that $G_{(p)} = G_{(J_p)}$, suppose at first that $g \in G_{(J_p)}$, and $a \in p^M$. We must show that $g(a) = a$. Obviously, we may assume $J_p \neq M$, equivalently $a \in I_0$, and by re-indexing the formulas ϕ_i we may assume that $\mathrm{card}\,\phi_0^M < \infty$. Then there is $S \in M$ coding the sequence of sets

$$S = \langle \phi_0^M, \phi_1^M, \ldots, \phi_n^M, \ldots \rangle$$

to nonstandard length. Since each $S_n = \phi_n^M$ is definable without parameters, $g(S_n) = S_n$ for each $n \in \mathbb{N}$. Putting $S' = g(S)$, and using overspill, $S'_i = S_i$ for all $i < \nu$, where ν is a nonstandard element of J_p. Hence $g(S_i) = S_i$ for all $i < \nu$ since $g(S_i) = S'_{g(i)} = S'_i$ for $i < \nu$. Now $a \in S_n$ for all standard n, so $a \in S_i$ for some nonstandard $i < \nu$. But for some $k \in J_p$, a is the kth element of S_i, $g(a)$ is the $g(k)$th element of $g(S_i) = S_i$, and $k = g(k)$, so $g(a) = a$.

For the converse, suppose $g \notin G_{(J_p)}$, $j \in J_p$ is moved by g, and suppose without loss that $g(j) > j$. Let x_n be the jth element of ϕ_n^M, so by saturation the sequence

$$x = \langle x_0, x_1, \ldots, x_n, \ldots \rangle$$

is coded to nonstandard length. Since $(g(x))_n > x_n$ for all $n \in \mathbb{N}$ there is $\nu > \mathbb{N}$ with $(g(x))_n > x_n$ for all $n < \nu$. Also, for all standard $n' > n$, $x_{n'} \geqslant x_n$ so by another overspill (changing ν if necessary) we may assume

$$M \vDash \forall n < n' < \nu \; x_n \leqslant x_{n'}.$$

We now consider the case when there is nonstandard $\mu < \nu$ with $g(\mu) \geqslant \mu$. In this case,

$$g(x_\mu) = (g(x))_{g(\mu)} \geqslant (g(x))_\mu > x_\mu$$

so x_μ is moved, and of course $x_\mu \in p^M$ since $\mu > \mathbb{N}$. In the other case, we have $g(\mu) < \mu$ for all nonstandard $\mu < \nu$. By replacing j with ν and considering g^{-1} instead of g we obtain a similar conclusion.

$G_{(>p)} = G_{(>J_p)}$ is similar, but a little easier. For one direction, if $g \in G_{(>J_p)}$, then g fixes $\{x : x < j_n\}$ pointwise, for some $n \in \mathbb{N}$. If $a \in \phi_n^M$, then a is the kth element of ϕ_n^M for some $k < j_n$, hence $g(a)$ is the $g(k)$th, i.e. the kth element of ϕ_n^M, hence $g(a) = a$. Conversely, if for each $n \in \mathbb{N}$ there is $k_n < j_n$ moved by g, then, for all n, the k_nth element of ϕ_n^M is moved by g, hence $g \notin G_{(>p)}$. □

REMARK. The proof in Kaye *et al.* (1991) of the inclusion $G_{(p)} \subseteq G_{(J_p)}$ is much more involved than the new proof given above. It also, unfortunately, contains a small error: in the notation of that paper, Lemma 6.6 requires the cut I_a^+ to be closed under all the functions G_n, and this does *not* follow from the other hypotheses. However, Corollary 6.8 can still be proved by considering $F(a)$ instead of a for a suitably fast-growing Skolem-term F, noting that $I^+_{F(a)}$ is closed under such functions and that $G_{(p)} \subseteq G_{(q)}$, where $p = \mathrm{tp}(a)$, $q = \mathrm{tp}(F(a))$.

4 The main theorem

For a given structure M, normal subgroups of $\mathrm{Aut}(M)$ often arise by considering notions of definability, such as first-order definability, $\mathscr{L}_{\omega_1\omega}$-definability (with or without parameters). Scott's Theorem (see p. 20) shows that, for countable M, the set of automorphisms of M that are $\mathscr{L}_{\omega_1\omega}$-definable without parameters forms the centre of $\mathrm{Aut}(M)$, and the automorphisms g that are $\mathscr{L}_{\omega_1\omega}$-definable with finitely many parameters are precisely those with only countably many conjugates g^h.

If we are to classify closed normal subgroups we will have to understand such notions of definability. In fact, for models of arithmetic, most of these definitions trivialize. For example, no non-trivial automorphism g is definable, for if it were, by induction on x in $g(x) = x$ one could show that $g = 1$. Tzouvaras (1991) has given an elegant and simple argument that uses a lemma due independently to Gaifman (1976, Theorem 4.1, p. 267) and Ehrenfeucht (1973) to show that in fact no non-trivial $g \subset G$ can have only countably many conjugates. Here are the details.

LEMMA 4.1. (Ehrenfeucht–Gaifman) Let $M \vDash \mathrm{PA}$, let $t()$ be a Skolem term and let $a \in M$. If $\mathrm{tp}(a) = \mathrm{tp}(t(a))$ then $t(a) = a$.

Proof. Suppose first that $t(a) < a$ and $\mathrm{tp}(a) = \mathrm{tp}(t(a))$. Let $s(a)$ be the length $n \in M$ of the longest sequence

$$y_0 = a > y_1 = t(y_0) > \cdots y_{i+1} = t(y_i) > \cdots y_n = t(y_{n-1}).$$

Then '$s(a)$ is odd' is an $\mathscr{L}_{\mathrm{A}}$-formula $\psi(a)$ and $s(a)$ is odd iff $s(t(a))$ is even.

If $t(a) > a$ argue similarly, but this time consider $r(a)$, the length of the longest sequence

$$a = y_0 = t(y_1) > y_1 = t(y_2) > \cdots > y_i = t(y_{i+1}) > \cdots > y_{n-1} = t(y_n). \qquad \square$$

COROLLARY 4.2. (Tzouvaras) Let $g \in G \setminus \{1\}$. Then for each $a \in M$ there is $h \in G_a$ with $gh \neq hg$.

Proof. If $ga = a$ and $gx \neq x$ then $gx \notin \mathrm{cl}_M(a, x)$ by the previous lemma, hence there is $h \in G_{a,x}$ with $hgx \neq gx = ghx$. If $ga \neq a$ then there is $h \in G_a$ with $hga \neq ga = gha$. $\square$

Next, we give two useful variations on Tzouvaras' result.

Lemma 4.3. Let M be a countable recursively saturated model of PA. Let $g \in G$, $a \in M$ and $b \in M \setminus I_{\mathrm{fix}}(g)$. Then there is $x \leqslant b$ in M and $h \in G_a$ such that $hgx \neq ghx$.

Proof. We find $x \leqslant \min(b, gb)$ such that either $gx \notin \mathrm{cl}_M(a,x)$ or $g^{-1}x \notin \mathrm{cl}_M(a,x)$. This suffices for, given x with $gx \notin \mathrm{cl}_M(a,x)$ there is $h \in G_{a,x}$ with $hgx \neq gx = ghx$, and if $g^{-1}x \notin \mathrm{cl}_M(a,x)$ and $hg^{-1}x \neq g^{-1}hx$ then $g^{-1}hg(g^{-1}x) \neq g^{-1}gh(g^{-1}x)$ so $y = g^{-1}x$ satisfies $ghy \neq hgy$.

Without loss, we may assume that $gb, b \in \mathrm{cl}_M(a)$ (otherwise replace a with $\langle a, b, gb \rangle$). So assume that for all $x < b$ there are Skolem terms t_x and s_x with

$$gx = t_x(a,x) \quad \text{and} \quad g^{-1}x = s_x(a,x).$$

Using this we prove that

$$\forall x < b\, \exists n \in \mathbb{N} \left(\left\lfloor \frac{x}{n} \right\rfloor \leqslant g(x) \leqslant nx \right)$$

Indeed, suppose $x < b$ and $g(x) > x \cdot \alpha$ for some $\alpha > \mathbb{N}$. Then for each $n \in \mathbb{N}$

$$\mathrm{card}\{t(a,y) : \ulcorner t \urcorner < n, y < x\} \leqslant xn.$$

By overspill, using a partial nonstandard satisfaction class, there is a nonstandard $\beta < \alpha$ in M such that

$$\mathrm{card}\{t(a,y) : \ulcorner t \urcorner < \beta, y < x\} \leqslant x \cdot \beta < g(x)$$

which is impossible since this set contains all elements of M less than $g(x)$. The other inequality is similar, using $s_n()$ and g^{-1}.

Now, if $x < \lfloor \log_2 b \rfloor$ is moved by g, $g(x) > x$ say, then $g(x) > x + n$ for all $n \in \mathbb{N}$ and $g(2^x) \geqslant 2^x \cdot 2^n > 2^x \cdot \mathbb{N}$. Similarly, if $x < \lfloor \log_2 b \rfloor$ and $g(x) < x$ then $g(2^x) < \lfloor 2^x / \mathbb{N} \rfloor$. But $b \notin I_{\mathrm{fix}}(g)$ and $I_{\mathrm{fix}}(g)$ is closed under exponentiation, so there is always such $x \in M$, by Proposition 2.2 for instance. □

In this paper, h^g will denote the conjugate $g^{-1}hg$ of h by g, and h^{-g} denotes $g^{-1}h^{-1}g$. Also, $\mathrm{fix}(g)$ denotes the fixed-point set of g, $\mathrm{fix}(g) = \{x \in M : g(x) = x\}$.

Lemma 4.4. Let $h \in G$ and $a \in M \setminus I_{\mathrm{fix}}(h)$ Then there are $g_1, g_2 \in G$ so that

$$\mathrm{fix}(h^{g_1} h^{-g_2}) \neq I_{\mathrm{fix}}(h^{g_1} h^{-g_2}) > \mathbb{N}$$

and

$$a > I_{\mathrm{fix}}(h^{g_1} h^{-g_2}).$$

Proof. The idea is to have $g_1 {\restriction} {\leqslant} c = g_2 {\restriction} {\leqslant} c$ for some $c > \mathbb{N}$. For then, for each $x < \min(c, g_2^{-1}h(c))$, $h^{g_1}h^{-g_2}(x) = g_1^{-1}hh^{-1}g_2(x) = g_1^{-1}g_2(x) = x$. Also, if $s \in M$ and $s, h^{-1}(s) \in \mathrm{fix}(g_1) \cap \mathrm{fix}(g_2)$ then $h^{g_1}h^{-g_2}(s) = s$.

We will take $g_1=\sigma g$ and $g_2=g$ for some g and some $\sigma\in G_{(>\mathbb{N})}$. A simple calculation shows that $h^{g_1}h^{-g_2}(x)=x$ if and only if $\sigma^{-1}h\sigma h^{-1}(gx)=gx$.

Fix s so that $s,h^{-1}(s)>\mathrm{I}_{\mathrm{fix}}(h)$. It suffices to find $\sigma\in G_{(>\mathbb{N})}\cap G_{s,h^{-1}(s)}$ and some $x<\min(a,s)$ such that $\sigma h\sigma^{-1}h^{-1}(x)\neq x$, for then any g fixing $x,s,h^{-1}(s)$ will work. By the last lemma there is $\tau\in G_{s,h^{-1}(s)}$ and $x<\min(a,s)$ such that

$$h^{-1}\tau^{-1}(x)\neq\tau^{-1}h^{-1}(x)$$

and by resplendency (or otherwise) there is $\sigma\in G_{(>\mathbb{N})}$ such that

$$\begin{aligned}\sigma(s) &= s\\ \sigma(h^{-1}(s)) &= h^{-1}(s)\\ \sigma(\tau^{-1}h^{-1}(x)) &= h^{-1}(x)\\ \sigma(\tau^{-1}(x)) &= x\end{aligned}$$

Then $\sigma h\sigma^{-1}h^{-1}(x)\neq x$, and we are finished. □

THEOREM 4.5. Let M be a countable recursively saturated model of PA (or PA*), let $h\in G$, $a,b\in M$, $I=\mathrm{I}_{\mathrm{fix}}(h)$ and $g\in G_{(I)}$ satisfy $g(a)=b$. Then there are $g_1,g_2,g_3,g_4,\in G$ such that either

$$h^{-g_1}h^{g_2}h^{-g_3}h^{g_4}(a)=b$$

or

$$h^{g_1}h^{-g_2}h^{g_3}h^{-g_4}(a)=b.$$

The proof will take the rest of this section. Throughout this proof we fix g,h,I,a,b as in the theorem. J will denote the initial segment $J=\inf_n j_n$ where

$$j_n=\max\left\{j\in M:\forall\bar{x}<j\bigwedge_{\ulcorner\theta\urcorner<n}(\theta(\bar{x},a)\leftrightarrow\theta(\bar{x},b))\right\}.$$

Since $g(a)=b$ and $g\in G_{(I)}$ we have $I\subsetneq J$ by Lemma 2.7. Let $p(x)=\mathrm{tp}_M(a)$ and $q(x,y)=\mathrm{tp}_M(a,b)$. These types are enumerated as

$$p(x)=\{\varphi_n(x):n\in\mathbb{N}\}$$

and

$$q(x,y)=\{\psi_n(x,y):n\in\mathbb{N}\}.$$

Without loss, we may assume that $a\neq b$ and that both sequences $(\varphi_n)_{n\in\mathbb{N}}$ and $(\psi_n)_{n\in\mathbb{N}}$ are coded in M,

$$M\models\forall x\,(\varphi_{n+1}(x)\rightarrow\varphi_n(x))$$

and

$$M\models\forall x,y\,(\psi_{n+1}(x,y)\rightarrow\psi_n(x,y))$$

for all $n \in \mathbb{N}$.

We will also need a notion of 'neighbourhood' of a point $x \in M$. This is the key notion for this argument.

DEFINITION 4.6. We say that $v, w \in M$ are *non-adjacent* if there are $u \in M$ and $\alpha, \beta \in G$ such that either

$$\alpha: u, w \mapsto v, w \text{ and } \beta: a, b \mapsto u, v$$

or

$$\alpha: u, v \mapsto w, v \text{ and } \beta: a, b \mapsto u, w.$$

Note in particular that if v, w are non-adjacent, and $h: v \mapsto w$ or $h^{-1}: v \mapsto w$, then taking α, β and u as in the definition just given and considering

$$g_1 = [h, \alpha]^\beta = (\beta^{-1} h^{-1} \beta)(\beta^{-1} \alpha^{-1} h \alpha \beta)$$

or

$$g_2 = [h^{-1}, \alpha]^\beta = (\beta^{-1} h \beta)(\beta^{-1} \alpha^{-1} h^{-1} \alpha \beta)$$

we have $g_1(a) = b$ or $g_2(a) = b$ as can be readily verified.

Our first task is to obtain a more convenient formulation of this adjacency relation, described in terms of first-order formulas.

Define $\mu_n(x, y)$ to be the formula

$$\forall z \, (\psi_n(z, x) \to \bigvee_{\ulcorner \theta \urcorner < n} (\theta(x, y) \not\leftrightarrow \theta(z, y)))$$

so $\bigvee_n \mu_n(x, y)$ says that 'whenever $k \in G$ moves x to z and $\mathrm{tp}(z, x) = q$ then k must also move y'. (This statement can be proved precisely using saturation: if $\bigwedge_n \neg \mu_n(x, y)$ then there exists $z \in M$ with $\mathrm{tp}_M(z, x) = q$ and $k \in G_y$ with $k(x) = z$ since q is coded in M.) We also use the notation $x \rightsquigarrow_n y$ for $\mu_n(x, y)$, $x \leftrightsquigarrow_n y$ for $\mu_n(x, y) \wedge \mu_n(y, x)$, $x \rightsquigarrow y$ for $\bigvee_{n \in \mathbb{N}} x \rightsquigarrow_n y$, and $x \leftrightsquigarrow y$ for $\bigvee_{n \in \mathbb{N}} x \leftrightsquigarrow_n y$. Note in particular that $x \leftrightsquigarrow y$ iff $x \rightsquigarrow y$ and $y \rightsquigarrow x$, since $M \vDash \forall x, y \, (\mu_n(x, y) \to \mu_{n+1}(x, y))$. Also, x and y are non-adjacent in the sense defined above iff $x \not\leftrightsquigarrow y$.

The first lemma required to prove Theorem 4.5 is an analogue of Theorem 3.6 in the previous section concerning $G_{(p)}$ and J_p. This can be stated in a general form for a (possibly incomplete) type p and a general adjacency relation $\leftrightsquigarrow$ defined in a way similar to that above.

DEFINITION 4.7. An *adjacency relation* is a coded family of formulas

$$\{\eta_n(x, y) : n \in \mathbb{N}\}$$

such that

$$M \vDash \forall x \, \eta_n(x, x) \wedge \forall x, y \, (\eta_n(x, y) \to \eta_n(y, x) \wedge \eta_{n+1}(x, y))$$

for all n. We denote $\eta_n(x,y)$ by $x \leftrightsquigarrow_n y$ and $\bigvee_n \eta_n(x,y)$ by $x \leftrightsquigarrow y$, etc., as before.

LEMMA 4.8. Let $p(x) = \{\varphi_n(x) : n \in \mathbb{N}\}$ be consistent, coded and satisfying

$$M \models \forall x\, (\varphi_{n+1}(x) \to \varphi_n(x))$$

for all n. (The set p need not be complete.) Let $(\eta_n(x,y) : n \in \mathbb{N})$ be an adjacency relation and define

$$k_n = \max_{E_n} \operatorname{card} E_n$$

where E_n ranges over all M-finite subsets of φ_n^M such that $\forall x,y \in E_n\, (x \neq y \to \neg x \leftrightsquigarrow_n y)$. (Some of these values may be ∞.) Hence define

$$K = \inf_{n \in \mathbb{N}} k_n.$$

Then if $h \notin G_{(K)}$ then there are $g_1, g_2 \in G$ and $u, v \in M$ such that, for $k = h^{-g_1} h^{g_2}$, $k(u) = v$ and $\neg u \leftrightsquigarrow v$.

The other important lemma is the following one, which enables us to estimate the values k_n.

LEMMA 4.9. If K is the initial segment of Lemma 4.8, and the adjacency relation is that defined by $\mu_n(x,y) \wedge \mu_n(y,x)$, then $K \geqslant J$.

Proof. [That the lemmas imply the theorem.]

I is closed under exponentiation, and $J \supsetneq I$, so $K \supsetneq I$ by Lemma 4.9. Thus there is $j \in K \setminus I$. By the definition of I, h moves some element of M smaller than j, i.e., $h \notin G_{(K)}$. By Lemma 4.8 there are $u, v \in p^M$ with $\neg u \leftrightsquigarrow v$ and $k \in G$ (which is the product of a conjugate of h with a conjugate of h^{-1}) such that $k(u) = v$. We may assume $\bigwedge_n \neg \mu_n(v,u)$, by swapping k and k^{-1} and u and v if necessary.

Since $\bigwedge_n \neg \mu_n(v,u)$ there is w in M such that (w,v) satisfies q and

$$\operatorname{tp}_M(v,u) = \operatorname{tp}_M(w,u).$$

Also, since $(w,v) \in q^M$, there is $g_1 \in G$ such that $g_1(a,b) = (w,v)$; and there is $g_2 \in G$ such that $g_2(v,u) = (w,u)$. Then it follows that

$$b = g_1^{-1} k g_1 g_1^{-1} g_2 k^{-1} g_2^{-1} g_1(a)$$

as required. □

It remains to prove the lemmas.

Proof of Lemma 4.8. By Lemma 4.4 there are $g_1, g_2 \in G$ such that for $k = h^{-g_1} h^{g_2}$, $k \notin G_{(K)}$, $\operatorname{fix}(k) \neq \mathrm{I}_{\mathrm{fix}}(k)$, and $\mathrm{I}_{\mathrm{fix}}(k) > \mathbb{N}$. Let $r < s$ be elements of M such that $r \in K \setminus \operatorname{fix}(k)$ and $s \in \operatorname{fix}(k)$. For each $n \in \mathbb{N}$ let E_n be the least M-finite set (in terms of its code) such that $E_n \subseteq \varphi_n^M$, $\forall x,y \in E_n\,(x \neq y \to \neg\, x \leftrightsquigarrow_n y)$, and $\operatorname{card} E_n = \min(s,k_n)$. Let $t_n(i,s)$ be the Skolem term enumerating E_n, i.e., $t_n(i,s)$ is the ith element of E_n, provided $i < \operatorname{card} E_n$, and $t_n(i,s) = 0$ otherwise. Note that this sequence of Skolem terms t_n is coded in M.

Now consider r. Since $r < \operatorname{card} E_n$ for all n and s is fixed by k,

$$t_n(r,s) \neq t_n(k(r),s) = t_n(k(r),k(s)) = k(t_n(r,s)).$$

But the sequence $(t_n(r,s))_{n\in\mathbb{N}}$ is coded in M, by t say. Thus

$$\forall l \in \mathbb{N}\; M \vDash \forall n < l\,(\neg\,(t)_l \leftrightsquigarrow_n (k(t))_l \wedge (t)_l \neq (k(t))_l).$$

So, by overspill, there is $\nu > \mathbb{N}$ such that

$$M \vDash \forall l \leqslant \nu\, \forall n < l\,(\neg\,(t)_l \leftrightsquigarrow_n (k(t))_l \wedge (t)_l \neq (k(t))_l).$$

Let $\mu < \nu$ be nonstandard and sufficiently small so that it is fixed by k and $(t)_\mu$ satisfies the type $\{\varphi_n : n \in \mathbb{N}\}$. Then $\forall n < \mu\, \neg\,(t)_\mu \leftrightsquigarrow_n (k(t))_\mu$, and $(k(t))_\mu = (k(t))_{k(\mu)} = k((t)_\mu)$, as required. □

Proof of Lemma 4.9. We must show

$$k_n = \max_{E_n} \operatorname{card} E_n > J$$

for all standard n. Suppose otherwise, that is $k_n \in J$, and let $E_n \subseteq \varphi_n^M$ be the least (in the sense of 'least code') M-finite set X with $\operatorname{card} X = k_n$ satisfying

$$\forall x,y \in X\,(x \neq y \to \neg\, x \leftrightsquigarrow_n y). \tag{‡}$$

Obviously $k_n \geqslant 1$. Note also that $E_n \in M_0$ since its definition only depends on the formulas ϕ_n, ψ_n and the θ with $\ulcorner\theta\urcorner < n$. We shall derive a contradiction by showing $b \notin E_n$ and that $X = E_n \cup \{b\}$ also satisfies (‡).

To see that $b \notin E_n$, suppose otherwise. Then there is $x < k_n$ such that

$$b \text{ is the } (x+1)\text{st element of } E_n,$$

i.e., there is a term t such that $t(x) = b$. Since $x \in J$ we have $M \vDash b = t(x) \leftrightarrow a = t(x)$, i.e., $b = a$, a contradiction.

To see that $X = E_n \cup \{b\}$ satisfies (‡), let $y \in E_n$ and $x \in J$ be such that y is the $(x+1)$st element of E_n, i.e., $y = t(x)$ where t is as in the last paragraph. We show $M \vDash \neg\,\mu_n(b,y)$. Indeed, $M \vDash \psi_n(a,b)$ and, for any formula $\theta(v,w)$,

$$\begin{aligned} & M \vDash \theta(a,y) \\ \Leftrightarrow\quad & M \vDash \theta(a,t(x)) \\ \Leftrightarrow\quad & M \vDash \theta(b,t(x)) \qquad \text{since } x \in J \\ \Leftrightarrow\quad & M \vDash \theta(b,y) \end{aligned}$$

and this completes the proof. □

REMARK. This proof of Lemma 4.9 is due essentially to Roman Kossak, and is considerably simpler than my original argument.

5 A Galois correspondence

Combining the results of the last section, we have:

THEOREM 5.1. Let M be a countable recursively saturated model of PA* and $G = \mathrm{Aut}(M)$ considered as a topological group with the usual topology.

(i) For initial segments I, J of M: $G_{(I)}$ and $G_{(J)}$ are closed; $G_{(I)} = G_{(J)}$ iff $\exp(I) = \exp(J)$; and $G_{(I)}$ is normal iff $\exp(I)$ is invariant (see p. 299 for a model-theoretic description of 'invariant'). Also, for $N \lhd G$, $\mathrm{I_{fix}}(N)$ is invariant and closed under exponentiation.

(ii) The operations

$$I \mapsto G_{(I)}$$

and

$$N \mapsto \mathrm{I_{fix}}(N)$$

defined on invariant initial segments I of M closed under exponentiation, and closed normal subgroups N of G are inverse to each other.

From this one can derive some interesting conclusions, such as: the closed normal subgroups of $G = \mathrm{Aut}(M)$ are linearly ordered in (the reversal of) one of the orderings given in Proposition 2.9. Non-trivial closed normal subgroups exist if and only if there are nonstandard definable elements in the model M. (For a model of PA in the original language this is equivalent to $M \models \mathrm{Th}(\mathbb{N})$.) If non-trivial closed normal subgroups do exist, there is a minimal one, namely $G_{(I_0)}$. There is a maximal proper closed normal subgroup if and only if $M_0 \setminus \mathbb{N}$ is not downward cofinal in $M \setminus \mathbb{N}$ if and only if $M_0 \setminus \mathbb{N}$ is coded in M.

Perhaps the most important question left open is

PROBLEM 5.2. Classify all normal subgroups, showing that they are all $G_{(I)}$ or $G_{(>I)}$ for invariant I.

The idea should be to build g_1 and g_2 such that $g = h^{-g_1} h^{g_2}$, say, by a back-and-forth argument using the ideas in the above theorem as the basis of a single step. Such a result would give the corollaries that (1) the normal subgroup generated by all $g \in G$ for which (M, g) is recursively saturated is precisely $G_{(>\mathbb{N})}$; and (2) the derived group G' of G is in fact the whole of G.

PROBLEM 5.3. Give a group-theoretic characterization of the closed subgroups that occur as $G_{(I)}$ for some initial segment I of M. Given such a characterization, extend the Galois correspondence in Theorem 5.1 to all such groups.

There are two separate classes of $G_{(I)}$ here. The first (and presumably easier case) is when I is *almost invariant* by which I mean, I is G_a-invariant for some $a \in M$. Here Smoryński's result (Lemma 2.6) is useful in showing that the normalizer of $G_{(I)}$ is $G_{\{I\}}$ when I is an initial segment closed under exponentiation. Thus such a group $H = G_{(I)}$ has the property that its normalizer $N_G(H)$ is open and for no closed $K \lneq H$ is $N(K) = N(H)$. I would conjecture that this characterizes such subgroups. The other $G_{(J)}$ (when J is not almost invariant) would probably have to be treated as the limit of such $G_{(I)}$.

Much more difficult, and I suspect beyond current methods is

PROBLEM 5.4. State and prove a Galois correspondence for arbitrary closed subgroups and the fix operation.

Here there are surprising difficulties, even in properly understanding those substructures K for which $K = \mathrm{fix}(G_{(K)})$. Kotlarski (1984) studies *initial segments* with this property, and a result in Kaye *et al.* (1991) characterizes those initial segments I for which $I = \mathrm{fix}(g)$ for a single $g \in G$. This seems to be just about all that is known on the matter at present.

PART III

Permutation groups of finite Morley rank

Stable groups

Groups (or model-theoretic expansions of groups) whose first-order theory is stable have been studied intensively lately. Many of the main results were first proved by Zil'ber and can be found in his papers in the bibliography to this volume. The general theory is developed in Poizat (1987) and there are several important articles in Nesin and Pillay (1989), including a useful survey there by Pillay. Groups whose first order theory has finite Morley rank have had particular attention, and a book by Borovik and Nesin on this subject is in preparation. In the present book, stable groups are considered only as permutation groups, though as mentioned in the introduction, the distinction is often spurious. The three chapters in this part of the book are all on permutation groups of finite Morley rank, and the subject is not too far away in the article by Cameron. Here we give a brief, and biased, resumé on what is known on permutation groups of finite Morley rank. This topic is interesting in its own right and is of importance in general stability theory since group actions frequently arise in the analysis of a structure. For permutation groups which are merely stable, the results in the literature are not nearly so strong.

The typical context is that M is a structure of finite Morley rank containing two disjoint 0-definable subsets, denoted Ω and G. There is a 0-definable group structure on G, and a 0 definable faithful group action of G on Ω. Since the G-orbits of Ω are definable, we may as well assume that G is transitive on Ω.

Clearly results on (abstract) groups of finite Morley rank are relevant. Here, the first and crucial observation, which is a direct consequence of the definition of Morley rank, is that in a group of finite Morley rank there are no infinite descending chains of definable subgroups. It follows easily that if G is such a group and $A \subseteq G$ then there is a finite $F \subseteq A$ such that the centralizer $C_G(A)$ is equal to $C_G(F)$ (in fact, this is true for stable groups, by a result of Baldwin and Saxl (1976)). Similarly, if (G, Ω) is a permutation group of finite Morley rank then for any $\Gamma \subset \Omega$ there is finite $\Delta \subset \Gamma$ such that $G_{(\Gamma)} = G_{(\Delta)}$. In particular, this observation already makes Frobenius groups of finite Morley rank (the subject of the articles by Nesin and by Epstein and Nesin) natural objects to investigate.

The dominant conjecture on groups of finite Morley rank is that of Cherlin (1979), that any non-abelian simple group of finite Morley rank is a Chevalley group over an algebraically closed field. (It is easy to see that any such group has finite Morley rank, since it is interpretable in the algebraically closed field, and algebraically closed fields have Morley rank one). As indicated in the paper by Epstein and Nesin, questions about Frobenius groups of finite Morley rank are closely related to this conjecture.

We now describe some of the established evidence for Cherlin's Conjecture.

First, any group interpretable in an algebraically closed field is an algebraic group over that field. This is a model-theoretic version of a theorem of Weil (1955), and is due independently to van den Dries and Hrushovski. A proof can be found in the article by Bouscaren in Nesin and Pillay (1989).

Second, there is a notion of *connected component* of a group G of finite Morley rank, analogous to the connected component (in the Zariski topology) of an algebraic group. The *connected component* of G, denoted G°, is the intersection of the definable subgroups of G of finite index, and by the descending chain condition on definable subgroups it too has finite index in G. The group G is said to be *connected* (even though there may be no topology around) if $G = G^\circ$.

Third, although there is no good structure theory for nilpotent groups of finite Morley rank, there are suggestive results for connected soluble groups of finite Morley rank which are not nilpotent. Zil'ber (1977) showed that in any such group there is an interpretable algebraically closed field. Closely related to this is a result of Zil'ber (1973) and Nesin (1989) that any connected soluble group of finite Morley rank has nilpotent derived subgroup. This is an analogue of the Lie–Kol'chin Theorem for linear algebraic groups (Humphreys, 1981). As an example of the Zil'ber–Nesin result, consider the group of non-singular upper-triangular $n \times n$ matrices over an algebraically closed field. This is a soluble algebraic group (so has finite Morley rank), and its derived subgroup is the group of unitriangular matrices, which is nilpotent.

Fourth, we mention the very powerful Zil'ber Indecomposability Theorem for groups of finite Morley rank. We shall not state it here, as it can be found in the chapter by Nesin.

Fifth, results on groups of Morley rank one and two back up the conjecture. Reineke (1975) showed that connected groups of Morley rank one are abelian, and classified them. In particular, any such group is elementary abelian or divisible abelian, as in the classification of connected one-dimensional algebraic groups. Cherlin (1979) proved his conjecture for groups of Morley rank two. Obstructions arise in Morley rank three, however, and the conjecture remains wide open.

A natural first step in examining transitive *permutation* groups (G, Ω) of finite Morley rank is to classify them in the case when Ω is strongly minimal. A classification was given by Hrushovski in his thesis (Hrushovski, 1986; see also Poizat, 1987, Théorème 3.27). A version of his theorem, appropriate for our context, is as follows.

THEOREM 0.5. (Hrushovski) Let (G, Ω) be a transitive permutation group of finite Morley rank, and suppose that Ω is strongly minimal. Then one of the following holds:

(a) G has Morley rank one, and the connected component G° is abelian and acts regularly on Ω;

(b) G has Morley rank two, is connected and sharply 2-transitive, and is isomorphic to the affine group $\mathrm{AGL}(1, K)$ (K an algebraically closed field). Also, Ω can be identified with K, and a typical element (g, h) of G ($g \in K$, $h \in K \setminus \{0\}$) acts on K by the rule $x^{(g,h)} \mapsto h.x + g$;

(c) G has Morley rank three, is connected and sharply 3-transitive, and is isomorphic to $\mathrm{PSL}(2, K)$ for some algebraically closed field K. The set Ω can be identified with the projective line $\mathrm{PG}(1, K)$, and G has the natural action on Ω.

It follows easily from this that there is an absolute bound (of 3) on the Morley rank of any permutation group of finite Morley rank which acts transitively on a set of Morley rank one. It is natural now to ask whether there is a function $f: \mathbb{N} \to \mathbb{N}$ such that if (G, Ω) is a transitive permutation group of finite Morley rank then $\mathrm{RM}(G) \leqslant f(\mathrm{RM}(\Omega))$. The answer is negative, and it was shown by Gropp (1991) that there is no bound on $\mathrm{RM}(G)$ even when $\mathrm{RM}(\Omega) = 2$. Examples to prove this are given in the chapter by Borovik and Thomas in this book. There, the authors ask whether there is such a function f if 'transitive' is replaced by 'primitive'. We should mention at this point the detailed analysis by Gropp of connected groups acting faithfully and transitively on a set of Morley rank two and Morley degree one. She shows (Gropp, 1992) that such a group cannot have a sharply 6-transitive action on the set of generic 6-tuples. Here, 'generic' is used in the standard sense of stability theory; a *generic 6-tuple* is just a 6-tuple from Ω, chosen to have maximal Morley rank, or, equivalently, chosen so that each element has maximal Morley rank in Ω, and the type of any one element over the other 5 does not fork over the empty set.

The papers below use facts about generic elements of groups of finite Morley rank. For more detail on this, see the texts on stable groups already referred to. We should mention that if G is a group definable in a structure of finite Morley rank, then a *generic type* of G is any type in G of maximal Morley rank over a tuple from which G is definable (the choice of the defining tuple is unimportant). It can be shown that the number of distinct generic types in G is equal to its Morley degree and to the index $|G : G^\circ|$, and hence that G is connected if and only if it has a unique generic type.

To fill out further the picture for permutation groups of finite Morley rank, we mention a generalization of the O'Nan–Scott Theorem for primitive such groups (Macpherson and Pillay, 1994). A brief account of the usual form of the O'Nan–Scott Theorem can be found in the chapter by

Cameron in this book. The statement of the finite Morley rank version is very close to that of the finite version, except that the twisted wreath case does not arise. If Cherlin's Conjecture were proved, then this theorem would almost make possible a classification of primitive permutation groups of finite Morley rank. For example, one of the families which arises is the affine case, where Ω is identified with a vector space V over a field F, and $G = V.H$ where H is an irreducible subgroup of $\mathrm{GL}(V)$. If V has infinite dimension over $C_{\mathrm{End}(V)}(FG)$ then very little can be said (except that the latter centralizer is a finite field); it is an interesting question whether this can arise. On the other hand, if the dimension is finite then we can reduce to a situation where F is algebraically closed and $\dim_F(V)$ is finite. In this latter case, using Clifford's theorem and an argument developed by Aschbacher for examining maximal subgroups of $\mathrm{GL}(n, q)$, one can reduce to the case when H is a central extension of a non-abelian simple group. This simple group is definable, so has finite Morley rank. There are results in (Macpherson and Pillay, 1994) which suggest that if there are no interpretable fields admitting a definable automorphism then H should be definable in F, that is, it should be algebraic over F.

On generic normal subgroups

Alexandre V. Borovik

Department of Mathematics,
UMIST,
PO Box 88,
Manchester,
M60 1QD,
U.K.
Electronic mail address: sasha@lanczos.ma.umist.ac.uk

Simon R. Thomas

Department of Mathematics,
Rutgers University,
New Brunswick,
NJ 08903,
U.S.A.
Electronic mail address: sthomas@math.rutgers.edu

1 Introduction

Uniformly definable families of subgroups. Let G be an ω-stable group of finite Morley rank. We say that the family $\mathscr{F} = \{F_i : i \in I\}$ of subgroups of a group G of finite Morley rank is *uniformly definable* if the index set I is definable in G and there exists a formula $\phi(x, y)$ of the first-order language, such that for any $i \in I$

$$F_i = \{g \in G : G \vDash \phi(g, i)\}.$$

For example, the family of centralizers $\{C_G(g) : g \in G\}$ is uniformly definable.

Generic normal subgroups. The following theorem is the main result of this paper.

THEOREM 1.1. Let $\mathscr{N} = \{N_i : i \in I\}$ be a uniformly definable family of normal subgroups of a connected group G of finite Morley rank. Assume that $\bigcap_{i \in J} N_i = 1$ for any generic subset $J \subseteq I$. Then for some generic subset $I_0 \subseteq I$ the subgroups N_i for $i \in I_0$ are nilpotent.

Notation. We shall use the following notation. $R(X)$ denotes the Morley rank of a definable set X. If G is a group of finite Morley rank and $X \subseteq G$, dc(X) is the least definable subgroup of G containing X (which exists by Theorem 1.8).

Permutation groups. Let G be a connected group of finite Morley rank acting faithfully and transitively on a set Ω. Assume that this action is definable. We say that a definable set Ω_0 is *generic* in Ω if Ω_0 has the same Morley rank as Ω.

THEOREM 1.2. Under the assumptions of Theorem 1.1, the action of G on Ω is generically faithful, i.e., $R(\mathrm{fix}(g)) < R(\Omega)$ for all $g \in G$, $g \neq 1$.

Let H be a connected definable subgroup of G. It easily follows from Theorem 1.2 that H acts faithfully on any generic H-invariant subset $\Omega_0 \subseteq \Omega$. Obviously $\Omega_0 = \bigcup_{i\in I} {\alpha_i}^H$ is the union of a definable family $\{{\alpha_i}^H\}$ of H-orbits, and the kernels N_i of the action of H on the orbits ${\alpha_i}^H$ constitute a uniformly definable family of subgroups satisfying the conditions of the Theorem 1.1. So Theorem 1.1 yields the following property of permutation groups.

COROLLARY 1.3. Let G be a connected group of finite Morley rank acting on a set Ω. Assume that this action is definable, transitive, and faithful. Let H be a connected definable subgroup in G. Then there exists an H-invariant definable generic subset $\Omega_0 \subseteq \Omega$ such that the kernels of the action of H on the H-orbits in Ω_0 are nilpotent.

Maybe we are mistaken, but it seems that this result was previously unknown even in the most natural of its settings:

COROLLARY 1.4. Let G be a connected algebraic group over an algebraically closed field K. Assume that G acts rationally on a variety X and this action is faithful and transitive. If H is a closed connected subgroup of G, then there exists an H-invariant open dense subset $Y \subseteq X$ such that all the kernels of the action of H on the H-orbits in Y are unipotent.

Proof. In view of Corollary 1.3 we can assume that the kernels K_i, $i \in I$, of the action of H on the H-orbits in Y are nilpotent. Let T_i be the maximal torus of K_i. Then $T_i \lhd H$ and so $\{T_i : i \in I\}$ is a uniformly definable family of subtori in a maximal torus $T \leqslant H$. But it is not so difficult to prove that any definable family of subtori in an algebraic torus is finite. So we can assume without loss that all T_i are equal and thus lie in the kernel of the action of H on Y. By Theorem 1.2 this kernel is trivial and $T_i = 1$. □

So our results may be stated in plain English as follows: *the kernel of the action of a subgroup of transitive permutation group on an orbit in general position is nilpotent.*

EXAMPLE 1.5. The following example is due to Gropp (1991). Let K be an algebraically closed field and $G = K \oplus \cdots \oplus K$ be a direct sum of n

copies of the additive group of K. Consider the following action of G on the affine space $\mathbb{A}^2$.

$$(a_1, \ldots, a_n) : \begin{pmatrix} x \\ y \end{pmatrix} \mapsto \begin{pmatrix} x \\ y + a_1x + \ldots + a_nx^n \end{pmatrix}.$$

Obviously this action is faithful and its orbits have a natural parametrization by the first coordinate $x \in K$ of a point $\binom{x}{y} \in \mathbb{A}^2$. It is also easy to see that the kernels of the orbit actions are hyperplanes

$$N_x = \{(a_1, \ldots a_n) : a_1x + \ldots + a_nx^n = 0\}.$$

It is well-known that generic subsets in K are cofinite, so obviously

$$\bigcap_{x \in J} N_x = 0$$

for any generic $J \subseteq K$.

This example shows that Morley rank of a group acting faithfully on a set of rank 2 is unbounded. So we cannot skip the transitivity assumption from the following problem.

PROBLEM 1.6. If a group G of finite Morley rank acts on a set Ω transitively and primitively and the action is ω-stable, is it true that $R(G) \leqslant f(R(\Omega))$ for some function $f\colon \mathbb{N} \to \mathbb{N}$?

The next example shows that the primitivity condition is also necessary.

EXAMPLE 1.7. Let K be an algebraically closed field and $A = K \oplus \ldots \oplus K$ be a K-vector space of dimension n. Let $T \simeq K^*$ be a one-dimensional torus acting on A by the matrices $\mathrm{diag}(t, t^2, \ldots, t^n)$ in a fixed basis $e_1, \ldots, e_n$ of A. Let B be a hyperplane in A trivially intersecting the subspaces $Ke_1, \ldots, Ke_n$. Then $\bigcap_{t \in T} B^t = 0$ and the right coset action of the natural semidirect product $G = AT$ on the coset space $\Omega = G/B$ is faithful and transitive. Since the stabilizer of the point B is not maximal in G (because $B < A$), this action is imprimitive.

Baldwin–Saxl Theorems. For convenience we include here the statements of the following two well-known theorems (Baldwin and Saxl, 1976; Baur *et al.*, 1979).

THEOREM 1.8. Let $\mathscr{F} = \{F_i : i \in I\}$ be a uniformly definable family of subgroups of a group G of finite Morley rank. Then there exists a natural number $n \in \mathbb{N}$ such that any intersection $\bigcap_{i \in J} F_i$, $J \subseteq I$, of subgroups from $\mathscr{F}$ can be written as an intersection of only n subgroups:

$$\bigcap_{i \in J} F_i = F_{i_1} \cap \ldots \cap F_{i_n},\ i_1, \ldots, i_n \in J.$$

THEOREM 1.9. Let $\mathscr{F} = \{F_i : i \in I\}$ be a uniformly definable family of subgroups of a group G of finite Morley rank. Then the lengths of chains of subgroups from $\mathscr{F}$ are uniformly bounded: there exists a constant m such that the length n of a chain of subgroups

$$F_{i_1} < F_{i_2} < \ldots < F_{i_n}$$

from $\mathscr{F}$ is less than m.

2 Generic faithfulness of a transitive action

This section contains the proof of Theorem 1.2.

Notation. Let us fix our notation. If $q \in S_1(\varnothing)$ is a stationary type, then $q{\restriction}A$ denotes the unique non-forking extension of q over A, and q^n denotes the type realized by n independent realizations of q.

If a, b are elements in a stable structure, $a \perp b$ means that a and b are independent.

Proof of Theorem 1.2. Let $R(\Omega) = n$. Since G is connected, it follows that $\deg(\Omega) = 1$. Let q be the unique type such that $\Omega \in q \in S_1(\varnothing)$ and $R(q) = n$. Suppose that there exists $1 \neq g \in G$ such that $R(\text{fix}(g)) = n$. Define

$$N = \{g \in G : R(\text{fix}(g)) = n\}.$$

It is clear that N is a normal subgroup of G. We now show that N is 0-definable. Notice that

$$\begin{array}{lll} g \in N & \text{iff} & \text{there exists } a \models q \text{ such that } a \perp g \text{ and } ga = a \\ & \text{iff} & gx = x \in q{\restriction}g. \end{array}$$

Since q is stationary, there exists a parameter-free formula $d(y)$ such that

$$gx = x \in q{\restriction}g \ \text{ iff } \ \models d(g).$$

Thus N is indeed 0-definable.

Choose a Morley sequence $\langle a_k : k < \omega \rangle$. Since N is ω-stable, there exists an integer m such that

$$N_{a_1,\ldots,a_m} = N_{a_1,\ldots,a_{m+1}}.$$

Now consider $1 \neq g \in N$. Since G acts faithfully on Ω, there exists $b \in \Omega$ with $gb \neq b$. Since G acts transitively on Ω and $N \lhd G$, we may assume that $b \models q$. Choose $b_1, \ldots, b_m \models q^m$ with $b_1, \ldots, b_m \perp g, b$. Then $b_1, \ldots, b_m, b \models q^{m+1}$ and

$$g \in N_{b_1,\ldots,b_m} \setminus N_{b_1,\ldots,b_m,b},$$

which contradicts the fact that $\text{tp}(b_1, \ldots, b_m, b) = \text{tp}(a_1, \ldots, a_{m+1})$. $\square$

3 Proof of Theorem 1.1

We shall prove Theorem 1.1 by induction on $n = R(G)$. If $n = 1$, then G is abelian (Reineke, 1975) and so the result clearly holds. Suppose then that $R(G) = n > 1$, and that the result holds for all $m < n$. Let

$$\mathscr{N} = \{N_i : i \in I\}$$

satisfy the hypothesis of Theorem 1.1. Clearly, we may assume that I has degree 1.

It follows from Theorem 1.8 that the family $\mathscr{L}$ of intersections of subgroups from $\mathscr{N}$ is also uniformly definable; and hence by Theorem 1.9 the lengths of chains in $\mathscr{L}$ are uniformly bounded. Let $\mathscr{M}$ be the definable family of minimal non-trivial subgroups from $\mathscr{L}$. Clearly we can suppose that $\mathscr{N}$ is infinite. Since the intersections of generic families of N_i are trivial it follows easily that $\mathscr{L}$ and $\mathscr{M}$ are also infinite.

For each $i \in I$, let

$$\mathscr{M}_i = \{M \in \mathscr{M} : M \leqslant N_i\}.$$

Define

$$J = \{i \in I : R(\mathscr{M}_i) = R(\mathscr{M})\}.$$

Claim. $R(J) < R(I)$.

Proof. For each $M \in \mathscr{M}$, let

$$J_M = \{i \in I : M \leqslant N_i\}.$$

Then $R(J_M) < R(I)$, since $M \leqslant \bigcap_{i \in J_M} N_i$. Suppose that $i \in I$ satisfies $R(i) = R(I)$. Whenever $M \in \mathscr{M}$ satisfies $M \leqslant N_i$, we have that $i \in J_M$ and hence $R(i/M) < R(i)$. By forking symmetry (see for example Pillay, 1983), $R(M/i) < R(M)$. It follows that $R(\mathscr{M}_i) < R(\mathscr{M})$ and so $i \notin J$. □

Let $K = I \setminus J$. If $i \in K$, then $[N_i, M] \leqslant N_i \cap M = 1$ for all M from the generic subset $\mathscr{M} \setminus \mathscr{M}_i$ of $\mathscr{M}$. Define

$$\tilde{N} = \langle N_i : i \in K \rangle$$

and $\tilde{M} = \langle M : M \in \mathscr{M}_i,\ i \in K \rangle$. Notice $\tilde{M} \leqslant \tilde{N}$. Since $\tilde{M}$ and $\tilde{N}$ are (not necessary definable) subgroups of the group G of finite Morley rank, they satisfy the descending chain conditions for centralizers. Thus

$$\mathrm{C}_{\tilde{M}}(\tilde{N}) = \mathrm{C}_{\tilde{M}}(N_{i_1}) \cap \ldots \cap \mathrm{C}_{\tilde{M}}(N_{i_s})$$

for some finite set of indices $\{i_1, \ldots, i_s\} \subseteq K$. Since $R(\mathscr{M}_i) < R(\mathscr{M})$, $i = i_1, \ldots, i_s$ by definition of K, the set

$$\bigcap_{i = i_1, \ldots, i_s} (\mathscr{M} \setminus \mathscr{M}_i)$$

is a generic subset of $\mathscr{M}$. Hence $C = \mathrm{C}_{\tilde{M}}(\tilde{N})$ is an infinite subgroup of $Z(\tilde{N})$. Thus $A = \mathrm{dc}(Z(\tilde{N})))$ is an infinite normal abelian subgroup of G satisfying $A \leqslant \mathrm{C}_G(\tilde{N})$. In particular, $R(G/A) < R(G)$.

If H is a subgroup of G, then we define $\overline{H} = HA/A$. In order to apply the inductive hypothesis to $\overline{G}$, we need to check that $\bigcap_{l \in L} \overline{N_l} = \overline{1}$ for any generic subset $L \subseteq K$. Set $\overline{H}_L = \bigcap_{l \in L} \overline{N_l}$. Let H_L be the full preimage of $\overline{H}_L$ in G. Then

$$H_L = \bigcap_{l \in L} N_l A.$$

Now if $i \in K$ is arbitrary, then by elementary properties of group commutators we have for each $l \in L$ that

$$[N_i, N_l A] = [N_i, N_l] \leqslant N_l$$

and so

$$[N_i, H_L] = [N_i, \bigcap_{l \in L} N_l A] \leqslant \bigcap_{l \in L} N_l = 1.$$

Thus $H_L \leqslant \mathrm{C}_G(\tilde{N})$. On the other hand, $H_L \leqslant \tilde{N}A$, so $H_L = A$ and $\overline{H}_L = \overline{1}$. Hence by inductive hypothesis, there exists a generic subset $J \subseteq K$ such that

$$\overline{N_i} = N_i A/A \cong N_i/N_i \cap A$$

is nilpotent for all $i \in J$. Since $N_i \cap A \leqslant Z(N_i)$, it follows that N_i is nilpotent for all $i \in J$.

On Frobenius groups of finite Morley rank I

Ali Nesin

Department of Mathematics,
University of California at Irvine,
Irvine,
California, CA 92717,
U.S.A.
Electronic mail address: anesin@math.uci.edu

1 Introduction

A *Frobenius group* is a group B with a proper and non-trivial subgroup T—called a *Frobenius complement*—such that for any $b \in B$, $T^b \cap T \neq 1$ implies $b \in T$. In this article, the phrase '$T < B$ is a Frobenius group' will mean that B is a Frobenius group and T is a Frobenius complement of B. Clearly if $T < B$ is a Frobenius group, then so is $T^b < B$ for any $b \in B$.

Frobenius groups play an important role in the theory of permutation groups because a group is a Frobenius group if and only if it is a transitive and non-regular permutation group where only the identity element fixes two distinct points. For example if G is a Zassenhaus group (a doubly transitive group where only the identity element fixes three distinct points) which is not a sharply 2-transitive group, then a one-point stabilizer B of G is a Frobenius group. Since the subgroup B of a Zassenhaus group of finite Morley rank is known to be definable (Delahan and Nesin, in press), the results of this article and of the next one in this volume can be useful in the classification of Zassenhaus groups of finite Morley rank.

Finite Frobenius groups have been investigated extensively. The finite Frobenius groups $T < B$ split, i.e., for some normal subgroup $U \lhd B$ (called *the Frobenius kernel*), $B = U \rtimes T$. This result is due to Frobenius and, as far as the author knows, all the known proofs make use of character theory (unless T has an involution in which case an easy counting argument suffices to prove the splitting (Passman, 1968)). Thompson (1959) proved that when B is finite, the Frobenius kernel U is necessarily nilpotent. This result was well-known when B is solvable (Higman, 1957). The structure of T was investigated by Zassenhaus (see Passman, 1968, §18).

The infinite Frobenius groups do not seem to have any common features unless additional hypotheses are assumed, like local finiteness or algebraicity (Collins, 1990; Hertzig, 1961). In this article we investigate Frobenius groups of finite Morley rank. Among other results we show that if $T < B$

is a Frobenius group of finite Morley rank, then the following statements hold.

1. The Frobenius complement T is necessarily definable in the pure group structure of B.
2. If $B = U \rtimes T$ is split, then U is also definable. Furthermore if U is solvable then U is nilpotent and the solvable subgroups of T are abelian-by-finite.
3. If B is solvable, then $B = U \rtimes T$ is split. Furthermore U is nilpotent and T is abelian-by-finite. Also, if B is connected, then $U = B'$.

In the next article (in this volume), we prove that a Frobenius group of finite Morley rank where the Frobenius complement T is finite, splits. We also consider minimal counterexamples to two conjectures regarding Frobenius groups: (1) we show that if B is a non-split Frobenius group of finite Morley rank whose proper and definable sections which happen to be Frobenius groups split then G is a counterexample to the Cherlin–Zil'ber Conjecture (i.e., is simple but not algebraic); (2) we show that if B is a Frobenius group of finite Morley rank whose proper, simple and definable sections are algebraic then either B is split and solvable or B is a counterexample to the Cherlin–Zil'ber Conjecture.

We conjecture that all infinite Frobenius groups of finite Morley rank are solvable.

Acknowledgement. We would like to thank Alexandre Borovik for helpful conversations and Dugald Macpherson for valuable comments.

2 Preliminaries

Our notation is mostly standard. If X is a definable subset of a structure of finite Morley rank, we let $\mathrm{rk}(X)$ and $\deg(X)$ be the Morley rank and degree of X. G always denotes a group. For $X \subseteq G$, we let $X^* = X \setminus \{1\}$. We use the notation $I(X)$ for the set of involutions of X, i.e., $I(X)$ is the set of elements of X^* whose square is 1. For $x, g \in G$ and $n \in \mathbb{N}$, we let $x^{ng} = g^{-1}x^n g$, $x^{-ng} = (g^{-1}x^n g)^{-1}$, and $\mathrm{ad}(g)(x) = [g, x] = g^{-1}x^{-1}gx$. We denote by x^G and x^{-G} the conjugacy classes of x and x^{-1} respectively. G' stands for the derived subgroup, $G^{(n+1)}$ is defined inductively as $G^{(1)} = G'$ and $G^{(n+1)} = (G^{(n)})'$ and $Z_n(G)$ stands for the nth centre of G, i.e., $Z_0(G) = 1$, $Z_1(G) = Z(G)$ is the centre of G and $Z_{n+1}(G)/Z_n(G) = Z(G/Z_n(G))$. If $X, Y \subseteq G$, then $[X, Y]$ denotes the subgroup generated by the elements of the form $[x, y]$ for $x \in X$ and $y \in Y$.

Let G be a group of finite Morley rank, and let $A \leqslant G$, $H \leqslant G$ be definable subgroups. Assume that A is infinite and normalized by H. Then, by the descending chain condition on definable subgroups, A contains a definable and infinite subgroup B which is normalized by H and which is minimal for these properties, i.e., if $C \leqslant B$ is infinite, definable and normalized by H then $C = B$. Such a subgroup B is called an *H-minimal subgroup of A*. Clearly an H-minimal subgroup is connected.

We will review a few facts used in the sequel. We start with the following fact.

FACT 2.1. (Gorenstein, 1980, Theorem 4.4, §3.4) Let $H = T \rtimes P$ be a group, where T is an elementary abelian normal q-subgroup of H, $|P| = p$, and p and q are two distinct primes. Assume further that $C_H(T) = T$ and that T is a minimal normal subgroup of H. Let V be a finite dimensional vector space over a field F of characteristic neither p nor q on which H acts faithfully. Then $C_V(P) \neq 0$.

The proof of this fact uses the following well-known result about representations of groups (Curtis and Reiner, 1988, Theorem 49.2, 49.6, 49.7; Gorenstein, 1980, §3 Theorem 4.1; Suzuki, 1986, §6, Theorem 1.5).

FACT 2.2. (Clifford's Theorem) Let G be any group and let $H \lhd G$ be a normal subgroup of G. Let K be any field and let V be an irreducible $K[G]$-module. Then V is a completely reducible $K[H]$-module and the irreducible $K[H]$-submodules of V are all conjugate[2] to each other under the action of G.

If G is finite then V is the direct sum of finitely many definable H-invariant subspaces V_i $(1 \leqslant i \leqslant r)$ which satisfy the following conditions:

1. There is a decomposition

$$V_i = W_{i1} \oplus W_{i2} \oplus \cdots \oplus W_{it}$$

where each W_{ij} is an irreducible $K[H]$-module $(1 \leqslant i \leqslant r)$, t is independent of i, and W_{ij}, $W_{i'j'}$ are isomorphic $K[H]$-submodules if and only if $i = i'$.
2. The group G acts on the set $\{V_1, \ldots, V_r\}$ and this action is transitive. Furthermore $HC_G(H)$ fixes each V_i.

Clifford's Theorem can be translated to groups of finite Morley rank quite easily (following the standard proof of the original statement) as follows.

FACT 2.3. (Clifford's Theorem for groups of finite Morley rank) Let G be a group acting irreducibly on an abelian group A. Let $H \lhd G$ be a normal subgroup. Assume that the subgroups A and H of $A \rtimes H$ are definable in $A \rtimes H$ and that $A \rtimes H$ has finite Morley rank. Assume further that $C_A(H^\circ) = 1$. Then A is the direct sum of finitely many definable H-invariant submodules A_i $(1 \leqslant i \leqslant r)$ which satisfy the following conditions.

[2]Let G be a group and let $H \lhd G$ be a normal subgroup of G. Let K be an arbitrary field, and finally let W_1, W_2 be two $K[H]$-modules. We say that W_1 and W_2 are *conjugate* under G if there is a $g \in G$ and a vector space isomorphism $\phi\colon W_1 \to W_2$ such that for $w \in W_1$ and $h \in H$ we have $h\phi(w) = \phi(h^g w)$.

1. There is a decomposition

$$A_i = B_{i1} \oplus B_{i2} \oplus \cdots \oplus B_{it}$$

   where each B_{ij} is an irreducible infinite H-module ($1 \leqslant i \leqslant r$), t is independent of i, and B_{ij}, $B_{i'j'}$ are isomorphic H-submodules if and only if $i = i'$. Furthermore for each i, i', j, j' there is a $g \in G$ such that $gB_{ij} = B_{i'j'}$.

2. For any H-submodule C of A, we have

$$C = C_1 \oplus C_2 \oplus \cdots \oplus C_r$$

   where $C_i = C \cap A_i$ ($1 \leqslant i \leqslant r$). In particular, any irreducible H-submodule lies in one of the A_i. Also, a direct sum of the H-submodules B_{ij} is a complement of C in A and C is completely reducible, definable (in $A \rtimes H$) and connected.

3. G permutes the submodules A_i transitively. If G is a definable and connected group then $r = 1$. Furthermore $H\mathrm{C}_G(H)$ fixes each A_i.

We will not give a proof of this fact here. A complete proof can be found in Borovik and Nesin (in press). The hypothesis that $\mathrm{C}_A(H^\circ) = 1$ is crucial. It is this hypothesis that makes the standard proof of Clifford's Theorem applicable to our context. For example, under this hypothesis, all the H-submodules of A are necessarily definable and connected, thanks to the following theorem of Zil'ber (1977) which is also applied several times in this paper.

FACT 2.4. (Zil'ber's Indecomposability Theorem) Let G be a group of finite Morley rank. Then the following hold:

1. If $H \leqslant G$ is a definable connected subgroup of G and X any subset of G, then the commutator subgroup $[H, X]$ is connected and definable;
2. The subgroup generated by any set of definable and connected subgroups of G is definable and connected.

In particular, if G is connected then G' is definable and connected. It also follows quite easily from Zil'ber's Indecomposability Theorem that a group of finite Morley rank is simple if and only if it has no proper, non-trivial, definable normal subgroups. Another (not quite immediate) consequence of this theorem is that the derived subgroup G' of any group G of finite Morley rank is always definable.

FACT 2.5. (Zil'ber, 1973; Nesin, 1989) Let G be a solvable and connected group of finite Morley rank. Then G' is nilpotent.

Now we list a few more facts about groups of finite Morley rank that may not be as well known as the other facts we will freely use.

We say that $\alpha \in \mathrm{Aut}(G)$ acts *freely* on G if $\alpha(g) = g$ implies $g = 1$. If $\alpha(g) = g$ we say that α *fixes* or *centralizes* g. If $T \leqslant \mathrm{Aut}(G)$, we say that T acts *freely* on G if every element of T^* acts freely on G. For $\alpha \in \mathrm{Aut}(G)$, we let

$$\mathrm{ad}(\alpha)(G) := \{\alpha(g)g^{-1} : g \in G\}.$$

The following is from Borovik *et al.*, in press*b*, Lemma 2.

FACT 2.6. (Borovik, DeBonis and Nesin) Let G be a nilpotent group of finite Morley rank and let $\alpha \in \mathrm{Aut}(G)$ be a definable automorphism that acts freely on G. Then $G = \mathrm{ad}(\alpha)(G)$.

FACT 2.7. (Nesin, 1990*a*, Proposition 4.1) Let G be a group of finite Morley rank and let α be a definable involutive automorphism of G.

1. If α centralizes only finitely many elements of G, then there is a definable abelian subgroup $B \lhd G$ of finite index which is inverted by α.
2. If α acts freely on G, then G is abelian, inverted by α and has no involutions.

LEMMA 2.8. Assume G is a group of finite Morley rank without elements of order p (a prime). Let $H \leqslant G$ be a definable subgroup. For $g \in G$, if $g^p \in H$ then $g \in H$.

Proof. $g^p \in Z(C_H(g))$ and the latter is an abelian group without elements of order p. Therefore, by Macintyre's theorem (Macintyre, 1971), $Z(C_H(g))$ is p divisible and so $g^p = c^p$ for some $c \in H$ that commutes with g. Hence $(gc^{-1})^p = 1$ and $g = c \in H$. □

FACT 2.9. (Borovik, 1982) Let G be a group of finite Morley rank and $i, j \in G$ be two involutions. Then either i and j are conjugate to each other or commute with a third involution.

If $X \subseteq G$ is a subset of a group G of finite Morley rank, by the descending chain condition on definable subgroups, there is a unique minimal definable subgroup of G that contains X. This subgroup is denoted by $d(X)$.

FACT 2.10. (Zil'ber, 1984*a*) Let G be a group of finite Morley rank and let $X, Y \leqslant G$. Then $[d(X), d(Y)] \leqslant d([X, Y])$. In particular, if $X \leqslant G$ is solvable (respectively nilpotent), then so is $d(X)$.

FACT 2.11. (Nesin, 1990*b*) Let G be a group of finite Morley rank and $H \lhd G$ be a definable normal nilpotent subgroup. Assume that G/H is abelian and that for some $g \in G$, $C_H(g) = 1$. Then $G = H \rtimes C_G(g)$ and any two complements of H in G are conjugate by an element of H.

3 Generalities

Our first lemma will show the reason for studying the Frobenius groups.

LEMMA 3.1. A group B is a Frobenius group if and only if it acts transitively but not regularly on a set in such a way that only the identity element of B fixes two distinct points.

Proof. Let $T < B$ be a Frobenius group. Let $X = B/T$ be the left coset space. Make B act on X by left multiplication. The converse is also easy. □

EXAMPLE 3.2. Let K be a field and let $T \leqslant K^*$ be a non-trivial subgroup. Then the standard semidirect product $K^+ \rtimes T$ is a Frobenius group.

EXAMPLE 3.3. A sharply 2-transitive group is a Frobenius group.

EXAMPLE 3.4. Let B be the free group on two generators x and y. Let $T = \langle [x, y] \rangle$. Then $T < B$ is a Frobenius group. T has no complement in B. This example is due to Kegel and Wehrfritz (1973, p. 51).

EXAMPLE 3.5. Let U be any group and let $T \leqslant \mathrm{Aut}(U)$ be an automorphism group that acts freely on U. Then the group $B = U \rtimes T$ is a split Frobenius group with T as a Frobenius complement. Conversely, if $B = U \rtimes T$ is a split Frobenius group with T as the Frobenius complement, then T acts freely on U.

Let us collect a few obvious properties of Frobenius groups.

LEMMA 3.6. Let $T < B$ be a Frobenius group.

1. If $B_1 \leqslant B$ is such that $T \cap B_1 \neq 1, B_1$, then $B_1 \cap T < B_1$ is a Frobenius group.
2. For $t \in T^*$, $\mathrm{C}_B(t) \leqslant T$.
3. For $1 \neq T_1 \leqslant T$, $\mathrm{N}_B(T_1) \leqslant T$.
4. $Z(B) = 1$. In particular if $B_1 \leqslant B$ is such that $T \cap B_1 \neq 1, B_1$, then $Z(B_1) = 1$.
5. For $a \in B \setminus \bigcup_{b \in B} T^b$, $\mathrm{C}_B(a)^* \subseteq B \setminus \bigcup_{b \in B} T^b$.
6. If $B = U \rtimes T$ is split, then $\mathrm{C}_B(U) \leqslant U$.

Proof. We only prove (6) and leave the rest as an exercise. Let $c \in \mathrm{C}_B(U)$. Write $c = ut$ where $u \in U$, $t \in T$. Since u and c commute, u and t commute also. Therefore, either $t = 1$ or $u = 1$. If $u = 1$, then $c = t \in \mathrm{C}_T(U) = 1$. If $t = 1$ then $c \in U$. □

4 Definability of the action

In this section we show that in a Frobenius group $T < B$ of finite Morley rank, the Frobenius complement T is necessarily definable. In other words the action of Lemma 3.1 is interpretable in the pure group structure of B.

PROPOSITION 4.1. Let $T < B$ be a Frobenius group of finite Morley rank. Then the Frobenius complement T is definable in the pure group structure of B.

Proof. Assume T is not definable. Then T is infinite.

CLAIM. For some $t \in T^*$, $C_B(t)$ is infinite.

Proof. Assume, in order to get a contradiction, that $C_B(t)$ is finite for all $t \in T^*$. We first show that T has no involutions. Assume T has an involution i. Then $C_B(i)$ being finite, by Fact 2.7, i inverts B° and so B° is abelian. Then for $t \in T \cap B^\circ \setminus \{1\}$, $T \cap B^\circ \leqslant C_B(t)$, contradicting our assumption that $C_B(t)$ is finite. Therefore T has no involutions.

Let $t \in T^* \cap B^\circ$. Since $C_B(t)$ is finite, t^B and t^{-B} are generic conjugacy classes in B°, therefore $t^B = t^{-B}$ and $t^b = t^{-1}$ for some $b \in B$. Since $t^{-1} = t^b \in T \cap T^b$, $b \in T$. In particular $b^2 \neq 1$. But now, $1 \neq b^2 \in C_B(t) \leqslant T$ and by Fact 2.8, $b \in C_B(t)$. Then $t^{-1} = t^b = t$ and t is an involution. This is a contradiction and the claim is proved. □

It follows from the claim that the subgroup $T_1 = \langle C_B(t)^\circ : t \in T^* \rangle$ is infinite. T_1 is clearly a normal subgroup of T. By Zil'ber's Indecomposability Theorem, T_1 is definable. Since $T = N_B(T_1)$, T is also definable. □

We will soon show that in a split Frobenius group of finite Morley rank, the Frobenius kernel is definable as well. But first we prove a few consequences of the definability of T.

COROLLARY 4.2. Let $T < B$ be a Frobenius group of finite Morley rank. Then $\bigcup_{b \in B} T^b$ is a generic subset of B.

Proof. The subgroups T^b and T^c are either equal or disjoint (except for the identity), and they are equal if and only if $c^{-1}b \in T$. Therefore,

$$\mathrm{rk}(\cup_{b \in B} T^b) = \mathrm{rk}(\bigsqcup_{\bar{b} \in B/T} T^b) = \mathrm{rk}(B/T) + \mathrm{rk}(T) = \mathrm{rk}(B).$$

The corollary is proved. □

LEMMA 4.3. (Delahan–Nesin) Let $T < B$ be a Frobenius group of finite Morley rank. Assume T has an involution. Then all the involutions of B are conjugate to each other and T has finitely many of them. Furthermore the involutions of T are conjugate by T and T° centralizes them. In particular if T is connected then T has a unique involution.

Proof. Let $i \in T$ and $j \notin T$ be involutions of B. By Fact 2.9, either they commute with a third involution k or they are conjugate to each other. Assume they commute with a third involution k. Then $k \in C_B(i) \leqslant T$ and therefore $j \in C_B(k) \leqslant T$, a contradiction. Since the conjugation relation is transitive, this proves the first assertion.

Now let $X = B/T$. We may assume that the set on which the group B acts as a Frobenius group is X. The Morley rank of the set of transpositions $A := \{(x, y) : x \neq y \in X\}$ is $2\,\mathrm{rk}(X)$. Now we calculate the Morley rank of transpositions involved in an involution. Let $I(B)$ and $I(T)$ denote the set of involutions of B and T respectively. For $i \in I(B)$, define the set $A(i) = \{(x, ix) : x \in X \text{ and } ix \neq x\}$. Since each element can fix at most one point of X, $\mathrm{rk}(A(i)) = \mathrm{rk}(X)$. Also, for two distinct involutions i and j, $A(i) \cap A(j) = \varnothing$, because if not ij would fix at least two points. Therefore the Morley rank of the set $\bigcup_{i \in I(B)} A(i)$ is $\mathrm{rk}(I(B)) + \mathrm{rk}(X)$. Since $\mathrm{rk}(A) = 2\,\mathrm{rk}(X)$, it follows that $\mathrm{rk}(I(B)) \leqslant \mathrm{rk}(X)$. On the other hand, for $i \in I(T)$, we have

$$\mathrm{rk}(I(B)) = \mathrm{rk}(i^B) = \mathrm{rk}(B) - \mathrm{rk}(C_B(i)) \geqslant \mathrm{rk}(B) - \mathrm{rk}(T) = \mathrm{rk}(X).$$

Therefore $\mathrm{rk}(X) = \mathrm{rk}(I(B))$.

Say that two involutions i and j of B are equivalent if $i, j \in T^b$ for some $b \in B$. Since $T < B$ is a Frobenius group, this is an equivalence relation, the rank of each class is $\mathrm{rk}(I(T))$ and the rank of the set of classes is equal to the rank of the set of conjugates of T, which is $\mathrm{rk}(B/T)$. Therefore $\mathrm{rk}(I(B)) = \mathrm{rk}(I(T)) + \mathrm{rk}(B/T)$. With the equality $\mathrm{rk}(I(B)) = \mathrm{rk}(X)$ proved above we get $\mathrm{rk}(I(T)) = 0$. This proves the second statement. The rest of the statements are easy and are left as an exercise. □

Now we prove that if T is finite and has an involution, then B splits. In the next article, we will see that the conclusion still holds even if T has no involution.

PROPOSITION 4.4. Let $T < B$ be a Frobenius group of finite Morley rank. Assume T is finite and has an involution i. Then $B = U \rtimes T$ for some definable, abelian and normal subgroup U which is inverted by i. Furthermore i is the unique involution of T.

Proof. All this is known when B is finite, so assume B is infinite. By Fact 2.7, there is a definable, normal subgroup $V \lhd B$ of finite index which is inverted by i. In particular V is abelian and $V \cap T = 1$. Take U to be a maximal normal subgroup containing V and inverted by i. Since U is abelian, $U \cap T = 1$. Assume T has another involution, say j. Then j inverts a subgroup W of finite index of B. Since B is infinite, $U \cap W \neq 1$ and ij centralizes $U \cap W$. But then $U \cap W \leqslant C_B(ij) \leqslant T$, a contradiction. So T has a unique involution, which of course must be central in T. Hence $i \in Z(T)$ and $T = C_B(i)$. We clearly have the following facts: $U \leqslant C_B(U)$,

$C_B(U) \cap T = 1$ and, by Fact 2.7(ii), i inverts $C_B(U)$. Now, the maximality of U gives $C_B(U) = U$. In particular U is definable.

Let $j \neq i$ be another involution of B. By Lemma 4.3, i and j are conjugate. Hence j inverts U as well. Therefore $ij \in C_B(U) = U$. Since U has no involutions (if not i would commute with this involution), U is 2-divisible and therefore there is a $v \in U$ such that $v^2 = ij$. Computing, $i^v = iiv^{-1}iv = iv^2 = iij = j$, we see that i and j are conjugate by an element of U.

Let $g \in B$ be any element. By above $i^g = i^v$ for some $v \in U$. Hence $g = gv^{-1}v \in C_B(i)U = TU = UT$. Therefore $B = UT = U \rtimes T$. □

Now we prove the definability of the Frobenius kernel in case it exists.

PROPOSITION 4.5. Let $B = U \rtimes T$ be a split Frobenius group of finite Morley rank. Then U is definable. Furthermore the following hold.

1. If B is infinite, then U is also infinite.
2. If T has an involution i then this involution is unique in T and i acts on U by inversion. In particular U is abelian.

Proof. Assuming U is definable, let us first prove the two items.

1. If U were finite, then T° would centralize U and T would be finite.
2. By Fact 2.7, i inverts U. The rest follows from this as in the above proposition.

Now we proceed to prove that U is definable. We may assume B is infinite. Recall that T is definable.

Assume U is not definable. Then $U < d(U) = U \rtimes (T \cap d(U))$ and $d(U)$ is a Frobenius group. Therefore, by induction we may assume that $B = d(U)$.

Case 1. T is finite. If T has an involution, this is Proposition 4.4. Assume T has no involution. By the Feit–Thompson Theorem, which states that a finite group without involutions is solvable (Feit and Thompson, 1963), the subgroup T is solvable.

For $t \in T^* \cap B$, the map

$$\mathrm{ad}(t) : B/C_B(t) \longrightarrow B$$

given by

$$C_B(t)b \longmapsto [b, t]$$

is well-defined and one-to-one. Since $C_B(t) \leqslant T$ and T is finite, $\mathrm{ad}(t)(B)$ is a generic subset of B. It follows that B' has finite index in B. Since B' is definable, UB' is also definable as a finite union of cosets of B'. Note that $UB' = UT' < UT = B$. If $U = UB'$, then there is no problem. If $U < UB'$, then UB' is a Frobenius group, and by induction U is definable.

Case 2. T is infinite. Let $U_1 := [T^\circ, U] \leqslant U$. By Zil'ber (1977) U_1 is definable and connected. We will show that U_1 has finite index in U. We have

$$[B, T^\circ] = [d(U), d(T^\circ)] \leqslant d([U, T^\circ]) = U_1 \leqslant U$$

by Lemma 2.10. In particular T° is central in T. Let $1 \neq t \in T^\circ$. We easily get $\mathrm{rk}(\mathrm{ad}(t)(B)) = \mathrm{rk}(B/\mathrm{C}_B(t)) = \mathrm{rk}(B/T)$. Also

$$\mathrm{ad}(t)(B) = \mathrm{ad}(t)(U) \subseteq [U, T^\circ] = U_1.$$

Therefore

$$\mathrm{rk}(B/T) = \mathrm{rk}(\mathrm{ad}(t)(B)) \leqslant \mathrm{rk}(U_1) \leqslant \mathrm{rk}(B/T).$$

So, $\mathrm{rk}(U_1) = \mathrm{rk}(B/T)$ and U_1T is a definable subgroup of finite index of B. Since $U_1 = U \cap U_1T$, it follows that U_1 is a subgroup of finite index of U and U is definable as a finite union of cosets of U_1. □

We end this section with one lemma and one corollary.

LEMMA 4.6. Let U be a group of finite Morley rank with a definable automorphism t that acts freely on U. Let $V \lhd U$ be a definable subgroup of U normalized by t. Then t acts freely on U/V.

Proof. Let $W = \{a \in U : a^{-1}t(a) \in V\}$. Then W is a definable subgroup containing V and the map $w \longmapsto w^{-1}t(w)$ from W into V is definable and one-to-one. Therefore V and W have the same Morley degree and rank. Hence $V = W$. The lemma follows easily. □

COROLLARY 4.7. Let $B = U \rtimes T$ be a split Frobenius group of finite Morley rank. Let $V \leqslant U$ be a B-normal subgroup of U. The following hold:

1. if V is definable and $V < U$ then $U/V \rtimes T$ is a Frobenius group;
2. if T is infinite then V is definable and connected. In particular U is connected.

Proof. Part 1 follows from Lemma 4.6. We prove part 2. We may assume that $V \neq 1$. The subgroup $[V, B^\circ]$ is definable, connected, non-trivial and is in V. By part 1, $(U/[V, B^\circ]) \rtimes T$ is a Frobenius group. Also $1 \neq [V, T^\circ] \leqslant [V, B^\circ]$. By induction $V/[V, B^\circ]$ is definable. Therefore V is definable.

Since $V \rtimes T$ is a Frobenius group of finite Morley rank, if V were not connected, by part (i), $V/V^\circ \rtimes T$ would be a Frobenius group, but as T is infinite, this is impossible. □

5 Solvable Frobenius groups

The following theorem is very important in the theory of finite groups (Passman, 1968, Theorem 17.4; Suzuki, 1986, Chapter 6, Theorem 2.6; or Gorenstein, 1980, §10.2, Theorem 2.1).

FACT 5.1. (Thompson, 1959) Let $B = U \rtimes T$ be a finite Frobenius group. Then U is nilpotent.

The analogue of this fact for groups of finite Morley rank is not known. Hertzig partially generalized Fact 5.1 to algebraic groups:

FACT 5.2. (Hertzig, 1961, Theorem 2) Let $T < B$ be a Frobenius group which is also an algebraic group over an algebraically closed field. Assume T is an algebraic subgroup of B. Then B splits, say as $U \rtimes T$. Furthermore U is abelian-by-finite.

When U is solvable, Fact 5.1 has a much simpler proof recorded by Higman (1957) that we were able to extend to our context.

THEOREM 5.3. Let $U \rtimes T$ be a split Frobenius group of finite Morley rank. Assume U is solvable. Then U is nilpotent. Furthermore the solvable subgroups of T are abelian-by-finite.

Proof. We first prove the second statement. Replacing U by U° we may assume that U is connected. If U is not nilpotent, replacing U by U', we may assume that U is nilpotent (Fact 2.5). Let $S \leqslant T$ be a solvable subgroup. Since the definable closure $d(S)$ of S is still solvable (Fact 2.10), replacing S by $d(S)$, we may assume that S is definable. We need to show that S is abelian-by-finite, i.e., that S° is abelian. Replacing S by S°, we may assume that S is connected. Assume S is not abelian. The group US is a solvable connected group and by Fact 2.6, $(US)' = US'$. On the other hand $(US)'$, being the derived subgroup of a connected solvable group, is nilpotent (Fact 2.5), and so has a non-trivial centre Z. Then $Z \leqslant C_B(S') \leqslant T$ and $U \leqslant C_B(Z) \leqslant T$, a contradiction.

To prove the first statement we will proceed in several steps and by induction on the Morley rank and degree of U. Recall that T and U are definable subgroups.

(a) We may assume U is infinite, T is abelian. If U is finite then T is finite and in this case the result is just Fact 5.1, or rather Higman's result that we have already mentioned. From now on we assume that U is infinite.

Replacing T by a definable abelian subgroup (for example by the centre of some centralizer), we may assume that T is abelian. Furthermore, if T is infinite, we may assume that it is connected.

(b) Remarks. We first show that there is an infinite, definable, B-normal, and proper subgroup of U. To show this, we may assume U' is finite. Let A be a U-minimal subgroup of U. Since A is connected, $[A, U]$ is a connected subgroup of U' by Zil'ber's Indecomposability Theorem. Hence $[A, U] = 1$ and A is central in U. Now consider the subgroup $V := \langle A^t : t \in T \rangle \leqslant Z(U)$. The subgroup V satisfies the requested properties.

Let V be a proper, infinite, definable and T-invariant subgroup of U whose existence has just been shown. By induction V is nilpotent (in particular U' is nilpotent). Also $V = \mathrm{ad}(t)(V)$ for any $t \in T^*$ (Fact 2.6). By Corollary 4.7, the group $(U/V) \rtimes T$ also satisfies the hypothesis of the theorem. Therefore, as V is infinite, U/V is nilpotent by induction and again $\mathrm{ad}(t)(U/V) = U/V$, i.e., $U = \mathrm{ad}(t)(U) \cdot V$. Since $V = \mathrm{ad}(t)(V)$, we get $U = \langle \mathrm{ad}(t)(U) \rangle$ for any $t \in T^*$. Since T is abelian, this shows that $B' = U$.

(c) The case when T is infinite. By Corollary 4.7, U is connected. It follows that B is a connected solvable group. Therefore B' is nilpotent. Since $B' = U$, we are done in this case by Fact 2.5.

From now on we assume that T is finite. Then we can also assume that $T = \langle \alpha \rangle$ is cyclic of prime order, say p. If for some n, $Z_n(U)$ is infinite, by taking $V = Z_n(U)$ in part b, we are done. Therefore, we assume that $Z_n(U)$ is finite for all n. We will obtain a contradiction.

(d) The case when the series $Z_n(U) \cap U'$ is stationary. We assume there is an n such that $U' \cap Z_n(U) = U' \cap Z_{n+1}(U)$. Dividing U by $Z_n(U)$, we may assume that $U' \cap Z(U) = 1$. Since U' is nilpotent, U' has a non-trivial centre. Let A be a U-normal, definable, α-invariant subgroup of $Z(U')$ which is minimal for these properties. Consider the group $V = A \rtimes U/\mathrm{C}_U(A)$ of finite Morley rank. Note that $U/\mathrm{C}_U(A)$ is abelian (because $A \leqslant Z(U')$) and non-trivial (because $Z(U) \cap U' = 1$). The automorphism α of U acts also on the group V by $\alpha(a, \overline{g}) = (\alpha(a), \overline{\alpha(g)})$. We claim that α does not fix any non-trivial elements of V. Assume $\alpha(a, \overline{g}) = (a, \overline{g})$. Then $a = 1$ and $\alpha(g)^{-1}g \in \mathrm{C}_U(A)$. But $\mathrm{C}_U(A)$ is nilpotent by induction. Therefore, by Fact 2.6, $\alpha(g)^{-1}g = \alpha(u)^{-1}u$ for some $u \in \mathrm{C}_U(A)$. But then $\alpha(ug^{-1}) = ug^{-1}$ and $g = u \in \mathrm{C}_U(A)$. So $\overline{g} = 1$. The claim is proved. Let $S = U/\mathrm{C}_U(A)$. Note that $\alpha(S) = S$ and $\mathrm{C}_S(A) = 1$.

Now we consider two subcases.

(d1) The case when S is finite. We will show that α fixes a non-trivial element of A. This will give the required contradiction. Assume first that the abelian group A has a non-trivial element a of finite order. Consider the group

$$C_1 := \langle a^S, \alpha(a)^S, \ldots, \alpha^{p-1}(a)^S \rangle.$$

Since C_1 is a finite group, it is definable. Furthermore C_1 is U-normal and α-invariant. Therefore $C_1 = A$. It follows that $A \rtimes S$ is a finite group with a fixed-point-free automorphism of prime order p. By Higman's result $A \rtimes S$ is nilpotent. But then $Z(A \rtimes S) \cap A \neq 1$ and so $Z(A \rtimes S) = A$ by the minimality of A. Hence S commutes with A, a contradiction. Therefore A is torsion-free and hence is divisible. If S had an element of order p, then α would normalize the unique p-Sylow subgroup S_p of the abelian group S

and so α would fix a non-trivial element of S_p, a contradiction. Therefore S is a p'-group (i.e. has no elements of order p). Let $s \in S^*$ be an element of prime order $q \neq p$. Replacing S by $\langle s, \alpha(s), \ldots, \alpha^{p-1}(s)\rangle$, we can assume that S is an elementary abelian q-group. By further reduction, we can also assume that S is a minimal subgroup of $S \rtimes \langle\alpha\rangle$. View A as a vector space over $\mathbb{Q}$. Let $a \in A^*$. Consider the $\mathbb{Q}$-subspace V of A generated by

$$\{a^S, \alpha(a)^S, \ldots, \alpha^{p-1}(a)^S\}.$$

Since S is finite, V is a finite dimensional $\mathbb{Q}$-vector space, and the finite group $S \rtimes \langle\alpha\rangle$ acts on it. We can apply Fact 2.1 to conclude that the group $\langle\alpha\rangle$ must centralize an element of V^*, a contradiction.

(d2) The case when S is infinite. $\mathrm{C}_A(S^\circ)$ is an $(S \rtimes \langle\alpha\rangle)$-invariant subgroup, so either $\mathrm{C}_A(S^\circ) = 1$ or $\mathrm{C}_A(S^\circ) = A$. But in the second case, we get $S^\circ \leqslant \mathrm{C}_S(A) = 1$, i.e., S is finite, a contradiction. Thus $\mathrm{C}_A(S^\circ) = 1$. Now we show that A is irreducible as an $(S \rtimes \langle\alpha\rangle)$-module. Let $C_1 \leqslant A$ be a non-trivial $(S \rtimes \langle\alpha\rangle)$-submodule. For $a \in C_1^*$, the subgroup $C_2 := [a, S^\circ]$ of C_1 is connected and definable by Zil'ber's Indecomposability Theorem. Since $\mathrm{C}_A(S^\circ) = 1$, $C_2 \neq 1$. Again by Zil'ber's Indecomposability Theorem, the subgroup

$$C_3 := \langle C_2^g : g \in S \rtimes \langle\alpha\rangle\rangle$$

of C_1 is definable. Since C_3 is normalized by $S \rtimes \langle\alpha\rangle$, by the minimality of A, $C_3 = A$. Therefore $C_1 = A$ and A is an irreducible $(S \rtimes \langle\alpha\rangle)$-module.

We have just seen that we can apply Fact 2.3 to the group $S \rtimes \langle\alpha\rangle$ as acting on the abelian group A. Hence $A = A_0 \oplus \cdots \oplus A_r$ where each A_i is S-invariant and $\langle\alpha\rangle$ permutes the subgroups A_i transitively. Since $o(\alpha) = p$ is a prime, either $r = 0$ or $r = p - 1$. Assume first $r = p - 1$. Reorder the subgroups A_i in such a way that $\alpha^i(A_0) = A_i$. Then for $a \in A_0^*$, the element $a\alpha(a)\cdots\alpha^{p-1}(a)$ is a non-trivial element fixed by α, a contradiction. Therefore $r = 0$. Continuing to apply Fact 2.3, we can write $A = B_1 \oplus \cdots \oplus B_t$ where the subgroups B_i are irreducible and definable S-modules which are isomorphic to each other. Since $S \leqslant \mathrm{Aut}(B_1) \subset \mathrm{End}(B_1)$, we can consider the subring R of $\mathrm{End}(B_1)$ generated by S. Recall the following: R is canonically isomorphic to the ring $\mathbb{Z}[S]/\mathrm{ann}_{\mathbb{Z}[S]}(B_1)$ where $\mathbb{Z}[S]$ is the group ring over $\mathbb{Z}$, an element $\gamma = n_1 s_1 + n_2 s_2 + \ldots + n_k s_k \in \mathbb{Z}[S]$ acts on B_1 by

$$\gamma(b) = \prod_{j=1}^{k} b^{n_j s_j}$$

and $\mathrm{ann}_{\mathbb{Z}[S]}(B_1)$ is the annihilator of B_1 in $\mathbb{Z}[S]$. By Zil'ber, R is an interpretable algebraically closed field. But the S-modules B_i are isomorphic. Therefore the annihilators $\mathrm{ann}_{\mathbb{Z}[S]}(B_i)$ are all equal and A is an R-vector space. Since $\alpha(S) = S$, the action of α on S can be extended to an action on R: if $\sum_{i=1}^{k} n_i s_i = 0$ in R, then (with the multiplicative notation) $\prod_{i=1}^{k} a^{n_i s_i} = 1$ for all $a \in A$, and applying α to this last equality we get $\prod_{i=1}^{k} \alpha(a)^{n_i \alpha(s_i)} = 1$, i.e., $\sum_{i=1}^{k} n_i \alpha(s_i) = 0$. Clearly, this action of $\langle\alpha\rangle$ on R gives rise to an interpretable group of field automorphisms. Therefore, α acts identically on R (Poizat, 1987, Corollaire 3.6) and hence on S, a contradiction.

(e) The case when U is connected. If U is connected, then the series $Z_n(U)$ is necessarily stationary[3] and therefore, by part (d), the theorem is proved in this case.

End of the proof. By part (e), we may assume that $U^\circ \neq U$. Then $U'U^\circ \neq U$ and is α-invariant. Let V be a maximal α-invariant subgroup containing $U'U^\circ$. Since $U' \leqslant V$, V is a normal subgroup. Also, since $U'U^\circ$ has finite index in V, V is definable. By induction V is nilpotent. Also, as $U' \leqslant V$, U/V is abelian.

Since V is infinite and nilpotent, its centre is infinite (Lemma 1 of (Borovik *et al.*, in press*a*)). Let A be a $(U \rtimes \langle\alpha\rangle)$-minimal subgroup of $Z(V)$. We want to show that U has infinite centre. This will prove the theorem, because, by part (b), $U/Z(U)$ is nilpotent. Assume $Z(U)$ is finite. Let $U_1 = A \rtimes S$ where $S := U/V$ is a finite group stabilized by α. Note that U_1 is interpretable. Since A is minimal, U has finite centre and $[A, S]$ is connected, we have $[A, S] = A$. Thus $A = U_1'$. By part (d), if the series $A_n = Z_n(U_1) \cap A$ stabilizes, then U_1 is nilpotent. Since $Z(U_1)$ is infinite and A has finite index in U_1, $A \cap Z(U_1)$ is infinite. But $A \cap Z(U_1) \leqslant Z(U)$, a contradiction. Therefore the series A_n is strictly increasing and each of them is finite. Let $B = \bigcup_{n\in\mathbb{N}} A_n$. Then the group $B \rtimes S$ is a locally nilpotent group which has a fixed-point-free automorphism α of prime order. Now we need the following fact.

FACT 5.4. (Higman, 1957, Theorem 3) There is a function k such that if a locally nilpotent group has an automorphism of prime order p which leaves fixed no elements except the identity, then it is nilpotent of class at most $k(p)$.

It follows that $B \rtimes S$ is nilpotent. Let $B_n := Z_n(B \rtimes S) \cap B$, and let n be such that B_{n+1} is infinite but B_n is finite. Since $[S, B_{n+1}] \leqslant B_n$, $C_{B_{n+1}}(S)$ is infinite (in fact has finite index in B_{n+1}). But $C_{B_{n+1}}(S) \leqslant C_A(S) \leqslant Z(U)$, a contradiction. □

COROLLARY 5.5. Let G be a solvable group of finite Morley rank. Let $T \leqslant \mathrm{Aut}(G)$ be a definable group of automorphisms of G. Assume T acts freely on G. Then G is nilpotent.

---

[3]For any connected group G of finite Morley rank, if $Z(G)$ is finite, then $G/Z(G)$ is centreless (Cherlin, 1979).

This result generalizes the second part of Hertzig's result quoted above as Fact 5.2. The subgroup U of that fact is in fact nilpotent.

Our final result of this section will show that solvable Frobenius groups of finite Morley rank are well-understood:

THEOREM 5.6. Let $T < B$ be a solvable Frobenius group of finite Morley rank. Then B is split and T° is abelian. Furthermore the following hold:

1. The Frobenius kernel is $(B \setminus \bigcup_{b \in B} T^b) \cup \{1\}$ and is nilpotent;
2. If B is connected, then $B = B' \rtimes T$;
3. All the complements of the Frobenius kernel are conjugate to each other.

Proof. By Theorem 5.3, if B splits then T° is abelian. We next show that if $B = U \rtimes T$ for some (necessarily definable) subgroup U, then the properties 1–3 hold. Note that U is nilpotent by Theorem 5.3. By Fact 2.6, $U = \mathrm{ad}(t)(U)$ for any $t \in T^*$. Let $b = ut \in B \setminus U$ where $u \in U$, $t \in T^*$. Let $v \in U$ be such that $u = [v^{-1}, t^{-1}]$. Then $b^v = t \in T$. This is (1).

Let us prove (2). Since B is connected, B' is nilpotent. Also, by Fact 2.6, $U \leqslant B'$. Therefore $B' = U \rtimes (B' \cap T)$. Assume $B' \cap T \neq 1$. Then $Z(B') \leqslant C_B(B' \cap T) \leqslant T$. But then $U \leqslant B' \leqslant C_B(Z(B')) \leqslant C_B(T) \leqslant T$, a contradiction. Therefore $B' \cap T = 1$ and $B' = U$.

Now we prove (3). Let S be another complement of U in B. Let r be such that $S^{(r)} \neq 1$ but $S^{(r+1)} = 1$. Let $1 \neq s \in S^{(r)}$. By part (1) we may assume that $s \in T$. Then $S^{(r)} \leqslant C_B(s) \leqslant T$ and $S \leqslant N_B(S^{(r)}) \leqslant T$.

Now we proceed in proving the splitting. We will proceed by induction on the Morley rank and degree. We take a smallest counterexample B. Therefore B is infinite. Let $1 \neq H \lhd B$ be a nilpotent definable subgroup. We claim that $H \cap T = 1$. Indeed, if $1 \neq t \in H \cap T$, then $Z(H)$ commutes with t, and so $Z(H) \leqslant T$. But then $H \leqslant C_B(Z(H)) \leqslant T$, and so for all $g \in B$, $H \leqslant T \cap T^g$. It follows that $T = B$, a contradiction.

First assume B is connected. Then B' is nilpotent, and $T \cap B' = 1$ by above. Therefore, T is abelian and $C_{B'}(t) = 1$ for any $t \in T^*$. By Fact 2.11, $B = B' \rtimes C_B(t)$. But then $C_B(t) = T$. The theorem is proved in this case. From now on we assume that B is not connected. Therefore $B'B^\circ < B$.

If $B'B^\circ \cap T \neq 1$, then we can apply the induction hypothesis to $B'B^\circ$: $B'B^\circ = U \rtimes (B'B^\circ \cap T)$ for some definable nilpotent subgroup U. Then $B'B^\circ T = U \rtimes T$. Since $B' \leqslant UT$, $UT \lhd B$. Let $g \in B$ and $t \in T^*$. Then $t^g \in UT$. By part (1), $t^g = t^u$ for some $u \in U$. Hence $gu^{-1} \in T$ and $g \in UT$. This proves that $B = UT = U \rtimes T$.

Assume now $B'B^\circ \cap T = 1$. Then T is abelian and, by Theorem 5.3, $B'B^\circ$ is nilpotent. Now we can apply Fact 2.11: $B = B'B^\circ \rtimes C_B(t)$ for any $t \in T^*$. Again $C_B(t) = T$ and we are done. $\square$

On Frobenius groups of finite Morley rank II

David Epstein and Ali Nesin

Department of Mathematics,
University of California at Irvine,
Irvine,
California,
CA 92717,
U.S.A.
Nesin's electronic mail address: anesin@math.uci.edu

1 Introduction

This is a continuation of the previous article by Nesin in this volume. The reader is invited to read the introduction of that article. We shall adopt the same notation throughout.

In this article, we will first show that if $T < B$ is a Frobenius group of finite Morley rank where T is finite, then B is split as $U \rtimes T$ (Theorem 3.1). Then we will deal with the minimal counterexamples to the following two conjectures.

CONJECTURE 1.1. (Cherlin–Zil'ber Conjecture) A simple group of finite Morley rank is an algebraic group over an algebraically closed field.

CONJECTURE 1.2. Let $T < B$ be a Frobenius group of finite Morley rank. Then $B = U \rtimes T$ for some (necessarily definable) normal subgroup U.

Before stating our next result we need a definition. Let G be a group of finite Morley rank. Let A and B be definable subgroups of G, and assume that $A \lhd B$. Then the group B/A is called a *definable section* of G. If either $\mathrm{rk}(B/A) < \mathrm{rk}(G)$ or $\mathrm{rk}(B/A) = \mathrm{rk}(G)$ and $\deg(B/A) < \deg(G)$, then B/A is called a *proper definable section* of G.

We show that if $T < B$ is a non-split Frobenius group of finite Morley rank whose proper sections which are Frobenius groups split (i.e., $T < B$ is in some sense a minimal counterexample to Conjecture 1.2) then B is a counterexample to Conjecture 1.1 (Theorem 4.1). We also show that if all the proper and simple sections of an infinite Frobenius group $T < B$ of finite Morley rank satisfy Conjecture 1.1, then either B is simple and a counterexample to Conjecture 1.1, or B is solvable, in which case $B = U \rtimes T$ with U nilpotent and T is abelian-by-finite (Theorem 4.2). Since the hypothetical bad groups are Frobenius groups, we cannot expect to do better than these results (at least not in such a short article). On the other hand, bad groups do not have involutions, and so one may think that there may be some hope in classifying Frobenius groups of finite Morley rank that contain an involution. However, sharply 2-transitive groups are Frobenius groups with involutions, and the problem of classifying these groups poses serious problems.

Acknowledgement. The authors would like to thank Dugald Macpherson for various valuable comments.

2 Preliminaries

A group of finite Morley rank may not have minimal normal subgroups (e.g., the group $\mathbb{Q}$ does not have minimal normal subgroups), but whenever the minimal normal subgroups exist, they are definable. On the other hand, a group of finite Morley rank whose connected component is centreless has minimal normal subgroups; in fact, every normal subgroup of such a group contains a minimal normal subgroup.

We recall that if G is a group of finite Morley rank, $S(G)$ and $R(G)$ stand for the socle and the radical of G, i.e., $S(G)$ is the subgroup generated by all the minimal normal subgroups and $R(G)$ is the subgroup generated by all the solvable normal subgroups. By Nesin (1991), $R(G)$ is definable and solvable. A *G-minimal* subgroup of G is a definable, normal and infinite subgroup of G which is minimal with respect to these properties. We also recall the following fact which can be found in Borovik and Nesin (in press) and which is part of the folklore.

FACT 2.1. Let G be a group of finite Morley rank. If $Z(G^\circ) = 1$ then $S(G)$ is definable and the following hold:

1. Assume G is connected.
   (a) $S(G)$ is the direct sum of finitely many G-minimal subgroups, say,
   $$S(G) = A_1 \oplus \cdots \oplus A_n,$$
   where each A_i is either abelian or simple.
   (b) If G is solvable, then each A_i is abelian.
2. Assume $R(G) = 1$. Then $S(G)$ is a direct sum of finitely many non-abelian simple (necessarily definable) subgroups. More precisely, we have
   $$S(G) = A_1 \oplus \cdots \oplus A_n$$
   where each A_i is G-normal and is a direct sum of finitely many isomorphic (in fact G-conjugate) simple definable subgroups B_{ij} which are normalized by G° and permuted by G transitively. If G is also connected then each A_i is an infinite simple subgroup of G.

We end this section with a lemma which is well-known among model theorists. We supply an algebraic proof of it.

LEMMA 2.2. Let $\mathscr{G} = G \rtimes H$ be a group of finite Morley rank where G and H are definable, G is an infinite simple (as an abstract group) algebraic group over an algebraically closed field, and $C_H(G) = 1$. Then, viewing H as a subgroup of $\mathrm{Aut}(G)$, we have $H \leqslant \mathrm{Inn}(G)\Gamma$ where $\mathrm{Inn}(G)$ is the inner automorphisms of G and Γ is the (finite) group of graph automorphisms of G.

Proof. Fix a Borel subgroup B and a maximal torus T in B. By Poizat (1987, Corollaire 4.16), every algebraic subgroup of G is definable in the pure group structure of G (so also in $\mathscr{G}$); but since in this proof we will only consider algebraic subgroups which have been extensively studied, we do not really need to refer to Poizat's corollary. In each instance where an algebraic subgroup appears in the proof, the reader can easily show that the subgroup is in fact definable in the pure group language.

Our proof is inspired from the Theorem 27.4 of Humphreys (1981).

We view H as a subgroup of $\mathrm{Aut}(G)$. We first show that, without loss of generality, we may assume that H normalizes the subgroups B and T. Consider the subgroup $\mathrm{Inn}(G)H$ of $\mathrm{Aut}(G)$. This group is interpretable: first consider the Cartesian product $G \times H$ and then the equivalence relation $\equiv$ on $G \times H$ given by

$$(g, \alpha) \equiv (g_1, \alpha_1) \text{ if and only if } \mathrm{Inn}(g) \circ \alpha = \mathrm{Inn}(g_1) \circ \alpha_1.$$

Now $G \times H/{\equiv}$ is a group isomorphic to $\mathrm{Inn}(G)H$. Consider the group

$$H_1 := \{\phi \in \mathrm{Inn}(G)H : \phi(B) = B \text{ and } \phi(T) = T\}.$$

The group H_1 is interpretable because the subgroups B and T are definable in the pure group language of G. We claim that $\mathrm{Inn}(G)H = \mathrm{Inn}(G)H_1$, i.e., that $H \leqslant \mathrm{Inn}(G)H_1$. Indeed, if $\alpha \in H$, then $\alpha(B)$ is another Borel subgroup of G, therefore $\alpha(B) = B^g$ for some $g \in G$ (the Borel subgroups of G are conjugate in G). Therefore, replacing α by $\mathrm{Inn}(g) \circ \alpha$, we may assume that $\alpha(B) = B$. Now $\alpha(T)$ is another torus of B, and so $\alpha(T) = T^b$ for some $b \in B$ (the maximal tori of B are conjugate in B). Hence, replacing α by $\mathrm{Inn}(b) \circ \alpha$, we may assume that α fixes B and T. Therefore $\alpha \in H_1$ and the claim is proved. Since the claim holds, we may assume that $H = H_1$, i.e., that the elements of H fix B and T.

The subgroups B and T give rise to a root system Φ and a basis Δ of Φ. Since H fixes B and T, each element of H induces an action on the root system Φ and under this action the basis Δ is fixed *setwise* as well. Now, we show that without loss of generality we may assume that H fixes the root system Φ pointwise. For this purpose, we consider the interpretable subgroup

$$H_2 = \{\alpha \in H : \alpha = \mathrm{Id} \text{ on } \Phi\}$$

of H. Since Φ is finite and H acts on Φ, H_2 has finite index in H. In fact we have $H_2 \leqslant H \subseteq H_2\Gamma$. Therefore we can assume that $H = H_2$, i.e., that H normalizes the subgroups B and T and it acts trivially on Φ.

The subgroup H normalizes the one-dimensional root subgroups U_β for all $\beta \in \Delta$. Furthermore, the root subgroups U_β are also normalized by T. Since H normalizes T, H normalizes also the subgroup $T_\beta := C_T(U_\beta)$ for $\beta \in \Delta$. Therefore H acts on the groups U_β, T/T_β and $U_\beta \rtimes T$. By §26.2 Corollary B, parts (c,d) and §26.3 Theorem, part (a) of Humphreys (1981), T/T_β has dimension 1, and $G_\beta := U_\beta \rtimes T/T_\beta \cong K^+ \rtimes K^*$ where K is the base field.

Consider the canonical map $\theta: T \longrightarrow \prod_{\beta\in\Delta} T/T_\beta$ given by $\theta(t) = (t_\beta)_\beta$ where t_β denotes the image of $t \in T$ in T/T_β. Clearly $\ker(\theta) = \bigcap_{\beta\in\Delta} T_\beta$, Since G is simple as an abstract group, by Corollary 26.2 (B, part e) of Humphreys (1981), $\ker(\theta) = 1$ (see also Theorem 27.3 of Humphreys, 1981). Therefore θ is one-to-one. On the other hand, $\dim(\prod_{\beta\in\Delta} T/T_\beta) = |\Delta| = dim(T)$ by Theorem 27.1 of Humphreys (1981). Since $\prod_{\beta\in\Delta} T/T_\beta$ is connected (as an algebraic group), the map θ is onto by dimension considerations.

The subgroup $\mathrm{Inn}(T)H$ of $\mathrm{Aut}(B)$ is interpretable (as above). Furthermore, $\mathrm{Inn}(T)H$ normalizes the root subgroup U_β. For each $\beta \in \Delta$, choose an element $u_\beta \in U_\beta \cong K^+$ which represents the element 1 of K. Thus $u_\beta \in U_\beta^*$ and $U_\beta^* = u_\beta^T = u_\beta^{T/T_\beta}$. Let $\alpha \in H$. Then $\alpha(u_\beta) = u_\beta^{t(\beta)}$ for some $t(\beta) \in T/T_\beta$ (because $U_\beta \rtimes T/T_\beta \cong K^+ \rtimes K^*$). Since the map θ is onto, there is a $t \in T$, such that $\alpha(u_\beta) = u_\beta^t$ for all $\beta \in \Delta$. Now consider the interpretable subgroup

$$H_3 = \{\alpha \in \mathrm{Inn}(T/T_\beta)H : \alpha(u_\beta) = u_\beta \text{ for all } \beta \in \Delta\}.$$

The consideration above shows that $\mathrm{Inn}(T)H = \mathrm{Inn}(T)H_3$. Therefore we may assume that $H = H_3$, i.e., that H normalizes the subgroups B and T, acts trivially on Φ and $\alpha(u_\beta) = u_\beta$ for all $\beta \in \Delta$.

Define the addition on $T/T_\beta \sqcup \{0\}$ by

$$t_1 + t_2 = t \text{ if and only if } u_\beta^{t_1} u_\beta^{t_2} = u_\beta^t$$

where $u_\beta^0 = 1$ by convention. Now the set T/T_β together with $+$, $\cdot$ (the group operation of T extended to $T/T_\beta \sqcup \{0\}$ in the obvious way) and the constants 0 and 1 is a field (in fact isomorphic to the base field K), that we will call F. Now let $\alpha \in H$. Since α acts on T, α acts on F also (by defining $\alpha(0) = 0$). We claim that α is a field automorphism of F. It is clearly multiplicative. Let us show that it is also additive: Assume $t_1 + t_2 = t$, i.e., that $u_\beta^{t_1} u_\beta^{t_2} = u_\beta^t$. Since $\alpha(u_\beta) = u_\beta$, we get $u_\beta^{\alpha(t_1)} u_\beta^{\alpha(t_2)} = u_\beta^{\alpha(t)}$, i.e., $\alpha(t_1) + \alpha(t_2) = \alpha(t)$. This proves that α gives rise to a field automorphism of F. By Poizat (1987, Corollaire 3.6), there is only one interpretable group of field automorphisms, namely the trivial group. Therefore H acts trivially on F, i.e., on T/T_β. Let $x \in U_\beta^*$. Let $t \in T/T_\beta$ be such that $u_\beta^t = x$. For $\alpha \in H$, we have

$$\alpha(x) = \alpha(u_\beta^t) = \alpha(u_\beta)^{\alpha(t)} = u_\beta^t = x,$$

and so H acts also trivially on U_β for all $\beta \in \Delta$. By Humphreys (1981, Theorem 26.3.b), H also acts on U_β where $\beta \in -\Delta$. By Humphreys (1981, Theorem 27.5(e)), α acts as identity on G. The lemma is proved. □

COROLLARY 2.3. Let $B = U \rtimes T$ be a split Frobenius group of finite Morley rank. Then U cannot be a simple algebraic group over an algebraically closed field.

Proof. If U is a simple algebraic group, by the lemma, T acts on U by rational automorphisms. By Fact 5.2, on p. 335, U is solvable. □

3 The case when T is finite

In this section we prove the following result which is an extension of Proposition 4.3 of the previous paper.

THEOREM 3.1. A Frobenius group $T < B$ of finite Morley rank, where T is finite, splits.

We first need the following lemma.

LEMMA 3.2. Let $B = U \rtimes T$ be a split Frobenius group of finite Morley rank where T is finite. If $1 < S < B$ is a subgroup such that $S \cap U = 1$ and $S < B$ is a Frobenius group, then S is conjugate to a subgroup of T by an element of U.

Proof. If S has an involution j, then T has also an involution i. By Lemma 4.3 on p. 331, we may assume that $i = j$. By Proposition 4.4 on p. 332, $T = C_B(i) = C_B(j) \leqslant S$. Since $S \cap U = 1$, this implies that $S = T$. From now on we assume that S has no involutions. By the Feit–Thompson Theorem (Feit and Thompson, 1963) (see also p. 333), S is then solvable.

Note that $\mathrm{rk}(U) = \mathrm{rk}(B)$ and $\deg(B) = \deg(U)|T|$. Set $n = |T|$ and $r = \deg(U)$. Let $T = \{1 = t_0, t_1, \dots, t_{n-1}\}$. The conjugacy classes t_i^U are disjoint for $i = 1, \dots, n-1$ and each of them has Morley rank equal to the Morley rank of B and they all have the same Morley degree r (because the conjugation gives rise to a definable one-to-one map between U and t_i^U). If $s \in S^*$ is an element not conjugate to an element of T, then each of the $n+1$ sets $U, t_1^U, \dots, t_{n-1}^U, s^U$ has Morley rank equal to the Morley rank of B and they are disjoint of the same degree r. This contradicts the fact that $\deg(B) = rn$. Therefore every element of S can be conjugated to an element of T.

Let $1 \neq s \in S^{(k)}$ where k is such that $S^{(k+1)} = 1$ and $S^{(k)} \neq 1$. By conjugating, we may assume that $s \in T$. Then $S^{(k)} \leqslant \mathrm{C}_B(s) \leqslant T$. But now $S \leqslant \mathrm{N}_B(S^{(k)}) \leqslant T$. □

Proof of Theorem 3.1. By Proposition 4.4 on p. 332, we may assume that T does not have an involution. Therefore by the Feit–Thompson Theorem, T is solvable. We claim that $T \cap B^\circ = 1$. Assume not. Then for $t \in T^* \cap B^\circ$, t^{B° is a generic conjugacy class in B°. Therefore the elements of $T^* \cap B^\circ$ are in a single B°-conjugacy class. But the elements of T^* can only be conjugated by the elements of T. Therefore $T^* \cap B^\circ$ is one conjugacy class under $T \cap B^\circ$. By a result of Reineke (Reineke, 1975; Cherlin, 1979), $|T \cap B^\circ| \leqslant 2$. Since T has no involutions, we get $T \cap B^\circ = 1$.

Since $B^\circ T = B^\circ \rtimes T$ and $B^\circ T$ is a Frobenius group, if $B = B^\circ T$ we are done. Assume $B^\circ T \neq B$. We claim that $TB^\circ/B^\circ < B/B^\circ$ is a Frobenius group. To prove this, we have to show that if $t \in T$ and $b \in B$ are such that $t^b \in TB^\circ$, then $b \in TB^\circ$. Choose such elements. Then $1 \neq T^b \cap TB^\circ < B^\circ T$ is a Frobenius group by Lemma 3.6 (1) on p. 330. By Lemma 3.2, $(T^b \cap TB^\circ)^c \cap T \neq 1$ for some $c \in B^\circ$. Then $T^{bc} \cap T \neq 1$ and $bc \in T$ and $b \in TB^\circ$. Thus $TB^\circ/B^\circ < B/B^\circ$ is a finite Frobenius group. Therefore $B/B^\circ = U/B^\circ \rtimes TB^\circ/B^\circ$ for some U containing B°. Now we have $B = UT = U \rtimes T$ and we are done. □

Corollary 3.3. Let $T < B$ be a Frobenius group of finite Morley rank. If $T \cap B^\circ$ has a normal complement in B°, then $T < B$ is split.

Proof. We may assume B is not connected and that T is infinite (Theorem 3.1). By assumption, $B^\circ = W \rtimes (T \cap B^\circ)$ for some W. Let $b \in B$. Then $(T \cap B^\circ)^b < B^\circ$ is a Frobenius group by Lemma 3.6 (1) on p. 330, and the set $\bigcup_{c \in B^\circ} (T \cap B^\circ)^{bc}$ is generic in B° (Corollary 4.2, p. 331). Since the set $\bigcup_{w \in W} (T \cap B^\circ)^w$ is also generic in B°, $(T \cap B^\circ)^{bc} \cap (T \cap B^\circ)^w \neq 1$ for some $w \in W$ and $c \in B^\circ$. Then $bcw^{-1} \in T$ and $b \in Twc^{-1}$. Thus $B = TWB^\circ = TB^\circ = B^\circ T = WT = W \rtimes T$. □

4 Minimal counterexamples

In this section, we prove the following two results.

Theorem 4.1. Let $T < B$ be a Frobenius group of finite Morley rank. Assume that every proper and definable section of B which is a Frobenius group splits. Then either B splits or B is simple and a counterexample to the Cherlin–Zil'ber Conjecture.

Theorem 4.2. Let $T < B$ be an infinite Frobenius group of finite Morley rank. Assume that every simple, proper and definable section of B satisfies the Cherlin–Zil'ber Conjecture (i.e., is an algebraic group over an algebraically closed field). Then either (1) B is simple and a counterexample to the Cherlin–Zil'ber Conjecture, or (2) B splits as $U \rtimes T$ for some nilpotent subgroup U and T is abelian-by-finite.

We need the following lemmas which are interesting in their own right.

LEMMA 4.3. Let $B = U \rtimes T$ be a split Frobenius group of finite Morley rank. Let $V \lhd B$ be a definable subgroup such that $V \cap T = 1$. Then $V \leqslant U$.

Proof. Assume not and take a minimal counterexample B. Then B is infinite (exercise). If V is not in U, then $UV = U \rtimes (UV \cap T)$ is a split Frobenius group with $UV \cap T$ as the Frobenius complement. Therefore, we may assume that $B = UV$ by Lemma 3.6 (1), p. 330.

Assume there is a definable, proper, infinite, B-normal subgroup W of U. We may take W to be B-minimal. Then either $W \leqslant V$ or $W \cap V$ is finite. In the second case, since W is connected, $[V, W]$ is a connected subgroup of the finite group $V \cap W$ and so $[V, W] = 1$. Therefore either $W \leqslant V$ or $[V, W] = 1$. By Corollary 4.7, p. 334, $TW/W < B/W = U/W \rtimes TW/W$ is a split Frobenius group. We claim that $VW \cap TW = W$, i.e., that $VW \cap T \subseteq W$. In case $W \leqslant V$ this is clear. Assume $[V, W] = 1$. Let $t = vw \in T \cap VW$. Then, since v and w commute, t and w commute also. Therefore either $t = 1$ or $w = 1$. If $w = 1$, then $t = v \in V \cap T = 1 \leqslant W$. If $t = 1$, then $v = w \in W$. The claim is proved. The claim says that the hypotheses of our theorem hold in the split Frobenius group B/W (with V replaced by VW/W). By induction, $VW \leqslant U$ and we are done. We assume from now on that U has no proper, infinite, definable, B-normal subgroups. In particular U is connected and $U \cap V$ is finite. As before, U and V commute. But then $V \leqslant C_B(U) \leqslant U$ by Lemma 3.6 (6), p. 330. □

LEMMA 4.4. Let $B = U \rtimes T$ be a split Frobenius group of finite Morley rank where either T is connected or has order p for some prime p. Assume also that U has no proper, non-trivial, definable and B-normal subgroups. Then U is either abelian or simple.

Proof. If $R(U) \neq 1$, then $R(U) = U$ and U is solvable. Then the lemma is quite trivial. Assume $R(U) = 1$ from now on. In particular, $Z(U^\circ) = 1$, $S(U) \neq 1$ and $S(U)$ is definable (Fact 2.1). Since $S(U)$ is characteristic in U, we have $S(U) = U$. Note also that $[R(B), U] \leqslant R(U) = 1$ and so $R(B) \leqslant C_B(U) \leqslant U$, and $R(B) \leqslant R(U) = 1$. Therefore $R(B) = 1$, $Z(B^\circ) = 1$ and $S(B)$ is definable.

If U were finite, B would be finite also, and then U would be nilpotent by Thompson's Theorem (Fact 5.1 on p. 335), a contradiction. So U is infinite. By hypothesis, U is connected. Now, Fact 2.1 (1) gives the result in case B is connected. Assume B is not connected. Then $|T| = p$. Applying Fact 2.1 (2), we have $U = U_1 \oplus \cdots \oplus U_k$ where each U_i is a definable, normal and simple subgroup of U and T acts on the set $\{U_1, \ldots, U_k\}$ transitively. Since $|T|$ is a prime number, either $k = 1$ or $k = p$. Assume $k = p$, and reorder the components in such a way that, for some fixed $t \in T^*$, $t^i U_1 = U_{i+1}$ for $i = 0, \ldots, p-1$. Let $a \in U_1^*$. Then the element $aa^t \cdots a^{t^{p-1}}$ is a non-trivial element of U fixed by t, a contradiction. Therefore $k = 1$ and we are done. □

LEMMA 4.5. Let $T < B$ be a Frobenius group which is also an algebraic group over an algebraically closed field. Then T is a closed subgroup and B is split, say $B = U \rtimes T$. Furthermore, U is a closed and nilpotent subgroup and T is abelian-by-finite.

Proof. As in Proposition 4.1 on p. 331, one can prove that T is an algebraic subgroup of B (use Proposition 7.5 of Humphreys (1981) instead of Zil'ber's Indecomposability Theorem). Thus $T < B$ is an *algebraic Frobenius group*. By Hertzig (1961, Theorem 4), B is split. Say $B = U \rtimes T$. By Hertzig (1961, Theorem 5), U is solvable. By Theorem 5.3 on p. 335, U is nilpotent.

Now we prove the last statement. By Theorem 5.3 on p. 335, the connected solvable subgroups of T are abelian. By Humphreys (1981, §21.4), T° is a nilpotent group, and so is abelian. □

LEMMA 4.6. Let $B = U \rtimes T$ be a split Frobenius group of finite Morley rank. Let $S < B$ be a subgroup of B such that $S \cap U = 1$ and $S < B$ is a Frobenius group. Then some non-trivial element of S can be conjugated to an element of T.

Proof. If T is finite, we can apply Lemma 3.2. Assume T is infinite from now on.

Note that S is definable by Proposition 4.1 p. 331. If S is infinite then T is also infinite and U is connected by Corollary 4.7, p. 334. Therefore we can assume that B is connected. Now, $\bigcup_{b \in B} T^b$ and $\bigcup_{b \in B} S^b$ are generic subsets of the connected group B (Corollary 4.2, p. 331). Therefore they intersect non-trivially and we are done in this case. We assume S is finite from now on.

Since S is finite, by Theorem 3.1, $B = V \rtimes S$ for some V. By Lemma 4.3, $U \leqslant V$. Therefore $V = U \rtimes (V \cap T)$ is a split Frobenius group where the Frobenius complement $V \cap T$ is infinite. We may assume that S is cyclic of prime order, say p. Let $s \in S$ be an element of order p. The group $(V \cap T)^s < U(V \cap T)$ is a Frobenius group. By the paragraph above, there is a $u \in U$ such that $(V \cap T)^{su} \cap (V \cap T) \neq 1$. Then $su \in T$. Set $t = su$. Since, for all integers i, $t^i = s^i u_i$ for some $u_i \in U$, we have $t^p \in T \cap U = 1$. Thus t is an element of order p of T. Consider the group $B_1 = U \rtimes \langle t \rangle$. This is a Frobenius group with $\langle t \rangle$ as a Frobenius complement, contains S and $S < B_1$ is a Frobenius group. By Lemma 3.2, S can be conjugated onto $\langle t \rangle$. □

LEMMA 4.7. Let $T < B$ be a Frobenius group of finite Morley rank. Let $V \lhd B$ be a definable subgroup such that $V \cap T = 1$. Then either $B = V \rtimes T$ or B/V is a Frobenius group with TV/V as a Frobenius complement.

Proof. Assume $B \neq VT$. Then $1 \neq TV/V < B/V$. To prove that this is a Frobenius group, we need to show that if $t \in T^*$, $b \in B$ are such that $t^b \in TV$, then $b \in TV$. Assume t and b satisfy the premises. By Lemma 3.6, p. 330, $(T^b \cap TV) < TV$ is a Frobenius group. But $TV = V \rtimes T$ is also a Frobenius group. By Lemma 4.6, there is a $v \in V$ such that $(T^b \cap TV)^v \cap T \neq 1$. In particular $T^{bv} \cap T \neq 1$ and so $bv \in T$. This shows that $b \in TV$ and finishes the proof of the lemma. □

Proof of Theorem 4.1. Assume B does not split. By Lemma 4.5, it is enough to show that B is simple. Assume not. By Theorem 3.1, T is infinite. By Corollary 3.3, B is connected. Let $V \lhd B$ be a proper and non-trivial normal subgroup. Rplacing V by $[B, V]$, we may assume that V is definable.

We will show that without loss of generality we may assume $V \cap T = 1$. Note that T cannot have a normal subgroup, so that V is not in T. Assume $V \cap T \neq 1$. Then by assumption, $V = A \rtimes (V \cap T)$ for some subgroup A which is necessarily definable by Proposition 4.5, p. 333. For $b \in B$, $A^b \leqslant V^b = V$. It follows that $A^b \lhd V$. Indeed if $v \in V$, then $v^{b^{-1}} \in V$ and so $A^{bv} = A^{v^{b^{-1}}b} = A^b$. Therefore by Lemma 4.3, $A^b \leqslant A$. Hence $A^b = A$ and $A \lhd B$. Replacing V by A if necessary, we may assume that $V \cap T = 1$. Since T is infinite, V is infinite also.

Now by Lemma 4.7, B/V is a Frobenius group. But it also splits by induction as $U/V \rtimes TV/V$. Therefore $B = UT = U \rtimes T$, a contradiction □

Proof of Theorem 4.2. If B is simple, Lemma 4.5 gives the result. Assume therefore that B is not simple. Then as in the proof of the previous theorem, B has a definable, normal and proper subgroup $V \neq 1$. As in the proof of Theorem 4.1, we may assume that $V \cap T = 1$.

(1) Now we show that B splits. If V is finite, then T is finite also, and in this case we can apply Theorem 3.1. Assume now V is infinite. By Lemma 4.7, either $B = V \rtimes T$ or B/V is a Frobenius group. Assume we are in the latter case. By induction, either B/V is simple or B/V splits. If B/V is simple, by assumption, B/V is an algebraic group. Therefore we can apply Lemma 4.5: B/V splits. Therefore B/V splits in all cases. It follows easily that B splits.

(2) Therefore $B = U \rtimes T$ for some subgroup U which is necessarily definable by Proposition 4.5, p. 333. Now we show that U is nilpotent. For this purpose we may restrict T and assume that either T is infinite and connected or is cyclic of prime exponent. Assume U has a proper, definable, infinite B-normal subgroup V. Since the group $V \rtimes T$ satisfies

the hypotheses, V is nilpotent by induction. But $B/V \cong U/V \rtimes T$ also satisfies the hypothesis by Corollary 4.7, p. 334. Therefore U/V is nilpotent by induction, and so U is solvable and hence nilpotent by by Theorem 5.3, p. 335. Therefore we may assume that U has no infinite and proper definable B-normal subgroups. In particular U is connected and its B-definable proper subgroups are central. Replacing U by $U/Z(U)$ we may assume that U has no proper, definable B-normal subgroups at all. By Lemma 4.4, U is simple. But this is impossible by Lemma 2.3. Therefore U is nilpotent.

(3) We now show that T is abelian-by-finite, i.e., that T° is abelian. By Lemma 4.3, p. 331, T has finitely many involutions. This fact together with our hypothesis on the simple sections of T° yields that T° is solvable. Now, Theorem 5.6, p. 339, yields the result. □

Bibliography

Abramson, F. and Harrington, L. (1978). Models without indiscernibles, *The Journal of Symbolic Logic* **43**: 572–600.

Adeleke, S. A. (1992). Semilinear tower of Steiner systems. Preprint.

Adeleke, S. A. (to appear). Examples of irregular infinite Jordan groups. In preparation.

Adeleke, S. A. and Macpherson, H. D. (1994). A classification of infinite Jordan groups. Preprint.

Adeleke, S. A. and Neumann, P. M. (in press*a*). Infinite bounded permutation groups, *The Journal of the London Mathematical Society, Series 2*. To appear.

Adeleke, S. A. and Neumann, P. M. (in press*b*). Primitive permutation groups with primitive Jordan sets, *The Journal of the London Mathematical Society, Series 2*. To appear.

Adeleke, S. A. and Neumann, P. M. (to appear). Relations related to betweenness: their structure and automorphisms. In preparation.

Ahlbrandt, G. (1987). Almost strongly minimal totally categorical theories, *Logic Colloquium '85* (The Paris Logic Group, eds.), North Holland, Amsterdam, pp. 17–31.

Ahlbrandt, G. and Ziegler, M. (1986). Quasi-finitely axiomatizable totally categorical theories, *Annals of Pure and Applied Logic* **30**: 63–82.

Ahlbrandt, G. and Ziegler, M. (1991*a*). Invariant subspaces of ${}^V V$, *The Journal of Algebra* **151**: 26–38.

Ahlbrandt, G. and Ziegler, M. (1991*b*). What's so special about $(\mathbb{Z}/4\mathbb{Z})^\omega$?, *Archive for Mathematical Logic* **31**: 115–132.

Albert, M. and Burris, S. (1986). Finite axiomatisations for existentially closed posets and semilattices, *Order* **3**: 169–178.

Alper, T. L. (1987). A classification of all order-preserving homeomorphisms of the reals that satisfy finite uniqueness, *The Journal of Mathematical Psychology* **31**: 135–154.

Apps, A. (1982). Boolean powers of groups, *Mathematical Proceedings of the Cambridge Philosophical Society* **91**: 375–395.

Apps, A. (1983*a*). On $\aleph_0$-categorical class two groups, *The Journal of Algebra* **82**: 516–538.

Apps, A. (1983*b*). On the structure of $\aleph_0$-categorical groups, *The Journal of Algebra* **81**: 320–339.

Apps, A. (1983*c*). Two counterexamples in $\aleph_0$-categorical groups, *Proceedings of the London Mathematical Society, Series 3* **47**: 385–410.

Archer, R. (1993). *Omega-categoricity and Boolean powers*, Ph.D. thesis, University of London.

Artin, E. (1955). The orders of the classical simple groups, *Communications in Pure and Applied Mathematics* **8**: 455–472.

Artin, E. (1957). *Geometric Algebra*, Interscience Publishers Inc., New York.

Aschbacher, M. (1971). Doubly transitive groups in which the stabilizer of two points is Abelian, *The Journal of Algebra* **18**: 114–136.

Baldwin, J. T. (1988). *Fundamentals of Stability Theory*, Springer-Verlag, Berlin.

Baldwin, J. T. (in press). An almost strongly minimal non-Desarguesian projective plane, *Transactions of the American Mathematical Society*. To appear.

Baldwin, J. T. and Lachlan, A. H. (1971). On strongly minimal sets, *The Journal of Symbolic Logic* **36**: 79–96.

Baldwin, J. T. and Saxl, J. (1976). Logical stability in group theory, *Journal of the Australian Mathematical Society* **21**: 267–276.

Baldwin, J. T. and Shi, N. (to appear). Stable generic structures. Preliminary version, 1992.

Barwise, J. (1975). *Admissible Sets and Structures*, Perspectives in Mathematical Logic, Springer-Verlag, Berlin.

Baudisch, A. (1992). A new $\aleph_1$-categorical pure group. Preprint, number A93–40, Freie Universität Berlin.

Baur, W., Cherlin, G. L. and Macintyre, A. J. (1979). Totally categorical groups and rings, *The Journal of Algebra* **57**: 407–440.

Bell, J. L. (1985). *Boolean-valued Models and Independence Proofs in Set Theory*, number 12 in *Oxford Logic Guides*, second edn, Oxford University Press.

Bell, J. L. and Slomson, A. (1969). *Models and Ultraproducts*, North Holland, Amsterdam.

Bennett, J. (1993). *Reducts of some binary homogeneous structures*, Ph.D. thesis, Rutgers University.

Berline, C. and Cherlin, G. L. (1981). QE rings in characteristic *p*, *in* M. Lerman (ed.), *Logic year 1979-80*, Vol. 859 of *Lecture Notes in Mathematics*, Springer-Verlag, Berlin.

Birch, B. J., Burns, R. G., Macdonald, S. O. and Neumann, P. M. (1976). On the degrees of permutation groups containing elements separating finite sets, *Bulletin of the Australian Mathematical Society* **14**: 7–10.

Blaha, K. (1992). Minimum bases for permutation groups: the greedy approximation, *J. Algorithms* **13**: 297–306.

Borovik, A. V. (1982). Involutions in groups with a dimension. Preprint, 18 pages (in Russian).

Borovik, A. V. and Nesin, A. (in press). *Groups of Finite Morley Rank*, Oxford University Press. To appear.

Borovik, A. V., DeBonis, M. J. and Nesin, A. (in press*a*). CIT-groups of finite Morley rank (I), *The Journal of Algebra*. To appear.

Borovik, A. V., DeBonis, M. J. and Nesin, A. (in press*b*). On some doubly transitive ω-stable groups, *The Journal of Algebra*. To appear.

Bouscaren, E. and Laskowski, L. C. (1993). S-homogeneity and automorphism groups, *The Journal of Symbolic Logic* **58**: 1302–1322.

Burris, S. (1975). Boolean powers, *Algebra Universalis* **5**: 341–360.

Cameron, P. J. (1976). Transitivity of permutation groups on unordered sets, *Mathematisches Zeitschrift* **148**: 127–139.

Cameron, P. J. (1977). Cohomological aspects of two-graphs, *Mathematisches Zeitschrift* **157**: 101–119.

Cameron, P. J. (1981). Finite permutation groups and finite simple groups, *The Bulletin of the London Mathematical Society* **13**: 1–22.

Cameron, P. J. (1983*a*). Orbits of permutation groups on unordered sets, IV: Homogeneity and transitivity, *The Journal of the London Mathematical Society, Series 2* **27**: 238–247.

Cameron, P. J. (1983*b*). Permutation groups acting on unordered sets IV: Homogeneity and transitivity, *The Journal of the London Mathematical Society, Series 2* **27**: 238–247.

Cameron, P. J. (1987). Some treelike objects, *Quarterly Journal of Mathematics, Series 2* **38**: 155–183.

Cameron, P. J. (1989). Groups of order-automorphisms of the rational numbers with prescribed scale type, *The Journal of Mathematical Psychology* **33**: 163–171.

Cameron, P. J. (1990). *Oligomorphic Permutation Groups*, number 152 in *London Mathematical Society Lecture Notes*, Cambridge University Press.

Cameron, P. J. (1992). Infinite geometric groups of rank 4, *European Journal of Combinatorics* **13**: 87–88.

Cameron, P. J. and Kantor, W. M. (1979). 2-transitive and antiflag transitive collineation groups of finite projective spaces, *The Journal of Algebra* **60**: 384–422.

Cameron, P. J. and Szabó, C. (to appear). Independence algebras. In preparation.

Cameron, P. J., Deza, M. and Frankl, P. (1987). Sharp sets of permutations, *The Journal of Algebra* **111**: 220–247.

Cameron, P. J., Deza, M. and Singhi, N. M. (1988). Infinite geometric groups and sets, *in* M.-M. Deza, P. Frankl and I. G. Rosenberg (eds), *Algebraic, Extremal and Metric Combinatorics*, Vol. 131 of *London Mathematical Society Lecture Notes*, Cambridge University Press, Cambridge, pp. 54–61.

Camina, A. R. and Evans, D. M. (1991). Some infinite permutation modules, *Quarterly Journal of Mathematics, Series 2* **42**: 15–26.

Chang, C. C. and Keisler, H. J. (1990). *Model Theory*, third edn, North Holland, Amsterdam.

Cherlin, G. L. (1979). Groups of small Morley rank, *Annals of Mathematical Logic* **17**: 1–28.

Cherlin, G. L. (1980). On $\aleph_0$-categorical nilrings II, *The Journal of Symbolic Logic* **45**: 291–301.

Cherlin, G. L. (1984). Totally categorical structures, *Proceedings of the International Congress of Mathematicians, Warsaw, 1983*, Państwowe Wydawnictwo Naukowe, Warsaw, pp. 301–16.

Cherlin, G. L. (1988). Homogeneous tournaments revisited, *Geometriae Dedicata* **26**: 231–240.

Cherlin, G. L. and Felgner, U. (1992). Homogeneous solvable groups, *The Journal of the London Mathematical Society, Series 2* **44**: 102–120.

Cherlin, G. L. and Lachlan, A. H. (1986). Stable finitely homogeneous structures, *Transactions of the American Mathematical Society* **296**: 815–850.

Cherlin, G. L., Harrington, L. and Lachlan, A. H. (1985). $\aleph_0$-categorical, $\aleph_0$-stable structures, *Annals of Pure and Applied Logic* **28**: 103–135.

Cherlin, G. L., Saracino, D. and Wood, C. (in press). On homogeneous nilpotent groups and rings, *Transactions of the American Mathematical Society*.

Cohen, P. J. (1966). *Set Theory and the Continuum Hypothesis*, Benjamin, London.

Collins, M. (1990). Some infinite Frobenius groups, *The Journal of Algebra* **131**: 161–165.

Covington, J. A. (1989). A universal structure for N-free graphs, *Proceedings of the London Mathematical Society, Series 3* **58**: 1–16.

Creed, P. (to appear). *On o-amorphous and quasi-amorphous sets*, Ph.D. thesis, University of Leeds. To appear.

Curtis, C. W. and Reiner, I. (1988). *Representation Theory of Finite Groups and Associative Algebras*, Wiley-Interscience, New York.

Cutland, N. (1980). *Computability*, Cambridge University Press.

Delahan, F. and Nesin, A. (in press). On split Zassenhaus groups of finite Morley rank, *Communications in Algebra*. To appear.

Delsarte, P. (1978). Bilinear forms over a finite field, with applications to coding theory, *Journal of Combinatorial Theory, Series A* **25**: 226–241.

Dixon, J. D., Neumann, P. M. and Thomas, S. R. (1986). Subgroups of small index in infinite symmetric groups, *The Bulletin of the London Mathematical Society* **18**: 580–586.

Droste, M. (1985). *Structure of Partially Ordered Sets with Transitive Automorphism Groups*, Vol. 334 of *Memoirs of the American Mathematical Society*, American Mathematical Society, Providence, RI.

Droste, M. and Macpherson, H. D. (1991). On k-homogeneous posets and graphs, *Journal of Combinatorial Theory, Series A* **56**: 1–15.

Droste, M. and Truss, J. K. (1991). Subgroups of small index in ordered permutation groups, *Quarterly Journal of Mathematics, Series 2* **42**: 31–47.

Droste, M., Holland, W. C. and Macpherson, H. D. (1989*a*). Automorphism groups of infinite semilinear orders (I), *Proceedings of the London Mathematical Society, Series 3* **58**: 454–478.

Droste, M., Holland, W. C. and Macpherson, H. D. (1989*b*). Automorphism groups of infinite semilinear orders (II), *Proceedings of the London Mathematical Society, Series 3* **58**: 479–494.

Ehrenfeucht, A. (1973). Discernible elements in models of Peano arithmetic, *The Journal of Symbolic Logic* **38**: 291–292.

Engeler, E. (1959). Äquivalenzklassen von n-Tupeln, *Zeitschrift für Mathematische Logik und Grundlagen der Mathematik* **5**: 340–345.

Erdős, P. and Rényi, A. (1963). Asymmetric graphs, *Acta Math. Acad. Sci Hungar.* **14**: 295–315.

Evans, D. M. (1986*a*). Growth rates for algebraic closure. Unpublished notes.

Evans, D. M. (1986*b*). Homogeneous geometries, *Proceedings of the London Mathematical Society, Series 3* **52**: 305–327.

Evans, D. M. (1986*c*). Subgroups of small index in infinite general linear groups, *The Bulletin of the London Mathematical Society* **18**: 587–590.

Evans, D. M. (1987). Infinite permutation groups and minimal sets, *Quarterly Journal of Mathematics, Series 2* **38**: 461–471.

Evans, D. M. (1994). *Some subdirect products of finite nilpotent groups*, Vol. 411 of *NATO ASI Series C: Mathematical and Physical Sciences*, Kluwer, Dordrecht, pp. 117–24. Proceedings, Banff, 1991.

Evans, D. M. and Hewitt, P. (1990). Counterexamples to a conjecture on relative categoricity, *Annals of Pure and Applied Logic* **46**: 201–209.

Evans, D. M. and Hrushovski, E. (1993). On the automorphism groups of finite covers, *Annals of Pure and Applied Logic* **62**: 83–112.

Evans, D. M., Hodges, W. A. and Hodkinson, I. M. (1991). Automorphisms of bounded abelian groups, *Forum Mathematicum* **3**: 523–541.

Feit, W. and Thompson, J. G. (1963). Solvability of groups of odd order, *Pacific Journal of Mathematics* **13**: 775–1029.

Felgner, U. (1970). *Models of ZF-Set Theory*, number 223 in *Lecture Notes in Mathematics*, Springer-Verlag, Berlin.

Felgner, U. (1972). *Die Unabhängigkeit des Booleschen Primidealtheorems vom Ordnungerweiterungssatz*, Habilitationsschrift, Heidelberg.

Felgner, U. and Truss, J. K. (to appear). Independence of the prime ideal theorem from the order extension principle. In preparation.

Fountain, J. and Lewin, A. (1992). Products of idempotent endomorphisms of an independence algebra of finite rank, *Proceedings of the Edinburgh Mathematical Society* **35**: 493–500.

Fraenkel, A. (1922). Der Begriff 'definit' und die Unabhängigkeit des Auswahlsaxioms, *Sitzungsberichte der Preussischen Akademie der Wissenschaften, Physikalisch-mathematische Klasse* pp. 253–257. English translation in: *From Frege to Gödel* (ed. J. van Heijenoort) Harvard University Press, 1967, pp. 284–289.

Fraïssé, R. (1953). Sur certaines relations qui généralisent l'ordre des nombres rationnels, *Comptes Rendus de l'Académie des Sciences de Paris* **237**: 540–542.

Fraïssé, R. (1954*a*). Sur l'extension aux relations de quelques propriétés des ordres, *Annales Scientifiques de lÉcole Normale Supérieure* **71**: 363–388.

Fraïssé, R. (1954*b*). Sur quelques classifications des systèmes de relations, *Publications Scientifiques, Université d'Alger, Série A* **1**: 35–182.

Friedman, H. M. (1973). Countable models of set theories, *Cambridge Summer School in Mathematical Logic*, Vol. 337 of *Lecture Notes in Mathematics*, Springer-Verlag, Berlin, pp. 539–573.

Frobenius, G. (1901). Uber auslösbare Gruppen IV, *Sitzungsberichte der Königlich Preussichen Akademie der Wissenschaften zu Berlin* pp. 1216–1230. Reprinted in *Ferdinand Georg Frobenius Gesammelte Abhandlungen*, Vol. 3 (Ed. J.-P. Serre), 1968, Springer, Berlin, 189–209.

Fuchs, L. (1960). *Abelian Groups*, Pergamon Press, Oxford.

Furst, M. L., Hopcroft, J. and Luks, E. M. (1980). Polynomial-time algorithms for permutation groups, *Proceedings of the 21st IEEE FOCS*, pp. 340–352.

Gaifman, H. (1976). Models and types of Peano's arithmetic, *Annals of Mathematical Logic* **9**: 223–306.

Gauntt, R. (1970). Axiom of choice for finite sets. Preprint.

Givant, S. (1979). A representation theorem for universal Horn classes categorical in power, *Annals of Mathematical Logic* **17**: 91–116.

Glass, A. (1981). *Ordered Permutation Groups*, Vol. 55 of *London Mathematical Society Lecture Note Series*, Cambridge University Press.

Godsil, C. D. (1979). *Graphs with Regular Groups*, Ph.D., University of Melbourne.

Goode, J. (1989). Hrushovski's geometries, *in* B. Dahn and H. Wolter (eds), *Proceedings of 7th Easter Conference on Model Theory*, pp. 106–118.

Gorenstein, D. (1980). *Finite Groups*, Chelsea Publishing Company, New York.

Grätzer, G. (1978). *General Lattice Theory*, Birkhaüser, Basel.

Gropp, U. (1991). *Action d'un Groupe ω-Stable sur un Ensemble de rang de Morley 2*, Thèse de doctorat, Université Paris VII.

Gropp, U. (1992). There is no sharp transitivity on q^6 when q is a type of Morley rank 2, *The Journal of Symbolic Logic* **57**: 1198–1212.

Hall, Jr., M. (1954). On a theorem of Jordan, *Pacific Journal of Mathematics* **4**: 219–226.

Hall, P. (1959). Some constructions for locally finite groups, *The Journal of the London Mathematical Society, Series 2* **34**: 305–319.

Halpern, J. D. and Levy, A. (1971). The Boolean prime ideal theorem does not imply the axiom of choice, *Proceedings of Symposia in Pure Mathematics*, Vol. XIII part 1, American Mathematical Society, Providence, RI, pp. 83–134.

Haskell, D. and Macpherson, H. D. (1994). Cell decompositions of C-minimal structures, *Annals of Pure and Applied Logic* **66**: 113–62.

Henson, C. W. (1971). A family of countable homogeneous graphs, *Pacific Journal of Mathematics* **38**: 69–83.

Henson, C. W. (1972). Countable homogeneous relational structures and $\aleph_0$-categorical theories, *The Journal of Symbolic Logic* **37**: 494–500.

Hertzig, D. (1961). The structure of Frobenius algebraic groups, *Amer. J. Math.* **83**: 421–431.

Herwig, B. (1992). Hrushovski's pseudoplane. Preprint.

Herwig, B. (in press). Weight ω in stable theories with few types, *The Journal of Symbolic Logic*. To appear.

Higman, D. G. (1967). Intersection matrices for finite permutation groups, *The Journal of Algebra* **66**: 22–42.

Higman, G. (1957). On groups and rings having automorphisms without non-trivial fixed elements, *The Journal of the London Mathematical Society, Series 2* **32**: 321–334.

Hodges, W. A. (1985). *Building Models By Games*, number 2 in *London Mathematical Society Student Texts*, Cambridge University Press.

Hodges, W. A. (1989). Categoricity and permutation groups, *in* H. D. Ebbinghaus *et al.* (eds), *Logic Colloquium '87*, North Holland, Amsterdam, pp. 53–72.

Hodges, W. A. (1993). *Model Theory*, Cambridge University Press, Cambridge.

Hodges, W. A. and Pillay, A. (in press). Cohomology of structures and some problems of Ahlbrandt and Ziegler, *The Journal of the London Mathematical Society, Series 2*. To appear.

Hodges, W. A., Hodkinson, I. M., Lascar, D. and Shelah, S. (1993). The small index property for ω-stable, ω-categorical structures and for the random graph, *The Journal of the London Mathematical Society, Series 2* **48**: 204–218.

Hrushovski, E. (1986). *Contributions to Stable Model Theory*, Ph.D. thesis, University of California, Berkeley.

Hrushovski, E. (1988*a*). A new strongly minimal set. Unpublished notes.

Hrushovski, E. (1988*b*). A stable $\aleph_0$-categorical pseudoplane. Unpublished notes.

Hrushovski, E. (1989*a*). Almost orthogonal regular types, *Annals of Pure and Applied Logic* **45**: 139–155.

Hrushovski, E. (1989*b*). Totally categorical theories, *Transactions of the American Mathematical Society* **313**: 131–159.

Hrushovski, E. (1989*c*). Unidimensional theories. An introduction to geometric stability theory, *in* H.-D. Ebbinghaus *et al.* (eds), *Logic Colloquium '87*, North Holland, Amsterdam, pp. 73–103.

Hrushovski, E. (1992*a*). Strongly minimal expansions of algebraically closed fields, *Israel Journal of Mathematics* **79**: 129–151.

Hrushovski, E. (1992*b*). Unimodular minimal theories, *The Journal of the London Mathematical Society, Series 2* **46**: 385–396.

Hrushovski, E. (1993*a*). A new strongly minimal set, *Annals of Pure and Applied Logic* **62**: 147–166.

Hrushovski, E. (1993*b*). Strongly minimal expansions of algebraically closed fields, *Israel Journal of Mathematics*.

Hrushovski, E. (1994). *Finite Structures with few Types*, Vol. 411 of *NATO ASI Series C: Mathematical and Physical Sciences*, Kluwer, Dordrecht, pp. 175–187. Proceedings, Banff, 1991.

Hrushovski, E. and Pillay, A. (1987). Weakly normal groups, *Logic Colloquium '85* (The Paris Logic Group, eds.), North Holland, Amsterdam, pp. 233–244.

Hrushovski, E. and Shelah, S. (1989). A dichotomy theorem for regular types, *Annals of Pure and Applied Logic* **45**: 157–169.

Hughes, D. R. and Piper, F. C. (1973). *Projective Planes*, Springer-Verlag, Berlin.

Humphreys, J. E. (1981). *Linear Algebraic Groups*, second edn, Springer-Verlag, Berlin.

Jacobson, N. (1951). *Lectures in Abstract Algebra*, van Nostrand, New York. Three volumes.

Jech, T. (1970). *Lectures in Set Theory, with Particular Emphasis on the Method of Forcing*, number 217 in *Lecture Notes in Mathematics*, Springer-Verlag, Berlin.

Jech, T. (1973). *The Axiom of Choice*, North Holland, Amsterdam.

Jech, T. and Sochor, A. (1966). Applications of the θ-model, *Bulletin de l'Académie Polonaise des Sciences* **14**: 297–303 and 351–355.

Jensen, D. and Ehrenfeucht, A. (1976). Some problem in elementary arithmetics, *Fundamenta Mathematicæ* **92**: 223–245.

Jónsson, B. (1956). Universal relational systems, *Mathematica Scandinavica* **4**: 193–208.

Jónsson, B. (1957). On isomorphism types of groups and other algebraic systems, *Mathematica Scandinavica* **5**: 224–229.

Jónsson, B. (1960). Homogeneous universal relational systems, *Mathematica Scandinavica* **8**: 137–142.

Jordan, C. (1871). Théorèmes sur les groupes primitives, *Journal de Mathematiques Pures et Appliqués* **16**: 383–408. Reprinted in *Oeuvres de Camille Jordan* (ed. J. Dieudonné, Gauthier-Villars, Paris, 1961), Vol. 1, pp. 313–338.

Jordan, C. (1875). Sur la limite du degré des groupes primitifs qui contiennent une substitution donnée, *Journal fur reine und angewandte Mathematik* **79**: 248–258. Reprinted in *Oeuvres de Camille Jordan* (ed. J. Dieudonné, Gauthier-Villars, Paris, 1961), Vol. 1, pp. 485–495.

Kaluzhnin, L. and Klin, M. (1976). On certain numerical invariants of substitution groups, *Latvijskiy Matematicheskiy Ezhegodnik* **18**: 81–99. In Russian.

Kantor, W. M. (1979). Permutation representations of the finite classical groups of small degree or rank, *The Journal of Algebra* **60**: 158–168.

Kantor, W. M. (1985). Homogeneous designs and geometric lattices, *Journal of Combinatorial Theory, Series A* **8**: 64–77.

Kantor, W. M., Liebeck, M. W. and Macpherson, H. D. (1989). $\aleph_0$-categorical structures smoothly approximated by finite substructures, *Proceedings of the London Mathematical Society, Series 3* **59**: 439–463.

Karp, C. (1965). Finite quantifier equivalence, *The Theory of Models*, North Holland, Amsterdam, pp. 407–412.

Kaye, R. (1991). *Models of Peano Arithmetic*, Vol. 15 of *Oxford Logic Guides*, Oxford University Press.

Kaye, R. (1992). The automorphism group of a countable recursively saturated structure, *Proceedings of the London Mathematical Society, Series 3* **65**: 225–244.

Kaye, R., Kossak, R. and Kotlarski, H. (1991). Automorphisms of recursively saturated models of arithmetic, *Annals of Pure and Applied Logic* **55**: 67–99.

Kegel, O. and Wehrfritz, B. (1973). *Locally Finite Groups*, North Holland, Amsterdam.

Kikyo, H. and Tsuboi, A. (in press). On reduction properties, *The Journal of Symbolic Logic*. To appear.

Kirby, L. and Paris, J. (1977). Initial segments of models of Peano's axioms, *in* A. H. Lachlan *et al.* (eds), *Set Theory and Hierarchy theory V, Bierutowice, Poland, 1976*, Vol. 619 of *Lecture Notes in Mathematics*, Springer-Verlag, Berlin, pp. 211–226.

Kleene, S. C. (1952). Finite axiomatizablilty of theories is the predicate calculus using additional predicate symbols, *Memoirs of the American Mathematical Society* **10**: 27–66.

Klin, M. and Poschel, R. (1981). The König problem, the isomorphism problem for cyclic graphs and the method of Schur rings, *in* L. Lovasz and V. T. Sos (eds), *Algebraic Methods in Graph Theory*, North Holland, Amsterdam, pp. 405–434. Jointly published with the Janos Bolyai Mathematical Society, Budapest.

Knight, J. F., Pillay, A. and Steinhorn, C. (1986). Definable sets in ordered structures II, *Transactions of the American Mathematical Society* **295**: 593–605.

Kossak, R. (1989). Models with the ω-property, *The Journal of Symbolic Logic* **54**: 177–189.

Kossak, R., Kotlarski, H. and Schmerl, J. H. (1993). On maximal subgroups of the automorphism group of a countable recursively saturated model of PA, *Annals of Pure and Applied Logic* **65**: 125–148.

Kotlarski, H. (1984). On elementary cuts in recursively saturated models of Peano arithmetic, *Fundamenta Mathematicæ* **120**: 205–222.

Kotlarski, H. and Kaye, R. (in press). Automorphisms of models of true arithmetic: Recognizing some basic open subgroups, *Notre Dame Journal of Formal Logic.* To appear.

Krantz, D., Luce, R. D., Suppes, P. and Tversky, A. (1971, 1989, 1990). *Foundations of Measurement*, Academic Press, New York. Vols. I–III.

Kueker, D. W. and Laskowski, C. (1992). On generic structures, *Notre Dame Journal of Formal Logic* **33**: 175–183.

Kueker, D. W. and Steitz, P. W. (unpublished *a*). Definability over definable sets. Preprint.

Kueker, D. W. and Steitz, P. W. (unpublished *b*). Stabilisers of definable sets in homogeneous models. Preprint.

Kuratowski, K. (1966). *Topology*, Academic Press, New York and London. Jointly published by PWN, Warsaw. English edition translated by J. Jaworowski.

Lachlan, A. H. (1973). On the number of countable models of a countable superstable theory, *in* Suppes (ed.), *Logic, Methodology and Philosophy of Sciences*, North Holland, Amsterdam, pp. 45–56.

Lachlan, A. H. (1974). Two conjectures regarding the stability of ω-categorical theories, *Fundamenta Mathematicæ* **81**: 133–145.

Lachlan, A. H. (1984). Countable homogeneous tournaments, *Transactions of the American Mathematical Society* **284**: 431–461.

Lachlan, A. H. (1987). Homogeneous structures, *in* A. M. Gleason (ed.), *Proceedings of the International Congress of Mathematicians, Berkeley 1986*, Vol. 1, American Mathematical Society, Providence, RI, pp. 314–21.

Lachlan, A. H. and Woodrow, R. E. (1980). Countable ultrahomogeneous undirected graphs, *Transactions of the American Mathematical Society* **262**: 51–94.

Lang, S. (1965). *Algebra*, Addison-Wesley, Reading, Mass.

Lascar, D. (1982). On the category of models of a complete theory, *The Journal of Symbolic Logic* **47**: 249–266.

Lascar, D. (1987). *Stability in Model Theory*, Vol. 36 of *Pitman Monographs and Surveys in Pure and Applied Mathematics*, Longman Scientific and Technical, Harlow, Essex. Copublished in the U.S.A. by John Willey and Son, New York.

Lascar, D. (1991). Autour de la propriété du petit indice, *Proceedings of the London Mathematical Society, Series 3* **62**: 25–53.

Lascar, D. (1992). Les automorphisms d'un ensemble fortement minimal, *The Journal of Symbolic Logic* **57**: 238–251.

Lascar, D. (1994). *The group of automorphisms of a relational saturated structure*, Vol. 411 of *NATO ASI Series C: Mathematical and Physical Sciences*, Kluwer, Dordrecht, pp. 225–36. Proceedings, Banff, 1991.

Lascar, D. and Shelah, S. (1993). Uncountable saturated structures have the small index property, *The Bulletin of the London Mathematical Society* **25**: 125–131.

Läuchli, H. (1964). The independence of the ordering principle from a restricted axiom of choice, *Fundamenta Mathematicæ* **54**: 31–43.

Levy, A. (1958). The independence of various definitions of finiteness, *Fundamenta Mathematicæ* **46**: 1–13.

Levy, A. (1965). The Fraenkel–Mostowski method for independence proofs in set theory, *The Theory of Models*, North Holland, Amsterdam, pp. 221–228.

Levy, A. (1966). Definability in axiomatic set theory I, *Logic, Methodology and Philosophy of Science, Jerusalem Congress 1964*, North Holland, Amsterdam, pp. 127–151.

Levy, A. (1973). Axioms of finite choice. Preprint.

Liebeck, M. W. (1987). The affine permutation groups of rank three, *Proceedings of the London Mathematical Society, Series 3* **54**: 477–516.

Liebeck, M. W., Praeger, C. E. and Saxl, J. (1988). On the O'Nan–Scott theorem for finite primitive permutation groups, *The Journal of the Australian Mathematical Society* **44**: 389–396.

Macintyre, A. J. (1971). On ω_1-categorical theories of abelian groups, *Fundamenta Mathematicæ* **70**: 253–270.

Macintyre, A. J. and Rosenstein, J. G. (1976). $\aleph_0$-categoricity for rings without nilpotent elements and for boolean structures, *The Journal of Algebra* **43**: 129–154.

Macpherson, H. D. (1985). Orbits of infinite permutation groups, *Proceedings of the London Mathematical Society, Series 3* **46**: 246–284.

Macpherson, H. D. (1986*a*). Groups of automorphisms of $\aleph_0$-categorical structures, *Quarterly Journal of Mathematics, Series 2* **37**: 449–465.

Macpherson, H. D. (1986*b*). Homogeneity in infinite permutation groups, *Periodica Mathematica Hungarica* **17**: 211–233.

Macpherson, H. D. (1991). Finitely axiomatisable theories and the strict order property, *Notre Dame Journal of Formal Logic* **32**: 188–192.

Macpherson, H. D. and Pillay, A. (1994). Primitive permutation groups of finite Morley rank. Preprint.

Macpherson, H. D. and Praeger, C. E. (1990). Maximal subgroups of infinite symmetric groups, *The Journal of the London Mathematical Society, Series 2* **42**: 85–92.

Macpherson, H. D. and Praeger, C. E. (in press*a*). Cycle types in infinite permutation groups, *The Journal of Algebra*. To appear.

Macpherson, H. D. and Praeger, C. E. (in press*b*). Infinitary versions of the O'Nan–Scott theorem, *The Journal of the London Mathematical Society, Series 2*. to appear.

Macpherson, H. D. and Steinhorn, C. (to appear). On variants of o-minimality. Preprint.

Macpherson, H. D. and Woodrow, R. (1992). The permutation group induced on a moiety, *Forum Mathematicum* **4**: 243–255.

Makkai, M. (1984). A survey of basic stability theory, with particular emphasis on orthogonality and regular types, *Israel Journal of Mathematics* **49**: 181–238.

Mathias, A. R. D. (1974). The order-extension principle, *Proceedings of Symposia in Pure Mathematics*, Vol. XIII part 2, American Mathematical Society, Providence, RI, pp. 179–184.

Mathieu, E. (1871). Mémoire sur l'étude des fonctions de plusiers quantités, sur la manière de les former, et sur les substitutions qui les laissent invariables, *J. Math. Pures Appl. (Liouville)* **18**: 241–323.

Maund, T. C. (1989). D.Phil., Oxford University.

McLeish, S. (1994). *The Sufficiency of Going Forth in First-order Homogeneous Structures*, Ph.D., Queen Mary and Westfield College.

Möller, R. G. (1992). Ends of graphs, *Mathematical Proceedings of the Cambridge Philosophical Society* **111**: 255–266.

Morley, M. D. (1965). Categoricity in power, *Transactions of the American Mathematical Society* **114**: 514–538.

Morley, M. D. (1967). Countable models of $\aleph_1$-categorical theories, *Israel Journal of Mathematics* **5**: 65–72.

Mostowski, A. (1938). Über den Begriff einer endlichen Menge, *Comptes Rendus des Séances de la Société des Sciences et des Lettres de Varsovie* **Cl.III, 31**: 13–20.

Mostowski, A. (1939). Über die Unabhängigkeit des Wohlordnungssatzes vom ordnungsprinzip, *Fundamenta Mathematicæ* **32**: 201–252.

Mostowski, A. (1945). Axiom of choice for finite sets, *Fundamenta Mathematicæ* **33**: 137–168.

Myhill, J. and Scott, D. S. (1971). Ordinal definability, *Proceedings of Symposia in Pure Mathematics*, Vol. XIII part 1, American Mathematical Society, Providence, RI, pp. 271–278.

Nesin, A. (1989). Solvable groups of finite Morley rank, *The Journal of Algebra* **121**: 26–39.

Nesin, A. (1990*a*). On sharply n-transitive superstable groups, *The Journal of Pure and Applied Algebra* **69**: 73–88.

Nesin, A. (1990*b*). On solvable groups of finite Morley rank, *Transactions of the American Mathematical Society* **321**: 659–690.

Nesin, A. (1991). Generalized Fitting subgroup of a group of finite Morley rank, *The Journal of Symbolic Logic* **56**: 1391–1399.

Nesin, A. and Pillay, A. (eds) (1989). *The Model Theory of Groups*, number 11 in *Notre Dame Mathematical Lectures*, University of Notre Dame Press.

Neumann, B. H. (1954). Groups covered by permutable subsets, *The Journal of the London Mathematical Society, Series 2* **29**: 236–248.

Neumann, P. M. (1975). The lawlessness of finitary permutation groups, *Archiv der Mathematik* **26**: 561–566.

Neumann, P. M. (1976). The structure of finitary permutation groups, *Archiv der Mathematik* **27**: 3–17.

Neumann, P. M. (1985). Some primitive permutation groups, *Proceedings of the London Mathematical Society, Series 3* **50**: 265–281.

Neumann, P. M. (1988). Homogeneity of infinite permutation groups, *The Bulletin of the London Mathematical Society* **20**: 305–312.

Nešetřil, J. and Rödl, V. (1977). Partitions of finite relational and set systems, *Journal of Combinatorial Theory, Series A* **22**: 289–312.

Nešetřil, J. and Rödl, V. (1983). Ramsey classes of set systems, *Journal of Combinatorial Theory, Series A* **34**: 183–201.

Pasini, A. (1991). Diagram geometries for sharply n-transitive sets of permutations or of mappings, *Designs, Codes and Cryptography* **1**: 275–297.

Passman, D. (1968). *Permutation Groups*, Mathematics Lecture Series, Benjamin, New York.

Pillay, A. (1983). *An Introduction to Stability Theory*, Vol. 8 of *Oxford Logic guides*, Oxford University Press.

Pillay, A. and Steinhorn, C. (1986). Definable sets in ordered structures, *Transactions of the American Mathematical Society* **295**: 565–592.

Pincus, D. (1971). Support structures for the axiom of choice, *The Journal of Symbolic Logic* **36**: 28–38.

Pincus, D. (1972). Zermelo–Fraenkel consistency results by Fraenkel–Mostowski methods, *The Journal of Symbolic Logic* **37**: 721–743.

Pincus, D. (1976). Two model-theoretic ideas in independence proofs, *Fundamenta Mathematicæ* **92**: 113–130.

Poizat, B. (1987). *Groupes Stables*, Nur Al-Mantiq Wal-Ma'rifah, 82 rue Racine, 69100 Villeurbanne, France. Available from author at the University of Lyon.

Pontrjagin, L. (1966). *Topological Groups*, Gordon and Breach, New York.

Pouzet, M. (1978). *Sur la Theorie des Relations*, PhD thesis, Lyon.

Prömel, H. and Voigt, B. (1989). A short proof of the restricted Ramsey theorem for finite set systems, *Journal of Combinatorial Theory, Series A* **52**: 313–320.

Pyber, L. (1993). Asymptotic results for permutation groups, *Proceedings of the DIMACS Conference on Computational Group Theory, New Brunswick, 1991*, Vol. 11 of *DIMACS Series in Discrete Mathematics and Theoretical Computer Science*, American Mathematical Society, Providence, RI, pp. 197–219.

Pyber, L. (in press). *On the Minimal Degree of Primitive Permutation Groups*, North Holland, Amsterdam. To appear.

Reineke, J. (1975). Minimale Gruppen, *Zeitschrift für Mathematische Logik und Grundlagen der Mathematik* **21**: 357–379.

Rosenberg, A. (1958). The structure of the infinite general linear group, *Annals of Mathematics* **68**: 278–294.

Rosenstein, J. (1969). $\aleph_0$-categoricity of linear orderings, *Fundamenta Mathematicæ* **64**: 1–5.

Rubin, H. and Rubin, J. E. (1976). *Equivalents of the Axiom of Choice II*, North Holland, Amsterdam.

Rubin, M. (in press). On the reconstruction of $\aleph_0$-catgeorical structures from their automorphism groups, *Proceedings of the London Mathematical Society, Series 3*. To appear.

Ryll-Nardzewski, C. (1959). On categoricity in power $\leqslant \aleph_0$, *Bulletin de l'Académie Polonaise des Sciences* **7**: 545–548.

Saracino, D. (1983). Amalgamation bases for nil-2 groups, *Algebra Universalis* **16**: 47–62.

Saracino, D. and Wood, C. (1979). Periodic existentially closed nilpotent groups, *The Journal of Algebra* **58**: 189–207.

Saracino, D. and Wood, C. (1982). QE nil-2 groups of exponent 4, *The Journal of Algebra* **76**: 337–352.

Schmerl, J. H. (1978). On the $\aleph_0$-categoricity of filtered Boolean extensions, *Algebra Universalis* **8**: 159–161.

Schmerl, J. H. (1979). Countable homogeneous partially ordered sets, *Algebra Universalis* **9**: 317–321.

Schmerl, J. H. (1980). Decidability and $\aleph_0$-categoricity of theories of partially ordered sets, *The Journal of Symbolic Logic* **45**: 585–611.

Schmerl, J. H. (1985). Recursively saturated models generated by indiscernibles, *Notre Dame Journal of Formal Logic* **26**: 99–105.

Schmerl, J. H. (1989). Large resplendent models generated by indiscernibles, *The Journal of Symbolic Logic* **54**: 1382–1388.

Scott, D. S. (1962). Algebras of sets binumerable in complete extensions of arithmetic, *in* J. C. E. Dekker (ed.), *Recursion Function Theory*, Vol. V of *Symposia in Pure Mathematics*, American Mathematical Society, Providence, RI, pp. 117–122.

Scott, D. S. (1964). Invariant Borel sets, *Fundamenta Mathematicæ* **56**: 117–128.

Scott, D. S. (1965). Logic with denumerably long formulas and finite strings of quantifiers, *The Theory of Models*, North Holland, Amsterdam, pp. 329–341.

Semmes, S. (1981). Endomorphisms of infinite symmetric groups, *Abstracts of the American Mathematical Society* **2**: 426.

Serre, J.-P. (1980). *Trees*, Springer-Verlag, Berlin.

Shelah, S. (1978). *Classification Theory and the Number of Nonisomorphic Models*, Vol. 92 of *Studies in Logic and the Foundations of Mathematics*, North Holland, Amsterdam. Revised edition, 1992.

Sikorski, R. (1964). *Boolean Algebras*, Springer-Verlag, Berlin.

Smoryński, C. (1982). Back and forth inside a recursively saturated model of arithmetic, *in* D. van Dalen *et al.* (eds), *Logic Colloquium '80*, North Holland, Amsterdam, pp. 273–278.

Solovay, R. (1970). A model of set theory in which every set of reals is Lebesgue-measurable, *Annals of Mathematics* **92**: 1–56.

Stevens, S. S. (1946). On the theory of scales of measurement, *Science* **103**: 677–680.

Suzuki, M. (1986). *Group Theory II*, Vol. 248, Springer-Verlag, Berlin.

Svenonius, L. (1959). $\aleph_0$-categoricity in first-order predicate calculus, *Theoria* **25**: 82–94.

Tarski, A. (1924). Sur les ensembles finis, *Fundamenta Mathematicæ* **6**: 45–95.

Tarski, A. (1949). Arithmetical classes and types of algebraically closed and real-closed fields, *Bulletin of the American Mathematical Society* **55**: 64. Abstract.

Thomas, S. R. (1986). Groups acting on infinite dimensional projective spaces, *The Journal of the London Mathematical Society, Series 2* **34**: 265–273.

Thomas, S. R. (1991). Reducts of the random graph, *The Journal of Symbolic Logic* **56**: 176–181.

Thomas, S. R. (to appear). Reducts of random hypergraphs. In preparation.

Thompson, J. G. (1959). Finite groups with fixed-point-free automorphisms of prime order, *Proceedings of the National Academy of Science* **45**: 578–581.

Tits, J. (1952). *Généralisations des Groupes Projectifs Basées sur leurs Propriétés de Transitivité*, Vol. 27 of *Classes des Mémoires, deuxième Serie*, Académie Royale de Belgique, Brussels.

Tits, J. (1974). *Buildings of Spherical Type and Finite BN-Pairs*, Vol. 386 of *Lecture Notes in Mathematics*, Springer-Verlag, Berlin.

Truss, J. K. (1973). Finite axioms of choice, *Annals of Mathematical Logic* **6**: 147–176.

Truss, J. K. (1974*a*). Classes of Dedekind finite cardinals, *Fundamenta Mathematicæ* **84**: 187–208.

Truss, J. K. (1974*b*). Models of set theory containing many perfect sets, *Annals of Mathematical Logic* **7**: 197–219.

Truss, J. K. (1976). Some cases of König's lemma, *Set Theory and Hierarchy Theory*, Vol. 537 of *Lecture Notes in Mathematics*, Springer-Verlag, Berlin, pp. 273–284.

Truss, J. K. (1978). The axiom of choice for linearly ordered families, *Fundamenta Mathematicæ* **99**: 133–139.

Truss, J. K. (1985). The group of the countable universal graph, *Mathematical Proceedings of the Cambridge Philosophical Society* **98**: 213–245.

Truss, J. K. (1989). Infinite permutation groups II. Subgroups of small index, *The Journal of Algebra* **120**: 494–515.

Truss, J. K. (in press). The structure of amorphous sets, *Annals of Pure and Applied Logic*. To appear.

Tzouvaras, A. (1991). A note on real subsets of a recursively saturated model, *Zeitschrift für Mathematische Logik und Grundlagen der Mathematik* **37**: 207–16.

Wagner, F. O. (1988). Relational structures and dimensions. Unpublished notes.

Warren, R. (1992). *The Structure of K-CS-Transitive Cycle-free Partial Orders*, Ph.D. thesis, Leeds University.

Waszkiewicz, J. and Węglorz, B. (1969). On ω_0-categoricity of powers, *Bulletin de l'Académie Polonaise des Sciences* **17**: 195–199.

Weil, A. (1955). On algebraic groups of transformations, *American Journal of Mathematics* **77**: 355–391.

Wielandt, H. (1959). Unendliche Permutationsgruppen. Unpublished typescript.

Wielandt, H. (1964). *Finite Permutation Groups*, Academic Press, New York.

Wilmers, G. M. (1975). *Some problems in set theory: non-standard models and their applications to model theory*, D.Phil., University of Oxford.

Yoshizawa, M. (1979). On infinite four-transitive permutation groups, *The Journal of the London Mathematical Society, Series 2* **19**: 437–438.

Zassenhaus, H. (1935*a*). Kennzeichnung endlicher linearer Gruppen als Permutationsgruppen, *Abh. Math. Sem. Hamburg* **11**: 17–40.

Zassenhaus, H. (1935*b*). Über endliche Fastkörper, *Abh. Math. Sem. Hamburg* **11**: 187–220.

Ziegler, M. (1992). Finite covers of disintegrated sets. Preprint.

Zil'ber, B. I. (1973). Groups with categorical theories, *Abstracts of Papers Presented at the Fourth All-Union Symposium on Group Theory*, pp. 63–68. In Russian.

Zil'ber, B. I. (1977). Groups and rings whose theory is categorical, *Fundamenta Mathematicæ* **55**: 173–188.

Zil'ber, B. I. (1979). Totally categorical theories: Structural properties and non-finite axiomatizability, *in* L. Pacholski *et al.* (eds), *Model Theory of Algebra and Arithmetic*, number 834 in *Lecture Notes in Mathematics*, Springer-Verlag, Berlin, pp. 381–410.

Zil'ber, B. I. (1984*a*). Some model theory of simple algebraic groups over algebraically closed fields, *Colloq. Math.* **48**: 173–180.

Zil'ber, B. I. (1984*b*). The structure of models of uncountably categorical theories, *Proceedings of the International Congress of Mathematicians, August 16–24, 1983, Warszawa*, Państwowe Wydawnictwo Naukowe, pp. 359–368.

Zil'ber, B. I. (1988). Finite homogeneous geometries, *in* B. I. Dahn and H. Wolter (eds), *Proceedings of the 6th Easter Conference on Model Theory*, Humboldt Universität zu Berlin, pp. 186–208.

Zil'ber, B. I. (1993). *Uncountably categorical theories*, Translations of Mathematical Monographs, American Mathematical Society, Providence, RI.

Zsigmondy, K. (1892). Zur Theorie der Potenzreste, *Monatshefte für Mathematik und Physik* **3**: 265–284.

Index of notation

Index

Page numbers in italics refer to the definition of the term given. Some terms are multiply defined—sometimes this is simply repetition, sometimes a later definition is an extension of an old one, but beware that definitions of the same term are occasionally inconsistent with each other. If a term has a variable hyphenated prefix, as for example *κ-categorical*, the term is indexed as if the prefix is absent. Some terms have prefixes which are not variable, *c-Jordan*, for example, and these are indexed in the usual way.